POSITRON PHYSICS

This book provides a comprehensive and up-to-date account of the field of low energy positrons and positronium within atomic and molecular physics. It begins with an introduction to the field, discussing the background to low energy positron beams, and then covers topics such as total scattering cross sections, elastic scattering, positronium formation, excitation and ionization, annihilation and positronium interactions. Each chapter contains a blend of theory and experiment, giving a balanced treatment of all the topics.

The book will be useful for graduate students and researchers in physics and chemistry. It is ideal for those wishing to gain rapid, in-depth knowledge of this unique branch of atomic physics.

MICHAEL CHARLTON obtained his degree and Ph.D. from University College London. In 1983 he was awarded a Royal Society University Research Fellowship, held at UCL. From 1991 to 1999 he was a Reader in Physics at UCL and was appointed to the Chair in Experimental Physics at the University of Wales, Swansea in 1999. Professor Charlton has published over one hundred research articles and written several reviews, notably for *Reports on Progress in Physics* and *Physics Reports.*

JOHN WATKIN HUMBERSTON was an undergraduate at Manchester University and obtained his Ph.D. from University College London. From 1965 to 1966 he taught at Trinity College Dublin and became a Lecturer at UCL in 1966, where he subsequently became Senior Lecturer, Reader and Professor. During this time he had sabbatical leave at the Goddard Space Flight Center, USA, and at York University, Toronto, Canada. Professor Humberston has written numerous research articles, published mainly in *Journal of Physics B*, as well as several review articles in *Advances in Atomic and Molecular Physics* and in *Physics Reports.*

CAMBRIDGE MONOGRAPHS ON ATOMIC, MOLECULAR AND CHEMICAL PHYSICS: 11

General editors: A. Dalgarno, P. L. Knight, F. H. Read, R. N. Zare

For Lucy and Bettina

Positron Physics

M. Charlton
University of Wales Swansea

J. W. Humberston
University College London

CAMBRIDGE UNIVERSITY PRESS
Cambridge, New York, Melbourne, Madrid, Cape Town, Singapore, São Paulo

Cambridge University Press
The Edinburgh Building, Cambridge CB2 2RU, UK

Published in the United States of America by Cambridge University Press, New York

www.cambridge.org
Information on this title: www.cambridge.org/9780521415507

First published 2001
This digitally printed first paperback version 2005

A catalogue record for this publication is available from the British Library

Library of Congress Cataloguing in Publication data

Charlton, M. (Michael)
Positron physics / M. Charlton, J.W. Humberston
p. cm. – (Cambridge monographs on atomic, molecular,
and chemical physics ; 11)
Includes bibliographic references
ISBN 0 521 41550 0
1. Positrons. 2. Positronium. I. Humberston, John W.
II. Title. III. Series.
QC793.5.P62 C48 2000
539.7′214–dc21 00-028909 CIP

ISBN-13 978-0-521-41550-7 hardback
ISBN-10 0-521-41550-0 hardback

ISBN-13 978-0-521-01939-2 paperback
ISBN-10 0-521-01939-7 paperback

Contents

Preface

This book is concerned mainly with the interactions of positrons and positronium with individual atoms and molecules in gases. Brief mention is also made of positrons interacting with bulk matter but this is in the context of describing the slowing down of positrons in solids and the subsequent ejection of low energy positrons and positronium from the surface of the solid. A technique using the angular correlation of annihilation radiation, which is widely used in studies of electron momentum distributions and defects in condensed matter, is also described but again the emphasis is mainly on positron annihilation in gases.

Theoretical studies of positron collisions with atomic and molecular systems have been made for many years, as also have both theoretical and experimental studies of the lifetimes of positrons diffusing in gases. Only since the development of energy-tunable monoenergetic positron beams in the early 1970s, however, has it been possible to make detailed comparisons between theoretical predictions and the increasingly accurate experimental measurements of total, partial and differential scattering cross sections. These experimental developments have in turn stimulated renewed interest in theoretical studies of systems containing positrons. In this book we have attempted to integrate both theoretical and experimental aspects of the field into a reasonably coherent whole, although some sections are predominantly either experimental or theoretical.

Positron physics has undergone very rapid development during the past several years. Accordingly, there has developed a need for a comprehensive up-to-date review of the field, which we hope this book will satisfy. No other extensive review of both experimental and theoretical aspects of the field has been published previously and therefore we believe it is timely to publish this book now.

We are indebted to the following people for providing information and permitting us to reproduce figures from their published work: E.A.G.

Armour, K.F. Canter, R.J. Drachman, D.W. Gidley, T.W. Hänsch, Y.K. Ho, W.E. Kauppila, R.P. McEachran, A.P. Mills Jr, W. Raith, H. Schneider, D.M. Schrader, A.D. Stauffer, T.S. Stein, C.M. Surko and H.R.J. Walters. Thanks are also due to the publishers of the journals from which these figures have been taken, namely the American Physical Society, the American Institute of Physics, Baltzer Science Publishers, Elsevier, and the Institute of Physics. Particular thanks are due to several of our immediate colleagues: to Dr G. Laricchia and Dr P. Van Reeth for their assistance and numerous helpful discussions, to Dr A. Garner for producing many of the figures, and to Mr P.A. Donnelly for his assistance in preparing the bibliography. Above all, however, we wish to thank Mrs Carol Broad for so ably, and with such patience, preparing the final version of the typescript and dealing with numerous modifications to the text. Also, we are indebted to the staff of Cambridge University Press for the care with which the final stages of the book's publication have been completed.

The experimental positron physics group at University College London was initiated by Professor T.C. Griffith and Dr G.R. Heyland at the instigation of the late Sir Harrie Massey. We wish to record our gratitude to these three pioneers for their seminal contributions to positron collision physics and for introducing us to this fascinating subject.

M. Charlton
J.W. Humberston

1
Introduction

1.1 Historical remarks

The prediction, and subsequent discovery, of the existence of the positron, e^+, constitutes one of the great successes of the theory of relativistic quantum mechanics and of twentieth century physics. When Dirac (1930) developed his theory of the electron, he realized that the negative energy solutions of the relativistically invariant wave equation, in which the total energy E of a particle with rest mass m is related to its linear momentum p by

$$E^2 = m^2c^4 + p^2c^2, \tag{1.1}$$

had real physical significance. He therefore postulated that the 'sea' of electron states with negative energies between $-mc^2$ and $-\infty$ was normally fully occupied in accordance with the Pauli exclusion principle, and would be unobservable. A vacancy in this ensemble, however, would manifest itself as a positively charged particle with a positive rest mass which, on the basis of uncalculated Coulomb energy corrections and the particles then known, Dirac assumed to be the proton. It was soon realized that this was not the case and that the theory actually predicted the existence of a new particle with the rest mass of the electron and an equal but opposite charge – the positron.

The positron was subsequently discovered by Anderson (1933) in a cloud chamber study of cosmic radiation, and this was soon confirmed by Blackett and Occhialini (1933), who also observed the phenomenon of pair production. There followed some activity devoted to understanding the various annihilation modes available to a positron in the presence of electrons; radiationless, single-gamma-ray and the dominant two-gamma-ray processes were considered (see section 1.2). The theory of pair production was also developed at this time (see e.g. Heitler, 1954).

In 1934 Mohorovičić proposed the existence of a bound state of a positron and an electron which, he (incorrectly) suggested, might be responsible for unexplained features in the spectra emitted by some stars. However, as summarized by Kragh (1990), Mohorovičić's ideas on the properties of this new atom were somewhat unconventional, and the name 'electrum' which he gave to it did not become widespread but was later replaced by the present appellation, positronium (Ruark, 1945), with the chemical symbol Ps.

Other significant developments took place in the 1940s. In 1949 DeBenedetti and coworkers discovered that the two gamma-rays emitted following positron annihilation in various solids deviated from precise collinearity, i.e. the angle between them was not exactly 180°, as would be expected from the annihilation of an electron–positron pair at rest. Although this deviation amounted to only a few milliradians, it was correctly interpreted as being due mainly to the effect of the motion of the bound electrons in the material, the positron having essentially thermalized. Somewhat earlier, DuMond, Lind and Watson (1949) had made an accurate measurement of the energy and width of the annihilation gamma-ray line using a crystal spectrometer. They found the width to be greater than that associated with the instrumental resolution, and they attributed this to Doppler broadening arising predominantly from electronic motion. These investigations laid the foundations for later advances in positron solid state physics, which were themselves to underpin the development of low energy positron beams.

In 1946 Wheeler undertook a theoretical study of the stability of various systems of positrons and electrons, which he termed polyelectrons. He found, as expected, that positronium was bound, but that so too was its negative ion ($e^-e^+e^-$). This entity, Ps^-, was not observed until much later (Mills, 1981), after the development of positron beams.

Positronium itself was eventually discovered in 1951 by Deutsch and its properties were investigated in an elegant series of experiments based around positron annihilation in gases. Many of the techniques developed then are still in use today. This advance stimulated further experimental and theoretical studies of the basic properties of the ground state of positronium (particularly the triplet 1^3S_1 state, ortho-positronium), including the hyperfine structure, the annihilation lifetime, elucidation of the selection rules governing annihilation and the calculation of the spectrum of photon energies emitted in the three-gamma-ray annihilation mode. Some of these topics are described in detail elsewhere in this book.

The recent production of relativistic antihydrogen (Baur *et al.*, 1996; Blanford *et al.*, 1998), and the prospect of its formation at very low energies (see Chapter 8), when detailed spectroscopic and other studies of this system should become possible, makes it appropriate to mention the antiproton. This particle, whose existence had been predicted by analogy with the positron, was discovered in 1955 by Chamberlain, Segrè, Weigand and Ypsilantis using the 6.2 GeV Bevatron accelerator at the Lawrence Berkeley Laboratory, California, USA.

For positron collision physics, a revolutionary advance came with the discovery and development of low energy positron beams. In a study of secondary electron emission by positrons, Cherry (1958) found that 'positrons in the energy interval 0–5 eV, very numerous in comparison to those in equal intervals at somewhat higher energies, were emitted from a chromium-on-mica surface when it was irradiated by a ^{64}Cu positron beta spectrum'. However, the efficiency of conversion from fast to slow positrons was only approximately 10^{-8}. This work was, in fact, predated by that of Madansky and Rasetti (1950), who unsuccessfully searched for low energy positron emission from a variety of samples. These experiments were largely ignored until the late 1960s and the work of Groce *et al.* (1968).

The decisive breakthrough in the development of positron beams probably came with the work of Canter *et al.* (1972) who discovered the smoked MgO moderator. Although only a very small fraction, 3×10^{-5}, of the incident β^+ activity was converted into a usable low energy beam, this advance paved the way for the ensuing rapid progress. Later in the same decade, the phenomenon of positron emission and re-emission from various surfaces, carefully prepared under ultra-high vacuum conditions, was investigated, mainly by Mills and his coworkers (see e.g. Mills, 1983a), and a physical understanding was obtained of the processes involved. As this understanding grew, so too did the efficiency of moderation (as the conversion process from fast to slow positrons is known); this culminated in the solid neon moderator (Mills and Gullikson, 1986) and variants thereof, which have moderation efficiencies close to 10^{-2}, fully six orders of magnitude greater than that in the seminal observation by Cherry (1958).

The mechanisms involved in the emission and re-emission of positrons from surfaces, and the attendant formation of beams with well-defined energies, are central to the main theme of this book and are described in greater detail in section 1.4.

1.2 Basic properties of the positron and other positronic systems

1 Positrons

The positron has an intrinsic spin of one half and is thus a fermion. According to the CPT theorem, which states that the fundamental laws of physics are invariant under the combined actions of charge conjugation (C), parity (P) and time reversal (T), its mass, lifetime and gyromagnetic ratio are equal to those of the electron, and it has the same magnitude of electric charge, though of opposite sign. There are at present no known exceptions to the CPT theorem.

Experimentally it has been shown from studies involving trapped particles that the gyromagnetic ratios of the electron and the positron are equal to within 2 parts in 10^{12} (Van Dyck, Schwinberg and Dehmelt, 1987). The magnitudes of the charges of the electron and the positron have been found by Hughes and Deutch (1992) to be equal to 4 parts in 10^{8} in an analysis of the measured charge-to-mass ratios and the values of the Rydberg constant derived from the energy spectra of hydrogen and positronium. A more stringent, though indirect, limit of 1 part in 10^{18} for the difference in charge magnitude was derived by Müller and Thoma (1992), in a method based on limits for the neutrality of atomic matter. They concluded that, because equal numbers of electrons and positrons contribute to the vacuum polarization of atoms, there would be an overall net charge on matter unless the charges of the two particles balanced precisely.

Current theories of particle physics predict that, in a vacuum, the positron is a stable particle, and laboratory evidence in support of this comes from experiments in which a single positron has been trapped for periods of the order of three months (Van Dyck, Schwinberg and Dehmelt, 1987). If the CPT theorem is invoked then the intrinsic positron lifetime must be $\geq 4 \times 10^{23}$ yr, the experimental limit on the stability of the electron (Aharonov *et al.*, 1995).

When a positron encounters normal matter it eventually annihilates with an electron after a lifetime which is inversely proportional to the local electron density. In condensed matter lifetimes are typically less than 500 ps, whilst in gases this figure can be considered as a lower limit, found either at very high gas densities or when the positron forms a bound state or long-lived resonance with an atom or molecule.

Annihilation of a positron with an electron may proceed by a number of mechanisms, and the Feynman diagrams for the radiationless process, which results in electron emission, and for the single-, two- and three-gamma processes are given in Figure 1.1. The positron can also annihilate with an inner shell electron in a radiationless process, the consequent energy release giving rise to nuclear excitation (see Saigusa and Shimizu, 1994, for a summary). The most probable of these annihilation processes, when the positron and electron are in a singlet spin state, is the two-gamma process, the cross section for which was derived by Dirac (1930) to be

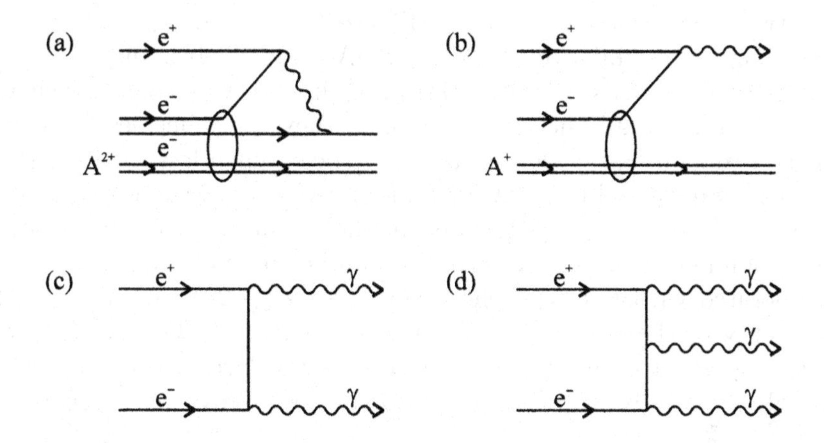

Fig. 1.1. Feynman diagrams of the lowest order contributions to (a) radiationless, (b) one-gamma, (c) two-gamma, (d) three-gamma-ray annihilation. A^{2+} and A^+ denote the charge states of the remnant atomic ion.

$$\sigma_{2\gamma} = \frac{4\pi r_0^2}{\gamma+1}\left[\frac{\gamma^2+4\gamma+1}{\gamma^2-1}\ln\left(\gamma+\sqrt{\gamma^2-1}\right) - \frac{\gamma+3}{\sqrt{\gamma^2-1}}\right], \tag{1.2}$$

where $r_0 = e^2/(4\pi\epsilon_0 mc^2)$ is the classical radius of the electron, $\gamma = 1/\sqrt{(1-\beta^2)}$, $\beta = v/c$, and v is the speed of the positron relative to the stationary electron. Of most relevance for our discussion is annihilation at low positron energies, where $v \ll c$, so that equation (1.2) reduces to the familiar form

$$\sigma_{2\gamma} = 4\pi r_0^2 c/v. \tag{1.3}$$

Note that $\sigma_{2\gamma} \to \infty$ as $v \to 0$, although the annihilation rate, which is proportional to the product $v\sigma_{2\gamma}$, remains finite. At low incident positron energies the two gamma-rays are emitted almost collinearly, the energy of each being close to mc^2 (= 511 keV). Annihilation of a small fraction of the positrons emanating from the radioactive source can occur at relativistic speeds and then it is necessary to use the full equation (1.2).

Annihilation can also occur with the emission of three (or more) gamma-rays, and Ore and Powell (1949) calculated that the ratio of the cross sections for the three- and two-gamma-ray cases is approximately 1/370. Higher order processes are expected to be further depressed by a similar factor. A case in point is the four-gamma-ray mode, for which the branching ratio with the two-gamma-ray mode was shown by Adachi *et al.* (1994) to be approximately 1.5×10^{-6}, in accord with QED calculations.

The two other processes shown in Figure 1.1 are the radiationless and single quantum annihilations (RA and SQA respectively), and both need to involve the nucleus or the entire atom in order to conserve energy and momentum simultaneously. As such, they are much less probable than the two-gamma-ray process and have been much less studied. Both processes are expected to involve mainly inner shell electrons. In the RA case shown here the energy released in the annihilation of the positron with a bound electron is transferred to another bound electron, which is then liberated with a kinetic energy of $E + mc^2 - 2E_b$, where E is the total energy of the positron as defined in equation (1.1) and E_b is the binding energy of each of the two electrons involved (assumed here to be equal). Similarly in SQA, the emitted gamma-ray has an energy of $E + mc^2 - E_b$.

The Born approximation for the cross section for SQA predicts a Z^5 dependence, where Z is the atomic number of the atom involved in the annihilation (e.g. Bhabha and Hulme, 1934), and its maximum value is approximately 5×10^{-29} m^2 at kinetic energies of the order of a few hundred keV; at these energies the positron can penetrate deep into the electronic core of the atom. The most recent experimental work by Palathingal *et al.* (1995), using a high-energy-resolution gamma-ray detector, has resolved the contributions to SQA from the K-, L- and M-shells for a number of targets. They found that the annihilation cross section for the K-shell scaled as $Z^{5.1}$, whereas the L-shell had a characteristic exponent of 6.4. Further details on the theoretical and experimental situation are given by Palathingal *et al.* (1995) and Bergstrom, Kissel and Pratt (1996).

The experimental evidence for radiationless annihilation is not very convincing and, indeed, the only claim to have observed this phenomenon is that of Shimizu, Mukoyama and Nakayama (1965, 1968), who used a β-ray spectrometer to fire 300 keV positrons into a lead foil. The emitted electrons were recorded using a silicon detector which allowed some energy selection. An excess of measured counts was found in the energy region to be expected for the target, and the derived cross section was approximately 10^{-30} m^2. According to theoretical work on radiationless annihilation by Mikhailov and Porsev (1992), in which the strong Coulomb repulsion experienced by the positron was taken into account, the cross section should scale as Z^8, with a value of approximately 10^{-32} m^2 at a positron kinetic energy of 500 keV and for a target with $Z = 80$. This is nearly two orders of magnitude lower than the value obtained by Massey and Burhop (1938), the discrepancy being attributed to the use of a plane wave representation of the electron state by Massey and Burhop. In the light of the more recent theoretical value, the experimental result appears too high, and further investigations are required.

Additional aspects of positron annihilation, with particular emphasis on the processes of relevance to atomic collisions at low energies, are described in Chapter 6.

2 Positronium

Positronium is the name given to the quasi-stable neutral bound state of an electron and a positron. It is hydrogen-like, but because the reduced mass is $m/2$ the gross values of the energy levels are decreased to half those found in the hydrogen atom, so that the binding energy of ground state positronium is approximately 6.8 eV. An energy level diagram of the ground and first excited states, with principal quantum numbers $n_{\rm Ps} = 1$ and 2 respectively, is given in Figure 1.2. Note that the fine and hyperfine separations are markedly different from the corresponding values for hydrogen, owing to the large magnetic moment of the positron (658 times that of the proton) and the presence of QED effects such as virtual annihilation (see e.g. Berko and Pendleton, 1980, and Rich, 1981, for summaries).

Positronium can exist in the two spin states, $S = 0, 1$. The singlet state ($S = 0$), in which the electron and positron spins are antiparallel, is termed para-positronium (para-Ps), whereas the triplet state ($S = 1$) is termed ortho-positronium (ortho-Ps). The spin state has a significant influence on the energy level structure of the positronium, and also on its lifetime against self-annihilation.

The need to conserve angular momentum and to impose CP invariance led Yang (1950) and Wolfenstein and Ravenhall (1952) to conclude that positronium in a state with spin S and orbital angular momentum L can only annihilate into n_γ gamma-rays, where

$$(-1)^{n_\gamma} = (-1)^{L+S}. \tag{1.4}$$

This selection rule does not appear to exclude radiationless annihilation and annihilation into a single gamma-ray, but these modes of annihilation are nevertheless forbidden for free positronium.

For ground state positronium with $L = 0$, annihilation of the singlet ($1^1\mathrm{S}_0$) and triplet ($1^3\mathrm{S}_1$) spin states can only proceed by the emission of even and odd numbers of photons respectively. Thus, in the absence of any perturbation the annihilation of para-Ps proceeds by the emission of two, four etc. gamma-rays, and the annihilation of ortho-Ps by the emission of three, five etc. gamma-rays. In both cases the lowest order processes dominate although observation of the five-photon decay of ortho-positronium has been reported (Matsumoto *et al.*, 1996). It is expected from spin statistics that positronium will in general be formed with a population

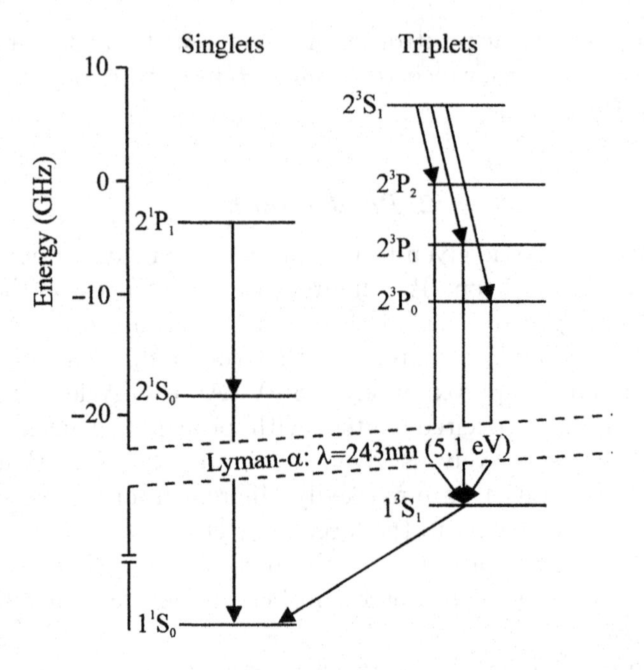

Fig. 1.2. Level diagram of the ground and first excited states of the positronium atom. The splittings are shown for the excited state. The Bohr energy level at $\frac{1}{8}$ ryd is chosen as the arbitrary zero and the $2^3\mathrm{P}_2$ and $2^1\mathrm{P}_1$ states are located approximately 1 GHz and 3.5 GHz respectively below that level. The frequencies in GHz are: $2^3\mathrm{S}_1 \rightarrow 2^3\mathrm{P}_2$, 8.62; $2^3\mathrm{S}_1 \rightarrow 2^3\mathrm{P}_1, 13.0$; $2^3\mathrm{S}_1 \rightarrow 2^3\mathrm{P}_0$, 18.5; $2^1\mathrm{P}_1 \rightarrow 2^1\mathrm{S}_0$, 14.6; $1^3\mathrm{S}_1 \rightarrow 1^1\mathrm{S}_0$, 203.4.

ratio of ortho- to para- equal to 3 : 1, and in the absence of any significant quenching (e.g. via the conversion of ortho-Ps to para-Ps considered in section 7.2), most of the ortho-Ps which is formed will eventually annihilate in this state. Thus, the three-gamma-ray annihilation mode will be much more prolific for positronium than it is for free positron annihilation. The three gamma-rays are emitted in a coplanar fashion, with predicted energy distributions (Ore and Powell, 1949; Adkins, 1983) shown in Figure 1.3(a) along with a recent experimental observation (Chang, Tang and Yaoqing, 1985). The difference between this and the near-monochromatic 511 keV radiation characteristic of the dominant two-gamma-ray annihilation of free positrons provides one way in which to distinguish between these two annihilation modes. This is emphasized in Figure 1.3(b), which shows gamma-ray energy spectra obtained using a high resolution detector under conditions of 0% and 100% positronium formation (Lahtinen *et al.*, 1986).

The lowest order contributions to the annihilation rates for the $n_{\mathrm{Ps}}{}^1\mathrm{S}_0$ and $n_{\mathrm{Ps}}{}^3\mathrm{S}_1$ states of positronium were first calculated by Pirenne (1946)

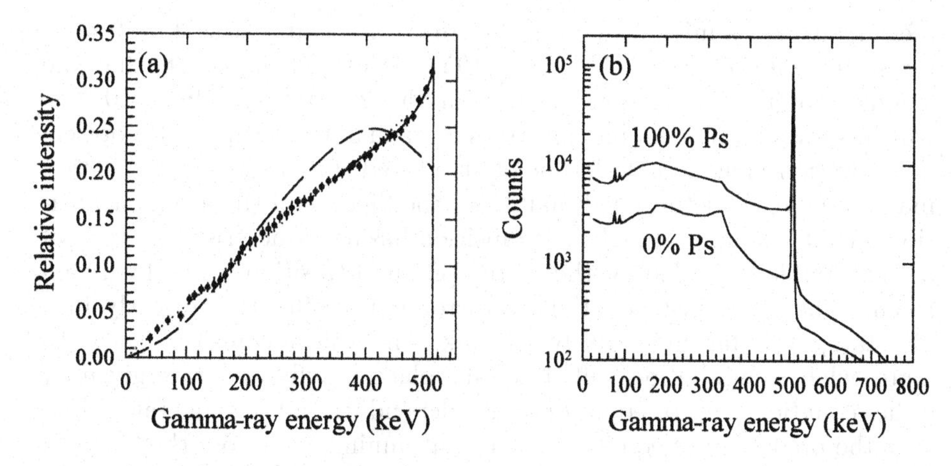

Fig. 1.3. (a) The gamma-ray energy spectrum for the three-photon decay of ortho-positronium. The broken curve is from the theoretical work of Ore and Powell (1949) whilst the dotted line shows the theory of Adkins (1983); the solid line includes an $O(\alpha)$ QED correction. The experimental points are from Chang *et al.* (1985). (b) Schematic gamma-ray energy spectra taken using a high resolution detector under conditions where the fraction of positrons forming positronium is 0% and 100% (e.g. Lahtinen *et al.*, 1986).

and Ore and Powell (1949) respectively, and are given by

$$\Gamma_{2\gamma}\left(n_{\rm Ps}{}^1{\rm S}_0\right) = \frac{1}{2}\frac{mc^2}{\hbar}\frac{\alpha^5}{n_{\rm Ps}^3} \tag{1.5}$$

and

$$\Gamma_{3\gamma}\left(n_{\rm Ps}{}^3{\rm S}_1\right) = \frac{2}{9\pi}(\pi^2-9)\frac{mc^2}{\hbar}\frac{\alpha^6}{n_{\rm Ps}^3}, \tag{1.6}$$

where $\alpha \approx 1/137.036$ is the fine structure constant. Inspection of these two expressions reveals that, owing to the extra power of α and to the numerical factor in equation (1.6), the two-gamma-ray annihilation rate is much greater than that for the three-gamma-ray process. For $n_{\rm Ps} = 1$, it is found that $\Gamma(1^1{\rm S}_0) \approx 8$ GHz whereas $\Gamma(1^3{\rm S}_1) \approx 7$ MHz. The lifetimes against annihilation of the $1^1{\rm S}_0$ and $1^3{\rm S}_1$ states, being the reciprocals of their annihilation rates, are therefore around 1.25×10^{-10} s and 1.4×10^{-7} s respectively. Further details of higher order contributions to the annihilation rates can be found in the review of Rich (1981) and in Chapter 7, where relevant experimental work is also described.

Considering the $n_{\rm Ps} = 2$ states ($2^1{\rm S}_0$, $2^1{\rm P}_1$, $2^3{\rm S}_1$, and $2^3{\rm P}_J$, with $J = 0, 1, 2$), the S-state lifetimes display the $n_{\rm Ps}^3$ scaling law of equations (1.5) and (1.6). For a given value of $n_{\rm Ps}$ the probability that the positron and electron will be found very close together is much lower, and therefore the lifetime against annihilation is much greater for states with $L \neq 0$ than for states with $L = 0$. Alekseev (1958, 1959) calculated the lifetime against annihilation for positronium in the 2P states to be $> 10^{-4}$ s, which is several orders of magnitude greater than the mean life for optical de-excitation. The actual lifetime of an excited state against annihilation may therefore be determined mainly by the lifetime of the atomic transition. As an example, the 2P–1S transition has a characteristic lifetime of 3.2 ns, double the value for the corresponding transition in the hydrogen atom. Therefore, instead of the positronium annihilating directly in a 2P state, it is far more likely to make an optical transition to a 1S state, where annihilation will then take place rapidly at a rate given by either equation (1.5) or equation (1.6), depending on the spin state. Note that the prediction of equation (1.4), that annihilation from the 2^3P and 2^1P states is predominantly into two and three gamma-rays respectively, only applies to direct annihilation. If the positronium first undergoes the optical 2P–1S transition, then the annihilation mode in the lower state is determined by the quantum numbers of that state.

3 Other bound states involving positrons

The next most complex bound state after positronium, and one that has now been observed (Mills, 1981), is the positronium negative ion, Ps^-, which consists of two electrons and a positron. This entity (and its charge conjugate counterpart, the positive ion consisting of two positrons and an electron) has a total spin $S = 1/2$ and a ground state configuration of $^1S^e$. It has no long-lived excited states, but there are several autoionizing resonant states. The most recent calculation of its binding energy with respect to break-up into an electron and positronium gives 0.326 68 eV (Ho, 1993; Frolov and Yeremin, 1989), and Ho (1993) obtained a value of 2.086 1222 ns^{-1} for its annihilation rate, in agreement with the experimental result of Mills (1983b). This value is also close to the spin-averaged annihilation rate of ground state positronium, i.e. one quarter of the rate for the 1^1S_0 state. Further discussion can be found in section 8.1.

Hylleraas and Ore (1947) first showed that the complex involving two electrons and two positrons, the positronium molecule Ps_2, is bound, and this was confirmed by the more accurate work of Ho (1986a), Kinghorn and Poshusta (1993) and Kozlowski and Adamowicz (1993). The later calculations gave the binding energy with respect to break-up into two positronium atoms as 0.435 eV, but a significantly larger value, 0.573 eV, has recently been obtained by El-Gogary *et al.* (1995). The system, with total spin $S = 0$, has no bound excited states, but several autodissociating states have been found (Ho, 1989b). Thus far, Ps_2 has not been observed in the laboratory because the large instantaneous positron densities required for its formation have not yet been achieved.

A related, but somewhat less exotic, complex is positronium hydride, PsH. This entity has been the subject of many theoretical studies, with the most recent predicting a binding energy with respect to break-up into hydrogen and positronium of approximately 1.067 eV (Ho, 1986b, Ryzhikh and Mitroy, 1997). Experimental evidence for its existence has been forthcoming from a recent study of positron–CH_4 collisions (Schrader *et al.*, 1992; see also Chapter 7), although its short lifetime against annihilation, 0.5 ns, makes further experimentation difficult. Again, the system has no bound excited states, but it possesses a Rydberg series of autodissociating resonances. Further discussion of these and other bound systems containing positrons is given in sections 1.6 and 7.5.

The last bound state to be introduced here is antihydrogen, consisting of a positron and an antiproton. From the CPT theorem this entity, which is stable in vacuum, is expected to have spectroscopic properties identical to those of atomic hydrogen. There is no evidence for the existence of bulk antimatter in the universe, but antihydrogen has recently been produced (Baur *et al.*, 1996; Blanford *et al.*, 1998), although at such high kinetic energies that no further investigation of its properties was possible. However, recent advances in the slowing down and trapping of antiprotons (Gabrielse *et al.*, 1990; Holzscheiter *et al.*, 1996), described in section 8.3, lend hope that the synthesis of very low energy antihydrogen will be achieved in the not too distant future.

Although there are rather few true bound systems containing positrons, numerous resonances are known to exist in the scattering of positrons and positronium by various target systems. Many of these are Feshbach resonances associated with the degenerate thresholds for excitation of positronium or of a hydrogenic target, but others arise from the Rydberg series of states of a positron 'bound' to the residual negative ion. For example, in PsH, excited states of the system consisting of a positron interacting with the H^- ion have energies in the continuum for positronium–hydrogen scattering, and they therefore manifest themselves as resonances in positronium–hydrogen scattering. All these resonances have very narrow energy widths, and they cannot yet be resolved experimentally.

1.3 Basic experimental techniques

In this section we introduce three techniques frequently encountered in positron physics, namely those used to measure annihilation lifetimes and the Doppler broadening (or Doppler shift) and angular correlation of the annihilation radiation. These techniques, or variants thereof, are encountered throughout the rest of this work, and here we briefly describe

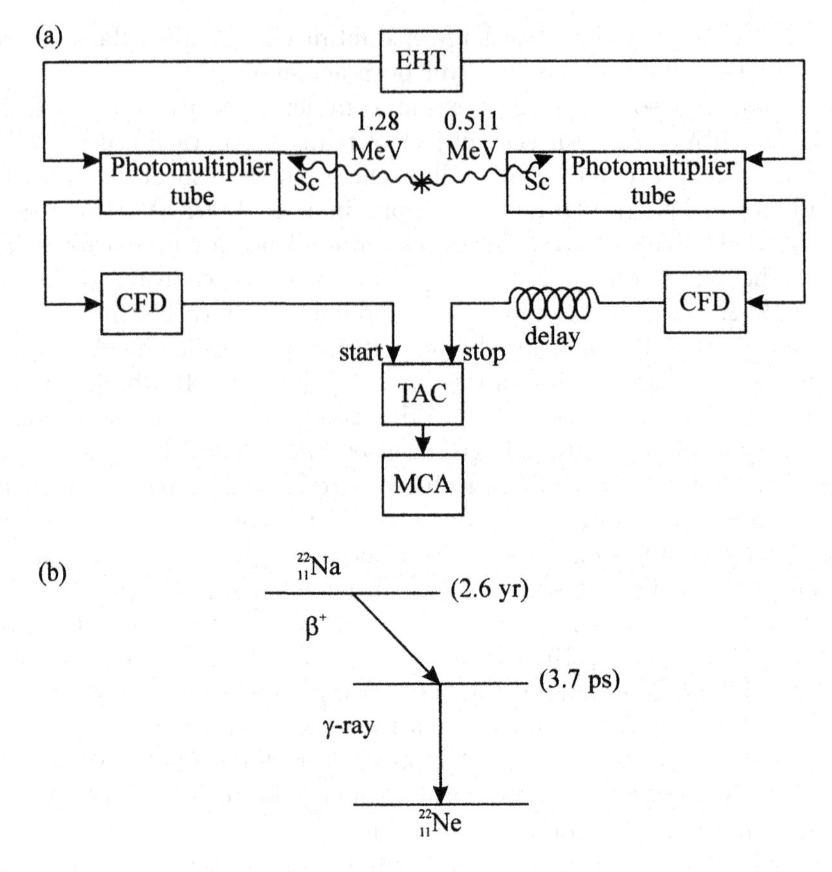

Fig. 1.4. (a) Schematic of a positron lifetime spectrometer; the star indicates the ^{22}Na source and only one of the 0.511 MeV annihilation photons is shown. Key: Sc, scintillator; CFD, constant fraction discriminator; TAC, time-to-amplitude converter; MCA, multichannel analyser. (b) Simplified level diagram of the ^{22}Na decay scheme. The β^+ fraction dominates that for electron capture for this isotope.

the principles behind them and give illustrations of the apparatus and methodology involved (see also Hautojärvi and Vehanen, 1979).

1 *Annihilation lifetimes*

The basic operating principle of all traditional positron lifetime systems, schematically illustrated in Figure 1.4(a), is to measure individual lifetimes of many positrons by suitably processing the timing signals derived from their birth and from their eventual annihilation with electrons in the medium in which the positrons are diffusing. In the apparatus shown, both the birth and annihilation signals are registered using gamma-ray scintillation counters (see e.g. Knoll, 1989 for a general discussion). One of the most commonly used sources of positrons in lifetime studies is the ^{22}Na radioisotope, a simplified decay scheme for which is shown in Figure 1.4(b). The β^+ branching ratio for this isotope is around 90% and the positron emission is followed promptly by a gamma-ray of energy 1.274 MeV. This gamma-ray is used to register the positron's birth by starting the timing sequence. In addition, ^{22}Na has a conveniently long half-life of around 2.6 years, and is commercially available in a form suitable for many applications.

The annihilation gamma-rays, at energies of 511 keV and below (see section 1.2 for a discussion of the various possible annihilation modes), are registered by a second scintillation counter. The start and stop signals are usually processed by a pair of discriminators, and the simplest arrangement, shown in Figure 1.4(a), consists of two constant-fraction discriminators, which combine good timing characteristics with the ability to set upper and lower limits on the pulse height accepted by the instrument. Thus, the higher energy 'start' gamma-ray can easily be selected. After the insertion of an appropriate delay, to introduce a minimum fixed time between the start and stop signals, the pulses are fed to the inputs of a time-to-amplitude converter. The output of this module, which is proportional to the length of time between the start and stop signals, and thus to the individual positron lifetime, is then stored using a multichannel analyser. A lifetime spectrum is thereby built up, frequently containing 10^6–10^7 events, from which various lifetimes, principally those due to free positron and ortho-Ps annihilation, along with several other parameters, can be extracted. Numerous examples of results obtained using this kind of instrumentation are described in Chapters 6 and 7.

Another useful lifetime system, though less frequently encountered, is the so-called β^+–gamma system. A variant of this technique was used in some of the pioneering measurements of positron–atom total collision cross sections described in Chapter 2. The annihilation gamma-ray still provides the stop signal, but the start signal is derived via the energy deposited by the positrons as they traverse a thin (typically 0.1–0.3 mm) scintillator. This method of start detection has a high efficiency, usually around 50%, which permits the use of a relatively weak radioactive source, resulting in a superior signal-to-background ratio.

Numerous technical descriptions of various aspects of lifetime apparatus can be found (e.g. MacKenzie, 1983, and references therein), and a number of sophisticated data analysis and fitting procedures have been developed to analyse the data collected (e.g. Coleman, Griffith and Heyland, 1974; Coleman, 1979; Kirkegaard, Pedersen and Eldrup, 1989), but detailed discussion of these topics is beyond the scope of the present treatment.

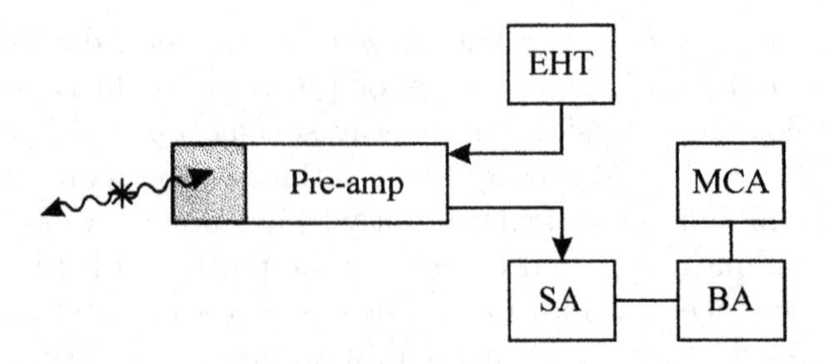

Fig. 1.5. Schematic of a gamma-ray energy spectrometer for Doppler broadening studies. The signal from the detector pre-amp is processed by spectroscopy and biassed amplifiers (SA and BA respectively) before being recorded in the multichannel analyser (MCA).

Technical advances in these areas have frequently been documented in the proceedings of the triennial series of Positron Annihilation conferences (see the appendix), which can serve as a useful starting point for a literature search.

2 Doppler shift or broadening of the annihilation radiation

In the frame of reference in which the centre of mass of an electron–positron pair is at rest, the two gamma-rays arising from their annihilation in a spin singlet state each have an energy of 511 keV, and they emerge in opposite directions; i.e. the angle between the two directions is π radians. However, the motion of the centre of mass creates a Doppler shift in the gamma-ray energies as measured in the laboratory frame of reference, and the angle between the gamma-rays is no longer π. When slowing down in matter, most positrons thermalize before annihilation, and the momentum of the centre of mass motion of an electron–positron pair is therefore predominantly that of the electron. The shift in the energy of the 511 keV line is given by

$$\Delta E_\gamma = mcv_{\rm cm}\cos\phi, \tag{1.7}$$

where $v_{\rm cm}$ is the speed of the centre of mass of the pair and ϕ is the angle between the direction of motion of the centre of mass and that of one of the emitted gamma-rays. Inserting typical values for $v_{\rm cm}$ ($\approx 10^6$ m s^{-1}) and $\cos\phi$ ($\simeq 1/\sqrt{2}$), we find that $\Delta E_\gamma \approx 1.2$ keV. This is comparable to the energy resolution of the commercially available high purity germanium gamma-ray detectors which, as illustrated schematically in Figure 1.5, are used in studies of this type. The amplification system following the detector is standard, usually consisting of a spectroscopy amplifier and a biassed amplifier; this allows the widened annihilation line to be examined

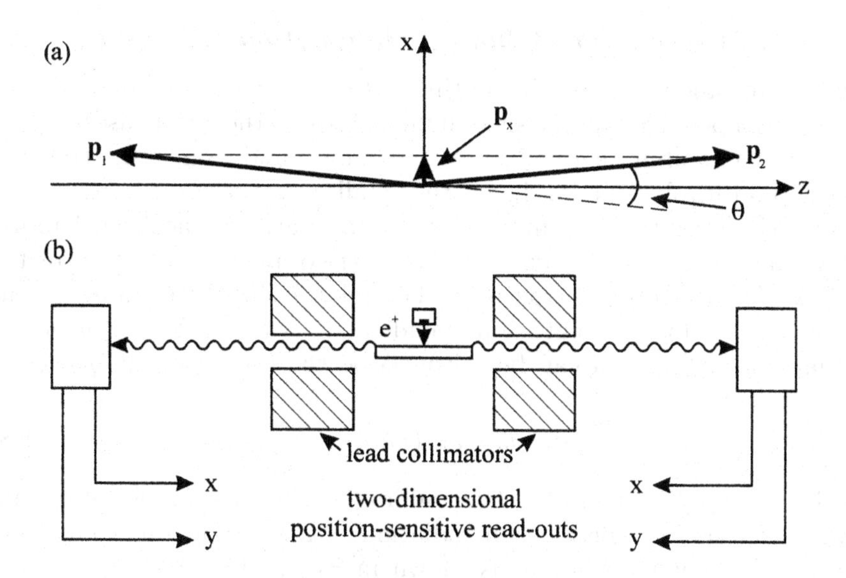

Fig. 1.6. (a) Illustration (exaggerated) of the relationship between the momenta $\boldsymbol{p}_1$ and $\boldsymbol{p}_2$ of the two annihilation photons, the angle θ between them and the x-component of the electron momentum prior to annihilation. (b) Schematic showing the principle of a two-dimensional ACAR apparatus. The source is shown in the small rectangle above the sample, from which the two 0.511 MeV gamma-rays are emitted.

in more detail. The gamma-ray energy distribution can then be stored in a multichannel analyser and processed in various ways depending upon the details of the study.

The widest field of application of this technique is in investigations of the solid state, where, in most cases, the geometry of the experiment and the random nature of the direction of motion of the electron–positron pairs means that the angle ϕ has a continuous distribution; consequently the 511 keV gamma-ray line is Doppler broadened by an amount related to the momentum distribution of the annihilating pair. This technique has only rarely been applied to the study of the behaviour of positrons and positronium in gases, although recent measurements of the Doppler broadening obtained from the annihilation of very low energy positrons confined in a trap containing helium gas have been found to be in excellent agreement with theoretical predictions (Van Reeth *et al.*, 1996; see also Chapter 6). Other studies using positron beams have also been reported in which the geometry allows a narrow band of values of ϕ to be selected and consequently a specific Doppler shift to be observed, as has been exploited in studies of positronium formation (Chapter 4) and in the detection of Ps^- (Chapter 8).

3 Angular correlation of annihilation radiation

The final technique we describe in this section has been used to study the behaviour of positrons and positronium in gases, in the latter case with the aim of understanding thermalization phenomena and momentum transfer. The method involves measuring θ, the small deviation from π radians in the angle between the two annihilation gamma-rays. As mentioned above, this deviation is a consequence of the centre-of-mass momentum of the annihilating electron–positron pair. The relationship between θ and the component of this momentum perpendicular to the direction of one of the gamma-rays, which can be taken to be the x-component, p_x, as in Figure 1.6(a), is

$$\theta = p_x/(mc); \tag{1.8}$$

this has a typical value of a few milliradians. An example of an angular correlation of annihilation radiation (ACAR) apparatus to detect such small angular deviations is shown in Figure 1.6(b). It consists of a pair of two-dimensional position-sensitive gamma-ray detectors, each typically located at a distance of 5 m from a radioactive source, which is immediately adjacent to the sample being studied. The field of view of each detector is limited by lead collimating slits. Using this type of arrangement, angular resolutions below one milliradian can be achieved routinely. In media in which there is no preferred axis of symmetry (e.g. in gases and liquids in the absence of external fields) it is not necessary to use a two-dimensional system, although it can sometimes be justified in terms of the improved count rate. Instead, the one-dimensional, or long-slit, technique can be used, in which the position-sensitive detectors are replaced by two single detectors, each with a long-slit collimator placed in front of it, giving integration over one of the components of the momentum. One of the detectors is fixed, whilst the other is scanned through the angle θ.

1.4 Slow positron beams

In this section we endeavour to describe the production of low energy, or slow, positron beams through a discussion of the following topics: the primary sources of high energy positrons and their slowing down, or moderation, through implantation into solids; the subsequent diffusion of the moderated positrons; the emission into vacuum of those positrons which, prior to annihilation, reach a suitable surface; and their subsequent manipulation to form a beam. It is hoped that this discussion will be of particular benefit to the non-specialist in illustrating some of the difficulties, limitations and new possibilities encountered when dealing with low

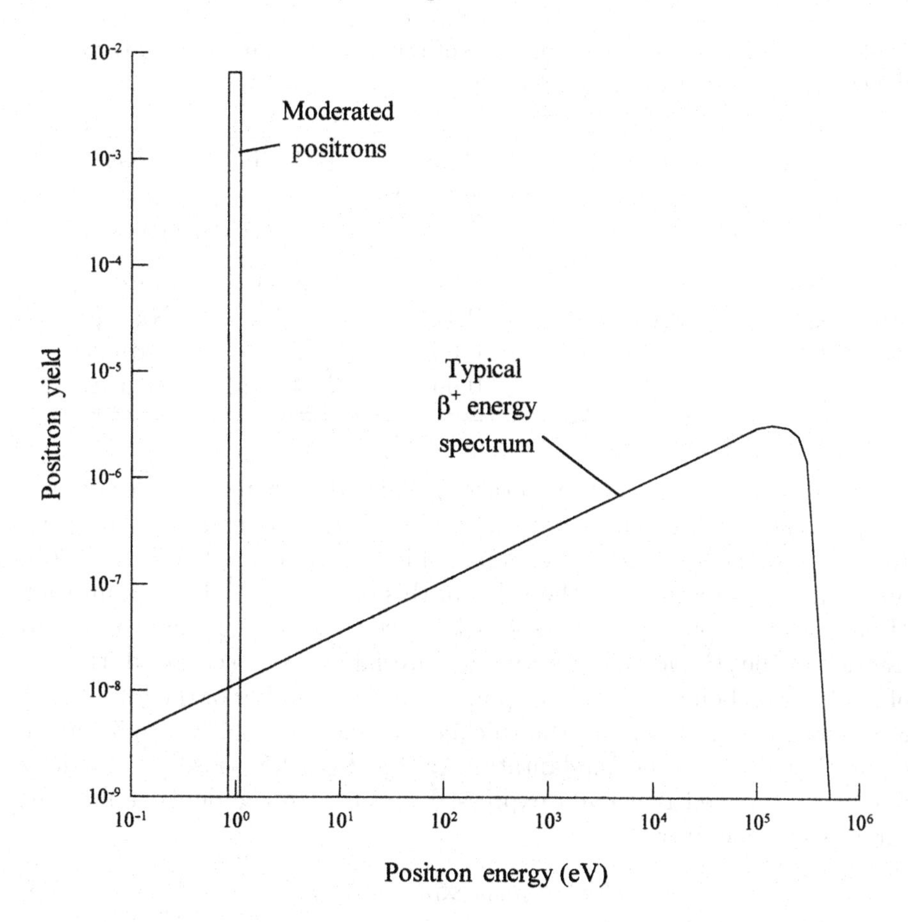

Fig. 1.7. Comparison of the energy spectrum of β^+ particles from a radioactive source with that for moderated positrons.

energy positron beams. A similar discussion has been given by Beling and Charlton (1987).

1 Introduction to positron moderation

In nature, high energy positrons are produced either as a result of nuclear decay, usually of an artificially produced radioactive isotope such as ^{22}Na, or by pair production from a photon of sufficiently high energy. Except near the threshold for the latter process, where the production cross section is very low, the positrons are produced with a very large energy spread (see Figure 1.7), typically of the order of MeV, which renders them unsuitable for most experiments in atomic collision physics.

Table 1.1 shows some of the properties of various β^+ emitters which have been used to create low energy beams. We adhere to the convention

Table 1.1. Selected properties of some of the radioisotopes commonly used in the creation of low energy positron beams

Isotope	β^+ Branching ratio	Endpoint energy (MeV)	Half-life	Typical production mechanism
^{22}Na	0.91	0.54	2.6 yr	^{24}Mg(d, α)
^{58}Co	0.15	0.47	70.8 d	^{58}Ni(n, p)
^{64}Cu	0.19	0.65	12.7 h	^{63}Cu(n, γ)
^{11}C	0.99	0.96	20.4 min	^{11}B(p, n)

of referring to a high energy positron emitted in a nuclear decay as a β^+ particle. Of particular note in table 1.1 are the end-point energies $E_{\max}$, since the average β^+ energy, and hence the implantation depth into matter, increases with the value of this parameter. The implantation of energetic β^+ particles into solids has been studied for many years. A fraction of the β^+ particles striking a material are back scattered, the size of the fraction being dependent upon the atomic number of the target (e.g. Mackenzie *et al.*, 1973) and the thickness of the material (up to a certain saturation value). The implantation profile $P(x)$ for those β^+ particles not back scattered is usually expressed in terms of the depth x into the target (or moderator) as

$$P(x) = \mu_{\mathrm{imp}} \exp(-\mu_{\mathrm{imp}} x), \tag{1.9}$$

where μ_{imp} is the absorption coefficient, which can be empirically expressed in terms of $E_{\max}$ and the target density (see e.g. Brandt and Paulin, 1977; Beling *et al.*, 1987). The positrons lose most of their kinetic energy in the material by ionizing collisions. In metals, where there are efficient energy-loss processes which can be excited by very low energy positrons, annihilation of the positrons occurs predominantly after their thermalization, which is typically reached a few picoseconds after implantation. However, this is not true of all materials, counter-examples being the rare gas solids, which are wide-band-gap insulators.

In solids the free positron lifetime τ lies in the approximate range 100–500 ps and is dependent upon the electron density. Following implantation, the positrons are able to diffuse in the solid by an average distance $L_+ = (D_+\tau)^{1/2}$, where D_+ is the diffusion coefficient. This quantity is usually expressed in cm^2 s^{-1} and is of order unity for defect-free metallic moderators at 300 K (Schultz and Lynn, 1988). The requirement of very low defect concentration arises because the value of D_+ is otherwise dramatically reduced owing to positron trapping at such sites.

Inserting values for D_+ and τ into the expression for L_+, a value of approximately 1000 Å is obtained, and this is typical for metals. One can then find an estimate of the efficiency ϵ of the moderator by multiplying the implantation profile $P(x)$, equation (1.9), by the probability that a positron reaches the surface from a depth x, $\exp(-x/L_+)$, and integrating over all values of x. Then, since $\mu_{\text{imp}}L_+ \ll 1$, the efficiency may be written as

$$\epsilon = y_0 \mu_{\text{imp}} L_+, \tag{1.10}$$

where $\mu_{\text{imp}}L_+$ is around 6×10^{-3} and y_0 is a surface-dependent quantity, the branching ratio for emission from the surface as a free positron. In general, y_0 is less than unity (and may even be zero if the positron work function ϕ_+ for the surface is positive), owing to the presence of competing processes at the surface. These are summarized in Figure 1.8(a), which shows the possible fates of positrons that have diffused to a metal surface.

Figure 1.8(b) illustrates a one-dimensional representation of the single-particle potential energy of a positron in the near-surface region for the important case where ϕ_+ is negative, so that escape of the thermalized positron from the solid into the vacuum is energetically allowed. Thus, a positron which has diffused to the surface may be emitted with the characteristic energy $-\phi_+$. Ideally, if the surface, at a temperature T, were atomically clean and flat, the positron would be ejected with a speed perpendicular to the surface of $(-2\phi_+/m)^{1/2}$ and a small transverse component approximately equal to $(2k_{\text{B}}T/m)^{1/2}$, where k_{B} is Boltzmann's constant, due to thermal smearing (Gullikson *et al.*, 1985). In reality, deviations occur which are thought to be due to the presence of adsorbed impurities and/or various irregularities on the surface; these cause an increase in the angular and energy spread of the positrons. At very low temperatures, positron emission is inhibited by quantum mechanical reflection at the surface (Britton *et al.*, 1989), although Jacobsen and Lynn (1996) have found that this effect can be reduced by the use of a thin moderator, of thickness $< L_+$, to encourage multiple encounters of the positrons with the surface.

The work function ϕ_+ of a particular surface has two contributing factors, namely, the relevant chemical potential μ experienced by the particle in the bulk material and the surface dipole potential D. Thus, for positrons and electrons the relevant work functions can be written (Tong, 1972; Hodges and Stott, 1973) as $-\phi_\pm = \mu_\pm \pm D$. The chemical potential contains terms due to the electron and positron interactions with the other electrons and with the ion cores. The surface dipole, which is attractive for positrons and repulsive for electrons, arises mainly from the tailing of the electron distribution into the vacuum for a distance of approximately

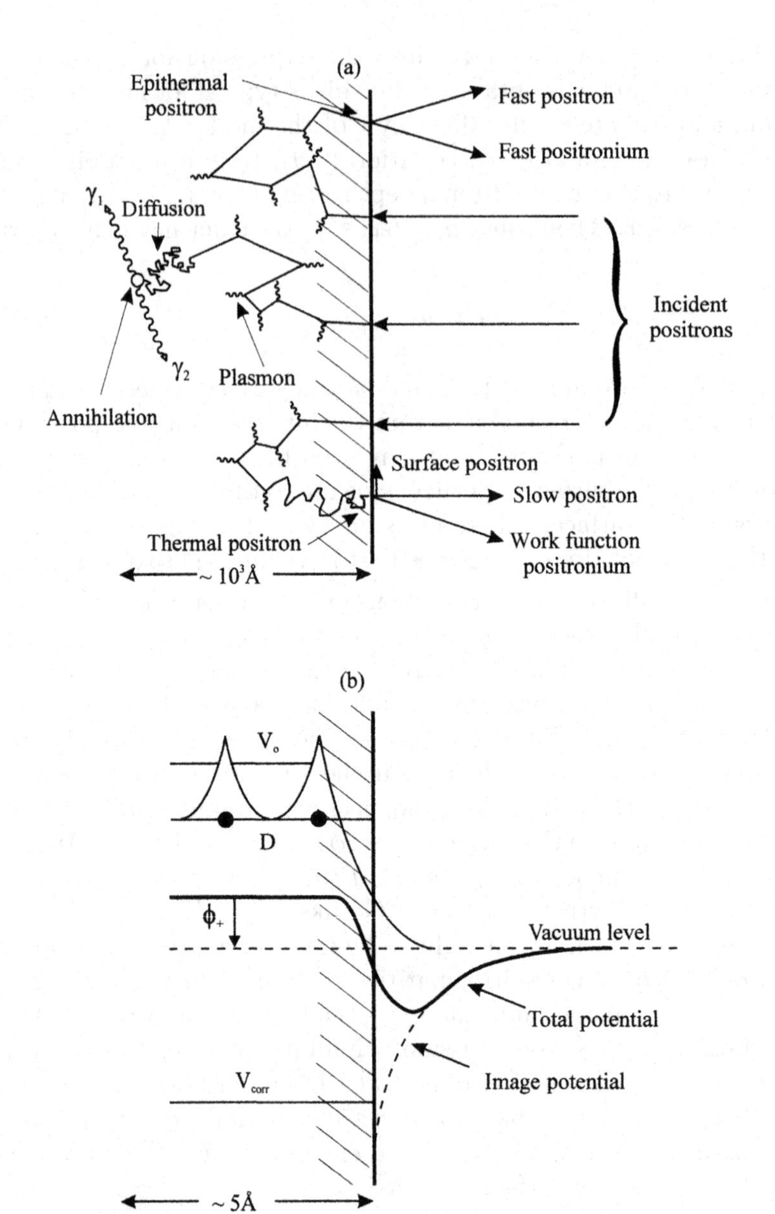

Fig. 1.8. (a) Simplified illustration of the interactions of positrons at a metal surface (after Mills, 1983a). An incident positron may return to the surface as a thermal or epithermal positron or it may be annihilated within the metal. (b) Representation of the one-dimensional potential for a thermalized positron near the surface of a metal (after Schultz and Lynn, 1988). The positron chemical potential contains a term V_{corr} due to correlation with the conduction electrons and a term V_0 due to the repulsive interaction with the ion cores, shown as black discs.

10^{-10} m. It is the presence of this effect which makes it possible for ϕ_+ to be negative.

The above discussion has been concerned with the energetics of positrons at surfaces with a negative positron work function. However, Mills and Gullikson (1986) have observed the copious emission of positrons from rare gas solids. Here, in contrast to metals, there are no free electrons near the surface so the dipole potential contribution is smaller and ϕ_+ is positive (Gullikson and Mills, 1986). However, this unfavourable circumstance is more than offset by the slow energy loss rate experienced by low energy positrons in these media, which results in many more reaching the surface epithermally. Some of these positrons are then able to overcome the positive ϕ_+ barrier and are emitted into the vacuum. Positron beams formed from solid rare gas moderators generally have inferior energy widths and angular properties compared with those formed from metal surfaces. They have, however, yielded efficiencies of around 1% (Khatri *et al.*, 1990; Mills and Gullikson, 1986; Greaves and Surko, 1996) and they can be fabricated in the unusual geometries suited to use at some high flux facilities (e.g. Weber *et al.*, 1992). Enhanced moderation efficiencies have been obtained by the electric-field-assisted drift of positrons in rare gas solids, achieved, as reported by Merrison *et al.* (1992), by deliberately charging the moderator surface. In many applications of low energy positron beams, much lower moderator efficiencies are used for practical reasons, typical values being in the range 10^{-3}–10^{-4}. Other considerations, such as the maximum radioactive source strength permissible and the self-absorption of the β^+ particles in the source, which reduces the number available for moderation, limit most laboratory beams to maximum intensities of around 10^7 s^{-1}, although much lower fluxes are often used.

Once the slow positrons are emitted into the vacuum surrounding the moderator they can be readily manipulated to form a beam and transported away from the region containing the radioactive source. Many different methods have been devised to achieve this, although they can be broadly divided into two classes, those using mainly magnetic fields, and those using electrostatic fields. These are usually termed B- and E- beams respectively, and examples of both are discussed below. So-called hybrid beams have also been developed, which usually employ an electrostatic field for positron extraction and focussing, with transport accomplished by an intermediary magnetic field.

2 Magnetically confined positron beams

A schematic illustration of a typical apparatus for the production of a B-beam is given in Figure 1.9 (Zafar *et al.*, 1992). The source-moderator region is shown in the inset: a commercially available encapsulated ^{22}Na source lies directly behind a metallic (typically tungsten or molybdenum) mesh or foil moderator. Low energy positrons are accelerated, as they leave the moderator, by a small potential difference (≈ 100 V) maintained between it and the surrounding vacuum chamber. Both the source and the moderator can be removed from their working positions remotely: movement of the moderator is needed to facilitate *in situ* treatment in an auxiliary chamber.

The positrons are constrained to follow helical trajectories by an axial magnetic field, typically in the range 5×10^{-3}–10^{-2} T, and they pass through a region containing a pair of electrostatically biassed plates. The electric field $\boldsymbol{E}$ perpendicular to $\boldsymbol{B}$ results in a drift speed $|\boldsymbol{E}|/|\boldsymbol{B}|$ perpendicular to both fields, which serves to deflect the low energy positrons without influencing the fast β^+ flux from the source. Thus, with the insertion of appropriate shielding inside the vacuum chamber, the remainder of the beam line can be removed from the line of sight of the high energy radiation from the source. In many cases the electrostatic plates are curved, resulting in a near distortion-free deflection of the beam (Hutchins *et al.*, 1986).

The entire region containing the source, the moderator and the deflector plate is electrically isolated from the rest of the beam line by a ceramic break, and it can be floated to an electrostatic potential of around 30 kV. Thus, the final beam energy can be varied up to 30 keV. Simpler systems, using only a carefully designed moderator region, can be used if lower beam energies (less than approximately 10 keV) are required.

Following acceleration, the beam, which is still confined by the magnetic field, traverses several pumping, scattering and other chambers before reaching the end of its flight path, where it is detected using a channel electron multiplier array (CEMA). Detection can also be accomplished using a gamma-ray counter located outside the vacuum chamber to register the annihilation photons, and such a method is frequently used, either alone or in conjunction with a CEMA or some other secondary electron multiplier. The scattering chambers shown in Figure 1.9 have been designed specifically for research into atomic collisions with low energy positrons. This is given only as an example, since similar devices using ultra-high vacuum instrumentation are frequently used for positron studies in surface and sub-surface physics (see e.g. Lahtinen *et al.*, 1986; Schultz, 1988).

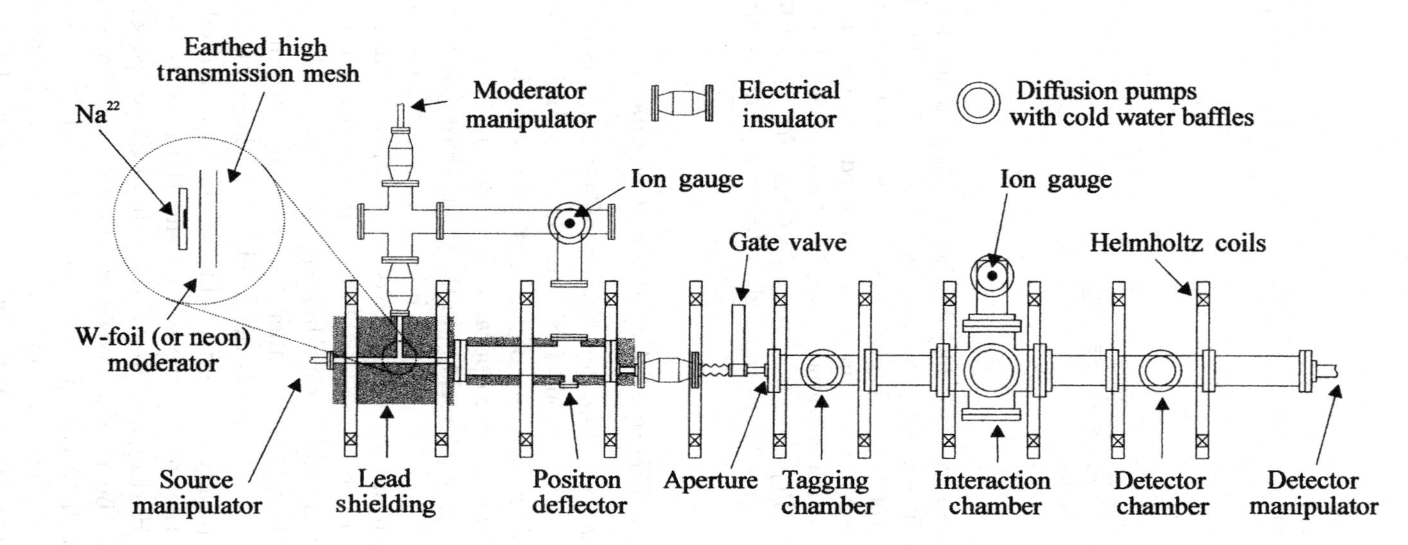

Fig. 1.9. Apparatus for the production of a magnetically confined positron beam in operation at University College London.

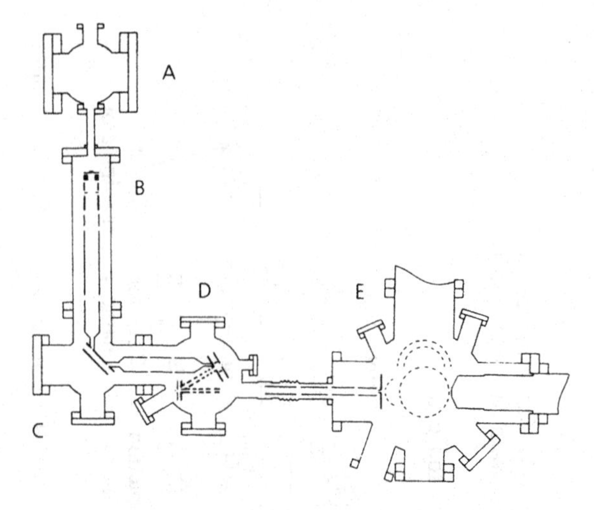

Fig. 1.10. The fully electrostatic high-brightness positron beam developed by the Brandeis group. The positron Soa gun is located near B. The beam is deflected at C using a cylindrical mirror analyser and focussed onto a remoderator in chamber D. The extracted beam is then focussed and remoderated at the lower left of D. The double brightness-enhanced beam is then transported into the target chamber, E. Reprinted from *Nucl. Instrum. Methods B* **143**, Charlton, Review of Positron Physics, 11–20, copyright 1998, with permission from Elsevier Science.

3 Electrostatic positron beams

The electrostatic beam system developed by Canter and coworkers (Canter, 1986; Canter *et al.*, 1986; 1987) at Brandeis University is shown in Figure 1.10. This was one of the first *E*-beams to be developed, and some of the features incorporated into this instrument have been duplicated by other groups working in the field. Here, positron extraction is accomplished using a so-called modified Soa gun, which is based upon a design well known in conventional electron optics but having an extra electrode (Canter *et al.*, 1986). This feature is necessary to prevent field penetration from the surrounding chamber walls, which could arise because of the relatively large inner diameter (35 mm in this case) of the lens elements. The latter is dictated by the large emitting area of the moderator (typically 0.5 cm^2), which is in turn limited mainly by the active area of the radioactive source and the need to keep the lens filling factors low (Harting and Read, 1976) in order to reduce the effects of aberrations. The beam is transported using a standard einzel lens to the entrance of a cylindrical mirror analyser (CMA). This device has an energy resolution of around 3% and a wide input angular acceptance; an analysed and focussed image of the beam is produced at the output (see e.g. Risley, 1972 for a more detailed discussion of the operation of this type of analyser). In this type of application the CMA is not usually employed for energy analysis of the slow positrons, which are all transmitted, but rather to separate them from the high energy β^+ particles and gamma-ray flux which is also present; it serves a similar purpose to the deflector described above in subsection 1.4.2. The beam can be transported and focussed onto the target or the first remoderator (located at the end of the upper lens stack, which provides beam transport between chambers C and D), by a variety of means; shown here is a further einzel lens with a final accelerating stage such that the beam strikes the target at an energy of 5 keV and with a diameter of approximately 1 mm. Depending upon the intended application of the beam, and bearing in mind the constraints thus imposed upon its energy and angular properties, this first section may be used directly for physics investigations.

The positrons re-emitted from the first remoderator, which is chosen to have ϕ_+ negative, form a brightness-enhanced beam (Mills, 1980) since, despite losing a fraction equal to $1 - \epsilon_{\rm rm}$, where $\epsilon_{\rm rm}$ is the re-emission efficiency, to other processes at the surface and to sub-surface annihilation, they originate from a smaller emitting area than that of the primary moderator and have a much reduced energy. The brightness gain, $G_{\rm B}$, is made possible by the non-conservative energy-loss processes during remoderation, which circumvent Liouville's theorem, and also the high efficiency of re-emission ($\epsilon_{\rm rm}$ is typically ≈ 0.2) following implantation at keV energies. Assuming the positron emission properties are the same for the moderator and the remoderator, and that lens aberrations can be neglected, the brightness gain is given by

$$G_{\rm B} = \epsilon_{\rm rm}(d_2/d_1)^2, \tag{1.11}$$

where d_1 and d_2 are the respective beam diameters at the primary moderator and the remoderator. This process of brightness enhancement can be continued, in principle, for several stages until the physical limit of the size of the beam is governed by the diffusion length of the implanted positrons. The system of Canter and his coworkers has a further stage of brightness enhancement: the beam is accelerated and focussed onto a second remoderator located at the end of the first small lens stack in chamber D. Here the re-emitted beam has a diameter ≈ 0.1 mm and an overall value of $G_{\rm B} \approx 500$ over that emitted from the primary moderator. Brandes *et al.* (1988) have described how this beam can be accelerated and focussed, to produce a microbeam, into the target chamber E for surface and sub-surface applications. Special experiments in the study of systems containing more than one positron can also be contemplated using such highly focussed beams (Mills, 1984; Platzman and Mills, 1994; see Chapter 8 for further discussion).

In practice, many atomic collision experiments can be performed without recourse to brightness enhancement, and the first generation of studies of various positron–atom (molecule) differential scattering cross sections were performed in this manner. The results of these investigations are reported in later chapters.

4 Facility-based beams

Almost since the earliest attempts to produce well-defined beams of low energy positrons, various types of accelerator have been used for this purpose, e.g. electron linear accelerators, microtrons and cyclotrons (see e.g. Dahm *et al.*, 1988; Itoh *et al.*, 1995). Positron beams have also been developed at nuclear reactors (Lynn *et al.*, 1987).

When using an electron accelerator, fast positrons are produced by pair production from bremsstrahlung gamma-rays generated as the high energy electrons from the accelerator slow down in matter, whereas with cyclotrons and reactors, very intense primary positron sources are produced directly. Slow positron beams are then produced and transported using similar techniques to those described previously in this section.

The main reason for using accelerator and reactor facilities is to produce positron beams with qualities which cannot easily be achieved using normal laboratory radioactive sources. One of the most important of these qualities is the beam intensity which, in most experiments, is limited either by the strength of the commercially available radioisotope or the amount of activity which can be safely handled. The intensity consideration may also be convoluted with other attributes of the beam, e.g. several stages of brightness enhancement (see subsection 1.4.3) may be needed, each involving a loss of slow positron flux. Some facility-based beams, particularly those located at electron linear accelerators, are naturally pulsed in nature, with pulse durations in the ns–μs range. Such beams can be used either for special applications in which large bursts of positrons are necessary or to interface with other pulsed sources (e.g. lasers for spectroscopic and other studies of the properties of positronium; see Chapter 7). Although pulsed beams can be produced in the laboratory, the instantaneous intensities available at electron linear accelerators can be much higher. Pulses containing around 10^6 positrons have been produced with nanosecond durations and at kHz frequencies (Howell, Alvarez and Stanek, 1982).

Although there have been many technical advances in this area, no one facility has yet emerged as significantly superior to any other. A brief overview of such facilities around the world can be found in the *Proceedings of the Sixth International Workshop on Slow Positron Beam*

Techniques for Solids and Surfaces (*Applied Surface Science*, Volume 85, 1995).

1.5 The production of positronium

1 Basic physics of positronium production

Of most relevance to us here is the production of positronium atoms in gases or at the surface of solids, and we restrict our discussion to these situations. In gases, positronium can be created in the collision of a positron with an atom or molecule according to

$$e^+ + X \to \mathrm{Ps} + X^+, \tag{1.12}$$

which has a formation threshold at

$$E_{\mathrm{Ps}} = E_{\mathrm{i}} - 6.8/n_{\mathrm{Ps}}^2, \tag{1.13}$$

where E_{i} is the ionization threshold of the atom or molecule and $6.8/n_{\mathrm{Ps}}^2$ is the binding energy for the positronium state with principal quantum number n_{Ps}, all energies being in eV. Formation into the ground state ($n_{\mathrm{Ps}} = 1$) is expected to dominate in most cases. For atomic targets E_{Ps} is the lowest inelastic threshold whereas for molecular gases rotational, vibrational and even some electronic states may have lower excitation thresholds, although positronium is still formed abundantly according to reaction (1.12). More detailed aspects of positronium formation are given in Chapter 4.

Owing to the high density of free electrons and the consequent screening of its positive charge, a positron cannot bind with an electron to form positronium in the bulk of a metal. Positronium can, however, be created as a positron passes through the outer, lower density, electron cloud at the surface. The positronium thus formed is then emitted into the vacuum with a kinetic energy $\leq -\epsilon_{\mathrm{Ps}}$, the positronium formation potential; this can be expressed, using energy conservation, in terms of the positron and electron work functions for the particular material, ϕ_+ and ϕ_-, as

$$\epsilon_{\mathrm{Ps}} = \phi_+ + \phi_- - 6.8/n_{\mathrm{Ps}}^2. \tag{1.14}$$

The formation potential is usually negative for $n_{\mathrm{Ps}} = 1$, and therefore positronium emission is allowed. However, with the possible exception of a diamond surface (Brandes, Mills and Zuckerman, 1992), it is positive for $n_{\mathrm{Ps}} \geq 2$, which therefore precludes the emission of excited state positronium following positron thermalization in the material.

Referring again to Figure 1.8(a), the surface-trapped positron shown there is bound by an energy E_{b}. It has been shown many times (e.g.

Mills and Pfeiffer, 1979; Lynn, 1979; Poulsen *et al.*, 1991; see also the discussion in subsection 1.5.3 below) that heating the metal surface can thermally activate positronium formation, the activation energy $E_{\rm a}$ being given by

$$E_{\rm a} = E_{\rm b} + \phi_{-} - 6.8/n_{\rm PS}^2, \tag{1.15}$$

which is typically less than 1 eV (Schultz and Lynn, 1988).

In contrast to the case for metals, positronium can be formed in the bulk of many insulators and molecular crystals, and any positronium which subsequently diffuses to the surface can be emitted into the vacuum with a kinetic energy $\leq -\phi_{\rm Ps}$, where $\phi_{\rm Ps}$ is the positronium work function. Its value can be expressed in terms of the binding energy of the positronium when in the solid, $E_{\rm B}$, and the positronium chemical potential, $\mu_{\rm Ps}$, as (Schultz and Lynn, 1988)

$$\phi_{\rm Ps} = -\mu_{\rm Ps} + E_{\rm B} - 6.8/n_{\rm PS}^2. \tag{1.16}$$

Note that the use of a work function here, rather than the formation potential used for metals, is appropriate since positronium can exist in the bulk of the material.

In what follows, we describe how positronium has been formed in gases and solids using low energy positron beams and also, in what is now regarded as the traditional way, using β^+ particles directly from radioactive sources.

2 Traditional methods

The discovery of positronium by Deutsch (1951) was accomplished by its formation in gases, a technique which has been used widely since then. The β^+ particles emitted from the radioactive source are moderated in the gas, which typically has a number density of atoms (or molecules) of approximately 10^{25} m^{-3}, and when losing their last few tens of eV of kinetic energy they may form positronium. At this density, the dominant formation mechanism is that given by equation (1.12); further discussion pertinent to dense gases is given in section 4.8. The positronium is formed with a range of kinetic energies and it, or more particularly the long-lived ortho-Ps component, slows down by collisions with other gas atoms. In contrast, the para-Ps, with its much shorter lifetime of 125 ps, is likely to annihilate before any substantial energy loss can occur. Eventually, if the ortho-Ps cannot break up in a subsequent collision then annihilation of the positron occurs, either with the electron to which it is bound or with an electron in an atom of the gas with which the positronium collides. These phenomena are discussed in more detail in Chapter 7.

An alternative means of positronium formation has been to use β^+ interactions with powders or aerogels of certain molecular solids (Paulin and Ambrosino, 1968; Brandt and Paulin, 1968; Chang *et al.*, 1985). Here positronium is formed inside small grains (usually less than 10^{-8} m in diameter) of the material packed in pellet form and having a mass density of approximately 100 kg m^{-3}, or less than 10% of the density of the bulk solid. As discussed by Brandt and Paulin (1968), positronium atoms which reach the grain surface may be emitted with a kinetic energy $\leq -\phi_{\text{Ps}}$ into the space between the powder or aerogel grains. If this space is evacuated, the positronium can survive many collisions with the grains, since the free electron concentration at the surface is very low. Indeed, such a system has been used to make measurements of the vacuum lifetime of ortho-Ps (e.g. Gidley, Marko and Rich, 1976). Before annihilation, the ortho-Ps slows down by inelastic collisions with the grains, possibly reaching eventual thermal equilibrium with its environment. This phenomenon has been of recent interest as a means of providing a source of very low energy positronium (see subsection 1.5.3 below).

The main disadvantages of studying positronium produced using the two techniques described above are that (i) the source of radiation is usually close by, (ii) the experimenter has no control over the energy of the positronium, so that only energy- or momentum-averaged values of parameters can be obtained and (iii) the positronium atoms are subject to perturbations caused by collisions in the medium, which can affect intrinsic properties such as annihilation and radiative lifetimes, ground state hyperfine splittings etc. Many of these complications have been overcome by new techniques developed using low energy positron beams.

3 Methods using positron beams

Investigations of low energy positron interactions with surfaces and gases have shown that both systems can be copious sources of positronium, and methods of production have been devised under well-controlled conditions. This section has been divided into two parts; this reflects a natural, though not strict, division between those methods which produce positronium at low energies (i.e. energies $\leq -\phi_{\text{Ps}}$ or $-\epsilon_{\text{Ps}}$) and those which are capable of producing it at higher energies and, in some instances, with energy tunability.

(a) Low energy positronium The most useful techniques to be considered here involve positronium production in vacuum by low energy positron bombardment of various surfaces. As discussed in subsection 1.5.1 above, positronium may be liberated with kinetic energies either $\leq -\epsilon_{\text{Ps}}$ or

$\leq -\phi_{\rm Ps}$, depending on whether the particular material is a metal or an insulator. Positrons which have been implanted into the material either form positronium, which subsequently diffuses to the surface, or thermalize and diffuse to the surface as free positrons, where they form positronium on emission. It should also be noted, as discovered by Howell, Rosenberg and Fluss (1986), that if the implantation energy is low then a fraction of the positrons return to the surface with epithermal energies; they now have a high probability of capturing electrons, which leads to the emission of positronium with energies greater than $-\phi_{\rm Ps}$ or $-\epsilon_{\rm Ps}$, for metals or insulators respectively. As described in subsection 7.1.2, this effect has been put to some use as a source of excited state ($n_{\rm Ps} = 2$) positronium for spectroscopic purposes. The incident positron energy needed to eliminate the effect is of the order of 2 keV but is dependent to an extent upon the atomic number of the target.

Perhaps of more general applicability for the study of the properties of positronium is its production by the desorption of surface-trapped positrons and by the interaction of positrons with powder samples. According to equation (1.15) it is energetically feasible for positrons which have diffused to, and become trapped at, the surface of a metal to be thermally desorbed as positronium. The probability that this will occur can be deduced (Lynn, 1980; Mills, 1979) from an Arrhenius plot of the positronium fraction versus the sample temperature, which can approach unity at sufficiently high temperatures. The fraction of thermally desorbed positronium has been found to vary as

$$F_{\rm s} = f_{\rm s}\Gamma \exp(-E_{\rm a}/k_{\rm B}T)/[\lambda_{\rm s} + \Gamma \exp(-E_{\rm a}/k_{\rm B}T)], \qquad (1.17)$$

where $E_{\rm a}$ is defined in equation (1.15), $\lambda_{\rm s}$ is the annihilation rate of the positrons trapped in the surface state and $f_{\rm s}$ is the fraction of positrons which reach the surface and become trapped in the surface potential well. Following Chu, Mills and Murray (1981) the parameter Γ can be thought of as an attempt rate for escape from the well and is of the order of the ratio of the thermal positron speed to the well dimension, around 10^{15} s^{-1}. Figure 1.11(a) shows the temperature dependence of the positronium fraction from an aluminium crystal when clean and when exposed to oxygen. A typical distribution of the component of the positronium energy normal to the surface of a heated aluminium single crystal is shown in Figure 1.11(b) (Mills and Pfeiffer, 1985). With a vacuum lifetime of 142 ns, typical flight distances of the thermal positronium component are 10^{-2} m, thus setting the scale for possible interactions of the positronium with various projectiles (photons, charged particles etc).

In some applications (e.g. precision laser spectroscopy, see section 7.1) it is the speed distribution of the positronium which deserves consideration. Owing to its small mass, the temperature usually required for

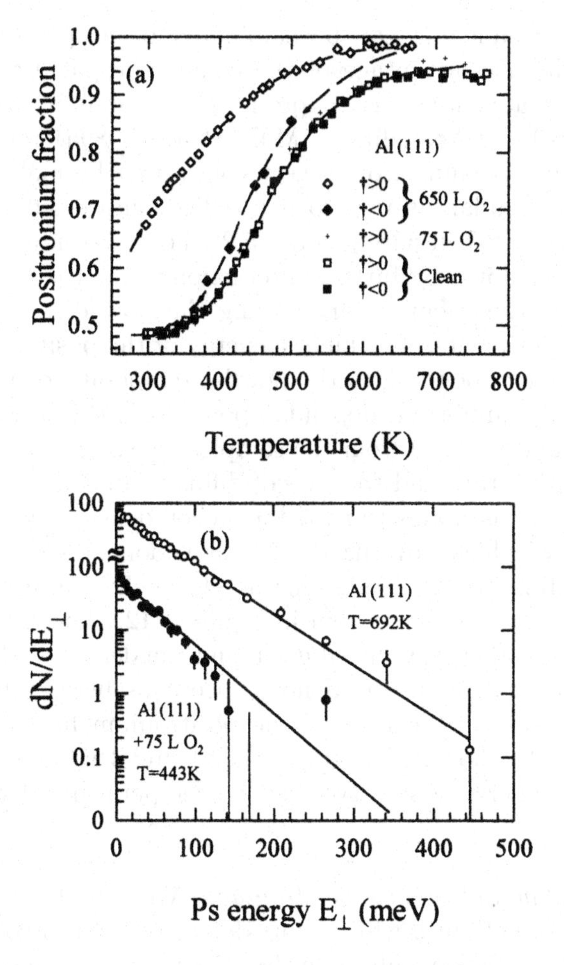

Fig. 1.11. (a) The variation of the positronium fraction versus temperature from clean and O_2-exposed Al[111] surfaces. 1 L is defined as 10^{-6} torr s. The arrows on the data and the dagger symbols refer to the direction of temperature change. The solid line is equation (1.17) fitted to the clean Al data. Note that there is a thermal hysteresis in the 650 L results. (b) Positronium energy spectra for clean and 75 L O_2-exposed Al[111] surfaces derived from the thermally desorbed positronium component only. The solid lines were fits to the data performed by Mills and Pfeiffer (1985) to deduce the temperature of the positronium: •, 443 K; ○, 692 K.

positronium activation from a metal surface can introduce substantial Doppler broadening of the spectral lines arising from transitions between the various states of positronium (e.g. for positronium with an average speed corresponding to a temperature of 600 K, the first order Doppler width of the 2P–1S transition is around 650 GHz). Although this first order Doppler broadening can be eliminated from spectroscopic investigations which take advantage of the counter-propagating two-photon technique, it is advantageous to produce positronium from as cold a surface as possible. Accordingly, Mills *et al.* (1989a) investigated the production of positronium from a very cold target by making use of low energy positron implantation into a powder compressed into the form of a pellet. As noted in subsection 1.5.2, the positronium produced in the sample slows down by further interactions with the grains and may be emitted into the vacuum surrounding the powder, where it can be accessed for experiments. The kinetic energy of the positronium tends to approach that corresponding to the powder temperature, and whether it reaches thermal equilibrium depends upon a balance between the rates of energy loss and removal from the sample. The latter is a combination of the annihilation rate and the rate of diffusion out of the entire pellet. Clearly the latter is a function of the pellet density and the incident positron energy, which governs the implantation depth. Results from the work of Mills *et al.* (1989a) for a SiO_2 pellet at temperatures of 300 K, 77 K and 4.2 K are shown in Figure 1.12. The actual details of the one-dimensional energy distribution and the degree of thermalization of the emitted positronium need not concern us here; suffice it to say that the average kinetic energy of the positronium has been markedly lowered over that shown in Figure 1.11(b) and that a progressive reduction in this energy is observed as the temperature of the sample is lowered.

(b) Higher energy and tunable positronium We now discuss methods of positronium production in which some degree of natural collimation and energy tunability exists, resulting in the production of what is now termed a positronium beam. All the techniques described here have analogues in hydrogen-atom beam production using protons, although for the much lighter positrons the effects of energy loss and scattering are more pronounced. This is particularly true of methods involving positron–solid interactions, and we treat these first.

Positronium production using a 'beam–foil' technique was demonstrated by Mills and Crane (1985). A pulsed beam of positrons with energies between 300 eV and 1550 eV was incident on a carbon foil 4 nm thick, and the resultant neutral particles, assumed to be positronium, were detected downstream by a channel-plate detector. The positronium energy distribution, attributed to formation solely in the ground state, was found, unexpectedly, to have a high yield at relatively low energies. This feature was the subject of close scrutiny for the possible presence of a long-lived excited state component, but the latter was eventually ruled out. In principle, some energy tunability of the beam would be feasible

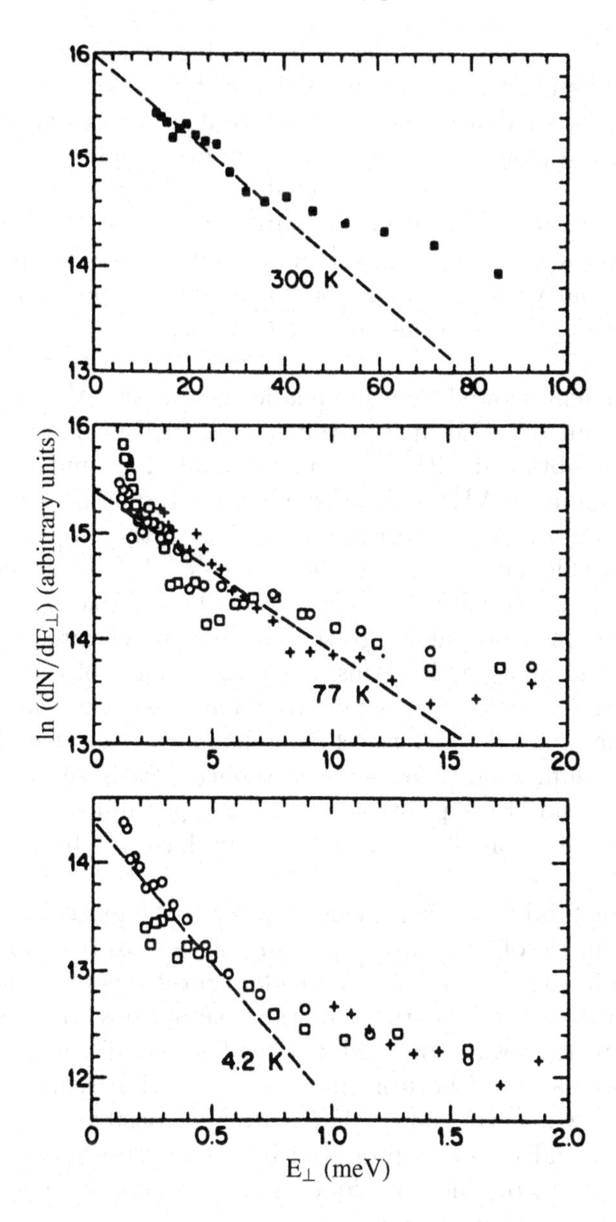

Fig. 1.12. Energy spectra of positronium produced from SiO_2 powders at 300 K, 77 K and 4.2 K. The broken lines correspond to fits to a Maxwellian distribution at the relevant temperature.

through changes in the energy of the incident beam, although this was not attempted. The interested reader is referred to the original work for further details.

Neutralization of a positron beam by grazing incidence on aluminium and copper crystal surfaces was investigated by Gidley *et al.* (1987). In order to work at shallow angles of incidence it was necessary to obtain a well-focussed positron beam (typically 1 mm in diameter with an angular divergence of 1° on target), which could only be obtained by a single stage of brightness enhancement (see subsection 1.4.3). Positronium was identified using a channel-plate–scintillator coincidence arrangement, and its position relative to the point of impact of the positron beam was determined using the position-sensitive read-out of the channel-plate detector.

The experiment showed that, at energies below 150 eV and at glancing angles in the range 5°–15°, a relatively well-defined beam of fast positronium could be obtained with an angular spread of around 20° full width at half maximum (FWHM). The maximum production efficiency (the ratio of the number of fast positronium atoms to the number of incident positrons) was deduced to be in the range 3%–5%. The positronium beam would be expected, to some extent, to be tunable with the energy of the incident positrons, although this was not investigated in the study. The results were analysed in terms of a simple model of the capture of a conduction band electron by the positron in an essentially elastic collision with the surface. To date, this technique has only been demonstrated in ultra-high vacuum conditions, and no applications have been reported. The energy spread of the positronium and the possibility that there may be several different states present in the beam have not been investigated in any detail.

The final method which is proving of value is the gas-cell technique, in which use is made of the natural peaking of the positronium formation cross section in the direction of the incident positrons (see Chapter 4 for further discussion of this feature) for the reaction described by equation (1.12). This method was pioneered independently by Brown (1985, 1986), and by Laricchia and Charlton and coworkers (Laricchia *et al.*, 1986, 1987b, 1988), who have shown that a tunable positronium beam with narrow energy width can be produced by the capture reaction in gases. Further discussion of this technique, and some applications in atomic physics, can be found in section 7.6.

In concluding this section we summarize by noting that a variety of techniques now exist for the controlled production of positronium atoms in vacuum in the kinetic energy range from meV up to keV and that these can be exploited for a number of studies in the fields of atomic and surface physics.

Table 1.2. Comparison of the main features of the interactions of electrons and positrons with atoms

	Electron	Positron
Static interaction	attractive	repulsive
Polarization interaction	attractive	attractive
Exchange with electrons	yes	no
Positronium formation	no	yes
Electron–positron annihilation	no	yes

1.6 The physical basis of the interactions of positrons and positronium with atoms and molecules

1 Positrons

Studies of positron collisions with atoms and molecules are of interest not only for their own sake but also because comparisons with the results obtained using other projectiles, such as electrons, protons and antiprotons, provide information about the effects on the scattering process of different masses and charges.

The most detailed of these comparisons has been made between positrons and electrons as projectiles, and the relevant features of their interactions with atoms and molecules are given in table 1.2. The main differences arise from the opposite sign of the charge on the two projectiles, and this has a very profound effect on the collision process. The positron is distinguishable from the electrons in the target atom or molecule, and therefore exchange effects with the projectile are absent. Also, the static interaction between a positron and an atom is equal in magnitude but opposite in sign to the attractive direct static interaction between an electron and the atom; however, the polarization potential, being quadratic in the charge of the projectile, is attractive and of the same magnitude for both positrons and electrons. Thus, two important components of the interaction between the projectile and the target are of opposite sign for positrons and therefore tend to cancel, making the overall interaction generally less attractive than for electrons. Consequently, at low projectile energies, when polarization effects are most pronounced, total scattering cross sections are usually significantly smaller for positrons than for electrons, as is clearly seen in Figure 2.1, which shows a schematic illustration of the total cross sections for electron and positron scattering by helium atoms (Stein and Kauppila, 1982). The main exceptions to this pattern are found with the alkali atoms, where the total positron scattering cross section includes a substantial contribution from the positronium formation channel, which is open even at zero incident energy.

A further consequence of the partial cancellation of the static and polarization potentials is that a positron is much less likely than an electron to be able to bind to an atom. It has been rigorously proved by Armour (1982, 1983) that a positron cannot bind to a hydrogen atom, nor can it bind to helium. However, it is reasonable to assume that binding is possible to a highly polarizable atom such as one of the alkalis. States of the positron–alkali atom system do indeed exist at energies below the positron scattering continuum, but, because the binding energy of ground state positronium is greater than the ionization energies of all the alkali atoms, these states are in the continuum of the corresponding positronium–ion system and are therefore not true bound states. Nevertheless, it has recently been proved by Ryzhikh and Mitroy (1997) that a positron can bind to a lithium atom, but with an energy of only 0.065 eV below the positronium–Li^+ scattering threshold. It is highly probable that a positron can bind to magnesium, and it has been plausibly argued by Dzuba *et al.* (1995) that positrons can bind to zinc, cadmium and mercury atoms, but the evidence is not conclusive. A positron can also bind to positronium to form the charge conjugate of Ps^-, provided the two positrons are in a singlet spin state.

Positrons exhibit resonance phenomena in collisions with some atomic and molecular targets and, as with electrons, an infinite series of resonances is expected to be associated with each degenerate excitation threshold (Mittleman, 1966). For electrons, such thresholds can only arise with hydrogenic targets, but for positrons there are also degenerate thresholds in the excitation of positronium. Several of these resonances have been identified theoretically for a few simple target systems, but they are too narrow to be observed experimentally with the presently available energy resolution of positron beams.

For many atoms the polarization potential at very low incident energies is sufficiently attractive that the s-wave elastic scattering phase shift is positive. As the positron energy is increased, however, this potential becomes less attractive because the target electrons then have less time to adjust to the influence of the positron, and the total interaction becomes repulsive, giving rise to a negative s-wave phase shift. The change in the sign of the s-wave phase shift typically occurs at a projectile energy between 1 eV and 3 eV, at which point the s-wave contribution to the total elastic scattering cross section is, of course, zero. At sufficiently low positron energies the higher-partial-wave phase shifts are determined by the polarizability of the target, and they are therefore all positive (see section 3.2). The zero in the s-wave phase shift at such a low energy gives rise to a prominent Ramsauer minimum in the total elastic scattering cross section for some atoms. A specific example of a Ramsauer minimum in positron–helium scattering is clearly seen in Figure 2.1, and a similar feature is found with several other targets, as described in Chapters 2 and 3. The Ramsauer effect is also observed in the low energy cross sections for electron scattering by some rare gas atoms, but it occurs then because the s-wave phase shift passes through π radians (or a multiple thereof) rather than zero.

The absence of exchange in positron–atom scattering might have been expected to lead to simplifications in the formulation of the scattering process, but unfortunately these are more than offset by the difficulties encountered in introducing an adequate representation of the strong electron–positron correlations, which must be taken into account if accurate theoretical results are to be obtained for the scattering parameters. The correlations arise from the attractive electrostatic interaction between the positron and the target electrons, and they can be considered as real or virtual states of positronium. They tend to be much more important than the corresponding electron–electron correlations in electron–atom collisions.

The calculated values of low energy positron scattering parameters are very sensitive to the inclusion of polarization and correlation terms in the wave function; differing methods of approximation yield a much wider range of results than for the corresponding electron case. The accurate determination of these parameters for positrons provides a particularly challenging test of approximation methods, and the most detailed theoretical studies have therefore been confined to simple atomic and molecular targets such as atomic hydrogen, helium, the alkali atoms (considered as equivalent one-electron atoms) and molecular hydrogen. Descriptions of some of the methods of approximation used, and the results obtained for various partial scattering cross sections, are given in the forthcoming chapters.

At sufficiently low positron energies, elastic scattering is usually the only open channel apart from annihilation. However, as the kinetic energy of the positron is increased various inelastic channels become accessible, including positronium formation, atomic excitation and ionization. Positronium may be formed in either the ground state or any one of the energetically available excited states, whereas, because of the absence of exchange between the positron and the atomic electrons, excitations of the target are restricted to those states which do not involve a spin flip. As an example, the lowest threshold for positron impact excitation of helium is that for the 2^1S_0 state, at 20.6 eV, rather than that for the 2^3S_1 state, at 19.8 eV.

Positron annihilation with one of the target electrons, introduced in subsection 1.2.1, is possible at all positron energies, but the annihilation cross section is usually much smaller than that for any other process. The rate of annihilation, whether for a scattering process (when it is usually expressed in terms of the annihilation cross section) or for a bound state or resonance, is a measure of the probability that the positron is at the same position as the electron with which it annihilates, and this may be calculated from the wave function of the total system. A detailed account of positron annihilation is given in Chapter 6.

Positronium formation is one of the simplest examples of a rearrangement collision and accordingly it has attracted considerable experimental and theoretical attention. Positronium can be formed provided the energy of the incident positron exceeds the difference between the ionization energy of the target atom and the binding energy of the positronium; see equation (1.13). If the ionization energy of the target atom is less than 6.8 eV, positronium formation into the ground state is possible even when the incident positron has zero kinetic energy, and the formation cross section for this exothermic reaction is then infinite. More commonly, however, the positronium formation threshold is at a positive projectile energy, e.g. it is 6.8 eV for atomic hydrogen and 17.8 eV for helium, the highest value for any atom. The lowest threshold for positron impact excitation is at a higher energy than the threshold for ground state positronium formation for all atoms, and therefore an energy interval exists between these two thresholds in which the only two scattering processes are elastic scattering and ground state positronium formation. It is in this energy interval, the so-called Ore gap, that the most detailed theoretical investigations of positronium formation have been made, as described in section 4.2.

2 Positronium

Positronium is electrically neutral, and its centre of mass is midway between the constituent electron and positron. Consequently, the interaction between positronium and any charged particle or atomic system, in the static approximation in which it is assumed that there is no distortion of either the positronium or any compound system with which it is interacting, is zero, provided exchange between the electron in the positronium and any electrons in the other system is ignored. It follows that the first Born approximation to the scattering amplitude for elastic positronium scattering by another system, assuming the neglect of exchange, is also zero and the leading term in the Born expansion of the elastic scattering amplitude is therefore the second order term. Consequently, at high energies, where the first Born approximation might have been expected to be valid and where exchange effects would in any event be small, the cross section for elastic positronium scattering by a target system should be very small.

At low energies, the absence of a direct static interaction between positronium and other systems gives an enhanced importance to polarization and exchange effects, and the relatively high value, $72a_0^3$, for its dipole polarizability enables positronium to bind to several other systems. Positronium can bind to a charged particle, either positive or negative, provided the mass of the charged particle is sufficiently small (Armour, 1983). Thus, it can bind to an electron or a positron if the two identical particles are in a singlet spin state, as has previously been mentioned, and also to a positive or negative muon, but it cannot bind to a proton or antiproton. Furthermore, positronium can bind to atomic hydrogen, provided the two electrons are in a singlet spin state, to form the positronium–hydride molecule PsH, and also to itself to form the positronium molecule Ps_2, in which the two electrons and the two positrons must separately be in singlet spin states. It has also been established that positronium can bind to the lithium atom and can probably bind to sodium (Ryzhikh, Mitroy and Varga, 1998a, b). A more detailed discussion of the binding of positronium to various systems is given in section 7.5.

The leading term in the interaction between positronium and a charged particle is the usual $1/R^4$ polarization potential, where R is the coordinate of the centre of mass of the positronium relative to the other particle. If, however, positronium interacts with a neutral atom or another positronium, the interaction is dominated by the van der Waals potential arising from the induced dipole–dipole and quadrupole–dipole terms, which has the long-range form

$$V = -A/R^6 + B/R^8, \tag{1.18}$$

where R is now the coordinate between the centre of mass of the positronium and that of the other system. Values of the coefficients A and B have been determined by Drachman (1987) for the positronium–hydrogen and positronium–positronium systems. Proper account must be taken of the long-range character of the van der Waals interaction if accurate results are to be obtained for low energy positronium–atom scattering. This topic is discussed further in section 7.2.

2
Total scattering cross sections

2.1 Introduction

The total positron scattering cross section, σ_T, is the sum of the partial cross sections for all the scattering channels available to the projectile, which may include elastic scattering, positronium formation, excitation, ionization and positron–electron annihilation. Elastic scattering and annihilation are always possible, but the cross section for the latter process is typically 10^{-20}–10^{-22} cm^2, so that its contribution to σ_T is negligible except in the limit of zero positron energy. All these processes are discussed in greater detail in Chapters 3–6.

Measurements of σ_T have been undertaken for more than two decades using low energy positron beams and the results, together with the theoretical predictions, are discussed in this chapter. The total cross section is defined by the Beer–Lambert law, in which projectiles scattered by a target are considered as having been being removed from the incident beam. Thus, the flux of projectiles, I, surviving from an initial flux I_0 with a projectile path length L, incident upon a target of number density n, is given by

$$I = I_0 \exp(-nL\sigma_T). \tag{2.1}$$

All measurements of σ_T for positrons have been made using the so-called attenuation method, in which a well-defined beam is passed through a gaseous target. The attenuation A is given by I_0/I. At the densities typical of this work the gases can be considered as ideal, and n can be written as

$$n = P/(k_B T) = 0.724 \times 10^{23} P/T \qquad (\mathrm{m}^{-3}), \tag{2.2}$$

where P and T are the pressure and absolute temperature of the gas in units of Pa and K respectively, and k_B is Boltzmann's constant.

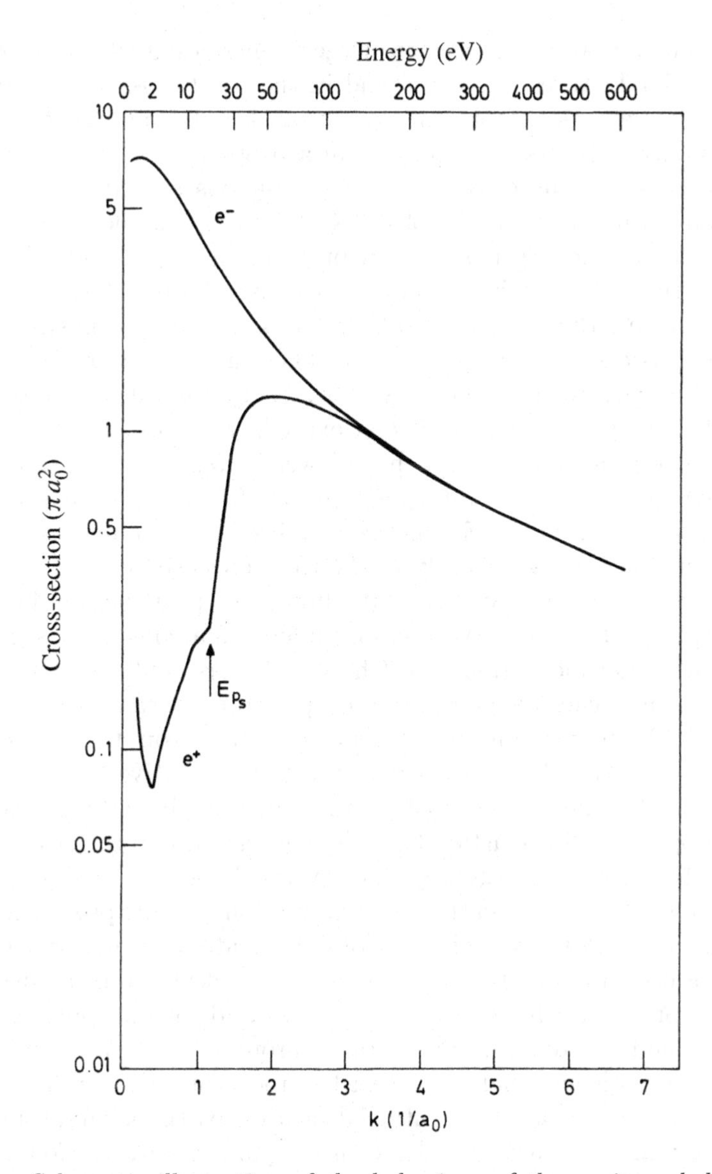

Fig. 2.1. Schematic illustration of the behaviour of the positron–helium and electron–helium total scattering cross sections. Notable are the large differences in magnitude of the cross sections at low energies, their merging at approximately 200 eV and the onset of inelastic processes at the positronium formation threshold E_{Ps} in the positron curve.

As an example of the energy dependence of the total cross section for positron–atom scattering, a schematic of the data for helium atoms is shown in Figure 2.1, together with the corresponding data for electrons. It is seen that $\sigma_T(e^-)$ is considerably larger than $\sigma_T(e^+)$ at low projectile energies, but the two cross sections merge at energies greater than 200 eV. Rather similar features are also found in the total cross sections for the other rare gas atoms except that, for argon, krypton and xenon, $\sigma_T(e^-)$ exhibits a narrow Ramsauer minimum at a very low energy, in the vicinity of which the scattering cross section for positrons exceeds that for electrons. Many other atoms and molecules also exhibit smaller cross sections for positrons than for electrons at low projectile energies, and this feature can be attributed directly to the partial cancellation of the attractive polarization and the repulsive static components of the positron–atom interaction, as discussed in section 1.6. There are, however, some exceptions to this pattern, most notably provided by the alkali atoms, where the total cross section for positrons exceeds that for electrons at low projectile energies. These atoms have lower ionization energies than the binding energy of ground state positronium (6.8 eV), and consequently the positronium formation channel is open even at zero incident positron energy, contributing significantly to the total cross section.

As the positron energy is raised above the positronium formation threshold, E_{Ps}, the total cross section undergoes a conspicuous increase. Subsequent experimentation (see Chapter 4) has confirmed that much of this increase can be attributed to positronium formation via the reaction (1.12). Significant contributions also arise from target excitation and, more importantly, ionization above the respective thresholds (see Chapter 5). In marked contrast to the structure in $\sigma_T(e^+)$ associated with the opening of inelastic channels, the electron total cross section has a much smoother energy dependence, which can be attributed to the dominance of the elastic scattering cross section for this projectile.

The example of helium given above illustrates the major trends in $\sigma_T(e^+)$, described in greater detail below. Our discussion is divided along somewhat pragmatic lines into low energies and intermediate and high energies. The loose dividing line is taken around the inelastic thresholds since the inelastic channels make such important contributions to the integrated cross section. As mentioned previously, the alkali atoms are a special case and they are considered in a separate section. Comparisons between experimental and theoretical results are made where possible, together with some comparisons with corresponding data for electrons. We first, however, consider some of the general theoretical features of total cross sections.

2.2 Theory

If elastic scattering is the only open channel (except for electron–positron annihilation with its very small cross section), the total and elastic scattering cross sections are identical, and the cross section may be calculated using the approximation schemes which have been developed for single-channel elastic scattering, some of which are described in Chapter 3. As the positron kinetic energy is increased, however, a succession of rearrangement and inelastic processes also become possible, and a proper formulation of the scattering process should then include the couplings between all the open channels.

In positron–helium scattering, for example, elastic scattering is the only open channel (apart from annihilation) up to an incident energy of 17.8 eV, the threshold for positronium formation, and thereafter positronium formation remains the only other open channel up to the threshold for the excitation of the 2^1S state of helium at 20.6 eV. The 2^3S state of helium has a lower excitation threshold at 19.8 eV, but the transition to this state from the singlet ground state is highly suppressed in positron impact because it would require an electron spin flip. Excitation of the helium atom to other states, and positronium formation into a range of excited states, all become possible within the next energy interval of 4 eV until, at an energy of 24.6 eV, ionization of the helium atom can also occur. As the energy is increased further, all these channels remain open, to be eventually augmented by others for double excitation and double ionization.

From such a multichannel formulation the required positron total scattering cross section may, in principle, be obtained as the sum of the partial cross sections for transitions between the input channel, describing positrons incident on the target system, and all possible open output channels. Details of the calculations of individual partial cross sections are not given here but in the relevant chapters devoted to specific processes, e.g. elastic scattering is discussed in Chapter 3, positronium formation in Chapter 4, excitation and ionization in Chapter 5 and annihilation in Chapter 6. In this section we are more concerned with the general properties of total scattering cross sections, although features of some specific systems are considered in more detail.

When only a few channels are open, it is feasible to calculate all the partial cross sections explicitly and sum them to obtain $\sigma_{\rm T}$, as has been done in some studies of positron scattering by atomic hydrogen (Humberston, 1986; Kernoghan, McAlinden and Walters, 1995; Kernoghan *et al.*, 1996), the alkali atoms (Hewitt, Noble and Bransden, 1993; McAlinden, Kernoghan and Walters, 1996, 1997; Kernoghan, McAlinden and Walters, 1996), and helium (Hewitt, Noble and Bransden, 1992a; Humberston and Van Reeth, 1996; Campbell *et al.*, 1998a). As the projectile energy continues to be raised, however, it ceases to be viable to calculate every possible partial cross section, although Kernoghan and coworkers, in the references cited above (and McAlinden, Kernoghan and Walters, 1997) have approximated to this task in their investigations of positron scattering by atomic hydrogen and the alkali atoms using the coupled-state approximation. These authors included up to nine states of the target atom (the alkali atoms being considered as equivalent one-electron systems) and an equal number of positronium states in their expansion of the wave function, and they made some allowance for ionization by including pseudostates. Further details of these calculations are given in subsection 4.2.3. Their 18-state results for atomic hydrogen, given in Figure 2.18 (see subsection 2.5.4), agree reasonably well with the experimental measurements of Zhou *et al.* (1994a), who obtained both upper and lower bounds on the total cross section. Even better agreement, however, is obtained between the 33-state results of Kernoghan *et al.* (1996) and the experimental results of Zhou *et al.* (1997) at all energies up to 100 eV. Below the positronium formation threshold, however, the experimental results fall below the accurate theoretical values. Investigations of positron–hydrogen scattering using the coupled-state approximation with the inclusion of several states of hydrogen and positronium have also been made by Mitroy and Ratnavelu (1995) and Gien (1997), but these authors did not extend their calculations to such high energies as did Kernoghan and coworkers.

Versions of the coupled-state method with all positronium terms omitted, such as the intermediate energy R-matrix method used by Higgins, Burke and Walters (1990) and the convergent close-coupling method used by Bray and Stelbovics (1994), when applied to positron–hydrogen scattering give similar results to those of Kernoghan *et al.* (1996) at the higher energies. However, they underestimate the total cross section below 40 eV, where positronium formation makes a significant contribution to it.

The total cross sections calculated by Kernoghan, McAlinden and Walters (1996) and Campbell *et al.* (1998a) for positron scattering by the alkali atoms sodium, potassium and rubidium are in good agreement with the slightly modified experimental results of Parikh *et al.* (1993) and Kwan *et al.* (1994), the modifications having been made to correct for the neglect of small-angle scattering. Figure 2.15 shows the experimental and theoretical results for sodium (see subsection 2.5.3). Coupled-state calculations of σ_T for the alkali atoms have also been performed by Ward *et al.* (1989) and McEachran, Horbatsch and Stauffer (1991), but these authors did not include any positronium states, although the positronium formation channel is open even at zero incident positron energy, where the cross section is infinite. Excluding the positronium formation channel has a very significant effect on the magnitude of the total cross section at low positron energies because the calculated elastic scattering cross section is then much larger than the true value, presumably in partial compensation for the absence of the positronium channel.

A more common means of calculating σ_T at intermediate and high energies is to use the optical theorem, which expresses the conservation of particle flux in the scattering process as the following relationship between the total cross section and the imaginary part of the elastic scattering amplitude in the forward direction:

$$\sigma_{\rm T} = (4\pi/k)\,{\rm Im}\, f_{\rm el}(\theta = 0), \tag{2.3}$$

where k is the wave number of the incident projectile. Winick and Reinhardt (1978a, b) used this relationship to calculate $\sigma_{\rm T}$ for positron–hydrogen scattering over the energy range 0–34 eV after first calculating the elastic scattering amplitude using their moment T-matrix method.

At intermediate and higher energies it is appropriate to expand the forward elastic scattering amplitude in a Born series:

$$f_{\rm el}(\theta = 0) = \sum_{n=1}^{\infty} f_{\rm el}^{{\rm B}_n}(\theta = 0), \tag{2.4}$$

and a reasonably good approximation to the forward scattering amplitude should then be given by the sum of the first few terms in the expansion. Because the first term in the Born expansion for $f_{\rm el}(\theta = 0)$ is real and therefore, according to the optical theorem, does not make a contribution to $\sigma_{\rm T}$, the lowest order contribution to the right-hand side of equation (2.4) is from the second order term, $f_{\rm el}^{{\rm B}_2}(\theta = 0)$. Calculations of the first few terms in the Born expansion of $f_{\rm el}(\theta = 0)$ have been made by Byron and Joachain (1977a). These and other authors have also used quite closely related approximation schemes such as the Glauber approximation and the eikonal-Born approximation. For positron–helium scattering these approximations are in reasonably good agreement with each other, and they also yield values of $\sigma_{\rm T}$ in reasonably good agreement with the experimental values for energies greater than 300 eV (Byron, 1982).

At sufficiently high positron energies, typically 1000 eV for helium, the first Born approximation yields reasonably accurate cross sections for elastic scattering and for the various inelastic processes. Furthermore, these cross sections have the same values for both electrons and positrons, except for positronium formation, which is, of course, absent in electron collisions. But at such high energies the positronium formation cross section is negligible, and therefore the first Born approximation to the total scattering cross section is essentially the same for both projectiles. This merging of the total cross sections for the two projectiles is indeed observed experimentally, as mentioned above, but it takes place at much lower energies than might have been expected. For helium, the two cross sections have already merged at an energy of 200 eV, as may be seen in Figure 2.1, and for the alkali atoms the merging occurs at energies as low as 50 eV (Stein *et al.*, 1990). The total cross section for positrons merges with that for electrons from above for atomic hydrogen and the alkali atoms, and from below for all the noble gases and many other atoms and molecules.

This phenomenon may be at least partially understood by reference to the optical theorem and the Born expansion of the imaginary part of the forward scattering amplitude, equation (2.4). If exchange is ignored when considering electron–atom scattering, the first non-zero contribution to the Born expansion is the second order term $f_{\rm el}^{{\rm B}_2}(\theta = 0)$, and this is the same for both electrons and positrons because it is quadratic in the projectile–target interaction potential. Similar identities exist for all the even-order terms in the Born series, and any differences between the total cross sections for electrons and positrons therefore arise from the odd-order terms in the series with $n \geq 3$. But Dewangan (1980) has shown that if the energy differences between atomic states are neglected in the calculation of the higher order terms in the Born series, then $f_{\rm el}^{{\rm B}_n}(\theta = 0) = 0$ for odd $n \geq 3$. Even though these odd higher order terms are not actually zero, they are likely to be small in magnitude compared with the adjacent even terms, as has been confirmed for hydrogen and helium by Byron and Joachain (1977a); see also Byron (1982).

Although the above argument is not rigorous, particularly in its neglect of electron exchange, it nevertheless provides a plausible explanation for the merging of the two total cross sections at much lower projectile energies than those for which the first Born approximation alone is valid.

The real part of the forward elastic scattering amplitude can be related to the total scattering cross section by means of the following dispersion relation:

$$\begin{aligned} {\rm Re}\, f_{\rm el}(k, \theta = 0) \; - \; & f_{\rm el}^{{\rm B}_1}(k, \theta = 0) \\ & = (1/2\pi){\rm P} \int_0^\infty [k'^2 \sigma_{\rm T}(k')/(k'^2 - k^2)]\, dk', \end{aligned} \tag{2.5}$$

or, in its subtracted form,

$$\begin{aligned} {\rm Re}\,[f_{\rm el}(k, \theta = 0) - f_{\rm el}(0, \theta = 0)] &= \left[f_{\rm el}^{{\rm B}_1}(k, \theta = 0) - f_{\rm el}^{{\rm B}_1}(0, \theta = 0) \right] \\ &= (k^2/2\pi) P \int_0^\infty [\sigma_{\rm T}(k')/(k'^2 - k^2)]\, dk', \end{aligned} \tag{2.6}$$

where P implies the principal value of the integral. This latter form, lacking the factor k'^2 in the numerator of the integrand, gives less weight to the total cross sections at high energies.

At zero incident energy equation (2.5) reduces to

$$a + a_{\rm B} = (-1/2\pi) \int_0^\infty \sigma_{\rm T}(k')\, dk', \tag{2.7}$$

where a is the scattering length, and $a_{\rm B}$ is the first Born approximation to it. Calculations of the scattering length are discussed further in subsection 3.2.1.

These equations provide useful checks on the consistency of the experimentally measured values of $\sigma_{\rm T}$ with the calculated values of the real part of the forward scattering amplitude, or the scattering length; they have been used as such by Bransden and Hutt (1975). For helium, these authors took the measured total cross sections of Coleman *et al.* (1976b) in the energy range 2–800 eV; below 2 eV they used an extrapolation based on a least-squares fit of the functional form

$$\sigma_{\rm T}(k) = b_0 + b_1 k + b_2 k^2 \ln k + b_3 k^2 + b_4 k^4 \tag{2.8}$$

to the measured elastic scattering cross sections below the positronium formation threshold. The first Born approximation was used to estimate $\sigma_{\rm T}$ above 800 eV. Using all this total cross section data, the value of $(-1.24 \pm 0.05)a_0$, where a_0 is the Bohr radius, was obtained for the right-hand side of equation (2.7), in very satisfactory agreement with the accurate value of $-1.28a_0$ for the left-hand side, as calculated by Humberston (1973). A similar analysis by Bransden, Hutt and Winters (1974), using the total cross section measurements of Canter *et al.* (1973), revealed that these earlier experimental data suffered from systematic errors which resulted in a serious underestimation of the true values of $\sigma_{\rm T}$ at higher positron energies.

Throughout the energy range up to the positronium formation threshold, Bransden and Hutt (1975) also obtained good agreement between the right-hand side of equation (2.6), as obtained from the measured total cross sections, and the left-hand side, as calculated from accurate theoretical elastic scattering phase shifts (η_l for the lth partial wave), using the relationship

$$\mathrm{Re}\, f_{\rm el}(\theta = 0) = (1/2k) \sum_{l=0}^{\infty} (2l+1) \sin 2\eta_l. \tag{2.9}$$

The theoretical results depend quite sensitively on the quality of the phase shifts, and good agreement with the values derived from the experimental cross sections was only obtained using the accurate phase shifts of Humberston (1973).

A similar comparison of measured total cross sections and calculated values of the elastic scattering parameters was also made by Bransden and Hutt (1975) for positron–neon collisions, using the theoretical polarized-orbital phase shifts of Montgomery and LaBahn (1970). These are almost certainly less accurate than the corresponding results for helium, and there is poorer agreement between the values of the two sides of equation (2.6).

2.3 Experimental techniques

In this section we describe the various experimental techniques which have been used to measure σ_T and we include a critical evaluation of their limitations so as to aid in the intercomparison of the various sets of data. Many experimental groups have contributed to this field, but the aim here is not to discuss every technique in great detail but rather to select a representative subset in order to illustrate the most interesting features. Discussions of the results obtained are given in sections 2.5 and 2.6.

Figure 2.2 shows the localized scattering system employed by the University College London group for their total cross measurements after 1976. The apparatus was based around a time-of-flight (TOF) arrangement where one timing trigger was provided by the amplified signal from a thin plastic scintillator traversed by the β^+ particles from the radioactive source and in which they deposited some of their kinetic energy before striking the moderator. The detection of a low energy positron by the channeltron electron multiplier (CEM) at the end of the flight path provided the other timing signal. In order to prevent loss of data this latter signal, with a low count rate of 1–10 s^{-1}, was used to start the timing sequence since the delayed scintillator output (stop-signal) rate, which was governed by the activity of the radioactive source, was typically 10^6 s^{-1}. In this system the axial magnetic field used to transport the positrons was generated by coils wound directly on to the beam vacuum pipeline. Initially a smoked MgO moderator was used, but later this was replaced by either tungsten vanes or a mesh.

Gas could be admitted to a localized region of the flight path, close to the moderator, which was differentially pumped by a system of three diffusion pumps. The total length of the gas cell was 10 cm, including 6 mm diameter apertures, each 1 cm long, at either end. The gas pressure was measured at the centre of the cell, on the side opposite to the inlet, using a capacitance manometer, whilst the temperature measurement was effected on the outside of the beam pipe using a simple mercury-in-glass thermometer. Thus, an estimate of the gas density n_0 at the centre of the scattering cell could be obtained from equation (2.2). It was found that the pressure gradients introduced by the use of a short cell could be accounted for using n_0 and a single normalization constant k_n, which was found to be nearly independent of the nature and pressure of the gas and to have the value 1.275 ± 0.020. In terms of the gas density $n(x)$ at point x along the positron flight path of length L,

$$\int_0^L n(x)\,dx = n_0 l_0 / k_n, \tag{2.10}$$

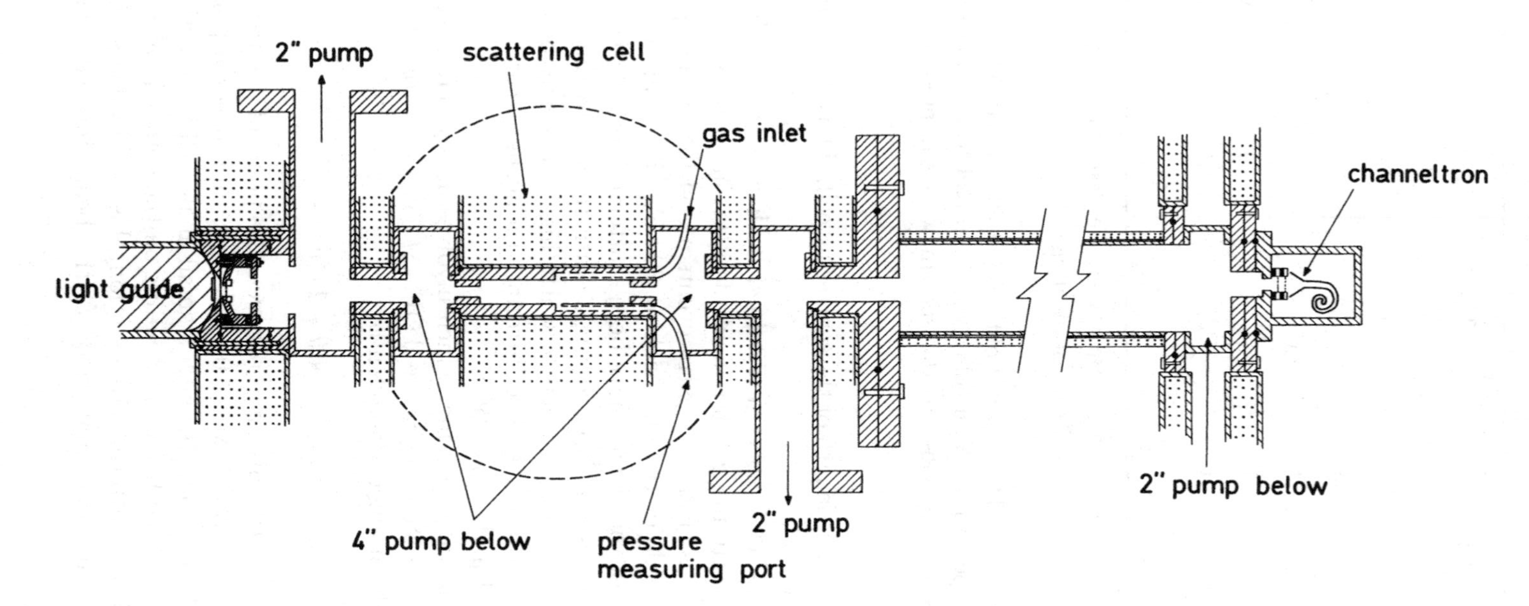

Fig. 2.2. Schematic diagram, not to scale, of the time-of-flight localized scattering system used by the UCL group for total scattering cross section measurements.

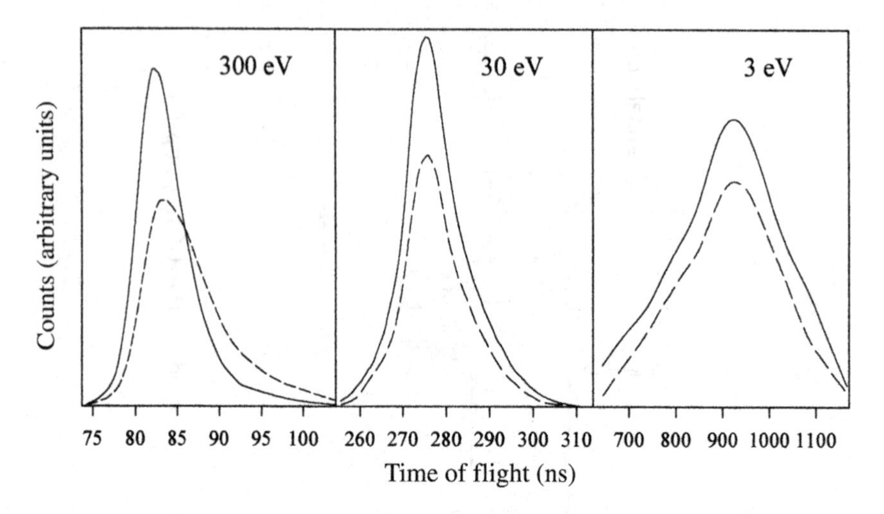

Fig. 2.3. The form of the TOF spectra obtained at 300 eV, 30 eV and 3 eV by the UCL group for positron–helium collisions. The solid lines are the vacuum spectra, whilst the broken lines are the spectra with gas present; they are discussed further in the text.

where l_0 is the geometric length of the scattering cell. Thus, from equation (2.1), $\sigma_{\rm T} = k_n \ln A/(n_0 l_0)$.

Figure 2.3 shows TOF spectra obtained both with and without gas admitted to the scattering cell for incident positron energies of 300 eV, 30 eV and 3 eV. It is clear that extra events are visible in the spectrum on the long TOF side of the gas peak at 300 eV. These are caused by the detection of positrons which have been scattered, both elastically and inelastically, through small forward angles with respect to the initial direction of the positron beam. Inelastic scattering could easily be resolved from the unscattered beam, and so introduced a negligible error into the determination of A and, hence, of $\sigma_{\rm T}$.

In the case of elastic scattering, which involves nearly zero energy loss for the positron, the increase in the TOF, Δt, of a positron of speed v scattering through an angle θ is given by

$$\Delta t = l_2(\sec\theta - 1)/v, \tag{2.11}$$

where l_2 is the geometric distance between the point of scattering and the CEM detector. Clearly $\Delta t \to 0$ as $\theta \to 0$, and at some point any positron elastically scattered through a small angle will be indistinguishable from the unscattered beam. Depending on the detailed behaviour of the differential elastic scattering cross section, this may lead to systematic overestimates in the determination of I, the beam intensity with gas present.

In most of the total cross section measurements made by the UCL group a magnetic field gradient existed along the positron beam flight path, which facilitated the distinction between scattered and unscattered particles. In particular, the magnetic fields in the scattering and detection regions were held in the ratio $1:8$ respectively, thereby producing a magnetic mirror. Consequently, from the adiabatic conservation law $B/\sin^2\theta$ = constant (see e.g. Griffith *et al.*, 1978a; Charlton *et al.*, 1984, for a more complete discussion of the experiment and e.g. Jackson, 1975, Chapter 12, for a treatment of the underlying principle), a positron initially travelling parallel to the axis of the magnetic field and then scattered through an angle greater than 20° would have been unable to reach the detector. A combination of the magnetic mirror and TOF techniques was used in an attempt to discriminate fully against small-angle positron scattering. Further details of the methods used to extract the true attenuation from the TOF spectra can be found in the works of Charlton *et al.* (1980a, 1984) for energies below approximately 20 eV, and Coleman *et al.* (1976b) and Griffith *et al.* (1979a) at higher energies. The apparatus shown in Figure 2.2 has also been used for determinations of σ_T for electron impact in the energy range 2–50 eV. The targets studied were helium, neon, argon and a number of simple molecules (Charlton *et al.*, 1980a, 1984; Griffith *et al.*, 1982). Very reasonable agreement (usually to better than 10%) was found with previously available cross sections, particularly for the noble gases.

Another system for measuring total cross sections, which has also been used for both positrons and electrons, is that developed by the Detroit group (see e.g. Stein, Kauppila and Roellig, 1974). Their apparatus and method embody many features not present in the widely used TOF technique. In addition, this group has made the most comprehensive survey of targets using both projectiles.

In the Detroit apparatus, illustrated in Figure 2.4, the β^+ source is created *in situ* by bombarding a boron target with a 4.75 MeV proton beam emanating from a van der Graaf accelerator. An unstable carbon isotope is then produced via the reaction

$$^{10}_{5}\mathrm{B} + \mathrm{p} \rightarrow {}^{11}_{6}\mathrm{C}, \tag{2.12}$$

the β^+ being emitted with a 20.3 minute half-life via the reaction

$$^{11}_{6}\mathrm{C} \rightarrow {}^{11}_{5}\mathrm{B} + \mathrm{e}^+ + \nu_\mathrm{e}. \tag{2.13}$$

It was found that the boron target itself acted as a moderator with a low efficiency of 10^{-7}, but the emitted positrons had a low energy, and therefore a narrow energy width, of approximately 0.1 eV.

Referring again to Figure 2.4, the slow positrons emitted from the boron were accelerated and focussed by the electrostatic lens system

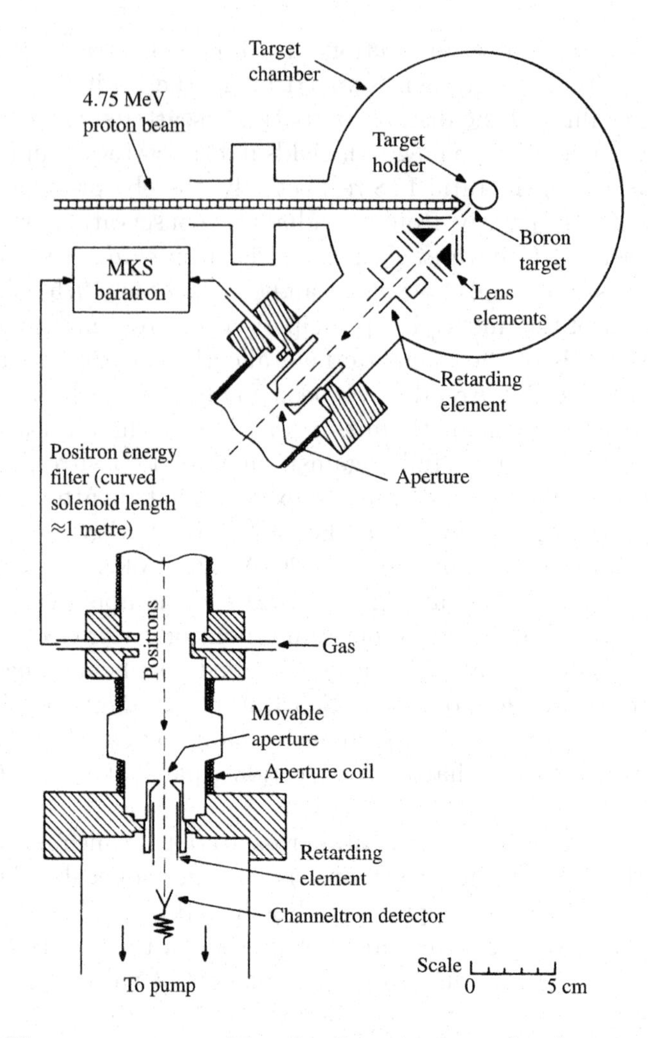

Fig. 2.4. The apparatus used by the Detroit group for the measurement of total cross sections for the scattering of positrons and electrons from noble and molecular gases. Linking the two parts of the apparatus is the curved positron energy filter.

into the magnetic guiding field (typically 10^{-3} T) produced by a long curved solenoid. This solenoid also served to remove the detector from the line of sight of the ^{11}C source. The positrons were detected by a CEM after passing through a 4.8 mm aperture and an electrostatic retarder arrangement.

The total cross section was measured by monitoring the attenuation of the beam as it passed through the curved solenoid, which could either be evacuated or filled with gas. The average pressure of the gas was

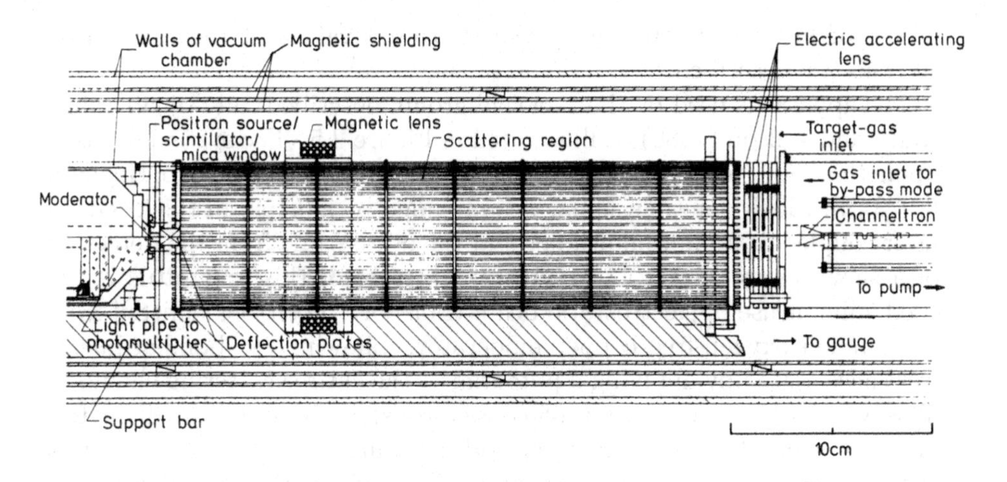

Fig. 2.5. Schematic illustration of the apparatus used by the Bielefeld group to measure total scattering cross sections. Reprinted from *Journal of Physics B* **13**, Sinapius, Raith and Wilson, Low-energy positrons scattering from noble gas atoms, 4079–4090, copyright 1980, with permission from IOP Publishing.

determined by direct measurements at either end of the solenoid using an absolute capacitance manometer. In order to obtain an absolute value for the product of n and L, see equation (2.1), the pressure measurement was corrected for thermal transpiration effects (section 2.4) using a direct temperature measurement on the inner surface of the wall of the solenoid of known geometric length. Thus, absolute values for σ_T could be obtained directly without the need for normalization. A further virtue of the Detroit apparatus was that a tungsten filament could be substituted for the boron moderator, thereby enabling cross sections for electrons to be measured in the same apparatus.

The narrow energy width of the positron and electron beams meant that a retarding electrode system located just before the CEM could be used to provide good angular discrimination against projectiles elastically scattered through small angles. The retarder voltage V_r was, as described by Kauppila *et al.* (1981), usually set at a level such that a small fraction of the incident beam was cut off, so that a particle with kinetic energy $E > eV_r$, scattered through an angle greater than $\cos^{-1}(eV_r/E)^{1/2}$, could not pass to the detector. The reader is referred to the aforementioned article, where the performance and importance of this device are discussed at greater length. In addition, the aperture beyond the scattering cell provided angular discrimination because scattered particles, having acquired larger components of velocity transverse to the magnetic field in the collision process, travelled in orbits with increased Larmor radii. These points are discussed further in section 2.4, where the systematic effects of forward scattering on the measured values of the total cross sections are described.

The apparatus used by the Bielefeld group, described fully by Sinapius, Raith and Wilson (1980), is illustrated in Figure 2.5. The system is based upon the TOF technique developed by Coleman, Griffith and Heyland (1973), in which one time signal was provided by a 0.15 mm thick scintillator coupled to a photomultiplier and the other came from the detection of a slow positron by a CEM. A novel feature of this apparatus is the predominant use of electrostatic beam transport. Positrons emitted from a grounded oxygen-free copper moderator, following β^+ bombardment, were accelerated by applying a potential to the entire scattering chamber. This was made from a cage of copper rods, each of 1 mm diameter, and was itself 70 mm in diameter and 280 mm long. The chamber was surrounded by magnetic shielding which reduced the magnetic field inside to approximately 35 nT. Adjustments to the positron beam could be made by two pairs of deflection plates at the entrance to the scattering chamber and by a weak magnetic lens located one third of the way along the length of the chamber, so as to facilitate focussing onto the cell exit aperture. A series of apertures beyond the scattering cell was used to accelerate and focus the positrons on to the channeltron.

The scattering chamber was pumped out through the apertures so that a pressure differential of around 80 was maintained between it and the remainder of the flight path. The pressure was recorded using a calibrated ionization gauge. The cross section was determined from measurements of the intensity of the beam with gas flow through the scattering chamber and with the gas flow bypassing the scattering chamber (see Figure 2.5) and directly entering the detector area. Since the gas flow was kept constant, the pressure in the detector region was equal in both cases. Thus, by simply subtracting the two signals, this method automatically accounted for collisions which occurred outside the scattering chamber.

Mizogawa *et al.* (1985) made measurements of positron–helium total cross sections at low energies using an apparatus based on the technique developed by the UCL group. The main differences were their use of low magnetic fields and small (2 mm radius) apertures in their scattering cell. This combination provided good angular resolution and also reduced the effect of spiralling. Further details are given in section 2.4.

It is also necessary to describe here the apparatus used by the Detroit group for measurements of total cross sections for positron and electron scattering from the alkali metals, since their special heated oven, which also served as the scattering cell, is significantly different from their long curved solenoid apparatus. The scattering cell, shown in Figure 2.6 (Kwan *et al.*, 1991), is located at the end of the curved solenoid shown in Figure 2.4, the continuation of the magnetic field being provided by the two

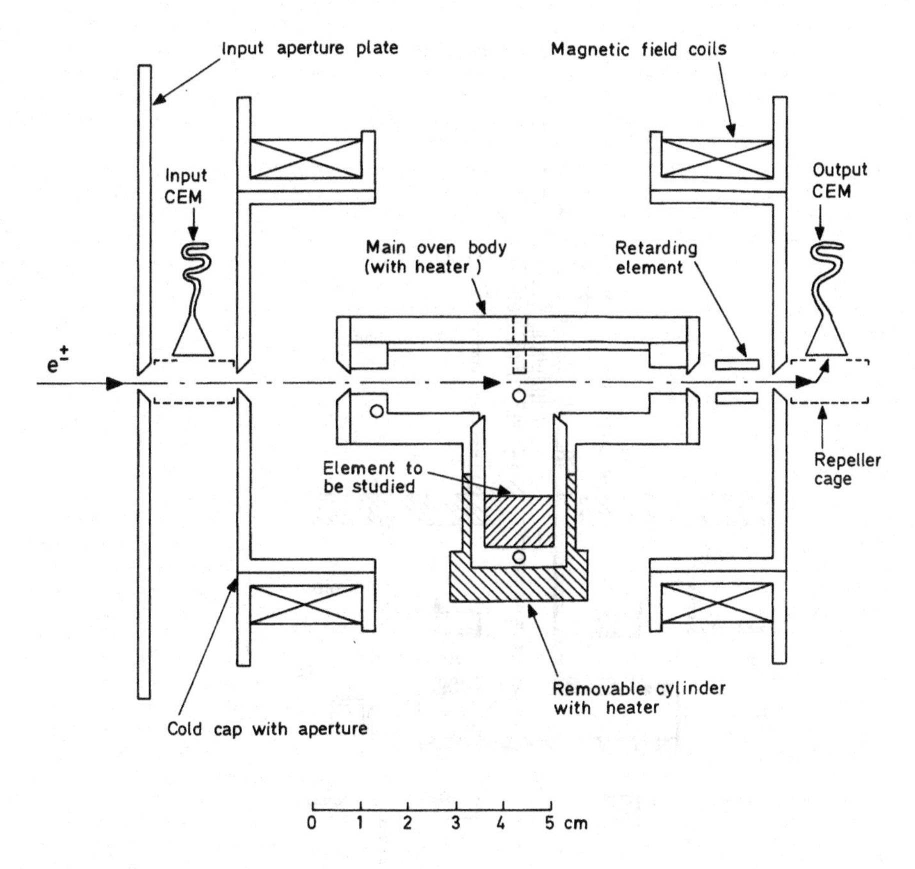

Fig. 2.6. The oven, which also served as the scattering cell, used by the Detroit group for studies of positron–alkali atom and electron–alkali atom total scattering cross sections. The open circles indicate thermocouples.

coils shown. Using appropriate voltages, the incident positron or electron beam could be deflected into a CEM at the input to the cell. This detector, and a similar one to record the beam transmitted through the cell, was further shielded by apertures from the alkali vapour which could emanate from the oven. A stainless steel element located on the output side of the cell (which became coated with alkali metal when the oven was hot) was used to provide some discrimination against forward-scattered projectiles. The oven was heated by coils wound in its walls, and the temperature was monitored at three different locations. During an experimental run these three temperatures were kept as close to one another as possible in order to reduce variations in the vapour pressure, which is a sensitive function of temperature. The vapour density, which is required in the determination of absolute values of σ_T, was calculated from standard vapour pressure tables using the average temperature. A full discussion

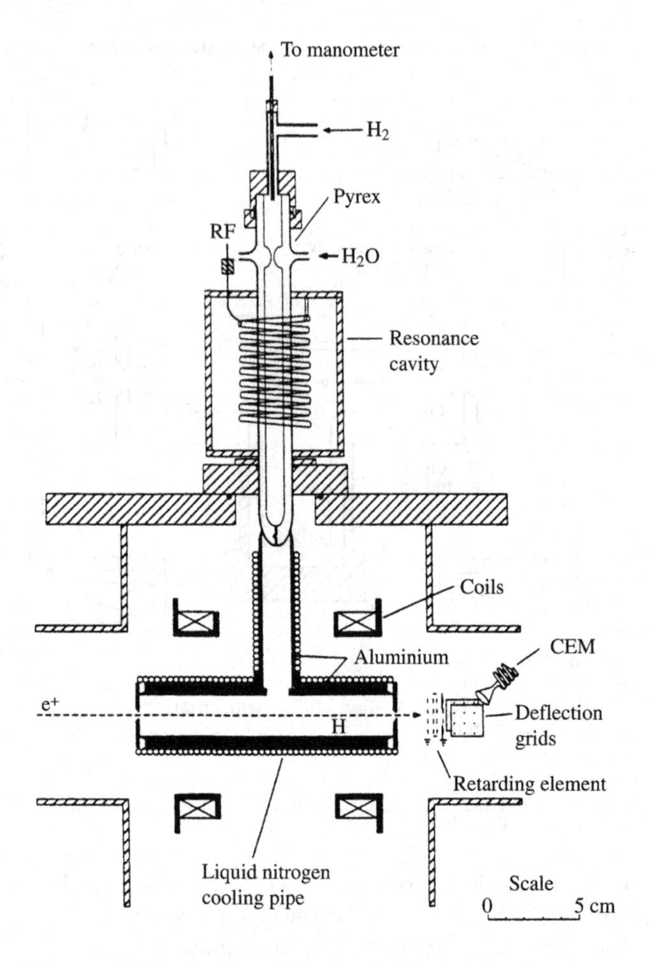

Fig. 2.7. Schematic illustration of the cooled-gas-cell arrangement used to study positron and electron collisions with atomic hydrogen (e.g. Zhou *et al.*, 1994a).

of the uncertainties involved in the latter procedure, which introduced the largest errors into the assignment of absolute values for σ_T ($\pm 20\%$), has been given by Kwan *et al.* (1991).

By measuring the intensity of the beam transmitted through the cell when it was hot, and therefore with alkali vapour present, and also when it was cold, when the vapour pressure was negligible, σ_T was deduced using

$$I_{\text{hot}} = I_{\text{cold}} \exp(-nL\sigma_T), \tag{2.14}$$

which is equivalent to equation (2.1).

The final apparatus we describe briefly here is that used by the Detroit group (Zhou *et al.*, 1994a) for the first measurements of the total cross section for positron scattering by atomic hydrogen, and again it was possible to study electron collisions with the same apparatus. The projectile beams were derived from the solenoid apparatus (see above) and the cross sections were measured by a transmission technique, using the special gas cell illustrated in Figure 2.7. A radio-frequency discharge source was used to produce hydrogen atoms, which were then fed into a small aluminium chamber; this, when cooled to 150 K, had a low recombination coefficient so that a sufficiently high density of hydrogen atoms (approximately 10^{19} m^{-3}) could be achieved for the attenuation measurement to be made. An estimate of the ratio of the number density of atomic hydrogen, $n(H)$, to the number density of atomic-plus-molecular hydrogen, $n(H) + n(H_2)$, in the cell was obtained by passing a sample of the output of the cell into a quadrupole mass spectrometer. In the earlier work this ratio, $f = n(H)/[n(H) + n(H_2)]$, was measured to be approximately 0.55. However, Zhou *et al.* (1994a) noted that, since some of the hydrogen atoms entering the cell recombined before leaving, the estimated value of f as encountered by the beam passing through the cell was likely to have been greater than 0.55, and possibly as high as unity. Thus, these two extreme values of f were used to calculate the cross section for atomic hydrogen using known values of the cross section for molecular hydrogen. The latter cross section was also measured using the same apparatus by turning the discharge off and normalizing to the earlier measurements by the same group (Hoffman *et al.*, 1982). Improvements reported by Stein *et al.* (1996) and Zhou *et al.* (1997), both in the degree of dissociation achieved and in its determination, mean that the results of the Detroit group can now be presented as definite values, rather than as upper and lower limits.

2.4 General discussion of systematic errors

Some of the systematic errors which can influence measurements of σ_T for low energy positron and electron scattering were introduced in the preceding section. We now proceed to a more general discussion of these errors and attempt to summarize the situation pertaining to all the groups who have made such measurements and whose data are presented later in this chapter. This summary is based on that given in the review of Charlton (1985a).

Equations (2.1) and (2.2) show that, in order to obtain accurate values of σ_T, it is necessary to make precise measurements of the incident and transmitted fluxes, I_0 and I respectively, the path length L in the scattering cell and the pressure P and temperature T of the gas. During the 1970s and 1980s it became technically feasible, using systems like those described in the previous section, to determine σ_T over a wide energy range, with quoted statistical errors usually less than $\pm 5\%$. It will become apparent in the following section that much greater discrepancies than this sometimes exist between the measurements from different groups. This implies that there are also systematic effects present in some of the measurements which cause erroneous determinations of one or more of the parameters listed above and which are peculiar to the method and apparatus employed by each of the groups. Following Charlton (1985a), the four main sources of systematic error can be summarized as follows:

(i) errors in L due to spiralling of the positrons and electrons in an axial magnetic field, if employed (in systems not using a magnetic field, other effects on the trajectories may be important);

(ii) errors in L caused by end effects in the gas cell;

(iii) errors in P due to incorrect determination of the absolute pressure;

(iv) errors in I, due predominantly to the detection of projectiles which have been scattered elastically, or with small energy loss, but which cannot be distinguished from the unscattered beam.

Any contribution from error (i) results in an overestimate of σ_T since the effective path length in the gas is greater by a factor $\sec\theta$ than the assumed value; see equation (2.11). Error (iv) also leads to an underestimate of σ_T since not all the scattered projectiles are so identified. All reported experimental values of σ_T may be subject to one or more of the errors (i)–(iv), and it is important to bear this in mind when making comparisons between the results from different groups.

Errors of type (iii) are probably less important than some of the others. Even so, as pointed out by Coleman *et al.* (1979) and Charlton (1985a), an absolute measurement of the pressure using a capacitance manometer can be complicated by thermal transpiration effects. Errors quoted by manufacturers on the pressure, as measured using such instruments, can be as low as 0.15%–0.25%. However, the sensor heads are usually electronically stabilized at a temperature T_s (typically ≈ 350 K) which, if not equal to the actual temperature T, can give rise to a thermal transpiration correction factor relating the true pressure P to that measured at the sensor, P_s, by $P = P_s(T/T_s)^{1/2}$. This is a 4% effect at $T = 297$ K. Such corrections were made by the Detroit group, but Coleman *et al.* (1979) pointed out that account should also be taken of the diameter of the tube or other arrangement which connects the scattering cell to the sensor, according to the method of Liang (1955). These authors, using a similar manometer to that of the Detroit group, found that a smaller adjustment should be applied, its value being dependent upon the gas under study (e.g. 0.25% for helium and 0.5% for neon). Mizogawa *et al.* (1985), who adopted the latter method, presented a detailed discussion of how the corrections were calculated.

An error of type (i), giving a systematic variation in the length of the flight path of the positron or electron in the scattering region, is a potentially serious contributor to the overall error in σ_T, particularly at low impact energies where the final beam energy is comparable to the emission energy from the moderator. At these energies the type of moderator used (the material and the geometry), and the quality of its surface, have some bearing on the angular properties of the beam. This affects the beam divergence in a system employing mainly electrostatic elements, such as that illustrated in Figure 2.5, and the pitch angle in a system similar to those shown in Figures 2.2 and 2.4, which incorporate axial magnetic guiding fields.

Considering first the latter type of instrument (which is used by the majority of workers), it is notable that no group has reported corrections for the effect of spiralling. No discussion of the possible effects of spiralling was given in the early UCL work, although Charlton *et al.* (1983a) claimed that their measurements were free from such errors down to an impact energy of 2 eV. These authors, and Coleman *et al.* (1979, 1980a), made use of the intrinsic energy spread of their beams (approximately 2 eV) to determine σ_T simultaneously at several energies across their TOF spectrum. Upon changing the mean energy of the beam by less than 2 eV it was possible to obtain several values of σ_T at a particular energy but from different parts of the TOF spectrum. If the time of flight had included a factor due to spiralling, then the values of these cross sections so determined would not have been in such good agreement. Total cross sections obtained using this technique were found to be self-consistent except on the long TOF side of the spectrum where, as described by Charlton *et al.* (1983a), small-angle elastic scattering was observed. A detailed discussion of effects at low positron energies was given by Coleman *et al.* (1979).

The positron–helium measurements of Mizogawa *et al.* (1985) were deemed to have small errors arising from positron spiralling, because of the use of low magnetic fields and the attendant limits on the transverse kinetic energy imposed by physical apertures in the system.

The Detroit and Toronto groups also used axial magnetic guiding fields, and they too reported negligible spiralling effects. Kauppila *et al.* (1977), who made total cross section measurements with their 0.1 eV emission energy beam, found that at energies above 2 eV the enhancement of the positron flight path in their curved solenoid gas cell was less than 1%. The Toronto group (Tsai, Lebow and Paul, 1976) found that spiralling in their 0.75 mT guiding field occurred at a pitch angle of typically 3°. In this case the details of the entire experimental arrangement are important since the low energy positrons were injected into the field generated by a solenoid, which also formed the scattering region, from an electrostatic extraction and deflection system.

Turning now to systems which do not employ axial magnetic fields, Sinapius *et al.* (1980), using trajectory simulations, found a maximum path-length spread of 3% due to their use of a weak magnetic lens in an otherwise electrostatic system. Finally, the Swansea group (Brenton *et al.*, 1977; Brenton, Dutton and Harris, 1978), who used a Ramsauer-type apparatus, did not give an estimate of the uncertainties in the semicircular trajectories arising from the angular divergence of their beam, although these are likely to have been small at the kinetic energies (usually > 20 eV) at which this instrument was used.

The second potential error listed above is that due to end effects in the gas cell. In section 2.3 the normalization method involving the use of a localized scattering cell, as employed by the UCL group, was described in some detail. In particular, we refer to equation (2.10) and the associated discussion. Further details can be found in the original works of Griffith *et al.* (1979a) and Charlton *et al.* (1983a, 1984), which have been summarized by Charlton (1985a). It is important to note that many gases were found to give the same value for the normalization constant k_n, 1.275 ± 0.020; this value was used for all the positron total cross sections measured using the apparatus shown in Figure 2.2. Of special note, though, is the study of low energy electron–helium scattering reported by Charlton *et al.* (1980a). Here k_n deviated from the above value and was found to be pressure dependent. This may have been due to the much lower helium pressure used in this work (because of the large difference in the cross sections for the two projectiles; see Figure 2.1) when compared with the original evaluation of k_n (Griffith *et al.*, 1979a) from studies of positron–helium scattering.

The early measurements of the Toronto group were also subject to corrections for end effects, and these were deduced by recording beam attenuations at two or more positions of the detector, which corresponded to varying the length of their total system. They found, by extrapolating to zero length, that equation (2.1) had to be modified in their case by a factor which was constant for a particular gas but which varied exponentially with gas density when the pressure was varied in the scattering region. Further details can be found in the review of Charlton (1985a) and in the works of Tsai *et al.* (1976) and Jaduszliwer, Nakashima and Paul (1975). Mizogawa *et al.* (1985) accounted for their end corrections by calculating the integral in equation (2.10) using direct measurements of the pressure at various points along the flight path.

In those experiments which employed long scattering cells, end effects were usually negligible. The Detroit and Swansea systems only required corrections for measured pressure gradients across the cell; this was also the case in the early UCL and Arlington experiments, both of which used the full geometric length of the flight path as the length of the scattering cell. An end, or background gas, correction was applied by Sinapius *et al.* (1980) to the results obtained with the system used at Bielefeld; this accounted for gas escaping from the scattering cell and entering the regions containing the moderator and the detector. This correction was applied directly to the data, as noted in section 2.3.

The final error, (iv), is that affecting the beam strength under gas flow conditions when positrons are detected which have undergone small-angle scattering but which cannot be distinguished experimentally from the unscattered beam. The largest contribution to the error is expected to arise from the elastic scattering channel although, for molecular targets, small-angle collisions in which there is rotational and vibrational excitation are also possible. This error, which is to some extent a feature of all experimental measurements of σ_T, is caused by the fact that the angle below which a scattered positron cannot be distinguished from the unscattered beam, the discrimination angle, θ_{disc}, is non-zero. Some workers have reported estimates of θ_{disc} which have enabled cross-evaluations of the size of the errors to be made.

The effect of forward scattering on the measured values of σ_T for positrons was first discussed briefly by Canter *et al.* (1973) in their investigations of positron scattering by the noble gases at low and intermediate energies. Detailed discussions of the effects of forward scattering have been given by Kauppila *et al.* (1981) and Kauppila and Stein (1982), and we now discuss the general features from a largely experimental viewpoint.

As mentioned previously, if scattered projectiles are detected as though they were unscattered then the transmitted intensity I will be overestimated, leading to a measured value of σ_T which is lower than the true value. In a beam system consisting entirely of electrostatic elements the angular discrimination can be set according to the geometry of the apparatus. When axial magnetic fields are employed, as has mostly been the case, some extra means of discriminating against small-angle elastic scattering must be sought. Here we briefly describe three techniques which have been applied in positron scattering studies.

The first technique, mentioned in section 2.3, uses a magnetic field gradient to produce a magnetic mirror effect on positrons with too high a pitch angle θ_p (e.g. Griffith *et al.*, 1978a). Note that, for projectiles initially propagating along the magnetic field axis, the pitch angle is equal to the scattering angle, i.e. $\theta_p = \theta$. On making a transition from a low magnetic field B_1 in the scattering region to a higher field B_2 in the region where the particle is detected the pitch angle of any scattered particle will increase, and when it becomes $90°$ the scattered particle can no longer reach the detector. From elementary considerations, θ_{disc} is given by

$$\theta_{\text{disc}} = \sin^{-1}(B_1/B_2)^{1/2}. \quad (2.15)$$

Note that θ_{disc} is independent of the total kinetic energy of the projectile provided the adiabatic criterion is fulfilled, namely that the magnetic field does not change appreciably over the axial distance travelled by the particle during one turn of its helical trajectory. A more detailed discussion of this point can be found in the work of Kruit and Read (1983).

The second method of discrimination uses apertures to intercept some of the scattered flux by virtue of its increased Larmor radius in the magnetic field (e.g. Kauppila *et al.*, 1981; Mizogawa *et al.*, 1985). The extent to which angular discrimination can be provided by this method depends not only on the radius of the aperture, r_{ap}, but also on the diameter of the initial beam. Assuming, for simplicity, a beam of particles, each with kinetic energy E, confined to the axis of the system by a uniform magnetic field B, then

$$\theta_{\text{disc}} = \sin^{-1}\left[r_{\text{ap}}eB/(2Em)^{1/2}\right]. \quad (2.16)$$

Using this formula, an average discrimination angle for a beam of finite diameter can easily be computed. At high speeds, however, where the positrons may not complete even one revolution about the magnetic field lines before leaving the scattering region, a trajectory calculation should be performed to deduce the effect of the aperture.

A third method of discrimination has been employed in those systems which combine the use of an axial magnetic field with the TOF technique (see section 2.3 and Figure 2.2). In this case it can be shown that

$$\theta_{\text{disc}} = \sec^{-1}\left[\tau_{\text{r}}(2E/m)^{1/2}/l_2 + 1\right], \quad (2.17)$$

where l_2 is again the geometric length of the flight path after scattering has taken place. Here, at high energies, τ_{r} can be interpreted as the intrinsic timing resolution of the TOF system. At low energies, where the TOF spread of the unscattered beam may be much greater than τ_{r}, it may be more suitable to use some other minimum resolvable time, greater than τ_{r} (see e.g. Charlton *et al.*, 1983a). This relationship is similar to that given as equation (2.11), except that here the significance of τ_{r} is emphasized. Clearly, if the increase in the time of flight, Δt, is less than τ_{r} then the scattered positrons cannot be distinguished from the unscattered beam. Note also that, using the TOF method, θ_{disc} decreases as the projectile kinetic energy increases.

The final method of discrimination, as pointed out above in relation to the Detroit apparatus, is the use of a retarding field analyser. Its effectiveness depends upon the incident energy according to (Kauppila *et al.*, 1981)

$$\theta_{\text{disc}} = \sin^{-1}\left[(\Delta E/E)(B_1/B_2)\right]^{1/2}, \tag{2.18}$$

where again B_1 and B_2 refer to the magnetic field strengths in the scattering region and in the retarding and detection region respectively. The quantity ΔE is effectively the energy resolution, either imposed on the beam using the retarding field analyser or of the beam itself.

Many detailed discussions have been given in the literature of the failure to discriminate adequately against small-angle forward scattering, particularly for the helium target at impact energies below the positronium formation threshold. This case is of great interest because it is the simplest target which is readily amenable to experimental investigation and for which benchmark calculations by Humberston and coworkers are available. Further details can be found in the works of Humberston (1978), Wadehra, Stein and Kauppila (1981) and the review of Charlton (1985a). These discussions have been mainly concerned with estimating the effect of forward scattering for each experiment, based upon the calculated differential cross sections for elastic scattering and the quoted values of θ_{disc}, when available. Unfortunately, the outcome of these analyses is that the discrepancies between the experiments, one with another and with theory, cannot be entirely explained by the neglect of forward scattering in the experiments. This implies that some other energy-dependent systematic errors were present in some of the data; these might be, for example, (i) above or perhaps (iii) if a wide range of pressures was used, although checks were usually performed to ensure that the measured values of σ_T were independent of the pressure. Notwithstanding this, there is generally good agreement in the values of σ_T between theory and most experiments for positron–helium scattering above 6 eV, as will be seen in the next section. Indeed, for many targets there is broad agreement between the experimental results to within the ±20% level, and sometimes much better. For the heavier noble gases and all molecules, however, theoretical uncertainties concerning the values of the differential elastic scattering cross sections mean that such analyses should be considered as merely offering a guide.

2.5 Results and discussion – atoms

Positron total scattering cross sections have been measured and calculated for a variety of atomic and molecular gases, and in this section we present a selection of results. In the light of the discussion given in section 2.4, particularly concerning small-angle elastic scattering, a critical evaluation

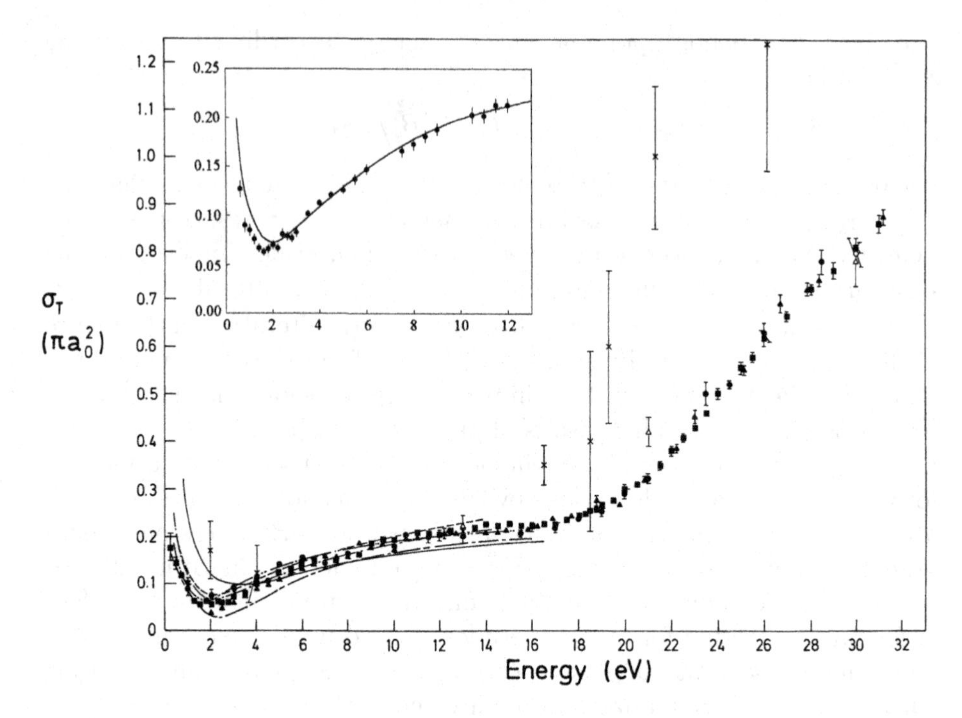

Fig. 2.8. Low energy positron–helium total scattering cross sections. Experimental data, main diagram: ×, Costello *et al.* (1972); •, Canter *et al.* (1972, 1973); ▼, Wilson (1978), after correction by Sinapius, Raith and Wilson (1980); ■, Stein *et al.* (1978); ▲, Coleman *et al.* (1979); △, Brenton *et al.* (1977); ○, Griffith *et al.* (1979a). The experimental data in the inset are from Mizogawa *et al.* (1985) and they are compared there with the theoretical work of Campeanu and Humberston (CH, see text). Theoretical curves, main diagram: — · —, CH; ——, McEachran *et al.* (1977); – – –, Schrader (1979); — · · —, Amusia *et al.* (1976); — - —, Aulenkamp, Heiss and Wichmann (1974).

of the available experimental data is attempted in some cases. Comparisons with relevant data for electron impact are also made.

1 Helium

This is the target most thoroughly studied experimentally, and it has also received considerable theoretical attention. Theoretical and experimental cross sections obtained by several groups are shown in Figure 2.8. Much consideration has been given to the low energy elastic scattering region; we now summarize the situation. Stein and Kauppila (1982) have calculated the extent to which their total cross section measurements underestimated the true values because of the neglect of small-angle scattering, assuming various values of the discrimination angle $\theta_{\rm disc}$.

Using the s-, p-, and d-wave phase shifts of Campeanu and Humberston (1977a), Humberston and Campeanu (1980) and Drachman (1966a) respectively, and all higher partial-wave phase shifts given by the formula of O'Malley, Spruch and Rosenberg (1961), equation (3.67), they found that the percentage discrepancy in σ_T, Δ, was largest in the vicinity of the Ramsauer minimum, at an energy of approximately 2 eV. At this energy they obtained $\Delta \approx 7\%$ for $\theta_{disc} = 10°$, rising to 41% for $\theta_{disc} = 30°$. However, at energies both above and below 2 eV the value of Δ falls sharply owing to the preponderance of the s-wave contribution to σ_T. Thus, at 13.6 eV the values of Δ are 3% and 8% for $\theta_{disc} = 10°$ and 30° respectively. In contrast, underestimates of the electron–helium scattering cross section due to the neglect of forward scattering are at a minimum at 2 eV, Δ being as small as 4% even for $\theta_{disc} = 30°$.

The reason for the large difference between the values of Δ for positrons and electrons at an energy of 2 eV is that for positrons the s-wave phase shift passes through zero at the Ramsauer minimum and the dominant contribution to the cross section therefore comes from the p-wave, which is quite strongly peaked in the forward and backward directions. In contrast, there is no Ramsauer minimum in electron–helium scattering, and the isotropic s-wave contribution to σ_T is dominant at this energy.

Let us now compare the experimental measurements of σ_T for low energy positron–helium scattering with the benchmark theoretical results of Humberston and Campeanu (1980) and Campeanu and Humberston (1975, 1977a), hereafter referred to collectively as CH, which are discussed in detail in section 3.2. It can be seen from Figure 2.8 that the data of Canter *et al.* (1972) are in good agreement with CH at energies close to 2 eV, but they fall slightly below these theoretical values above approximately 7 eV. The data of Wilson (1978) in the range 1–6 eV, corrected in the manner described by Sinapius, Raith and Wilson (1980), are also in good accord with CH, whereas the data from the Detroit and Arlington groups below 6 eV are seen to fall gradually below those of CH, with Δ amounting to approximately 15% and 40% respectively at 2 eV. This discrepancy has attracted considerable attention, and Humberston (1978, 1979) claimed that the experimental results from Detroit and Arlington at the lower energies were in error because of the neglect of forward scattering and that the data of Canter *et al.* (1972) and Sinapius *et al.* (1980) were more accurate. However, as pointed out by Wadehra *et al.* (1981), it is impossible to explain the differences between the data of Canter *et al.* (1972) and CH above 6 eV using these arguments. Canter *et al.* (1973) originally asserted that θ_{disc} for their earlier experiment was approximately 35°, which, using the analysis of Stein and Kauppila (1982), would have resulted in a value of $\Delta \geq 40\%$ at 2 eV. In a reappraisal of the experimental method of Canter *et al.* (1972, 1973), Griffith *et al.* (1978a) considered the magnetic field strengths employed in various regions along the flight path of the positron beam and concluded that 10° was a more realistic upper limit for $\theta_{\rm disc}$. These authors further stated that this value of $\theta_{\rm disc}$ implied that the measured value $(0.075 \pm 0.012)\pi a_0^2$ for $\sigma_{\rm T}$ at an energy of 2 eV is an underestimate by only $0.003\pi a_0^2$. This differs from the value $0.006\pi a_0^2$ deduced from the value of Δ given by Stein and Kauppila (1982), but is within the statistical uncertainty of the measurement. In a complementary fashion, Humberston (1978) and Stein and Kauppila (1982) assumed that the non-zero values for Δ were solely due to forward scattering errors and they obtained values of 12° and 20° for $\theta_{\rm disc}$ for the Detroit and Arlington systems respectively. The Detroit value is in agreement with estimates of this parameter made by Dababneh *et al.* (1980).

Stein and Kauppila (1982) suggested a possible error in the analysis of Griffith *et al.* (1978a) arising from positron scattering in the curved solenoid of length 0.15 m located at the end of the flight path in the original UCL experiment. They argued that, owing to the positioning of the coils, the value of 10° for $\theta_{\rm disc}$ only applied to the straight section, of length 0.7 m, of the flight path; positrons scattered in the curved section, which was maintained at a roughly uniform magnetic field, would be virtually indistinguishable from those in the unscattered beam. These comments are valid and, as such, it is appropriate to weight the value of $\theta_{\rm disc}$ given by Griffith *et al.* (1978a) according to the length of the flight path. A revised upper estimate of $\theta_{\rm disc}$ for the experiment of Canter *et al.* (1972, 1973) is then

$$(0.7 \times 10^\circ + 0.15 \times 90^\circ)/0.85 \approx 24^\circ,$$

which would mean, according to Stein and Kauppila (1982), that Δ is approximately 30% at an energy of 2 eV. This would be an overestimate of Δ since there would still be a TOF separation between positrons scattered in the final 0.15 m and those in the unscattered beam. At 2 eV this separation may be sufficiently large to reduce Δ substantially, and Griffith and Heyland (1978) presented convincing evidence (see their Figure 14) that scattered positrons are essentially absent from the TOF spectra obtained at low energies. A more detailed discussion of forward scattering errors has been given by Stein and Kauppila (1982), who pointed out that the cross sections of Sinapius, Raith and Wilson (1980), obtained with a system for which $\theta_{\rm disc} \approx 7^\circ$, are in best agreement with the results of CH in the energy region 1–6 eV.

The results of Mizogawa *et al.* (1985) are also shown in Figure 2.8. Their data were taken using a magnetic field of 0.8 mT at energies below 3 eV and 1.3 mT at higher energies. Above 10 eV their results are in very good agreement with those of the Detroit group. For smaller energies they lie above the latter, and in the energy region 1–6 eV they are close to the data of Sinapius, Raith and Wilson (1980) and are also in good agreement with CH, though slightly lower in the region of the Ramsauer–Townsend minimum. This discrepancy can, according to Mizogawa *et al.* (1985), be accounted for entirely by forward scattering errors.

Thus, despite some reservations concerning the role of forward scattering and other potential systematic errors, most of the experimental measurements of σ_T are in broad qualitative agreement with each other, and the data of Canter *et al.* (1972, 1973), Stein *et al.* (1978), Coleman *et al.* (1979) and Mizogawa *et al.* (1985) all exhibit a marked change of slope as the positron energy passes through the positronium formation threshold at 17.8 eV. The importance of this channel in positron collisions is clear from these data and is emphasized in the high resolution measurements of Stein *et al.* (1978).

One of the most striking features of Figure 2.8 is the presence of a Ramsauer–Townsend minimum in the vicinity of 2 eV. Similar features are also observed in low energy electron scattering by certain atoms and molecules, although the mechanism responsible for their existence is then different. For positrons, the minimum is caused by the partial cancellation of the attractive polarization and the repulsive static components of the interaction between the positron and the target atom, giving rise to a zero s-wave phase shift at a specific value of the energy. The corresponding two components in the electron–target interaction are both attractive, and the overall interaction may be sufficiently attractive to give an s-wave phase shift of π radians (or a multiple thereof). As mentioned in subsection 1.6.1 and in section 2.1, the difference in sign of the static interaction is also responsible for the large differences between the scattering cross sections for electrons and positrons, as highlighted in Figure 2.1, where cross sections for both projectiles are presented. At the energy of the Ramsauer minimum, the cross section for positrons is nearly two orders of magnitude smaller than that for electrons.

Another noteworthy feature of Figure 2.1, which is based upon the results of Kauppila *et al.* (1981) for electrons and positrons, is the merging of the cross sections for the two projectiles at the relatively low energy of approximately 200 eV; a plausible, rather than rigorous, explanation of this feature, exploiting the unitarity of the scattering process implied by the use of the optical theorem, has been given in section 2.2. The individual partial cross sections for the two projectiles are quite different at such a relatively low energy, and they only merge at much higher energies. Furthermore, there is no counterpart to positronium formation in electron scattering, although this process has a very small cross section at 200 eV. When considering the total cross section as the sum of the

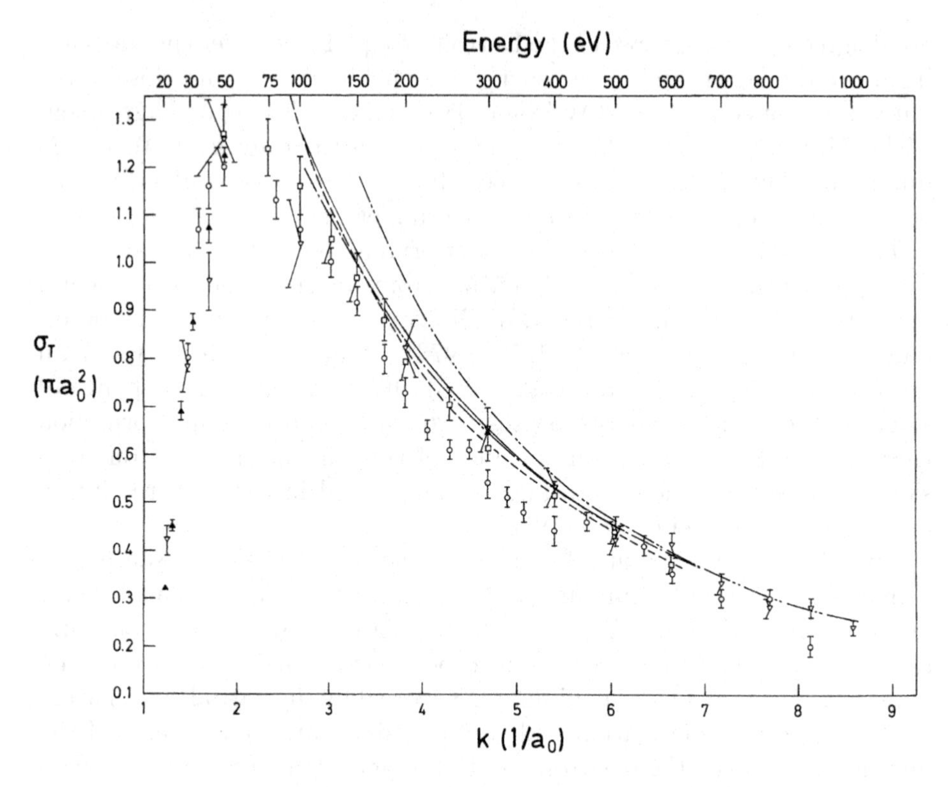

Fig. 2.9. Intermediate energy positron–helium total scattering cross sections. Experiment: ▲, Coleman *et al.* (1979); ○, Griffith *et al.* (1979a); ▽, Brenton *et al.* (1977); □, Kauppila *et al.* (1981). Experimental data for electrons (– – –) are shown for comparison. Theory: — · —, Dewangan and Walters (1977); ——, Byron and Joachain (1977a); — · · —, Bethe–Born calculations of Inokuti and McDowell (1974).

constituent partial cross sections, it may therefore seem surprising, and somewhat fortuitous, that the greater elastic scattering cross section for electrons is so well compensated by the greater inelastic scattering cross section for positrons. Similar merging of the total cross sections for the two projectiles at energies significantly lower than those for which the individual partial cross sections merge is encountered with several other targets. The partitioning of σ_T into its various constituent partial cross sections is described in greater detail in section 2.7.

The situation at intermediate energies is shown in Figure 2.9. All groups find that σ_T continues to rise rapidly with energy, reaching a maximum somewhere in the energy range 50–60 eV before falling gradually as the energy is increased up to 1 keV. The actual values are in tolerably good agreement (within approximately 10%) over most of the energy range, although the data of Griffith *et al.* (1979a) are 15% lower than the other results in the range 200–400 eV, and they exhibit a marked drop above 800 eV. This latter effect is almost certainly caused by failure to resolve at high energies the diminishing difference between the times of flight of positrons which have been scattered through small angles and those which are unscattered.

The early measurements of Costello *et al.* (1972) are included in Figure 2.8 even though they have now been superseded, but the results of Jaduszliwer and Paul (1973) are not because, as discussed by Charlton (1985a), they were subject to a substantial systematic error in the effective path length of the positrons in the gas. Neither are the intermediate energy data of Canter *et al.* (1973) and of Coleman *et al.* (1976b) included in Figure 2.9, because they suffered from large forward scattering errors.

Also shown in Figures 2.8 and 2.9 are results from various theoretical investigations. The methods used, and the results obtained, below the positronium formation threshold relate only to elastic scattering and are described in more detail in subsection 3.2.2. The most accurate results in this energy region have been obtained using variational methods (Humberston, 1979), and these are in good agreement with the experimental measurements throughout this energy range. Similar techniques have recently been used to obtain accurate total cross sections in the Ore gap, the energy range between the positronium formation threshold and the lowest positron-impact excitation threshold of the helium target 17.8–20.6 eV. The results of the polarized-orbital method, as used by Massey, Lawson and Thompson (1966), by Montgomery and LaBahn (1970) and, particularly, by McEachran *et al.* (1977) are in rather less good agreement but the method is of special significance because it is one of the few to have been applied to all the noble gases. A much simpler method of calculating the elastic scattering cross sections, which yields surprisingly good results, has been used by Schrader (1979). The positron–target interaction is represented by a simple central polarization potential containing an adjustable range parameter, and the scattering problem then reduces to solving a one-dimensional Schrödinger equation for each partial-wave phase shift. At intermediate and higher energies (>100 eV), total cross sections have been calculated using the optical theorem in conjunction with various approximation methods for determining the forward scattering amplitude. Among the most important of these studies have been those by Byron and Joachain (1977a), who used the eikonal Born series, and Dewangan and Walters (1977), who used the distorted-wave second Born approximation. Calculations have also been made by Byron and Joachain (1977b) using an optical potential formalism

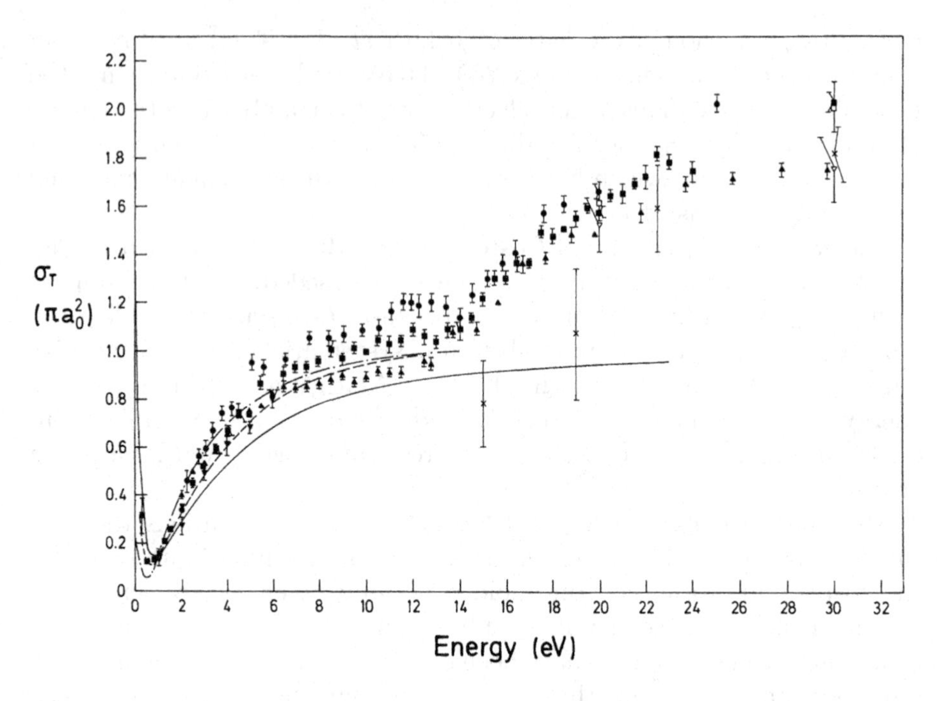

Fig. 2.10. Low energy positron–neon total scattering cross sections. Experiment: ■, Stein *et al.* (1978); ▲, Coleman *et al.* (1979); ▼, Sinapius *et al.* (1980); •, Charlton *et al.* (1984); □, Kauppila *et al.* (1981); ×, Tsai, Lebow and Paul (1976); ▽, Brenton, Dutton and Harris (1978). Theory: — · —, Campeanu and Dubau (1978); ——, McEachran, Ryman and Stauffer (1978); – – –, Schrader (1979).

in which the imaginary part of the potential represents absorption into channels not explicitly included in the formulation. The results of Dewangan and Walters (1977) are in particularly good agreement with the measured values of Kauppila *et al.* (1981) at all energies beyond 100 eV, but all methods agree well beyond 700 eV. Virtually no theoretical studies have been made of the total cross section between the upper limit of the Ore gap and 100 eV, although several calculations have been made of individual partial cross sections, details of which are given in later chapters.

2 Heavier noble gases

The values of σ_T for neon at low and intermediate energies are shown in Figures 2.10 and 2.11, which include experimental results obtained by a number of authors in the energy range 0–1000 eV. For the same reasons as outlined above when discussing helium, the data of Coleman *et al.*

(1976b) and Jaduszwiler and Paul (1974) are not included. Figure 2.10 shows that only the data of Stein *et al.* (1978) extend to sufficiently low energies to observe the deep, narrow Ramsauer minimum which exists at around 0.6 eV. All groups find that the cross section rises sharply with energy and then levels out around 10 eV before rising dramatically again above E_{Ps}. This latter effect is similar to that found in positron–helium scattering and is again due to positronium formation.

Although the data from all the experiments exhibit similar qualitative features, closer inspection reveals discrepancies which, according to the analysis of Kauppila and Stein (1982), cannot be explained by assuming that the main systematic error is due to the neglect of forward scattering. The results of Stein *et al.* (1978) are mostly higher than those of Coleman *et al.* (1979) above 6 eV, but they fall significantly below them at lower energies. Both measurements are lower than those of Charlton *et al.* (1984), except perhaps at the higher part of the energy range covered by the latter authors. As mentioned earlier, the system used by Sinapius *et al.* (1980), which has the lowest value of θ_{disc}, should produce the most reliable results. It is noticeable from Figure 2.10 that their data are significantly lower than those of the other groups. The agreement between the results of the Bielefeld and Detroit groups can be improved somewhat since the latter (Kauppila and Stein, 1982) have reported that their energy scale should be increased by approximately 0.2 eV. However, in the range 4–6 eV the data of Sinapius *et al.* (1980) are still approximately 7% lower than those of Stein *et al.* (1978) whereas, according to Stein and Kauppila (1982), they should agree to within 1%–2% at these energies.

Theoretical results are also included in Figure 2.10. Nothing comparable to the accurate variational calculations for positron–helium elastic scattering has been attempted for the other noble gases, and simpler methods of approximation have had to be used instead. The most notable of these has been the polarized-orbital method, which has been used by McEachran and coworkers (1978, 1979, 1980) for all the heavier noble gases. Their results are in reasonably good agreement with the experimental measurements for all the noble gases except argon, where the method fails to reproduce the flat energy dependence between 2 eV and the positronium formation threshold at 9 eV. The simple model-potential method of Schrader (1979) has also been applied to the heavier noble gases and, as with helium, the results are surprisingly good, particularly for neon and argon. Very few theoretical results have been obtained at energies beyond 100 eV; optical model calculations have been performed by Byron and Joachain (1977b) for neon and by Joachain *et al.* (1977) for argon, and the distorted-wave Born approximation has been used by Dewangan and Walters (1977) for neon.

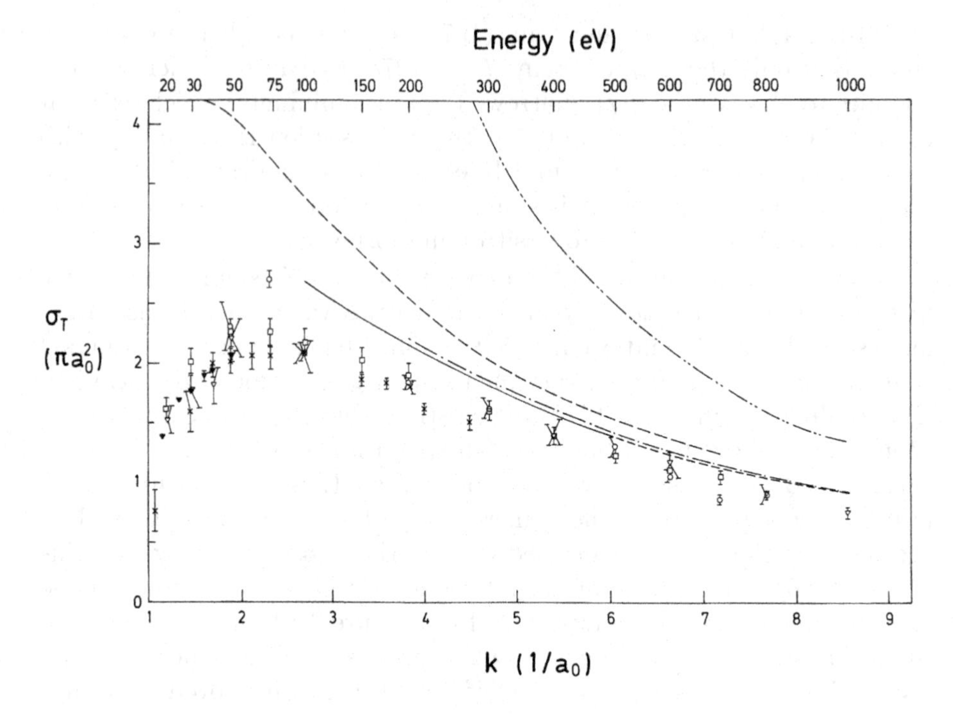

Fig. 2.11. Intermediate energy positron–neon total scattering cross sections. Experiment: ▼, Coleman *et al.* (1979); ▽, Brenton, Dutton and Harris (1978); ×, Tsai, Lebow and Paul (1976); ○, Griffith *et al.* (1979a); □, Kauppila *et al.* (1981). Theory: ——, Byron and Joachain (1977b); — · —, Dewangan and Walters (1977); — · · —, Bethe–Born calculation of Inokuti and McDowell (1974). The experimental electron data (– – –) are shown for comparison.

Theoretical and experimental data at intermediate energies are shown in Figure 2.11. The experimental measurements are in tolerable agreement with one another, but are everywhere lower than the theoretical results.

The available experimental and theoretical data for positron–argon scattering at low energies are shown in Figure 2.12, where it can be seen that the data of Kauppila, Stein and Jesion (1976) and Coleman *et al.* (1980a) are in reasonable agreement with each other, although both sets are lower than the results of Sinapius *et al.* (1980) and Charlton *et al.* (1984). As was the case with helium and neon, it is unlikely that these discrepancies are the result of forward scattering errors alone. Of particular note are the large differences between the data of Charlton *et al.* (1984), Kauppila, Stein and Jesion (1976) and Coleman *et al.* (1980a) at energies above E_{Ps}, where it is expected that forward elastic scattering will contribute only a small fraction to σ_T. Furthermore, the measurements

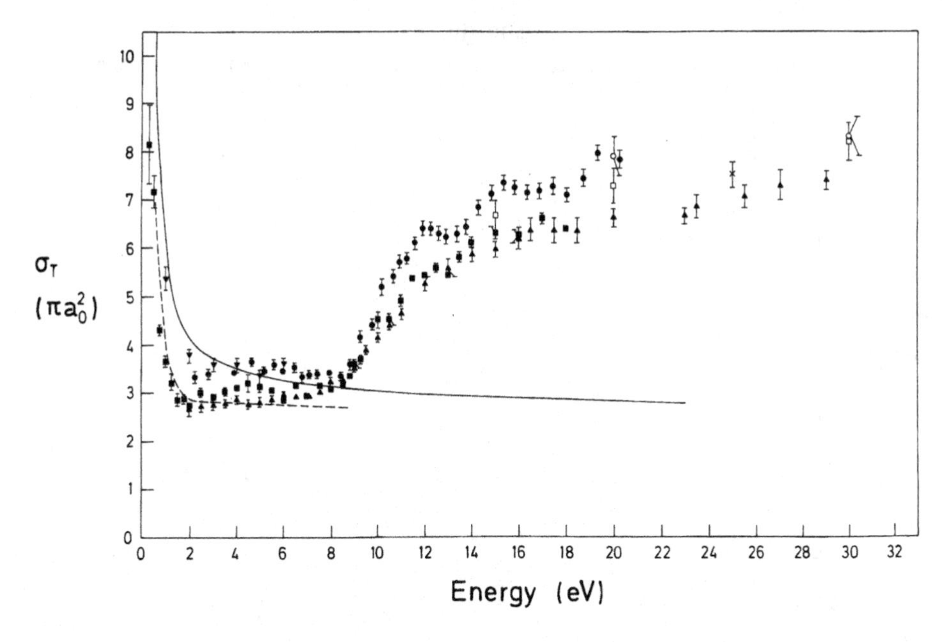

Fig. 2.12. Low energy positron–argon total scattering cross sections. Experiment: ■, Kauppila, Stein and Jesion (1976); ×, Tsai, Lebow and Paul (1976); ▼, Sinapius, Raith and Wilson (1980); ▲, Coleman *et al.* (1980a); •, Charlton *et al.* (1984); □, Kauppila *et al.* (1981); ○, Griffith *et al.* (1979a). Theory: ——, McEachran, Ryman and Stauffer (1979); – – –, Schrader (1979).

of Charlton *et al.* (1984) exhibit some structure in the 12–18 eV energy range, whereas none can be discerned in the results from the other groups. The calculations of McEachran *et al.* (1979) and Schrader (1979) for elastic cross sections are also included, but they do little to help resolve the experimental discrepancies.

Experimental data at intermediate energies are presented in Figure 2.13, together with the results of the optical model calculations of Joachain *et al.* (1977). The results of Kauppila *et al.* (1981), of Tsai, Lebow and Paul (1976) in the energy range 25–270 eV and of Brenton, Dutton and Harris (1978) in the energy range 200–1000 eV are in reasonable agreement, whereas the data of Griffith *et al.* (1979a) are 7%–13% higher below 150 eV and around 12% lower above this energy. The optical model results of Joachain *et al.* (1977) are somewhat higher than the experimental data.

Most experimental groups have performed measurements of σ_T for krypton and xenon although, as described by Stein and Kauppila (1982), the results are not in good agreement with each other. At the lowest energies the magnitude and the trend of σ_T, as observed by Dababneh

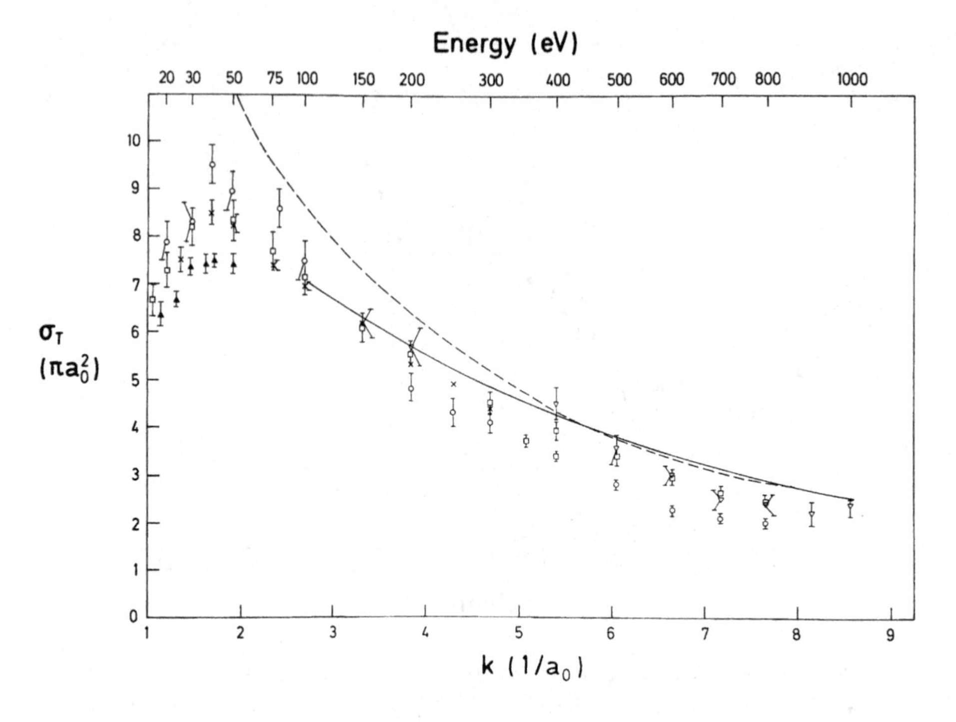

Fig. 2.13. Intermediate energy positron–argon total scattering cross sections. Experiment: ×, Tsai, Lebow and Paul (1976); ▲, Coleman *et al.* (1980a); ○, Griffith *et al.* (1979a); □, Kauppila *et al.* (1981); ▽, Brenton, Dutton and Harris (1978). Theory: Joachain *et al.* (1977). The experimental electron data (– – –) are shown for comparison.

et al. (1980), appear to be closest to the polarized-orbital calculations of McEachran, Stauffer and Campbell (1980) and the model potential approach of Schrader (1979). It is, however, notable that, in contrast to the case of neon, but like that for argon, the data of Sinapius, Raith and Wilson (1980) are much higher than the Detroit results (Dababneh *et al.*, 1980). It should be remembered that the Bielefeld data are expected to be the least prone to forward scattering errors. Stein and Kauppila (1982) argued that if the Detroit and Bielefeld results are each corrected for their respective forward scattering errors, as calculated using the differential cross sections of McEachran, Stauffer and Campbell (1980), then the two data sets are in much better agreement, leaving the results of Canter *et al.* (1973) and Coleman *et al.* (1980a) apparently too low, by as much as a factor of two around 2 eV. However, in view of the theoretical difficulty of treating such complex target atoms accurately, and the lack of overall consistency in the experimental data of the noble gases when taken together, this conclusion should be regarded as tentative.

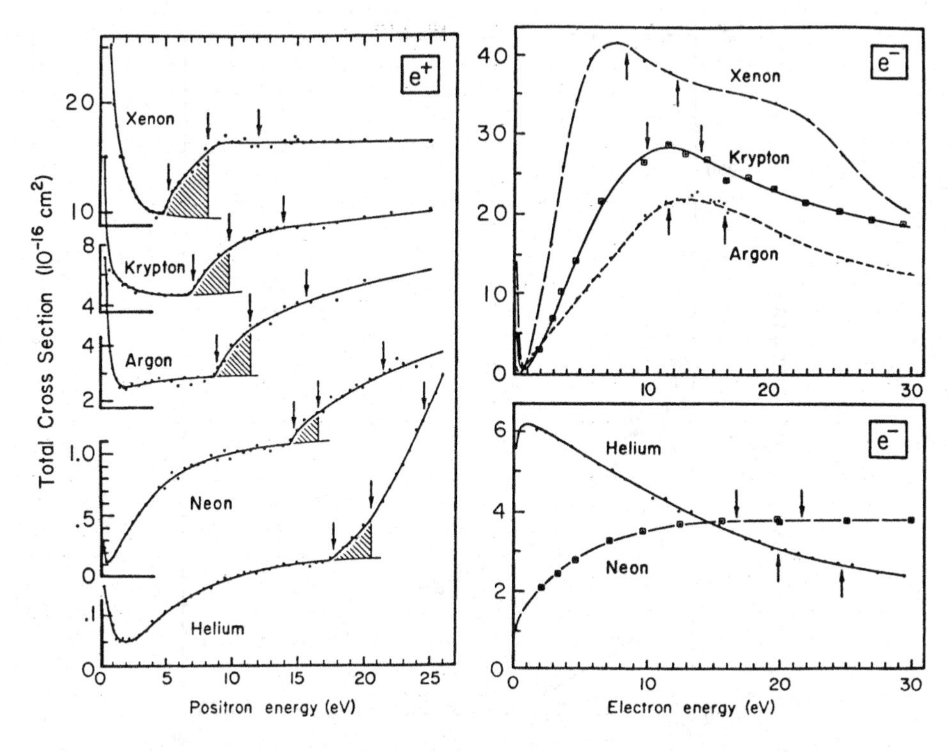

Fig. 2.14. Compendium of total cross section data for positron–noble gas and electron–noble gas scattering. The arrows refer to thresholds for (in order of increasing energy) positronium formation (positrons only), excitation and ionization. (From Kauppila and Stein, 1982.)

It is also instructive to compare the positron and electron scattering cross sections for the heavier noble gases, as was done for helium in subsection 2.5.1. Rather than present all the available data for both projectiles, we show in Figure 2.14 a compendium based upon the results of the Detroit group. This concentrates on the low energy range, where most of the structure in the cross sections can be found. The arrows on the curves indicate the major inelastic thresholds, E_{Ps} (positronium formation), E_{ex} (excitation) and E_{i} (ionization) for positrons, and E_{ex} and E_{i} for electrons. At higher impact energies the cross sections for both projectiles fall monotonically, with those for electrons tending to the positron values from above. The merging of these two cross sections for helium has been described previously, but it has not been observed for the heavier noble gases even at the highest energies investigated so far, typically 1 keV; nevertheless, it is expected to occur at sufficiently high energies.

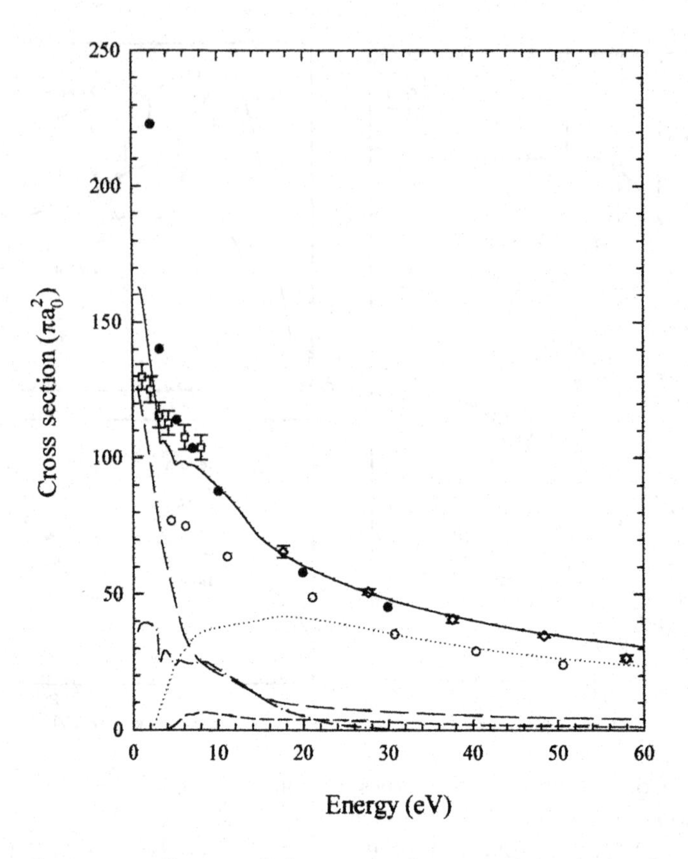

Fig. 2.15. Positron–sodium and electron–sodium total (and also for positrons partial) scattering cross sections. Experiment (positrons): ◇, Kwan *et al.* (1991); □, Kauppila *et al.* (1994). Experiment (electrons): ○, Kauppila *et al.* (1994). Theory: •, Hewitt, Noble and Bransden (1993); ——, Kernoghan (1996). Various partial cross sections from the work of Kernoghan (1996) are also shown: — —, elastic scattering; — · —, positronium formation; - - - -, excitation of sodium to the 3D level; · · · · ·, resonant excitation (3S–3P).

3 Alkali metals

The apparatus and technique used by Stein *et al.* (1985), Kwan *et al.* (1991), Parikh *et al.* (1993) and Kauppila *et al.* (1994) has already been discussed in section 2.3. Kwan *et al.* (1991) stated that their data for positron and electron scattering by potassium superseded those obtained by Stein *et al.* (1985), and consequently the earlier results are not described here.

In Figure 2.15 we present the total cross sections for positron–sodium scattering. Shown here are the experimental data of Kwan *et al.* (1991) and Kauppila *et al.* (1994), and the theoretical coupled-state results of

Hewitt, Noble and Bransden (1993) and Kernoghan (1996). As described in section 2.2, these latter authors are believed to have produced the most accurate theoretical results at low energies. The experimental results have been corrected by Kernoghan (1996) to account for the failure to discriminate completely against small-angle elastic scattering. Angular discrimination values given by Kwan *et al.* (1991) were used for this purpose, the largest corrections arising at impact energies below 10 eV. The theoretical results of Kernoghan (1996) and the corrected experimental data (which were further scaled upwards by 4% compared to the estimated systematic error of ±20% in order to achieve the best agreement in shape) are in reasonable accord, although less so below 10 eV. The two sets of theoretical results shown are in good agreement, except at the very lowest energies when the predictions of Hewitt, Noble and Bransden (1993) greatly exceed those of Kernoghan (1996). This discrepancy has yet to be explained satisfactorily.

The total cross section for electron–sodium scattering is also shown in Figure 2.15, where it can be seen to decrease monotonically with increasing energy. Considering the difficulty in obtaining an absolute scale for these cross sections (see section 2.3), the data of Kwan *et al.* (1991) are in good agreement with most of the other direct measurements and also with theoretical estimates derived from the integrated forms of several differential elastic scattering cross sections obtained by Srivastava and Vušković (1980), to which were added excitation and ionization cross sections obtained by other workers. The behaviour of the electron scattering cross sections for other alkali atoms is similar to that for sodium (see e.g. Kwan *et al.*, 1991; Parikh *et al.*, 1993).

The total cross sections for positron scattering by potassium and rubidium, shown in Figures 2.16 and 2.17, display a markedly different behaviour at low positron energies from the corresponding data for positron and electron scattering from sodium. Of particular note is the broad maximum in the cross sections around 6 eV, in both theory and experiment (after corrections for forward scattering errors, and an overall slight upward rescaling), for both targets. The behaviour of the partial cross sections as calculated in the coupled-state approximation by McAlinden, Kernoghan and Walters (1996) and Kernoghan, McAlinden and Walters (1996) is also illustrated in Figures 2.16 and 2.17. Again, the dominant contributions are resonant excitation, 4S–4P for potassium and 5S–5P for rubidium, at the higher energies and elastic scattering at lower energies, except in the limit of zero incident energy when positronium formation becomes dominant. The fall in σ_T at low energies is caused by a more rapid decrease of the resonant excitation cross section than is the case for sodium, and also the behaviour of the positronium formation cross section.

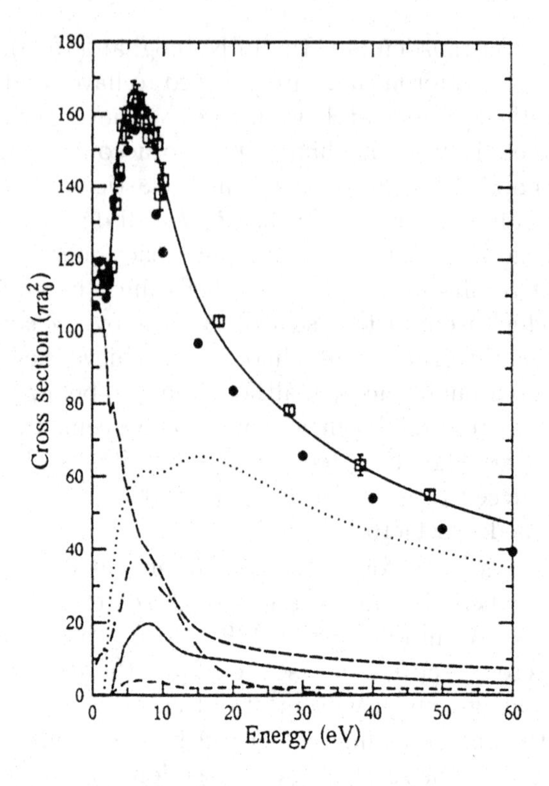

Fig. 2.16. Total (and partial) cross sections for positron–potassium scattering. Experiment: □, Parikh *et al.* (1993). Theory: •, Hewitt, Noble and Bransden (1993); upper solid line, McAlinden, Kernoghan and Walters (1996). Various partial cross sections from the work of the latter are also shown: — —, elastic scattering; · · · · ·, resonant excitation (4S–4P); lower solid curve, 4S–3D excitation; - - - -, excitation to the n = 5 levels of potassium; — · —, positronium formation.

This latter cross section peaks at a higher positron energy for potassium and rubidium than for sodium, and the calculations of Hewitt, Noble and Bransden (1993), McAlinden, Kernoghan and Walters (1996) and Kernoghan, McAlinden and Walters (1996) have established that this is due to large contributions from positronium formation into excited states, outweighing the contribution from ground state positronium formation. This situation is unique in targets studied so far and is discussed further in Chapter 4.

A further noteworthy feature of the total scattering cross sections for the alkali atoms, which is clearly illustrated in the data presented by Kwan *et al.* (1991), is the merging of the data for electrons and positrons at a projectile energy of approximately 50 eV. This is much lower than the energy at which the corresponding two cross sections merge for other atomic targets, most notably the noble gases. The energy at which

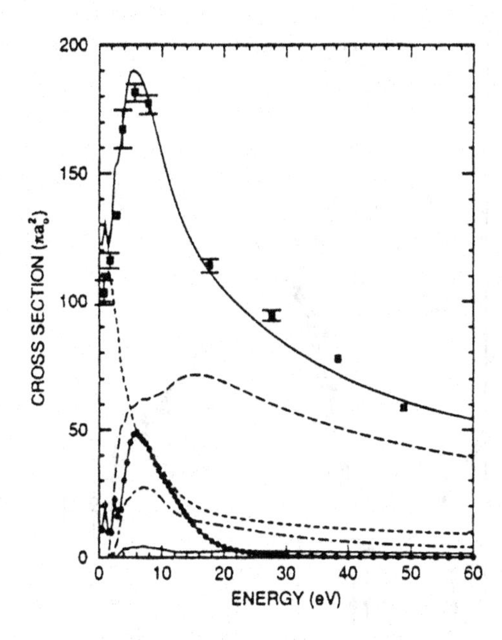

Fig. 2.17. Cross sections for positron–rubidium scattering. Experiment: ■, Parikh *et al.* (1993), total cross sections. Theory: upper solid curve, total cross sections from the work of Kernoghan, McAlinden and Walters (1996). Various partial cross sections from the latter are also shown: - - - -, elastic scattering; — —, 5S–5P excitation; — · —, 5S–4D excitation; uneven solid curve, at bottom of figure, sum of 5S–6S and 5S–6P excitation cross sections; –•–•–•–, positronium formation.

merging occurs is expected to be related to the mean kinetic energy of the electrons in the target. If the projectile energy is large compared to this mean energy, exchange effects will be small for electron projectiles and the conditions for merging outlined in section 2.2 will be satisfied. In the case of an alkali atom, probably only the mean kinetic energy of the weakly bound valence electron is relevant, and this is much smaller (typically a few electron volts) than the corresponding value for a noble gas. Merging is therefore expected to occur at significantly lower energies for the alkali atoms than for other atoms, as observed.

4 Atomic hydrogen

The total cross sections measured by Zhou *et al.* (1997) for positron and electron impact are illustrated in Figure 2.18. As noted by these authors, the positron total cross sections are in excess of those for electrons in the energy range 15–100 eV, a situation similar to that found for the alkali atoms, as described in subsection 2.5.3. The experimental electron–hydrogen data agree well with the semi-empirical values derived

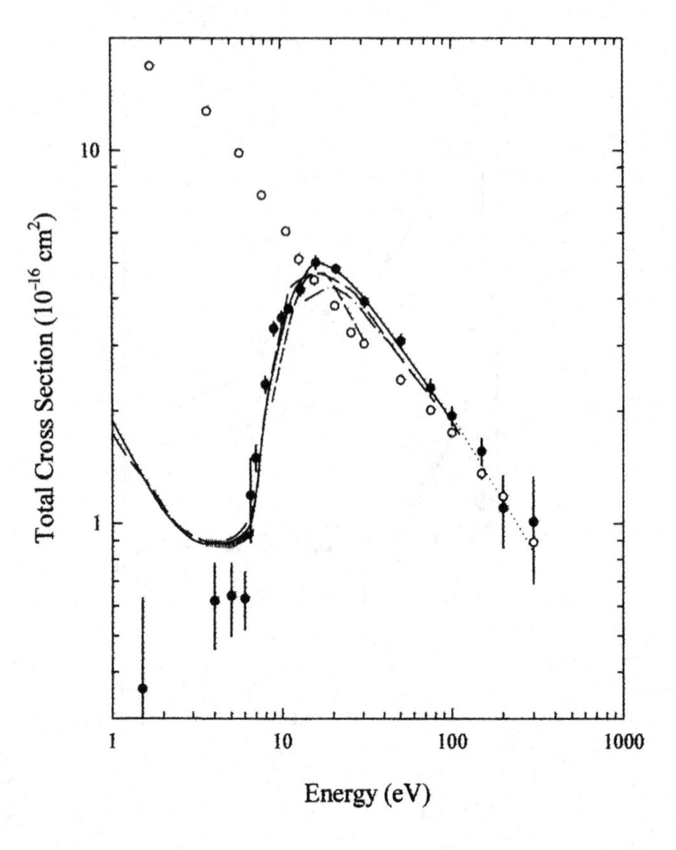

Fig. 2.18. Positron–atomic hydrogen and electron–atomic hydrogen total scattering cross sections. Experiment: •, positrons; ◦, electrons; both sets of data are from Zhou *et al.* (1997). Theory: ——, Kernoghan, McAlinden and Walters (1996); — —, Mitroy (1996); — · —, Higgins, Burke and Walters (1990); · · · · ·, Walters (1988); - - - -, Winick and Reinhardt (1978a, b).

by de Heer, McDowell and Wagenaar (1977) and also with the results of various theories; the positron data agree with the T-matrix calculations of Winick and Reinhardt (1978a, b) throughout the energy range 8–300 eV and those of Walters (1988) in the energy range 50–300 eV. At lower energies, and at least down to the positronium formation threshold, the experimental results of Zhou *et al.* (1997) are in very good accord with the results of the elaborate coupled-state calculations of Kernoghan, McAlinden and Walters (1995, 1996). At energies below 8 eV the experimental results fall below the calculated values of Kernoghan, McAlinden and Walters (1995, 1996) and the accurate variational results of Brown and Humberston (1985), though Stein *et al.* (1996) have attributed this to forward scattering errors.

It is interesting to note that the total cross sections, both theoretical and experimental, for the two projectiles are close to one another above about 12 eV, although true merging does not occur until above 100 eV. This energy, intermediate between the energies at which merging occurs for the alkali atoms and helium, is what would be expected, given the relationship previously stated in subsection 2.5.3 between the energy at which merging occurs and the mean kinetic energy of the target electrons. As emphasized by Stein *et al.* (1996) and Zhou *et al.* (1997), the merging occurs despite the large differences in the partial cross sections (e.g. the cross section for elastic scattering is estimated to be four times higher for electrons than for positrons at an energy of 30 eV, where the positronium formation cross section is almost 50% of the total positron cross section). Zhou *et al.* (1997) have speculated that this is due to some form of coupling between the various channels which results in the total cross sections for the two projectiles being very similar. These authors also alluded to the argument of Dewangan (1980) as summarized in section 2.2.

2.6 Results and discussion – molecules

Total scattering cross sections have been measured for a wide variety of molecules using low energy positron beams, but in this section we consider just four targets, H_2, N_2, CO_2 and H_2O, each of which exhibits intriguing and complex behaviour. Relevant theoretical work is also described, although it is in shorter supply than for the atomic targets described above. General reviews of positron–molecule theory were given by Armour (1988) and Ghosh, Sil and Mandal (1982) and references to other measurements of positron–molecule total scattering cross sections have been given by Kimura *et al.* (2000).

1 Hydrogen

Total cross sections for low energy positron and electron scattering by H_2 are presented in Figures 2.19(a), (b) respectively, and the experimental data at intermediate energies are shown in Figure 2.20. The results display a somewhat similar energy dependence to that found for helium, the electron data falling monotonically with increasing energy from a value of 2×10^{-15} cm^2 at 2 eV and the positron data displaying a Ramsauer minimum of approximately 10^{-16} cm^2 at 3 eV. As the positron energy is reduced below this value the cross section rises steeply, and there is also a pronounced rise as the energy is increased above the positronium formation threshold, indicating once again the importance of the positronium formation channel. It should be noted, however, that other channels, e.g.

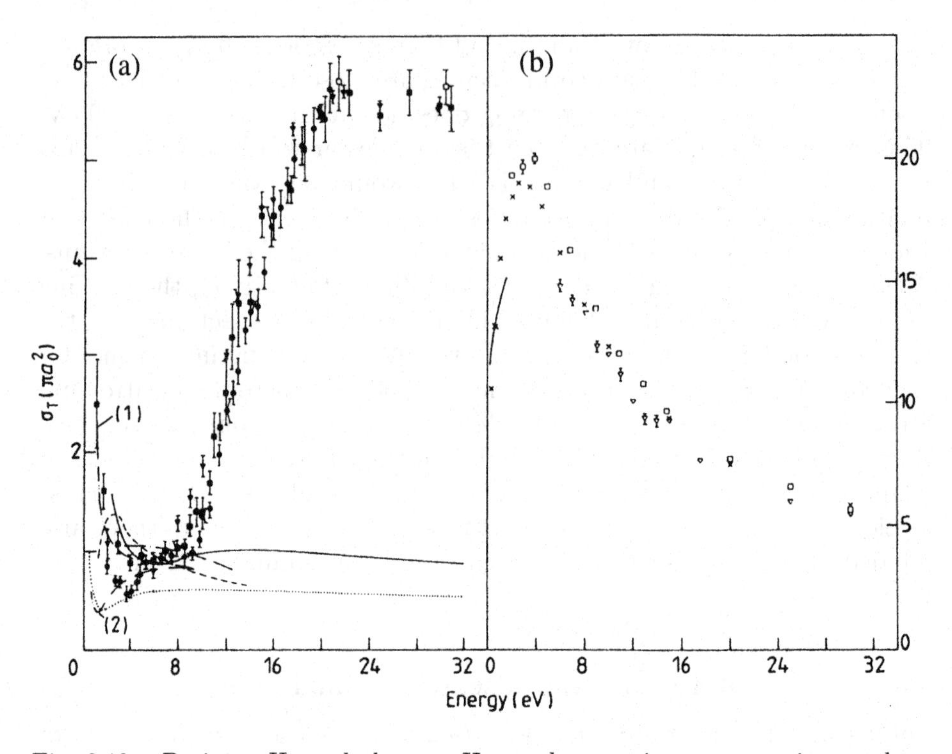

Fig. 2.19. Positron–H_2 and electron–H_2 total scattering cross sections at low energies. (a) Positron scattering. Experiment: ▪, Hoffman *et al.* (1982); •, Charlton *et al.* (1983a); ▼, Deuring *et al.* (1983). Theory: · · · · ·, Baille, Darewych and Lodge (1974); — · —, Hara, (1974); ——, Bhattacharyya and Ghosh (1975); - - - -, Armour (1984): — · · —, Morrison, Gibson and Austin (1984) (two such curves for different approximations used, labelled 1 and 2). Other recent theoretical work (e.g. Armour, Baker and Plummer (1990), Danby and Tennyson (1990) and Gianturco and Mukherjee (1997) is not shown – see the main text for a discussion. (b) Electron scattering, experiment only: □, Hoffman *et al.* (1982); ▽, Deuring *et al.* (1983); ×, Dalba *et al.* (1980); ——, Ferch, Raith and Schröder (1980).

vibrational and rotational excitation and molecular dissociation, may also have a minor influence on the behaviour of the total cross section in this energy region. Of these processes, one might expect molecular dissociation, which is energetically possible at a projectile energy of 4.48 eV, to be the most significant, but Massey (1969, Chapter 13) has shown, by appealing to the Frank–Condon principle, that the cross section for this process is very small for energies less than 8.8 eV. The total cross sections for the two projectiles merge at an energy of approximately 100 eV.

Above 9 eV all sets of experimental data are in reasonable agreement, although those of Deuring *et al.* (1983) are on average 15% higher than

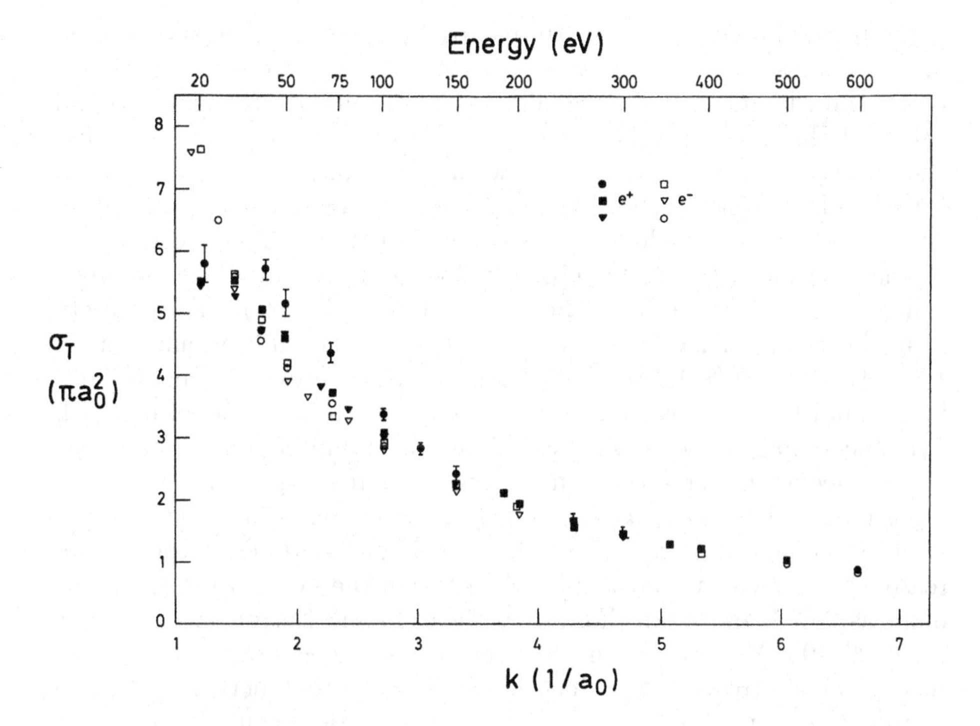

Fig. 2.20. Intermediate energy experimental results for positron–H_2 and electron–H_2 total scattering cross sections. Positrons: •, Charlton *et al.* (1980b); ■, Hoffman *et al.* (1982); ▼, Deuring *et al.* (1983). Electrons: ○, Van Wingerden *et al.* (1980); □, Hoffman *et al.* (1982); ▽, Deuring *et al.* (1983).

the Detroit and London results. At energies below approximately 5 eV, however, the results of Charlton *et al.* (1983a) fall below those of Hoffman *et al.* (1982) and exhibit a shallow minimum around 3.5 eV. This feature is not observed by Hoffman *et al.* (1982) nor is it present in any of the theoretical results.

The results of several calculations are also shown in Figure 2.19. Hara (1974) obtained fair agreement with experiment using a fixed-nuclei approximation for elastic scattering and rotational excitation. However, Baille, Darewych and Lodge (1974), who applied an adiabatic nuclei approximation for elastic scattering, obtained cross sections which are much lower than the experimental values at all energies. Better agreement with experiment was obtained by Bhattacharyya and Ghosh (1975) who used an eikonal approximation to calculate elastic cross sections in the energy range 2–32 eV. The most detailed investigations of low energy positron scattering by molecular hydrogen have been made by Armour, Baker and Plummer (1990), who used the Kohn variational method with elaborate trial functions. This method is similar to that used to obtain accurate results for elastic scattering by atomic hydrogen and helium, further details of which are given in section 3.2. The results of Armour, Baker and Plummer (1990) are in good agreement with the experimental values of Hoffman *et al.* (1982) up to an energy of 5 eV, but at higher energies they fall progressively below the experimental values. Armour, Baker and Plummer (1990) extended their calculations to an energy of 14 eV, but did not include positronium formation. Cross sections for elastic positron scattering by molecular hydrogen have also been obtained using the R-matrix method (Danby and Tennyson, 1990) and a variety of model potential methods (Hara, 1974; Morrison, Gibson and Austin 1984; Morrison, 1986). Recently, good accord with experiment has also been found by Gianturco and Mukherjee (1997) using a model in which dynamical vibrational coupling effects are taken into account.

Cross section measurements in the intermediate energy range have been reported by Charlton *et al.* (1980b), Hoffman *et al.* (1982) and Deuring *et al.* (1983), and their results are shown in Figure 2.20. These data are in reasonable agreement above 150 eV, although the London results are as much as 20% higher than those from Bielefeld and Detroit in the energy range 20–100 eV. Also shown are recent measurements of the total cross section for electrons, from which it can be seen that between approximately 30 eV and 150 eV the electron cross sections fall slightly below those for positrons. This may be due to the contribution from positronium formation in the positron case and to the fact that the ionization cross section for positrons has been found to exceed that for electrons in this energy range (see Chapter 5).

2 Nitrogen

The total cross sections for positron and electron scattering by N_2 at low energies are shown in Figure 2.21. In this case, as also for CO_2 and H_2O, see the discussion below, only low energy data are presented. The only groups to have published reliable low energy total cross sections for positron–N_2 scattering are those in London (Charlton *et al.*, 1983a) and Detroit (Hoffman *et al.*, 1982), and their results are shown together with a few points from the intermediate energy data of Charlton *et al.* (1980b) and Dutton, Evans and Mansour (1982).

The data of Charlton *et al.* (1983a) and Hoffman *et al.* (1982) are in reasonable agreement over the entire energy range, the latter showing that the total cross section rises sharply below 2 eV. Close inspection of the points reveals some interesting features. Both experiments indicate that the cross section grows steadily at higher energies, although Charlton *et al.* (1983a) find the rise beginning at approximately 7 eV, whilst the Detroit group find the rise starting at 8 eV. The reason for this difference

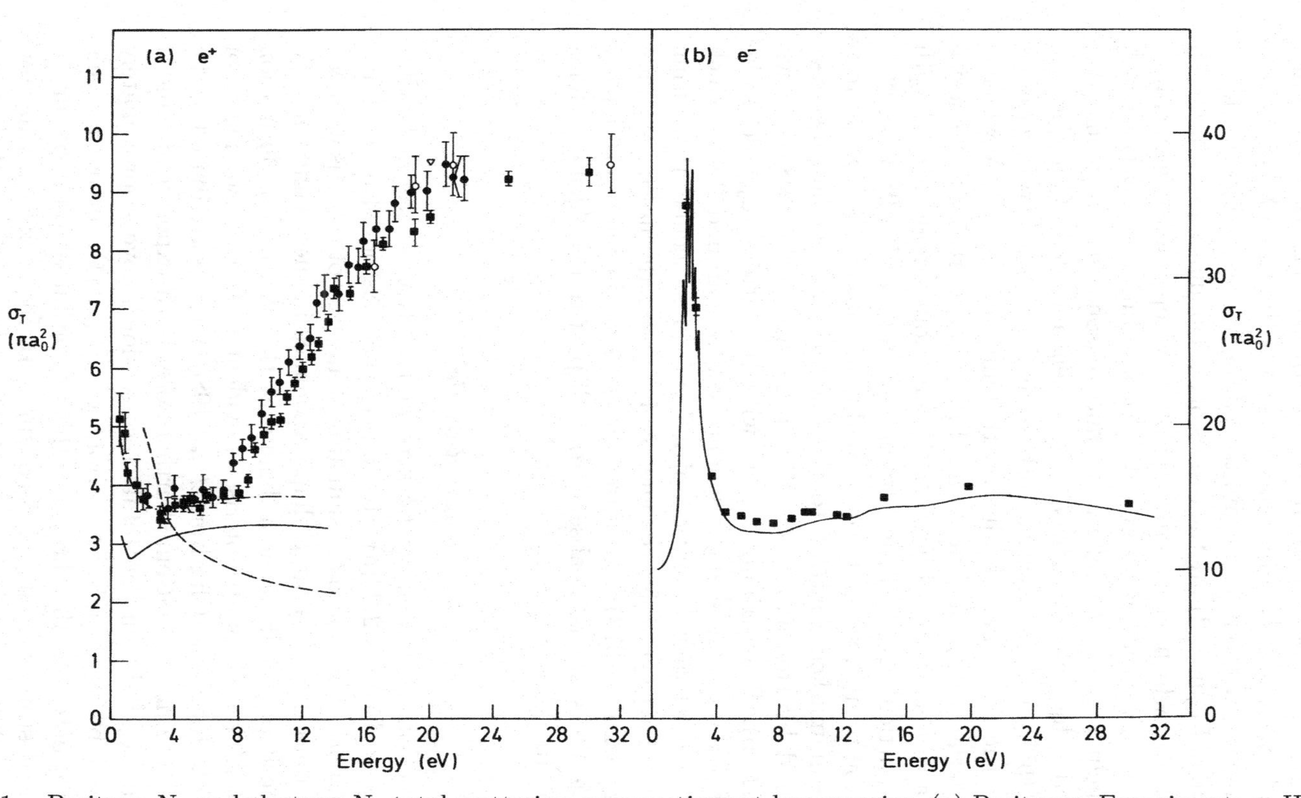

Fig. 2.21. Positron–N_2 and electron–N_2 total scattering cross sections at low energies. (a) Positrons. Experiment: ■, Hoffman *et al.* (1982); ●, Charlton *et al.* (1983a); ○, Charlton *et al.* (1980b); ▽, Dutton, Evans and Mansour (1982). Theory: – – –, Darewych and Baille (1974); ——, Gillespie and Thompson (1975); — · —, Darewych (1982). (b) Simplified experimental situation for low energy electron–N_2 scattering. The discrete points are those of Hoffman *et al.* (1982) whilst the full curve is taken from the detailed work of Kennerly (1980).

is not clear since neither energy coincides with the positronium formation threshold at 8.8 eV. The possibility that the increase is due to the onset of electronic excitation was suggested by Charlton *et al.* (1983a), though Schrader and Svetic (1982) noted that the lowest threshold for excitation of the N_2 molecule without a spin-flip transition is at an energy of 8.6 eV.

Also shown in Figure 2.21 are the results of the calculations of Darewych and Baille (1974), which have an incorrect energy dependence in the range 3–10 eV. These investigations were extended and improved by Darewych (1982) to include more partial waves, and reasonable agreement with the measured values of σ_T below E_{Ps} was then obtained. Gillespie and Thompson (1975), using a polarized-orbital method, obtained lower cross sections than the experimental values, with a minimum near 1 eV which is not present in the data of Hoffman *et al.* (1982). Gianturco and Mukherjee (1997) (data not shown on Figure 2.21) have found good accord with experiment, particularly with the data of Charlton *et al.* (1983a).

The positron results, which have a broad minimum in the energy range 2–7 eV, are in marked contrast to the low energy electron cross sections measured by Hoffman *et al.* (1982) and Kennerly (1980), shown in Figure 2.21(b). The striking feature with the latter projectile is the existence of a prominent shape resonance centred around 2 eV, which is due to the temporary formation of a negative molecular ion during the collision. This type of phenomenon has been discussed by many authors (e.g. Schulz, 1973) and, although observed in many low energy electron–molecule systems, is absent from all positron–molecule systems studied hitherto.

3 Carbon dioxide

Total cross sections for low energy positron and electron scattering by CO_2 are presented in Figures 2.22(a), (b) respectively. As noted by Hoffman *et al.* (1982), both sets of data are similar in shape except that, as with N_2, the electron data display a large shape resonance in the vicinity of 4 eV whilst the positron results show a ‘bump’ at the positronium formation threshold at 7.0 eV. This abrupt rise is evident in the measurements of Hoffman *et al.* (1982) and Charlton *et al.* (1983a) and is attributed to the onset of positronium formation. Furthermore, both experiments find a plateau or a slight dip in the total cross section above approximately 8 eV, followed by a second rise. Kwan *et al.* (1984) noted that this rise starts close to the threshold for the formation of positronium in its first excited state, although subsequent work by Laricchia, Charlton and Griffith (1988) and Laricchia and Moxom (1993) has found that the process responsible is the formation of ground state positronium with

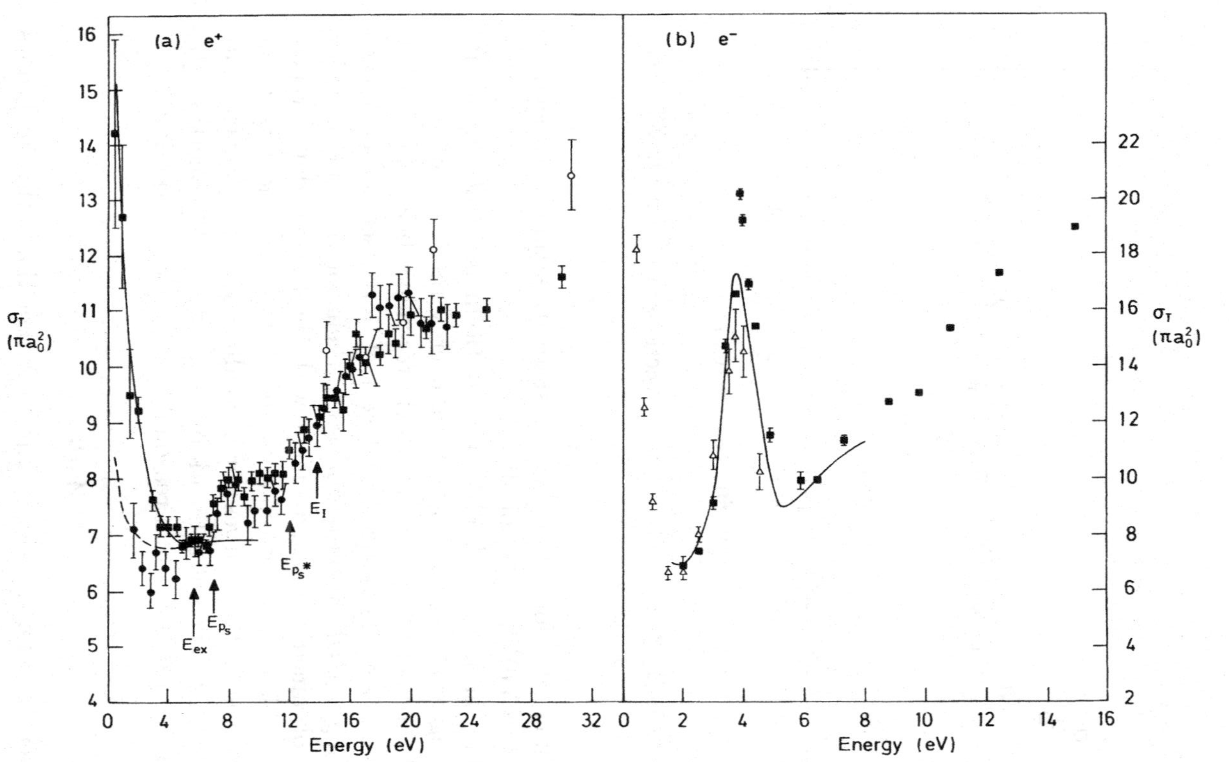

Fig. 2.22. Positron–CO_2 and electron–CO_2 total scattering cross sections at low energies. (a) Positrons. Experiment: ■, Hoffman *et al.* (1982); •, Charlton *et al.* (1983a); ○, Charlton *et al.* (1980b). Theory: Horbatsch and Darewych (1983). The two curves correspond to the use of a fixed cut-off (– – –) and an energy-dependent cut-off (——) parameter in the polarization potential. (b) Experimental situation for electrons: ■, Hoffman *et al.* (1982); △, Ferch, Masche and Raith (1981); ——, Szmytkowski and Zubek (1978).

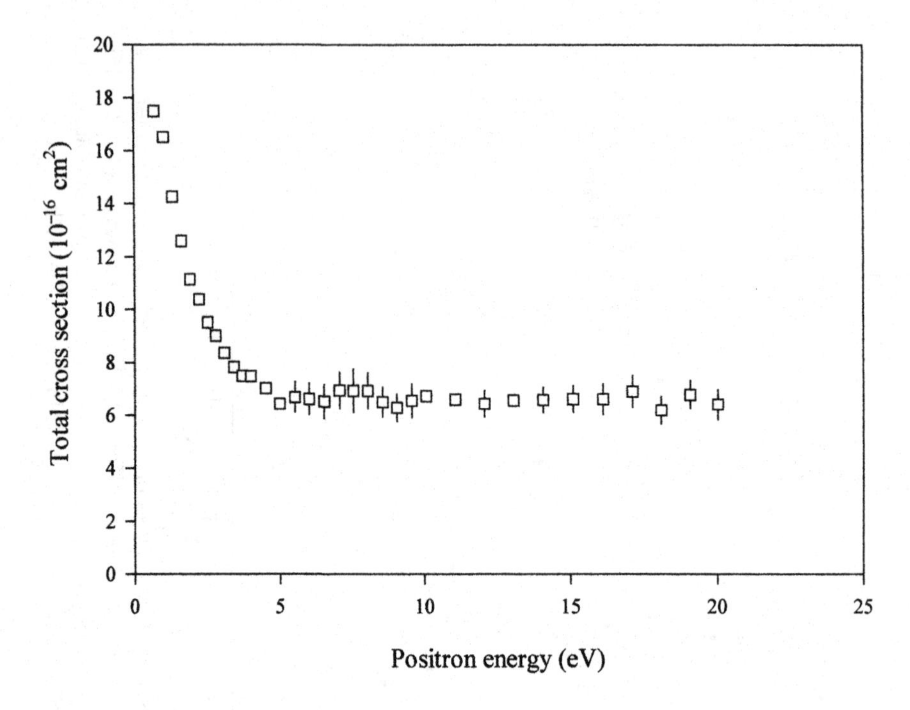

Fig. 2.23. Positron–H_2O total scattering cross sections at low energies (Sueoka, Mori and Katayama, 1987).

the residual molecular ion in an excited state, this threshold being at approximately 10.5 eV (see section 4.5).

The data of Charlton *et al.* (1983a) and Hoffman *et al.* (1982) are in reasonable agreement over most of the energy range presented here, except that the former are lower than the latter values below 4 eV. Comparison is also made with the theoretical work of Horbatsch and Darewych (1983). By using a fixed cut-off parameter in their polarization potential, these authors obtained better agreement with the London data. When, however, a variable-energy cut-off was used, better agreement with the Detroit results was found. Curves corresponding to each calculation are given in Figure 2.22. Theoretical work by Gianturco and Paioletti (1997) (not shown in Figure 2.22) are in reasonable accord with experiment.

4 Water

We have chosen H_2O for detailed comment rather than other molecules because it was, until recently, one of the few with a large permanent dipole moment to have been studied. The positron scattering data of Sueoka, Mori and Katayama (1987) are shown in Figure 2.23, and they were obtained using a similar TOF system to that described in section

2.3. It is immediately apparent that the total cross section does not show a dramatic rise or change of slope as the impact energy is raised through the threshold for positronium formation. This is in marked contrast to the results for the other gases discussed so far in this section and for the noble gases, discussed in subsections 2.5.1 and 2.5.2. In fact, following an initial decline from approximately 2×10^{-15} cm^2 at 1 eV to 7×10^{-16} cm^2 at 5 eV, σ_T remains nearly constant up to 20 eV (there is a gradual decline at higher energies). The apparent dearth of positronium formation may, however, be an illusion resulting from the behaviour of the elastic scattering cross section near this threshold (see sections 3.3 and 4.4 and the discussion of Meyerhof *et al.*, 1996). There have, thus far, been no direct studies of positronium formation in water, although a study of low energy positron impact with an ice surface found abundant positronium emission (Eldrup *et al.*, 1985). In positron lifetime studies in other gases with high dipole moments (e.g. NH_3, CH_3Cl), high positronium formation probabilities have been found (Heyland *et al.*, 1982).

2.7 Partitioning of the total cross section

In this, the concluding section of this chapter, we present a discussion of the partitioning of the positron and electron total cross sections for helium gas, based upon the work of Campeanu *et al.* (1987) for positron scattering and of de Heer and Jansen (1977) for electron scattering. This target has been chosen because it has been the subject of extensive theoretical and experimental study. A similar exercise has also recently been undertaken by Zhou *et al.* (1997) in connection with their measurements of σ_T for atomic hydrogen, and by Sueoka and Mori (1994) for helium, neon and argon gases.

The presentation of positron and electron data in sections 2.5 and 2.6 has illustrated that there are often large differences between the two total cross sections, both in magnitude and energy dependence, although sometimes the converse is true, with the behaviour being remarkably similar. In an effort to understand some of these effects it is necessary to probe deeper into the total cross section by splitting it up into its partial contributions, taking the most reliable data from both theory and experiment. In so doing, one may also hope to gain some degree of self-consistency. Detailed discussions of the processes which make the major contributions to σ_T are contained in Chapters 3–5, and one of our purposes here is to set the scene for these.

The earliest attempts to partition σ_T for positron–helium scattering were those of Griffith *et al.* (1979b), before the advent of any partial cross section measurements, and of Coleman *et al.* (1982), when the first inelastic scattering data became available. The most detailed study is that of Campeanu *et al.* (1987), who defined

$$\sigma_{\text{no ion}} = \sigma_{\text{T}} - \sigma_{\text{Ps}} - \sigma_{\text{i}}, \tag{2.19}$$

where σ_{Ps} is the positronium formation cross section, σ_{i} is the positron impact-ionization cross section and $\sigma_{\text{no ion}}$ is that part of the total positron cross section in which there is no ion left in the final state. The right-hand side of equation (2.19) may be rewritten as

$$\sigma_{\text{no ion}} = \sigma_{\text{el}} + \sum_{n=2}^{\infty} \sigma_{\text{ex}}(nLS), \tag{2.20}$$

where σ_{el} is the elastic scattering cross section and the summation is over the angle-integrated excitation cross sections for various specific bound electronic states of the target system. Similar equations would of course apply to electron scattering, but without the positronium formation term. Campeanu *et al.* (1987) justified the partitioning in this way by noting that $\sigma_{\text{no ion}}$, and the corresponding quantity for electrons, is relatively easy to obtain from various experimental measurements through equation (2.19). This can be seen from the discussion in this chapter and in Chapters 4 and 5. However, the measurement of $\sigma_{\text{no ion}}$ from equation (2.20) is not straightforward because σ_{el} can only be easily determined below the positronium formation threshold energy and measurements of the total excitation cross section are difficult (see e.g. the discussion of electron impact data given by Heddle, 1979). In a theoretical partitioning of the total scattering cross section the situation is reversed, and $\sigma_{\text{no ion}}$ is easier to determine via equation (2.20) than via equation (2.19) because the cross sections for positronium formation and ionization are the most difficult to compute accurately. Thus, by different routes it is possible to extract reliable values for $\sigma_{\text{no ion}}$ from both theory and experiment.

In order to test the general utility of their approach, Campeanu *et al.* (1987) performed an analysis of electron–helium scattering, for which several sets of theoretical and experimental data were available, relating to all the major partial cross sections. Their approach was similar to that of de Heer and Jansen (1977), except that these latter authors aimed to construct a semi-empirical σ_{T} for electrons by summing all available partial cross sections, many of which (e.g. the elastic and excitation contributions) were determined by integrating published differential cross sections. The cross sections obtained by de Heer and Janson agreed well with what were then the most recent experimental results of Blaauw *et al.* (1977), from 30 eV up to the experimental limit of 700 eV. Campeanu *et al.* (1987) used the data assembled and evaluated by de Heer and Jansen (1977), constructed $\sigma_{\text{no ion}}$ for electrons according to equation (2.20) and

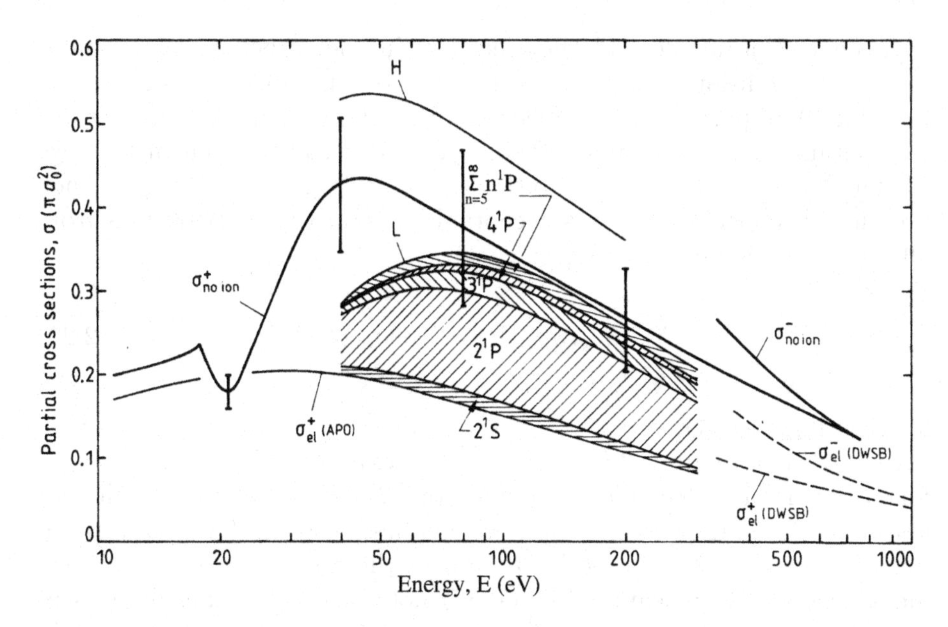

Fig. 2.24. The integrated cross sections for positron scattering from helium atoms without ionization ($\sigma^+_{\text{no ion}}$), obtained from experiment according to equation (2.19), and theoretical predictions derived from equation (2.20). The corresponding cross sections for electrons (superscript minus) are also shown on the high energy side. Curves H and L represent the highest and lowest theoretical estimates of $\sigma^+_{\text{no ion}}$ respectively. The σ^+_{el} (APO) curve is from Campeanu *et al.* (1987) – see the main text for the origins of the other theoretical and experimental data. The broken curves at the higher energies are for electron (−) and positron (+) elastic scattering cross sections from the distorted-wave second Born (DWSB) calculations of Dewangan and Walters (1977).

then compared 'theory' with experiment. They found reasonably good agreement between the two when the lowest theoretical values of $\sigma_{\text{no ion}}$ were used, and this led to some confidence being placed in the corresponding positron analysis.

Data for positrons are presented in Figure 2.24, which shows some of the major contributions to σ_{T}. The 'experimental' values of $\sigma_{\text{no ion}}$ for positrons were deduced from the σ_{T} measurements of Kauppila *et al.* (1981) by subtracting the cross sections for positronium formation and ionization as given by Fromme *et al.* (1986). The theoretical results, as for electrons, have been given with higher and lower values (curves H and L). For the excitation cross sections the distorted-wave approach of Parcell, McEachran and Stauffer (1983, 1987) was extended to higher impact energies and to higher values of the principal quantum number of helium, n_{He}, for the excited channels. (Note that since exchange between the projectile and the electrons in the target atom is absent in positron collisions, the collisional excitation of triplet states is strictly forbidden.) In particular, Campeanu *et al.* (1987) reported cross sections for excitation to the 3^1P and 4^1P states and showed that their ratio obeys the familiar $1/n_{\rm He}^3$ scaling law, so that $\sigma_{\rm ex}(4^1P)/\sigma_{\rm ex}(3^1P) = (3/4)^3$. They therefore assumed that it was acceptable to sum the contributions from all higher ^{1}P channels to give

$$\sum_{n_{\rm He}=5}^{\infty} \sigma_{\rm ex}(n_{\rm He}\,{}^1{\rm P}) = 1.56\sigma_{\rm ex}(4^1{\rm P}) \qquad (2.21)$$

for each impact energy.

Inspection of Figure 2.24 reveals that the 'experimental' values of $\sigma_{\rm no\,ion}$ for positrons, as calculated from equation (2.19), lie between the higher and lower theoretical estimates at all energies. Above 100 eV they are in better agreement with curve L, in view of the error bars on the experimental values. Campeanu *et al.* (1987) noted that the main differences between the two theoretical estimates derive entirely from differences in the cross sections for processes leaving the target atom in the final states 1^1S (elastic scattering), 2^1S and 2^1P. The upper curve, H, comes from the work of Willis and McDowell (1982); their results are markedly higher than those of Dewangan and Walters (1977), Campeanu *et al.* (1987) and Parcell, McEachran and Stauffer (1983, 1987), and are thought to be overestimates at all energies.

As noted above, the lower curve, L, is in good agreement with experiment above 100 eV but falls significantly below the experimental values at lower energies. This is attributed by Campeanu *et al.* (1987) to a failure of Parcell's polarized-orbital method. The experimental work of Coleman *et al.* (1982) and of Sueoka (1982) is cited in support of this, and the later results of Mori and Sueoka (1994), which supersede the earlier data of Sueoka (1982), do not alter this conclusion. Further discussion of elastic scattering can be found in Chapter 3; positron impact excitation is treated in Chapter 5.

Before leaving this section it is worth noting some other points raised in the work of Campeanu *et al.* (1987). One pertains to the established merging of the total cross sections for positrons and electrons at approximately 200 eV for helium, as previously mentioned in section 2.5. It can be seen from Figure 2.24, by subtracting the ionizing channels from the relevant $\sigma_{\rm T}$, that the merging of the remainder does not occur until 600 eV. Moreover, since the distorted-wave Born approximation calculations of Dewangan and Walters (1977) reveal that $\sigma_{\rm el}$ for electrons exceeds $\sigma_{\rm el}$ for positrons even out to 2 keV impact energy, the preponderance of the excitation cross section for positrons over that for electrons must be

responsible for the merging of the no-ionization cross sections. As noted by Campeanu *et al.* (1987), theoretical verification of this has not been made using the same model for both projectiles.

Campeanu *et al.* (1987) also discussed the behaviour of the ionization cross sections for positrons and electrons near to the ionization threshold, but our treatment of this topic is deferred until subsection 5.4.5. Furthermore, in subtracting σ_{Ps} from σ_T Campeanu *et al.* (1987) obtained an estimate of the behaviour of σ_{el} between the positronium formation threshold and the first excitation threshold of the helium atom. Their derived cross section appeared to contain a cusp or threshold anomaly around E_{Ps}, but more recent experimentation and theoretical analysis has cast some doubt on the existence of a feature of this size in helium. Further discussion of these interesting phenomena is given in Chapters 3 and 4.

3
Elastic scattering

3.1 Introduction

In this chapter we describe the elastic scattering of positrons by atoms and molecules over the kinetic energy range from zero to several keV, concentrating mainly on the angle-integrated cross section, $\sigma_{\rm el}$. However, reference is also made to differential cross sections, $d\sigma_{\rm el}/d\Omega$, which have recently become amenable to experimental measurement using crossed gas and positron beams.

Particular attention is given to relatively simple targets, e.g. atomic hydrogen, helium, the alkali and heavier rare gas atoms and small molecules, and some comparisons are made with the corresponding data for electron impact. This again highlights the differences and similarities in the scattering properties of the two projectiles, which have already been mentioned in subsection 1.6.1 and in Chapter 2.

At energies below the lowest inelastic threshold, elastic scattering is the only open channel (except for electron–positron annihilation, which is always possible but which usually has a negligibly small cross section). For all atoms, the lowest inelastic threshold is that for positronium formation, at an energy $E_{\rm Ps}$, but for the alkali atoms positronium formation is possible even at zero incident energy. Molecular targets usually have thresholds for rotational and vibrational excitation at energies below $E_{\rm Ps}$, although the elastic scattering cross section is nevertheless expected to dominate over the cross sections for these inelastic channels.

We continue this chapter with a detailed description of the theoretical models applied to the elastic scattering of positrons by atoms and molecules.

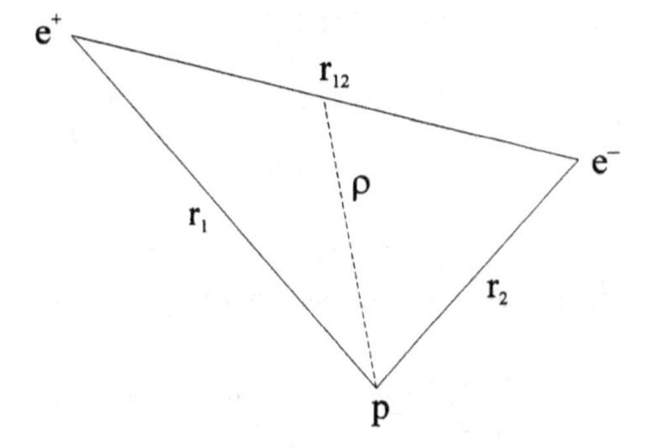

Fig. 3.1. The coordinates of the positron–hydrogen system.

3.2 Theory

A discussion of elastic positron–atom scattering is most conveniently introduced in the context of positron–hydrogen scattering, and we therefore describe this system in considerable detail and use it to illustrate some of the more important methods of approximation used in positron collision physics.

1 Positron–hydrogen scattering

The Hamiltonian of the positron–hydrogen system may be written, with reference to the nomenclature of Figure 3.1, as

$$H = -\frac{1}{2}\nabla^2_{r_1} + \frac{1}{r_1} - \frac{1}{r_{12}} + H_{\mathrm{H}}, \tag{3.1}$$

where

$$H_{\mathrm{H}} = -\frac{1}{2}\nabla^2_{r_2} - \frac{1}{r_2} \tag{3.2}$$

is the Hamiltonian of the hydrogen atom. If elastic scattering is the only open channel, the total wave function when the positron is far from the target atom has the product form

$$\Psi(\boldsymbol{r}_1, \boldsymbol{r}_2) \underset{r_1\to\infty}{\sim} F(\boldsymbol{r}_1)\Phi_{\mathrm{H}}(\boldsymbol{r}_2), \tag{3.3}$$

where $\Phi_{\mathrm{H}}(r_2)$ is the hydrogen target wave function, with form

$$\Phi_{\mathrm{H}}(r_2) = \frac{1}{\sqrt{\pi}}\exp(-r_2), \tag{3.4}$$

and

$$F(\boldsymbol{r}_1) \underset{r_1\to\infty}{\sim} e^{i\boldsymbol{k}\cdot\boldsymbol{r}_1} + f_{\mathrm{el}}(\theta)\frac{e^{ikr_1}}{r_1}. \tag{3.5}$$

A partial-wave expansion of the scattering function $F(\boldsymbol{r}_1)$ gives

$$F(\boldsymbol{r}_1) \underset{r_1\to\infty}{\sim} \sum_{l=0}^{\infty}(2l+1)i^l \left[\frac{\sin(kr_1 - \frac{1}{2}l\pi + \eta_l)}{kr_1}\right] P_l(\cos\theta), \tag{3.6}$$

where η_l is the lth partial-wave phase shift and $P_l(\cos\theta)$ is the corresponding Legendre function. The elastic scattering amplitude is then

$$f_{\mathrm{el}}(\theta) = \frac{1}{2ik}\sum_{l=0}^{\infty}(2l+1)(e^{2i\eta_l}-1)P_l(\cos\theta) \tag{3.7}$$

and the elastic scattering cross section is

$$\sigma_{\mathrm{el}} = 2\pi\int_0^{\pi}|f_{\mathrm{el}}(\theta)|^2\sin\theta\,d\theta = \frac{4\pi}{k^2}\sum_{l=0}^{\infty}(2l+1)\sin^2\eta_l. \tag{3.8}$$

The differential scattering cross section is

$$\frac{d\sigma_{\mathrm{el}}}{d\Omega}(\theta) = |f_{\mathrm{el}}(\theta)|^2 = A^2 + B^2, \tag{3.9}$$

where

$$A = \frac{1}{2k}\sum_{l=0}^{\infty}(2l+1)(\cos 2\eta_l - 1)P_l(\cos\theta) \tag{3.10}$$

and

$$B = \frac{1}{2k}\sum_{l=0}^{\infty}(2l+1)\sin 2\eta_l\, P_l(\cos\theta). \tag{3.11}$$

It readily follows from equations (3.7) and (3.8) that

$$\sigma_{\mathrm{el}} = \frac{4\pi}{k}\,\mathrm{Im}\, f_{\mathrm{el}}(\theta = 0). \tag{3.12}$$

This is the optical theorem, and it expresses the conservation of the number of particles in the scattering process. As already mentioned in section 2.2, it is valid even when inelastic processes can occur, although σ_{el} is then replaced by the total scattering cross section σ_{T}, which includes contributions from all open scattering channels.

As the positron approaches the target system it interacts with and distorts it, so that the total wave function no longer has the separable form of equation (3.3). Nevertheless, an equivalent Schrödinger equation can be derived for the positron, the solution to which is a function of the positron coordinate $\boldsymbol{r}_1$ only, with the correct asymptotic form but at the cost of introducing a non-local optical potential.

If one introduces the projection operator (Feshbach, 1962)

$$P = |\Phi_{\mathrm{H}}\rangle\langle\Phi_{\mathrm{H}}|, \tag{3.13}$$

which projects onto the ground state of the target atom, and

$$Q = 1 - P, \tag{3.14}$$

which projects on to all other states, then

$$P + Q = 1, \qquad P^2 = P, \qquad Q^2 = Q, \qquad PQ = QP = 0. \tag{3.15}$$

The Schrödinger equation for the total system can be expressed as

$$(H - E)\,|\Psi\rangle = (H - E)(P + Q)\,|\Psi\rangle = 0, \tag{3.16}$$

where $E = -\frac{1}{2} + \frac{1}{2}k^2$ is the total energy of the positron–hydrogen system. Applying the projection operators P and Q separately to equation (3.16) and eliminating $Q|\Psi\rangle$ by substituting from one equation into the other, the following equation for $P|\Psi\rangle$ is obtained:

$$\{P(H - E)P + PV_{\mathrm{int}}Q[Q(E - H)Q]^{-1}QV_{\mathrm{int}}P\}P\,|\Psi\rangle = 0, \tag{3.17}$$

where V_{int} is the interaction potential between the positron and the particles in the target. For the positron–hydrogen system

$$V_{\mathrm{int}} = \left(\frac{1}{r_1} - \frac{1}{r_{12}}\right). \tag{3.18}$$

In configuration space, $P|\Psi\rangle$ has the representation

$$\Phi_{\mathrm{H}}(r_2)\langle\Phi_{\mathrm{H}}|\Psi\rangle = \Phi_{\mathrm{H}}(r_2)F(\boldsymbol{r}_1), \tag{3.19}$$

and the equation satisfied by the positron function $F(r_1)$ is then

$$\left(-\tfrac{1}{2}\nabla^2_{r_1} + V_1 - \tfrac{1}{2}k^2\right)F(\boldsymbol{r}_1) + \int V_{\mathrm{opt}}(\boldsymbol{r}_1, \boldsymbol{r}_1')F(\boldsymbol{r}_1')\,d\boldsymbol{r}_1' = 0, \tag{3.20}$$

where

$$V_1 = e^{-2r_1}\left(1 + \frac{1}{r_1}\right) \tag{3.21}$$

is the static potential between the positron and the undistorted hydrogen atom and V_{opt} is the non-local optical potential. Thus, the Schrödinger equation for a positron–hydrogen system has been converted to an equation for the positron alone moving in a potential field. This may seem a significant simplification of the scattering problem, but the difficulties have merely been transferred to the task of calculating the optical potential. However, this formalism provides a convenient and consistent means of generating approximation schemes for solving the scattering problem; different approximate trial wave functions will correspond to different choices of the optical potential.

An approximation to the exact optical potential may be written as

$$V_{\text{opt}} = \sum_{\lambda} \left\{ \frac{|PHQ\Phi_\lambda\rangle\langle\Phi_\lambda QHP|}{E - E_\lambda} \right\}, \tag{3.22}$$

where the functions Φ_λ and energies E_λ are the eigenfunctions and associated eigenvalues of the matrix eigenvalue equation

$$QHQ\,|\Phi\rangle = E\,|\Phi\rangle \tag{3.23}$$

in whatever basis is being used.

It has been shown by Gailitis (1965) that the optical potential defined by equation (3.22) is less attractive than the exact optical potential for energies below the lowest eigenvalue of QHQ, and the resulting phase shifts are therefore lower bounds on the exact values. Furthermore, as the number of basis functions used in the matrix representation of the operator QHQ is increased, the optical potential becomes more attractive and the resulting phase shift therefore also increases, becoming closer to the exact value.

The simplest approximation is to retain only the local static potential V_1; see equation (3.20). This defines the static approximation, in which there is assumed to be no distortion of the target by the incident positron, so that the total wave function is

$$\Psi(\boldsymbol{r}_1, \boldsymbol{r}_2) = F(\boldsymbol{r}_1)\Phi_{\text{H}}(r_2) \tag{3.24}$$

for all values of $\boldsymbol{r}_1$. For positrons, the static potential is always repulsive, and all partial-wave phase shifts in the static approximation are therefore negative. Although these phase shifts are rigorous lower bounds, they are in poor agreement with the exact results, as may be seen in Figures 3.2 and 3.3, because the attractive polarization potential, which arises from the distortion of the target, has been neglected.

Account must therefore be taken of the distortion of the target if more accurate results are to be obtained. A relatively easy means of doing so, which has been quite widely used for investigating positron scattering by a variety of atoms, is the polarized-orbital method (Temkin, 1957, 1959; McEachran *et al.*, 1977). The method proceeds in two stages. First, the wave function of the distorted atom in the field of a stationary positron is expressed as $[1+G(\boldsymbol{r}_1, \boldsymbol{r}_2)]\Phi_{\text{H}}(r_2)$, and $G(\boldsymbol{r}_1, \boldsymbol{r}_2)$ is determined using either perturbation theory, as for hydrogen (Drachman, 1965), or a variational method in which the energy of the stationary positron–atom

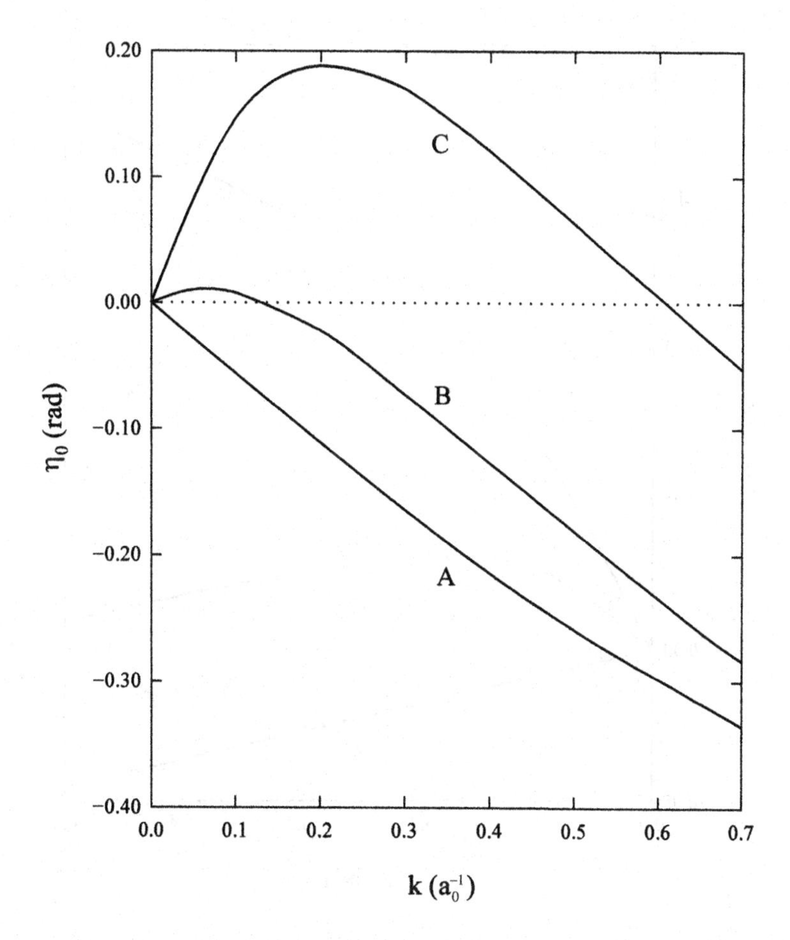

Fig. 3.2. The s-wave phase shift for positron–hydrogen scattering: A, static approximation; B, result for six-term coupled state (1s, 2s, 2p, 3s, 3p, 3d of H) (McEachran and Fraser, 1965); C, exact variational result (Schwartz, 1961b; Bhatia *et al.*, 1971).

system is minimized using some suitable trial function for each electron orbital (McEachran *et al.*, 1977). Usually, only the dipole component of the positron–atom interaction potential is considered. The total wave function is then expressed as

$$\Psi(\boldsymbol{r}_1, \boldsymbol{r}_2) = [1 + G(\boldsymbol{r}_1, \boldsymbol{r}_2)]\Phi_{\rm H}(r_2)F(\boldsymbol{r}_1), \tag{3.25}$$

and this is substituted into the Schrödinger equation. After projecting onto the undistorted target wave function, $\Phi_{\rm H}(r_2)$, the following equation is obtained for the scattering function, $F(\boldsymbol{r}_1)$:

$$\left(-\tfrac{1}{2}\,\nabla^2_{r_1} + V_1 + V_2\right) F(\boldsymbol{r}_1) = \tfrac{1}{2}k^2 F(\boldsymbol{r}_1), \tag{3.26}$$

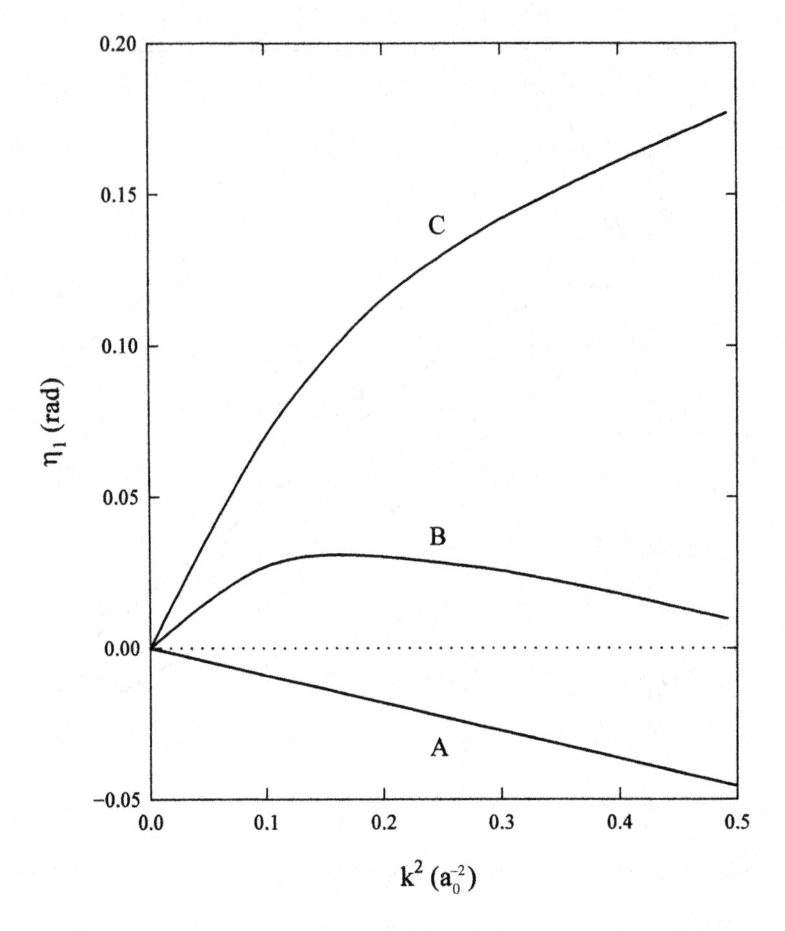

Fig. 3.3. The p-wave phase shift for positron–hydrogen scattering: A, static approximation; B, result for six-term coupled state (1s, 2s, 2p, 3s, 3p, 3d of H) (McEachran and Fraser, 1965); C, exact variational result (Armstead, 1968; Bhatia, Temkin and Eiserike, 1974).

where V_1 is the static potential and V_2 is the polarization potential arising from the distortion of the atom. In the investigations of Drachman (1965), who first applied the polarized-orbital method to positron–hydrogen scattering, V_2 was taken to be the second order adiabatic potential of Dalgarno and Lynn (1957). This potential has the correct long-range behaviour, namely

$$V_2 \underset{r_1 \to \infty}{\sim} -\frac{\alpha}{2r_1^4} \tag{3.27}$$

where α is the dipole polarizability of the hydrogen atom ($= 4.5a_0^3$), but it is too attractive for small values of r_1. In the adiabatic approximation the positron is assumed to be stationary whereas in reality it is moving, so that the target atom does not have unlimited time in which to adjust to the changing field created by the moving positron. Consequently the true positron–hydrogen interaction is less attractive than the adiabatic interaction, even for positrons with zero incident energy. Drachman (1965) attempted to compensate for this excessive attraction by reducing the monopole component of the distortion function $G(\boldsymbol{r}_1, \boldsymbol{r}_2)$, thereby modifying the form of the potential V_2. He found that almost total suppression of the monopole component gave excellent agreement with the very accurate variational results of Schwartz (1961b) for s-wave scattering. It has since been found that accurate variationally determined wave functions for low energy s-wave positron–hydrogen scattering (Humberston and Wallace, 1972) do indeed contain very little monopole distortion of the target.

Drachman (1965) also applied the polarized-orbital method to higher-partial-wave scattering, but his phase shifts are rather less positive than the accurate p-wave results of Armstead (1968) and the d-wave results of Register and Poe (1975), the discrepancies becoming greater as the positron energy is increased. In partial waves with $l > 0$ the adiabatic potential is less attractive than the exact interaction because the polarized-orbital method cannot adequately represent the most general forms of distortion of a system in which the orbital angular momentum is shared between the positron and the electron in the target.

One of the most commonly used approximation schemes in the study of both electron and positron collisions with atoms is the coupled-state, or close-coupling, approximation. As applied to positron–hydrogen scattering, the total wave function may be formally expanded in terms of complete sets of the discrete and continuum states of the hydrogen atom, Φ_{H_i}, and those of positronium, Φ_{Ps_j}. Such an expansion is doubly complete, and in principle the total wave function could be expressed solely in terms of a complete set of either hydrogen or positronium states. Higgins, Burke and Walters (1990) considered an expansion of the wave function in terms of states of the hydrogen atom alone, but the rate of convergence of the elastic scattering phase shifts with respect to adding more states is very slow. More usually, a truncated expansion involving states of both hydrogen and positronium is used, so that

$$\Psi(\boldsymbol{r}_1, \boldsymbol{r}_2) = \sum_{i=1}^{n} F_i(\boldsymbol{r}_1)\Phi_{\mathrm{H}_i}(\boldsymbol{r}_2) + \sum_{j=1}^{p} G_j(\boldsymbol{\rho})\Phi_{\mathrm{Ps}_j}(\boldsymbol{r}_{12}), \tag{3.28}$$

where, using the nomenclature of Figure 3.1, $\boldsymbol{\rho}$ is the position vector of the centre of mass of the positronium relative to the proton. Operating with $H - E$ on Ψ, and requiring that the projection of the resulting function on each of the hydrogen and positronium eigenfunctions, Φ_{H_i} and Φ_{Ps_j}, be zero, we obtain

$$\langle \Phi_{\mathrm{H}_i} | (H - E) | \Psi \rangle = 0 \qquad (i = 1, \ldots, n) \tag{3.29}$$

and

$$\langle \Phi_{\mathrm{Ps}_j} | (H - E) | \Psi \rangle = 0 \qquad (j = 1, \ldots, p). \tag{3.30}$$

Then the following equations are obtained for the functions $F_i(\boldsymbol{r}_1)$, $i = 1, \ldots, n$, and $G_j(\boldsymbol{\rho})$, $j = 1, \ldots, p$:

$$\left(-\tfrac{1}{2}\nabla^2_{r_1} - \tfrac{1}{2}k_i^2\right) F_i(\boldsymbol{r}_1) = \sum_{i'=1}^{n} V_{ii'}(\boldsymbol{r}_1) F_{i'}(\boldsymbol{r}_1) + \sum_{j'=1}^{p} \int K_{ij'}(\boldsymbol{\rho}, \boldsymbol{r}_1) G_{j'}(\boldsymbol{\rho})\, d\boldsymbol{\rho}, \tag{3.31}$$

$$\left(-\tfrac{1}{4}\nabla^2_{\boldsymbol{\rho}} - \tfrac{1}{4}k_j^2\right) G_j(\boldsymbol{\rho}) = \sum_{j'=1}^{p} U_{jj'}(\boldsymbol{\rho}) G_{j'}(\boldsymbol{\rho}) + \sum_{i'=1}^{n} \int K_{ji'}(\boldsymbol{\rho}, \boldsymbol{r}_1) F_{i'}(\boldsymbol{r}_1)\, d\boldsymbol{r}_1, \tag{3.32}$$

where

$$V_{ii'}(\boldsymbol{r}_1) = \left\langle \Phi_{\mathrm{H}_i} \left| \left(\frac{1}{r_1} - \frac{1}{r_{12}} \right) \right| \Phi'_{\mathrm{H}_i} \right\rangle, \tag{3.33}$$

$$U_{jj'}(\boldsymbol{\rho}) = \left\langle \Phi_{\mathrm{Ps}_j} \left| \left(\frac{1}{r_1} - \frac{1}{r_2} \right) \right| \Phi_{\mathrm{Ps}_{j'}} \right\rangle, \tag{3.34}$$

and

$$K_{ij} = 8\left\{ -\tfrac{1}{4}\nabla^2_{\boldsymbol{\rho}} [\Phi_{\mathrm{H}_i}(\boldsymbol{r}_2) \Phi_{\mathrm{Ps}_j}(\boldsymbol{r}_{12})] + \left[\left(\frac{1}{r_1} - \frac{1}{r_2} \right) - \tfrac{1}{4}\kappa_j^2 \right] \Phi_{\mathrm{H}_i}(\boldsymbol{r}_2) \Phi_{\mathrm{Ps}_j}(\boldsymbol{r}_{12}) \right\}. \tag{3.35}$$

Energy conservation gives the relationships

$$E = \tfrac{1}{2}k_i^2 + E_{\mathrm{H}_i} = \tfrac{1}{4}\kappa_j^2 + E_{\mathrm{Ps}_j} \qquad (i = 1, \ldots, n;\ j = 1, \ldots, p), \tag{3.36}$$

where E_{H_i} and E_{Ps_j} are the energy eigenvalues of hydrogen and positronium respectively.

If a positron channel labelled by i is open, then k_i^2 is positive and the corresponding positron function $F_i(\boldsymbol{r}_1)$ is oscillatory for large values of r_1. However, if the channel is closed, because the positron energy is below the relevant threshold, then k_i^2 is negative and the function $F_i(\boldsymbol{r}_1)$ decays exponentially for large values of r_1. Similarly, if a positronium channel labelled by j is open, then κ_j^2 is positive and $G_j(\boldsymbol{\rho})$ is oscillatory for large values of ρ, whereas for a closed positronium channel κ_i^2 is negative and $G_j(\boldsymbol{\rho})$ decays exponentially for large values of ρ.

When elastic scattering is the only open channel, k_1^2 is positive but all other values of k_i^2, and all values of κ_j^2, are negative. Consequently, all the functions $F_i(\boldsymbol{r}_1)$ and $G_j(\boldsymbol{\rho})$, except for $F_1(\boldsymbol{r}_1)$, decay exponentially for large values of r_1 and ρ. The resulting equation for $F_1(\boldsymbol{r}_1)$ is similar in form to equation (3.20), in which the optical potential $V_{\rm opt}$ was introduced; indeed a truncated coupled-state expansion essentially defines an approximation to the optical potential which satisfies the conditions for the phase shifts to be lower bounds on the exact values.

Including only the ground state of the target atom in the coupled-state expansion defines the static approximation, which has been mentioned previously as poor. Adding the ground state of positronium creates the coupled-static approximation (Cody *et al.*, 1964; Bransden and Jundi, 1967; Chan and Fraser, 1973) and introduces some positron–electron correlation into the wave function, but the results, although showing significant improvements over the static results, still differ substantially from the exact, variationally determined, values.

The rate of convergence of the phase shifts with respect to increasing the number of eigenstates in the close-coupling expansion is rather low, as may be seen in Figures 3.2 and 3.3, but faster convergence may be achieved by replacing some of the terms in the expansion by pseudostates, each of which is chosen so as to incorporate the important features of several eigenstates into one algebraic function. The projection of $(H-E)|\Psi\rangle$ onto each pseudostate is assumed to be zero, and the energy of the pseudostate is taken to be the expectation value of the relevant Hamiltonian, either of hydrogen or positronium, as calculated with the pseudostate.

Many different formulations of positron–hydrogen scattering based on the coupled-state approximation have been made, particularly where several channels are open. One of the most detailed studies was that of Kernoghan *et al.* (1995), who used 18 states, H(1s, 2s, $\overline{\rm 3s}$, $\overline{\rm 4s}$, 2p, $\overline{\rm 3p}$, $\overline{\rm 4p}$, $\overline{\rm 3d}$, $\overline{\rm 4d}$) and Ps(1s, 2s, $\overline{\rm 3s}$, $\overline{\rm 4s}$, 2p, $\overline{\rm 3p}$, $\overline{\rm 4p}$, $\overline{\rm 3d}$, $\overline{\rm 4d}$), where the bar above a term signifies a pseudostate, and solved the resulting integro-differential equations using an R-matrix technique (Burke and Robb, 1975) in the energy range 0–100 eV. Kernoghan *et al.* (1996) increased the number of states in the expansion to 33 and these data are very similar to the 18-state results shown in Figure 3.11. The elastic scattering cross sections change only slightly when going from 18 to 33 states, implying that they are then close to the exact values. Other calculations using similar methods have been made by Hewitt, Noble and Bransden (1990, 1991), Higgins and Burke (1991, 1993), Gien (1994, 1997) and Archer *et al.* (1990). A recently developed variant of the standard close-coupling method, which employs an expansion of the wave function in terms of states of the hydrogen target alone, is the convergent close-coupling method of Bray and Stelbovics (1993). These results are in good agreement with those of Kernoghan *et al.* (1995, 1996) at energies beyond 40 eV, although somewhat smaller in the energy interval 10–30 eV.

The formulation outlined above is in configuration space, but several authors, notably Ghosh and his collaborators (Chaudhury, Ghosh and Sil, 1974) and Mitroy (1993), also Mitroy, Berge and Stelbovics (1994) and Mitroy and Ratnavelu (1995), have preferred to work in momentum space with a set of coupled integral equations rather than the coupled integro-differential equations (3.31) and (3.32).

An alternative means of incorporating the elaborate correlations between the constituent particles into the formulation of positron–atom scattering is to use a flexible algebraic trial wave function in a variational method, somewhat similar to the way in which the Rayleigh–Ritz variational method is used for determining the energies of bound states. One of the simplest and most convenient computational methods for scattering problems is the Kohn variational method, which has been used quite extensively to obtain very accurate results for low energy positron scattering by several light atoms. Indeed, the first accurate values of the s-wave positron–hydrogen phase shifts were obtained in this way (Schwartz, 1961b), and subsequently the method has been used to obtain accurate values of the various partial-wave elastic scattering and positronium formation cross sections for positrons colliding with hydrogen (Armstead, 1968; Humberston and Wallace, 1972; Houston and Drachman, 1971; Stein and Sternlicht, 1972; Humberston, 1982, 1984; Brown and Humberston, 1984, 1985) and also for positrons colliding with helium (Houston and Drachman, 1971; Humberston, 1973; Campeanu and Humberston, 1975, 1977a; Van Reeth and Humberston, 1995b). Several references will be made to this method of approximation, and it is therefore appropriate to describe it in some detail in the context of positron–hydrogen scattering. Initially only elastic scattering will be considered, but the way in which the method can be extended to multichannel scattering, and particularly to positronium formation, will be outlined later, in section 4.2.

The basis of the Kohn variational method is the functional

$$\tan\eta_{\rm v} = \tan\eta_{\rm t} - \langle\Psi_{\rm t}|\, L\,|\Psi_{\rm t}\rangle, \tag{3.37}$$

where $L = 2(H - E)$ and $\Psi_{\rm t}$ is the trial wave function representing the scattering process. For the lth partial wave, the trial wave function must be chosen to have the asymptotic form (using the nomenclature of Figure 3.1)

$$\Psi_{\mathrm{t}} \underset{r_1 \to \infty}{\sim} Y_{l,0}(\theta_1, \phi_1)\sqrt{k}\,[\,j_l(kr_1) - \tan\eta_{\mathrm{t}}\, n_l(kr_1)]\,\Phi_{\mathrm{H}}(r_2), \tag{3.38}$$

where $j_l(kr_1)$ and $n_l(kr_1)$ are spherical Bessel and Neumann functions respectively and $Y_{l,0}(\theta_1, \phi_1)$ is the corresponding spherical harmonic. The function Ψ_{t} differs from the exact asymptotic form of the wave function only in the value of the trial phase shift η_{t}. The Kohn functional $\tan\eta_{\mathrm{v}}$ is stationary with respect to small variations in the trial wave function Ψ_{t} from the exact wave function Ψ, so that if the trial wave function is written as

$$\Psi_{\mathrm{t}} = \Psi + \delta\Psi, \tag{3.39}$$

where

$$\delta\Psi \underset{r_1 \to \infty}{\sim} (\tan\eta - \tan\eta_{\mathrm{t}}) n_l(kr_1)\Phi_{\mathrm{H}}(r_2), \tag{3.40}$$

η being the exact phase shift, then the error in the variationally determined phase shift η_{v} is given by

$$\delta(\tan\eta) = \tan\eta - \tan\eta_{\mathrm{v}} = \langle\delta\Psi|\, L\, |\delta\Psi\rangle, \tag{3.41}$$

which is of second order in the error in the trial wave function. Consequently, if the trial function is reasonably accurate, the error in $\tan\eta_{\mathrm{v}}$ should be much smaller than the errors in Ψ_{t} and $\tan\eta_{\mathrm{t}}$.

(a) s-wave scattering For s-wave scattering, a convenient and flexible choice of trial function, similar to that used by Schwartz (1961b), Stein and Sternlicht (1972) and Humberston and Wallace (1972), is

$$\Psi_{\mathrm{t}} = \sqrt{\frac{k}{4\pi}}\,\{j_0(kr_1) - (\tan\eta_{\mathrm{t}}) n_0(kr_1)[1 - \exp(-\lambda r_1)]\}\,\Phi_{\mathrm{H}}(r_2) + \sum_{i=1}^{n} c_i \exp\left[-(\alpha r_1 + \beta r_2 + \gamma r_{12})\right] r_1^{k_i} r_2^{l_i} r_{12}^{m_i}, \tag{3.42}$$

where the short-range Hylleraas terms, $\exp[-(\alpha r_1 + \beta r_2 + \gamma r_{12})] r_1^{k_i} r_2^{l_i} r_{12}^{m_i}$, represent the various interparticle correlations. Similar functions have been used extensively to represent correlations in many different three-body systems. In the summation in equation (3.42) it is usual to include all correlation terms such that

$$k_i + l_i + m_i \leq \omega, \tag{3.43}$$

where k_i, l_i, m_i and ω are non-negative integers. Increasing the value of ω then provides a convenient and systematic means of improving the trial wave function.

Because the orbital angular momentum of the positron–hydrogen system is zero for s-wave scattering, the total wave function is spherically symmetric and depends only on the three internal coordinates which specify the shape of the three-body system. The kinetic energy operator

$$T = -\tfrac{1}{2}\left(\nabla_{r_1}^2 + \nabla_{r_2}^2\right) \tag{3.44}$$

can therefore be expressed solely in terms of r_1, r_2 and the angle between them, θ_{12} (see Figure 3.1):

$$\begin{aligned} T = & -\frac{1}{2}\left[\frac{1}{r_1^2}\frac{\partial}{\partial r_1}\left(r_1^2\frac{\partial}{\partial r_1}\right) + \frac{1}{r_2^2}\frac{\partial}{\partial r_2}\left(r_2^2\frac{\partial}{\partial r_2}\right)\right. \\ & \left. + \left(\frac{1}{r_1^2}+\frac{1}{r_2^2}\right)\frac{1}{\sin\theta_{12}}\frac{\partial}{\partial\theta_{12}}\left(\sin\theta_{12}\frac{\partial}{\partial\theta_{12}}\right)\right]. \end{aligned} \tag{3.45}$$

Alternatively, it can be expressed in terms of the three interparticle distances r_1, r_2 and r_{12}, as

$$\begin{aligned} T = & -\frac{1}{2}\left[\frac{1}{r_1^2}\frac{\partial}{\partial r_1}\left(r_1^2\frac{\partial}{\partial r_1}\right) + \frac{1}{r_2^2}\frac{\partial}{\partial r_2}\left(r_2^2\frac{\partial}{\partial r_2}\right)\right. \\ & \left. + \frac{2}{r_{12}^2}\frac{\partial}{\partial r_{12}}\left(r_{12}^2\frac{\partial}{\partial r_{12}}\right) + \frac{r_1^2+r_{12}^2-r_2^2}{r_1 r_{12}}\frac{\partial^2}{\partial r_1 \partial r_{12}}\right]. \end{aligned} \tag{3.46}$$

The trial function, equation (3.42), may be written in an abbreviated form as

$$\Psi_t = S + K_t C + \sum_{i=l}^{n} c_i \phi_i \tag{3.47}$$

where

$$K_t = \tan\eta_t, \tag{3.48}$$

$$S = \sqrt{\frac{k}{4\pi}}\, j_0(kr_1)\Phi_H(r_2), \tag{3.49}$$

$$C = -\sqrt{\frac{k}{4\pi}}\, n_0(kr_1)[1-\exp(-\lambda r_1)]\Phi_H(r_2) \tag{3.50}$$

and

$$\phi_i = \exp\left[-(\alpha r_1 + \beta r_2 + \gamma r_{12})\right] r_1^{k_i} r_2^{l_i} r_{12}^{m_i}. \tag{3.51}$$

The requirement that the Kohn functional $K_v = \tan\eta_v$, equation (3.37), be stationary with respect to variations of the linear parameters K_t and c_i $(i = 1, \ldots, n)$, i.e. that

$$\frac{\partial K_v}{\partial K_t} = 0 \quad \text{and} \quad \frac{\partial K_v}{\partial c_i} = 0 \quad (i = 1, \ldots, n), \tag{3.52}$$

then results in the following set of linear simultaneous equations, which we express in matrix form:

$$\begin{bmatrix} \langle C|L|C\rangle & \langle C|L|\phi_1\rangle & \dots & \langle C|L|\phi_n\rangle \\ \langle \phi_1|L|C\rangle & \langle \phi_1|L|\phi_1\rangle & \dots & \langle \phi_1|L|\phi_n\rangle \\ \vdots & \vdots & & \vdots \\ \langle \phi_i|L|C\rangle & \langle \phi_i|L|\phi_1\rangle & \dots & \langle \phi_i|L|\phi_n\rangle \\ \vdots & \vdots & & \vdots \\ \langle \phi_n|L|C\rangle & \langle \phi_n|L|\phi_1\rangle & \dots & \langle \phi_n|L|\phi_n\rangle \end{bmatrix} \begin{bmatrix} K_{\rm t} \\ c_1 \\ \vdots \\ c_i \\ \vdots \\ c_n \end{bmatrix} = - \begin{bmatrix} \langle C|L|S\rangle \\ \langle \phi_1|L|S\rangle \\ \vdots \\ \langle \phi_i|L|S\rangle \\ \vdots \\ \langle \phi_n|L|S\rangle \end{bmatrix}. \tag{3.53}$$

Writing this matrix as $\mathbf{AX} = -\mathbf{B}$, where $\mathbf{X}$ is the column matrix of the unknown linear parameters, then $\mathbf{X} = -\mathbf{A}^{-1}\mathbf{B}$ and the stationary value of $\tan\eta_{\rm v}$ is obtained by substituting the elements of $\mathbf{X}$ back into the Kohn functional, equation (3.37). The optimum values of the non-linear variational parameters in the trial function are determined by repeating the entire calculation for a range of values of each non-linear parameter, thereby identifying the values which yield a maximum value of $\tan\eta_{\rm v}$. Further details of this procedure were given by Armour and Humberston (1991).

Although the Kohn variational method is not a bounded method (except at zero energy when, subject to certain conditions, it can yield an upper bound on the scattering length), it is found in practice that the phase shift usually becomes more positive as the flexibility of the trial function is enhanced by increasing ω; it converges towards what is assumed to be the exact value according to a pattern which is quite accurately represented by

$$K_{\rm v}(\omega) = K_{\rm v}(\infty) + \frac{b}{\omega^q}, \tag{3.54}$$

provided ω is not too small. Thus, a plot of $K_{\rm v}(\omega)$ against $1/\omega^q$ gives a straight line, which can be extrapolated to infinite ω to yield a good approximation to the exact result. An example of this convergence pattern for the s-wave phase shift is given in Figure 3.4.

At very low positron energies the rate of convergence of the phase shift with respect to ω deteriorates significantly because the exact wave function then resembles the form given by the adiabatic approximation, with long-range components varying as $1/r_1^2$. Such terms are not represented very efficiently by a finite sum of short-range Hylleraas terms; instead they need to be explicitly added to the trial function, as will be described in the following account of positron scattering at zero energy (see subsection 3.2.1(b) below). The convergence with respect to ω also

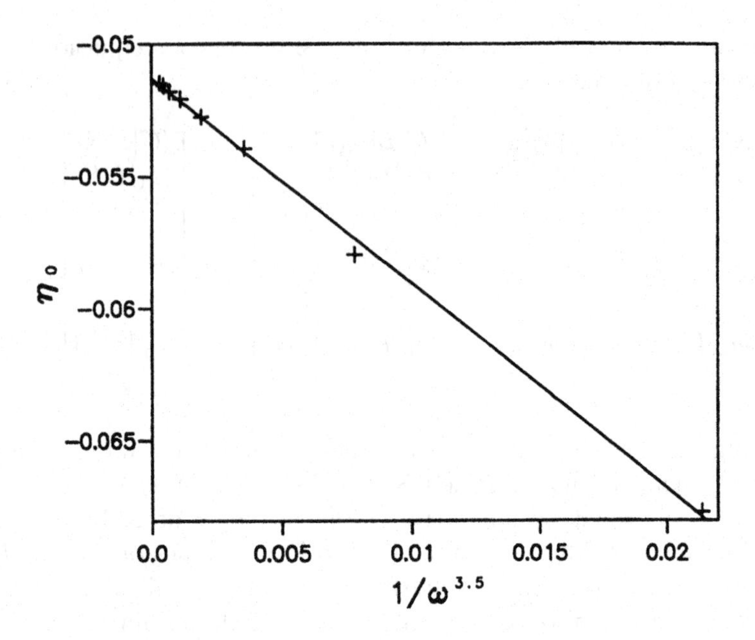

Fig. 3.4. The convergence of the positron–hydrogen s-wave phase shift (for $k = 0.7a_0^{-1}$) with respect to systematic improvements in the trial wave function; see equation (3.54).

deteriorates in the vicinity of the positronium formation threshold because the configuration of virtual positronium loosely bound to the residual ion is not easily represented by the Hylleraas terms either. Again, the addition to the trial function of a term explicitly representing this configuration improves the convergence.

Although most results conform to the convergence pattern described by equation (3.54), very erratic values of the phase shift are occasionally obtained which clearly contravene the empirical lower bound. These so-called Schwartz singularities (Schwartz, 1961b) arise because the matrix **A** defined as in equation (3.53), must formally be inverted in order to obtain the optimum values of the linear parameters in the trial function; however, **A** is not positive definite and may have an eigenvalue very close to zero, making the matrix ill-conditioned. It is quite easy to recognize when the results are influenced by such a singularity because they do not conform to the pattern obtained at slightly different energies or with slightly different trial functions.

Various techniques have been devised for coping with Schwartz singularities. They may either be ignored, or they can be avoided by using a modified form of the Kohn variational method in which the asymptotic form of the trial function has an amplitude different from that of equation (3.42);

see Armour and Humberston (1991). Alternatively, techniques can be used in which these singularities do not arise, such as the method devised by Harris (1967). This method assumes a trial function of the same general form as that used in the Kohn variational method, equation (3.42), and the eigenvalues and eigenvectors of the total Hamiltonian matrix are then calculated in the basis ϕ_i of the short-range correlation terms. Thus

$$(\mathbf{H} - E\mathbf{N})\mathbf{C} = 0, \tag{3.55}$$

where the matrix elements of $\mathbf{H}$ and $\mathbf{N}$ are $\langle \phi_i | H | \phi_j \rangle$ and $\langle \phi_i | \phi_j \rangle$ respectively. At one of the energy eigenvalues, E_p, the trial function is expressed, using a similar nomenclature to that in equation (3.47), as

$$\Psi_p = S + \tan\eta C + \sum_{i=1}^{N} c_{ip}\phi_i = S + \tan\eta C + \Phi_p, \tag{3.56}$$

where the c_{ip} are the elements of the eigenvector $\boldsymbol{C}$ corresponding to the eigenvalue E_p. The projection of $(H - E_p)\Psi_p$ onto Φ_p is then required to be zero,

$$\langle \Phi_p | H - E_p | \Psi_p \rangle = 0, \tag{3.57}$$

whence

$$\tan\eta = -\frac{\langle \Phi_p | H - E_p | S \rangle}{\langle \Phi_p | H - E_p | C \rangle}. \tag{3.58}$$

Schwartz singularities are avoided using the Harris method, but results can only be obtained at the discrete energies E_p (although the values of E_p can be altered by changing the values of the non-linear parameters in the trial function). Furthermore, the error in the phase shift is only of first order in the error in the trial wave function, and the results may therefore be less accurate than those of a well-behaved Kohn calculation.

Probably the most accurate positron–hydrogen s-wave phase shifts are those obtained by Bhatia *et al.* (1971), who avoided the possibility of Schwartz singularities by using a bounded variational method based on the optical potential formalism described previously. These authors chose their basis functions spanning the closed-channel Q-space, see equation (3.14), to be of essentially the same Hylleraas form as those used in the Kohn trial function, equation (3.42), and their most accurate results were obtained with 84 such terms. By extrapolating to infinite ω in a somewhat similar way to that described in equation (3.54), they obtained phase shifts which are believed to be accurate to within 0.0002 rad. They also established that there are no Feshbach resonances below the positronium formation threshold.

Several other accurate calculations have been made of s-wave elastic scattering phase shifts, the more recent ones usually being the byproduct of studies of positronium formation and other inelastic processes at higher energies. Among these have been the calculations of Doolen *et al.* (1971), who used the extrapolated T-matrix method, and Chan and Fraser (1973), who used the formalism of the coupled-static approximation with the addition of several short-range Hylleraas correlation terms. Worthy of particular mention is the moment T-matrix method developed by Winick and Reinhardt (1978a, b), who used it to calculate the various partial-wave elastic scattering amplitudes, t_l, from the off-shell elastic scattering T-matrix. In terms of these amplitudes, the lth partial-wave contribution to the elastic scattering cross section is then

$$\sigma_{\mathrm{el}}^{l} = \frac{4\pi}{k^2}(2l+1)|t_l|^2. \tag{3.59}$$

The most accurate values of the positron–hydrogen s-wave phase shifts, obtained using variational methods, are plotted in Figure 3.2, together with the results of other less accurate approximations. At low energies, the attractive polarization potential is dominant and the phase shift is positive. As the energy increases, however, the polarization potential becomes less attractive and the repulsive static potential begins to dominate, whereupon the phase shift changes sign and becomes negative.

(b) The scattering length According to Levinson's theorem, the s-wave phase shift tends to $n\pi$ as the energy of the projectile tends to zero, where n is the number of bound states of the composite projectile–target system. However,

$$\lim_{k\to 0}\left(-\frac{\tan\eta_0}{k}\right) = a, \tag{3.60}$$

where a is the scattering length, and the elastic scattering cross section at zero positron energy is therefore

$$\sigma_{\mathrm{el}}(k=0) = 4\pi a^2. \tag{3.61}$$

The scattering length can be calculated using the Kohn variational method in a similar manner to that employed for the phase shift, but the Kohn functional then becomes

$$a_{\mathrm{v}} = a_{\mathrm{t}} + \langle \Psi_{\mathrm{t}}|L|\Psi_{\mathrm{t}}\rangle. \tag{3.62}$$

For positron–hydrogen scattering the trial function must have the asymptotic form

$$\Psi_{\mathrm{t}} \underset{r_1\to\infty}{\sim} \frac{1}{\sqrt{4\pi}}\left(1 - \frac{a_{\mathrm{t}}}{r_1}\right)\Phi_{\mathrm{H}}(r_2). \tag{3.63}$$

If no bound states of the composite projectile–target system exist, which has been proved by Armour (1978, 1982) to be the case for the positron–hydrogen system, the Kohn variational method immediately gives a rigorous upper bound on the scattering length (Spruch and Rosenberg, 1960). If, however, there are bound states of the projectile–target system then the Kohn method will also yield an upper bound on the scattering length, provided that the matrix representation of the total Hamiltonian, in the basis of the short-range correlation terms in the trial function has as many eigenvalues below the energy of the target system as there are bound states of the composite system.

The existence of an upper bound ensures that the value of the scattering length becomes more negative as the flexibility of the trial function is increased by the addition of more short-range correlation terms. However, as already mentioned, the convergence of the scattering length with respect to the number of such terms is poor unless long-range correlation terms with a $1/r_1^2$ dependence are added to the trial wave function (Schwartz, 1961b; Humberston and Wallace, 1972). These terms arise from the long-range character of the interaction potential between the projectile and the target at zero projectile energy, which, for sufficiently large values of r_1, reduces to the adiabatic polarization potential whose asymptotic form is given by equation (3.27). The distortion of the hydrogen atom in the field of the stationary positron, which gives rise to the polarization potential, has the dipole form $(r_2 + r_2^2/2)\Phi_{\mathrm{H}}(r_2)\cos(\theta_{12})/r_1^2$ (Temkin and Lamkin, 1961). Thus, an appropriate form of zero energy trial function incorporating these long-range features is

$$\begin{aligned}\Psi_{\mathrm{t}} = \frac{1}{\sqrt{4\pi}}\Bigg\{ & 1 - \frac{a_{\mathrm{t}}}{r_1}[1-\exp(-\lambda r_1)] + \frac{b_1}{r_1^2}[1-\exp(-\lambda r_1)]^2 \\ & + b_2\left(r_2 + \frac{1}{2}r_2^2\right)\cos\theta_{12}\frac{[1-\exp(-\lambda r_1)]_3}{r_1^2}\Bigg\}\Phi_{\mathrm{H}}(r_2) \\ & + \sum_{i=1}^{n} c_i\phi_i, \end{aligned} \tag{3.64}$$

where the long-range monopole term multiplied by the linear parameter b_1 is also included, to represent more accurately the influence of the polarization potential on the positron.

This form of trial wave function has been used in several variational calculations (Schwartz, 1961b; Houston and Drachman, 1971; Humberston and Wallace, 1972) to obtain well-converged upper bounds on the scattering length, the most accurate value found being $-2.103a_0$. In its most flexible form the coefficients b_1 and b_2 multiplying the long-range correlation terms in equation (3.64) were left free, to be determined by the variational method (Humberston and Wallace, 1972), and their values were found to be very close to those predicted by the adiabatic approximation, namely -2.25 and 1.0 respectively.

(c) Higher partial waves Accurate calculations of higher-partial-wave phase shifts require the use of trial wave functions which incorporate the total angular momentum of the system in the most general manner. Asymptotically, the total orbital angular momentum of the positron–hydrogen system resides solely on the positron, but when the positron is close to the atom, and interacting with it, the total angular momentum is shared between the positron, with angular momentum l_1, and the electron, with angular momentum l_2. In principle, all combinations of l_1 and l_2 which couple together to form a total angular momentum of l should be included in the trial function, but Schwartz (1961a) showed that an equivalent formulation may be used which involves a summation over just $l+1$ rotational harmonics of the required parity, each one multiplied by a suitable spherically symmetric function of the three interparticle coordinates of the system.

For p-wave scattering, with $l = 1$, there are two such rotational harmonics of odd parity, $Y_{1,0}(\theta_1, \phi_1)$ and $Y_{1,0}(\theta_2, \phi_2)$, corresponding to $l_1 = 1$, $l_2 = 0$ and $l_1 = 0$, $l_2 = 1$ respectively; $Y_{1,0}(\theta, \phi) = \sqrt{\frac{3}{4\pi}} \cos\theta$. For d-wave scattering, with $l = 2$, there are three rotational harmonics of even parity, $Y_{2,0}(\theta_1, \phi_1)$ and $Y_{2,0}(\theta_2, \phi_2)$, corresponding to $l_1 = 2, l_2 = 0$ and $l_1 = 0, l_2 = 2$ respectively, where $Y_{2,0}(\theta, \phi) = \sqrt{\frac{5}{4\pi}}(\frac{3}{2}\cos^2\theta - \frac{1}{2})$, and

$$Y(\theta_1, \phi_1; \theta_2, \phi_2) = \frac{3}{4\pi\sqrt{6}}(3\cos\theta_1\cos\theta_2 - \cos\theta_{12}),$$

corresponding to $l_1 = 1$, $l_2 = 1$. A suitably flexible trial function for p-wave scattering, similar in form to equation (3.42) but now containing short-range correlation terms of both symmetries, is then

$$\begin{aligned} \Psi_t = {} & Y_{1,0}(\theta_1)\sqrt{k}\left\{j_1(kr_1) - \tan\eta_t\, n_1(kr_1)[1 - \exp(-\lambda r_1)]^3\right\}\Phi_H(r_2) \\ & + Y_{1,0}(\theta_1) r_1 \sum_{i=1}^{n_1} c_i \exp\left[-(\alpha r_1 + \beta r_2 + \gamma r_{12})\right] r_1^{k_i} r_2^{l_i} r_{12}^{m_i} \\ & + Y_{1,0}(\theta_2) r_2 \sum_{j=1}^{n_2} d_j \exp[-(\alpha r_1 + \beta r_2 + \gamma r_{12})] r_1^{k_j} r_2^{l_j} r_{12}^{m_j}, \end{aligned} \quad (3.65)$$

where the summations include all short-range correlation terms with

$$k_i + l_i + m_i \leq \omega_1 \quad \text{and} \quad k_j + l_j + m_j \leq \omega_2. \quad (3.66)$$

This form of trial function was used in the Kohn variational method by Armstead (1968) and Humberston and Campeanu (1980) to obtain well-converged p-wave phase shifts.

Humberston and Campeanu (1980) investigated the convergence of the p-wave phase shifts with respect to the number of short-range correlation terms of each symmetry separately, and they showed that at low positron energies the first-symmetry terms are the most important but that as the positron energy is raised the inclusion of the second-symmetry terms becomes increasingly significant.

Several partial-wave phase shifts have been calculated using a range of other techniques, notable examples being the Harris method, used by Register and Poe (1975) for the d-wave, the intermediate energy R-matrix method (Higgins, Burke and Walters, 1990), the convergent close-coupling method (Bray and Stelbovics, 1994) and various forms of the coupled-state method (Kernoghan *et al.*, 1995, 1996; Higgins, Burke and Walters, 1990; Kuang and Gien, 1997; Mitroy and Ratnavelu, 1995).

At sufficiently low positron energies the scattering in all partial waves with $l > 0$ is dominated by the polarization potential, which has the asymptotic form of equation (3.27); the phase shifts are given by the formula (O'Malley, Spruch and Rosenberg, 1962)

$$\eta_l = \frac{\pi \alpha k^2}{(2l-1)(2l+1)(2l+3)} + R, \tag{3.67}$$

where the remainder term R is of order k^3 for $l = 1$ and of order k^4 for $l > 1$. As can be seen in Figure 3.3, at very low positron energies the variational values of the p-wave phase shift display the linear increase with positron energy predicted by equation (3.67), but as the energy is increased the phase shifts fall progressively below the linear form. The d-wave phase shifts, however, are in reasonably good agreement with the linear energy dependence of equation (3.67) over a much wider range, almost up to the positronium formation threshold at 6.8 eV. This equation can therefore be expected to provide reasonably accurate phase shifts for all higher partial waves, and this has indeed been confirmed by the results of accurate coupled-state calculations.

2 Positron–helium scattering

Until quite recently, helium was the simplest atomic target used in experimental studies of positron collisions. It is also the simplest atom for which the wave function is not known exactly. Accordingly, positron–helium scattering has attracted considerable theoretical attention, and detailed comparisons have been made between the experimental measurements of the scattering parameters and the corresponding theoretical results obtained using a wide variety of approximation methods.

The use of an inexact target wave function in a scattering calculation inevitably introduces some inconsistencies into the formulation; this is

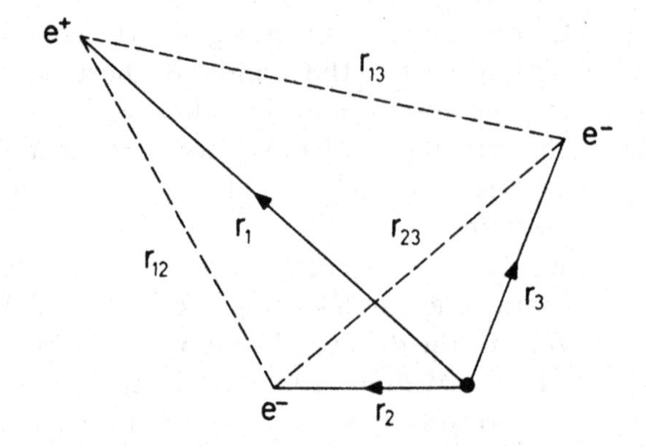

Fig. 3.5. The coordinates of the positron–helium system.

so even for elastic scattering, where the usual practice is to replace the exact energy of the target system by the expectation value of the target Hamiltonian as calculated using the chosen inexact target wave function. However, it has been shown by several authors that this procedure can produce quite inaccurate results unless very accurate target wave functions are used (Peterkop and Rabik, 1971; Houston, 1973; Page, 1975; Van Reeth and Humberston, 1995a). The most satisfactory means of avoiding this problem is to use the method of models (Drachman, 1972), but it can only be used for the elastic scattering of projectiles, such as positrons, which are distinguishable from the electrons in the target. The exact target wave function, Φ_0, although not known, satisfies the eigenvalue equation

$$H_0\Phi_0 = E_0\Phi_0, \tag{3.68}$$

where H_0 is the exact target Hamiltonian and E_0 is the exact, but also unknown, target ground state energy. It is assumed that a known approximation to the target wave function, $\Phi_{\rm m}$, is an exact eigenfunction of a model Hamiltonian $H_{\rm m}$, so that

$$H_{\rm m}\Phi_{\rm m} = E_{\rm m}\Phi_{\rm m}. \tag{3.69}$$

For the helium atom, with the two electron coordinates $\boldsymbol{r}_2$ and $\boldsymbol{r}_3$,

$$H_{\rm m} = -\tfrac{1}{2}(\nabla^2_{r_2} + \nabla^2_{r_3}) + V_{\rm m}, \tag{3.70}$$

where the model interaction potential $V_{\rm m}$ and the model energy $E_{\rm m}$ may be determined by substituting the approximate wave function into the Schrödinger equation for the model, equation (3.69).

The exact scattering process is now replaced by one in which the projectile is scattered by the model target system but it is assumed that the interactions between the projectile and the constituent particles in the target retain their exact Coulomb form. Thus, using the nomenclature of Figure 3.5, the interaction potential between the positron, with position vector $\boldsymbol{r}_1$, and the helium atom is

$$V_{\text{int}} = \left(\frac{2}{r_1} - \frac{1}{r_{12}} - \frac{1}{r_{13}} \right), \tag{3.71}$$

and the total potential function is therefore

$$V = V_{\text{int}} + V_{\text{m}}. \tag{3.72}$$

It is important to appreciate that any results obtained using the method of models are approximations to the exact results for scattering by the model target rather than the real target. This distinction is particularly significant when calculating a rigorous bound on a scattering parameter. If, however, the model target wave function is a good approximation to the exact target wave function, then an accurate scattering result for the model system is expected to be a good approximation to the exact result for the real system.

Although providing a self-consistent formulation for the elastic scattering of a distinguishable particle by an inexact target system, the use of the method of models fails to avoid inconsistencies in the formulation of rearrangement collisions. This may be readily appreciated by considering the example of positronium formation in positron–helium scattering, discussed in detail in subsection 4.2.2, where the interactions of the electrons with the nucleus, and with each other, in the initial helium target are given in terms of the model potential V_{m}, but the electron–nucleus interaction in the residual He^+ ion has the exact Coulomb form.

Let us now consider further the use of the method of models in elastic positron–helium scattering, which is the sole open channel for positron energies below 17.8 eV, the threshold for ground state positronium formation. The total Hamiltonian of the system is

$$H = -\tfrac{1}{2}(\nabla^2_{\boldsymbol{r}_1} + \nabla^2_{\boldsymbol{r}_2} + \nabla^2_{\boldsymbol{r}_3}) + V_{\text{int}} + V_{\text{m}} \tag{3.73}$$

and the total energy is

$$E = \tfrac{1}{2}k^2 + E_{\text{m}}. \tag{3.74}$$

Because the ground state helium model wave function is nodeless, the total positron–helium wave function may be written without any loss of generality as the product form

$$\Psi(\boldsymbol{r}_1, \boldsymbol{r}_2, \boldsymbol{r}_3) = F(\boldsymbol{r}_1, \boldsymbol{r}_2, \boldsymbol{r}_3)\Phi_{\text{He}}(\boldsymbol{r}_2, \boldsymbol{r}_3), \tag{3.75}$$

where $\Phi_{\rm He}(\mathbf{r}_2, \mathbf{r}_3)$ is the model helium wave function. Operating with $L = 2(H - E)$ on Ψ and using equation (3.69) yields

$$\begin{aligned} L\Psi &= LF\Phi_{\rm He} \\ &= \Phi_{\rm He}\left[-(\nabla^2_{r_1} + \nabla^2_{r_2} + \nabla^2_{r_3}) + 2V_{\rm int} - k^2\right]F \\ &\quad - 2\left(\nabla_{r_2}\Phi_{\rm He} \cdot \nabla_{r_2}F + \nabla_{r_3}\Phi_{\rm He} \cdot \nabla_{r_3}F\right), \end{aligned} \tag{3.76}$$

in which there is no longer any explicit reference to either the form of the model potential, $V_{\rm m}$, or the energy eigenvalue of the model Hamiltonian, $E_{\rm m}$. Thus, if the method of models is used with the product form of wave function, equation (3.75), nothing need be known about the model target except its wave function.

Several authors have used the method of models in conjunction with the Kohn variational method to obtain partial-wave phase shifts for elastic positron–helium scattering. The most detailed investigations were those of Humberston (1973), Campeanu and Humberston (1975) and Humberston and Campeanu (1980), who formulated the problem in a manner rather similar to that described previously for elastic positron–hydrogen scattering in subsection 3.2.1 (Humberston and Wallace, 1972). Here, however, the short-range correlation terms in the trial function involve all six interparticle distances between the four particles in the system (see Figure 3.5), and there is the additional complication of ensuring the spatial symmetry of the trial function with respect to the interchange of the two electrons, which are in a singlet spin state. For s-wave scattering, the total wave function used by Humberston (1973) had the product form of equation (3.75), with

$$\begin{aligned} F &= \sqrt{\frac{k}{4\pi}}\,\{j_0(kr_1) - \tan\eta_{\rm t}\, n_0(kr_1)[1 - \exp(-\lambda r_1)]\} \\ &\quad + (1 + {\rm P}_{23})\sum_i c_i \exp[-(\alpha r_1 + \beta r_2 + \beta r_3)]\, r_1^{k_i} r_2^{l_i} r_{12}^{m_i} r_3^{n_i} r_{13}^{p_i} r_{23}^{q_i}, \end{aligned} \tag{3.77}$$

where P_{23} is the space-exchange operator for the two electrons. The summation included all short-range correlation terms such that

$$k_i + l_i + m_i + n_i + p_i + q_i \le \omega, \tag{3.78}$$

where k_i, l_i, m_i, n_i, p_i, q_i and ω are all non-negative integers; furthermore, for numerical convenience, only even values of q_i were included. In order to avoid the same term being generated twice by the action of the exchange operator P_{23}, additional constraints were imposed, namely that $l_i \ge n_i$ and if $l_i = n_i$ then $m_i \ge p_i$. Further details of the formulation and a description of the numerical techniques developed to evaluate the various matrix elements were given by Armour and Humberston (1991).

Systematic improvements in the trial wave function were achieved by increasing the value of ω, and investigations of the convergence of the phase shifts revealed a similar pattern to that described earlier for positron–hydrogen scattering, equation (3.54), with extrapolation to infinite ω expected to yield essentially exact results for the particular helium model being used.

A suitable representation of the helium wave function was taken to be the Hylleraas form

$$\Phi_{\mathrm{He}} = \exp[-B(r_2 + r_3)] \sum_{j=1}^{n_{\mathrm{He}}} d_j (r_2 + r_3)^{L_j} (r_2 - r_3)^{M_j} r_{23}^{N_j}, \tag{3.79}$$

where the summation included all terms with non-negative integer powers such that

$$L_j + M_j + N_j \leq \omega_{\mathrm{He}}, \tag{3.80}$$

with M_j taking only even values. Again for numerical convenience, N_j was also limited to even values. The values of the parameters d_j ($j = 1, \ldots, n_{\mathrm{He}}$) and B were determined using the Rayleigh–Ritz variational method to minimize the expectation value of the ground state energy of the helium atom subject to its dipole polarizability having the correct value, $\alpha = 1.383a_0^3$ (Dalgarno and Kingston, 1960; Thomas and Humberston, 1972). This requirement was imposed because of the importance of the dipole polarizability in determining the values of the low energy phase shifts with $l > 0$; see equation (3.67).

With such an elaborate form for the helium wave function, the matrix elements of the operator L between any two short-range terms in the trial function of the product form $\Phi_{\mathrm{He}}\phi_i$ and $\Phi_{\mathrm{He}}\phi_j$ may be more conveniently expressed, after integrating by parts, as

$$\begin{aligned} \langle \Phi_{\mathrm{He}}\phi_i | L | \Phi_{\mathrm{He}}\phi_j \rangle &= \int \Phi_{\mathrm{He}}^2 [\nabla_{r_1}\phi_i \cdot \nabla_{r_1}\phi_j + \nabla_{r_2}\phi_i \cdot \nabla_{r_2}\phi_j + \nabla_{r_3}\phi_i \cdot \nabla_{r_3}\phi_j \\ &\quad + (V_{\mathrm{int}} - k^2)\phi_i\phi_j] \, d\boldsymbol{r}_1 \, d\boldsymbol{r}_2 \, d\boldsymbol{r}_3 \end{aligned} \tag{3.81}$$

rather than the form derived from direct use of equation (3.76). The helium wave function now enters only as a squared factor multiplying the remainder of the integrand and not in terms of its gradient.

Three helium models were generated (Thomas and Humberston, 1972) for these investigations by setting $\omega_{\mathrm{He}} = 0$, 2 and 4; the corresponding numbers of terms in the helium model wave function, according to the scheme in equation (3.80), are 1, 5 and 14 respectively. These models are

Table 3.1. Properties of three helium wave functions used in variational calculations of positron–helium elastic scattering. Each wave function is of the form given in equation (3.79), the number of terms n being determined by the value of ω_{He} according to the pattern specified in equation (3.80)

	H1	H5	H14	Exact
ω_{He}	0	2	4	
$n(\omega_{\mathrm{He}})$	1	5	14	
Energy (a.u.)	−2.840	−2.895	−2.900	−2.9037
Polarizability (a.u.)	1.376	1.372	1.387	1.3834
$\langle r^2 \rangle$ (a_0^2)	1.1730	1.2117	1.196 81	1.1935

therefore referred to as H1, H5 and H14 and their properties are given in table 3.1. Increasing the value of ω_{He} in this way makes it possible to investigate the convergence of the scattering parameters with respect to systematic improvements in the helium model wave function as well as with respect to ω for a given helium model. Results which are well converged with respect to increases in both ω and ω_{He} are then assumed to be very close to the exact results for positron scattering by a real helium atom.

Well-converged (with respect to ω) s-wave phase shifts for the three helium models defined above are plotted in Figure 3.6. As in positron–hydrogen scattering (see subsection 3.2.1), the rate of convergence with respect to ω deteriorates as the positron energy approaches zero. An accurate determination of the scattering length requires a zero energy trial wave function which is a suitably modified form of equation (3.77), with the addition of long-range polarization terms similar to those introduced into the zero energy positron–hydrogen trial wave function, equation (3.64).

Substantial differences exist between the results for the simple uncorrelated helium function H1 and those for the more elaborate correlated helium functions H5 and H14, but the very close agreement between these latter two sets of results strongly suggests that they are both close to the exact values. Further evidence in support of this claim has recently been provided by Van Reeth and Humberston (1995a), who obtained very similar results using even more accurate helium wave functions, both with and without the use of the method of models.

Similar variational techniques were employed by Campeanu and Humberston (1975) (see also Humberston, 1979) to determine accurate values of the p- and d-wave phase shifts. Trial wave functions were used which

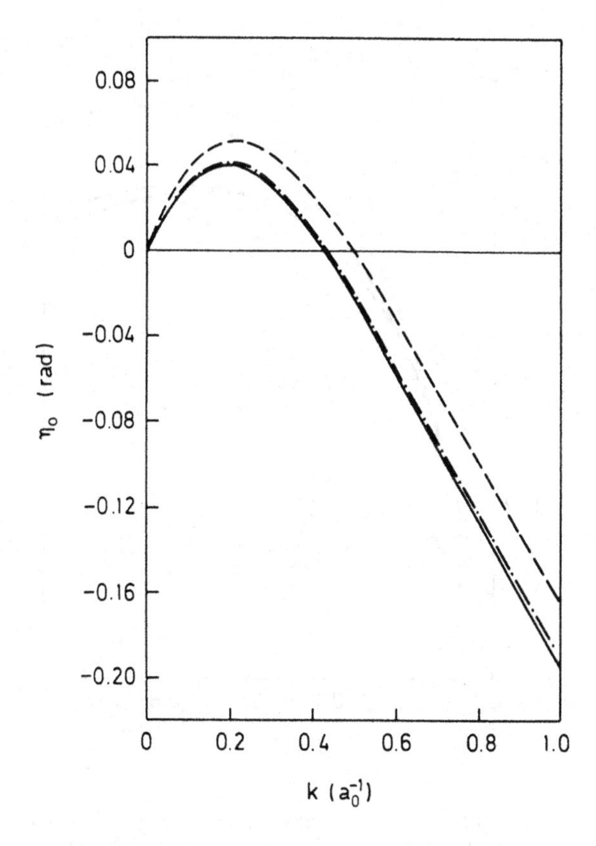

Fig. 3.6. Positron–helium s-wave phase shifts for three helium models: – – –, H1; ——, H5; — · —, H14.

are obvious generalizations to positron–helium scattering of the trial functions used in the determination of the higher-partial-wave phase shifts in positron–hydrogen scattering; see equation (3.65). As in hydrogen, the scattering at very low positron energies in all partial waves with $l > 0$ is dominated by the long-range polarization potential, and the phase shifts are therefore given by equation (3.67). This equation provides little more than the correct gradient of the p-wave phase shift in the limit of zero incident energy, but it gives a reasonably good approximation to the d-wave phase shift over quite a wide energy range. Well-converged results for all helium models with the same dipole polarizability should therefore agree at very low positron energies and this is clearly seen in Figures 3.7 and 3.8, where the p- and d-wave phase shifts for helium models H1 and H5 are displayed. As the positron energy increases, however, the results for the two models diverge, particularly for $l = 1$.

Phase shifts for partial waves with $l > 2$ are expected to be in even better agreement with the values given by equation (3.67) up to the

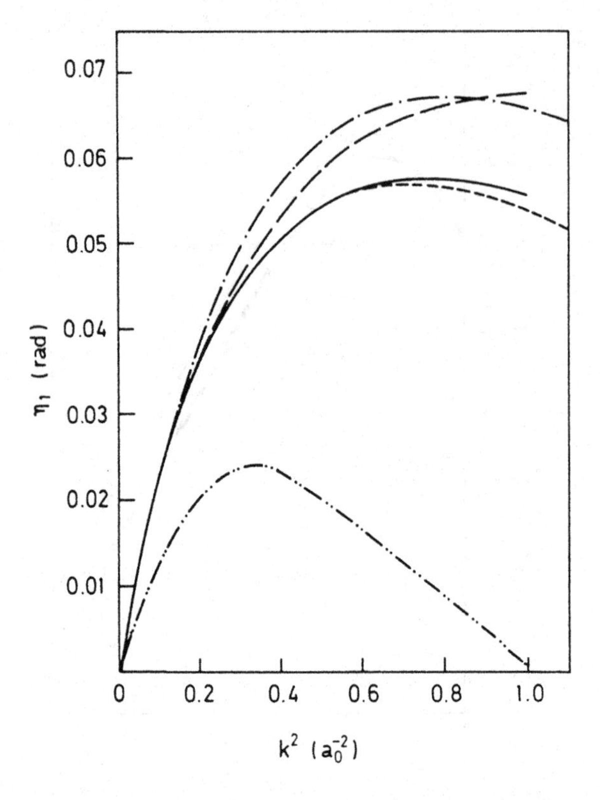

Fig. 3.7. Positron–helium p-wave phase shifts: ——, Campeanu and Humberston (1975) for model H5; — —, Campeanu and Humberston (1975) for model H1; - - - -, McEachran *et al.* (1977); — · —, Drachman (1966a); — · · —, Aulenkamp, Heiss and Wichman (1974).

positronium formation threshold at 17.8 eV and, consequently, no elaborate variational calculations have yet been performed for $l > 2$.

Several other calculations of the first few partial-wave phase shifts for positron–helium scattering have been carried out using a variety of approximation methods; in all cases, however, rather simple uncorrelated helium wave functions have been used. Drachman (1966a, 1968) and McEachran *et al.* (1977) used the polarized-orbital method, whereas Ho and Fraser (1976) used a formulation based on the static approximation, with the addition of several short-range correlation terms, to determine the s-wave phase shifts only. The only other elaborate variational calculations of the s-wave phase shift were made by Houston and Drachman (1971), who employed the Harris method with a trial wave function similar to that used by Humberston (1973, 1974), see equation (3.77), and with the same helium model H1. Their results were slightly less positive than Humberston's H1 values, and are therefore probably less

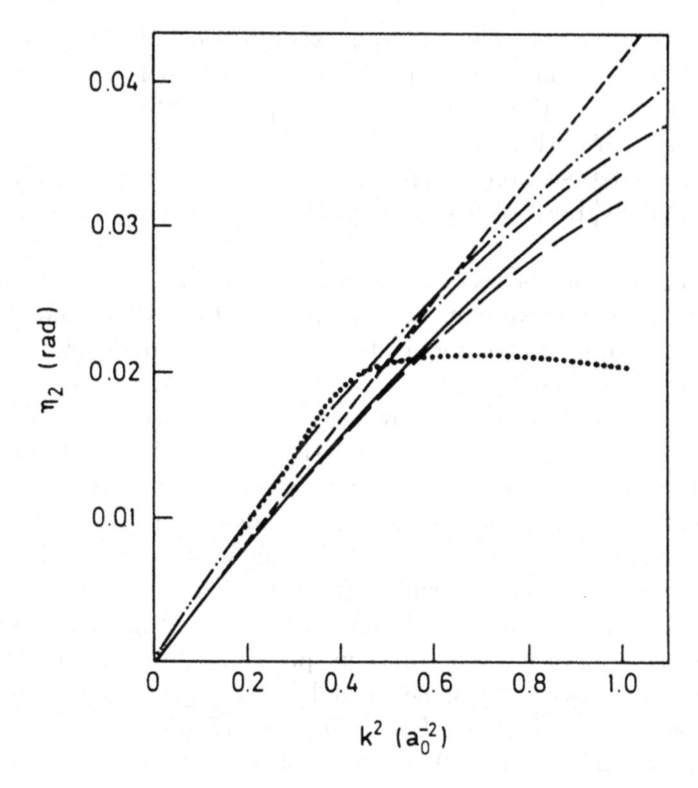

Fig. 3.8. Positron–helium d-wave phase shifts: ——, Campeanu (1977) for model H5; — —, Campeanu (1977) for model H1; — · —, Drachman (1966a); — · · —, Amusia *et al.* (1976); · · · · ·, Aulenkamp, Heiss and Wichman (1974); - - - -, equation (3.67).

accurate. Houston and Drachman also calculated the scattering length, using the Kohn variational method, and obtained essentially the same value as Humberston, $a = -0.524a_0$. A rather different method, the random-phase approximation derived from many-body theory, was used by Amusia *et al.* (1976) to calculate the phase shifts for $l \leq 3$. These authors used a more accurate, but still uncorrelated, Hartree–Fock helium wave function and obtained reasonably good agreement with the accurate variational results.

The total elastic scattering cross sections for the three helium models, calculated using the accurate variationally determined phase shifts for $l \leq 2$ and equation (3.67) for $l > 2$, are shown in Figure 2.8, together with several sets of experimental measurements. Excellent agreement is obtained between the results for the two helium models H5 and H14 and the experimental measurements of Canter *et al.* (1973) and, more recently, those of Mizogawa *et al.* (1985). Also, the accurate theoretical results are also in good agreement with an extrapolation of the experimental cross sections of Canter *et al.* (1973) to energies below 2 eV obtained by fitting to the functional form given in equation (2.8) (Bransden, Hutt and Winters, 1974; Humberston, 1974). Furthermore, the extrapolated experimental cross section at zero energy, $0.88\pi a_0^2$, agrees quite well with the value $0.92\pi a_0^2$ derived from the accurate calculation of the scattering length.

Other comparisons between experimental and theoretical values of the scattering parameters were made by Bransden *et al.* (1974) and Bransden and Hutt (1975), using techniques based on forward dispersion relations, as described in section 2.2.

Additional, but rather less direct, evidence for the accuracy of the variational results for models H5 and H14 is provided by the excellent agreement between the theoretical and experimental lifetime spectra for positrons diffusing in helium gas, where calculation of the theoretical spectrum requires a knowledge of the momentum transfer and annihilation cross sections, both of which are derived from the wave functions generated in the calculations of the elastic scattering phase shifts. A detailed discussion of positron lifetime spectra is given in Chapter 6.

Differential cross sections, see equation (3.9), for elastic positron–helium scattering, calculated for helium model H5 at several positron energies are given in Figure 3.9. At low positron energies, in the vicinity of the Ramsauer minimum, where the s-wave phase shift is zero, a significant fraction of the total elastic scattering cross section is seen to arise from scattering through small angles, $\theta < 30°$. It is precisely in this low energy region that the measurements of Stein *et al.* (1978) and Coleman *et al.* (1979) (see Figure 2.8) are significantly lower than the most accurate theoretical results. A discussion of differential cross sections at higher incident energies is given in subsection 3.2.5; see also section 3.4.

3 Positron scattering by alkali atoms

It is a reasonably good approximation to consider an alkali atom as a single electron moving in the modified Coulomb field of the ionic core, and this approximation has been made in almost all theoretical investigations of positron scattering by the alkali atoms. The interaction of the electron with the core is expressed as a local central potential of the general form

$$V_-(r) = -\frac{1}{r} + V_0(r), \tag{3.82}$$

where $V_0(r)$ is of short range. The positron–core potential has usually been taken to be $V_+(r) = -V_-(r)$. This form for $V_+(r)$ is not altogether appropriate, because the electron–core interaction implicitly takes

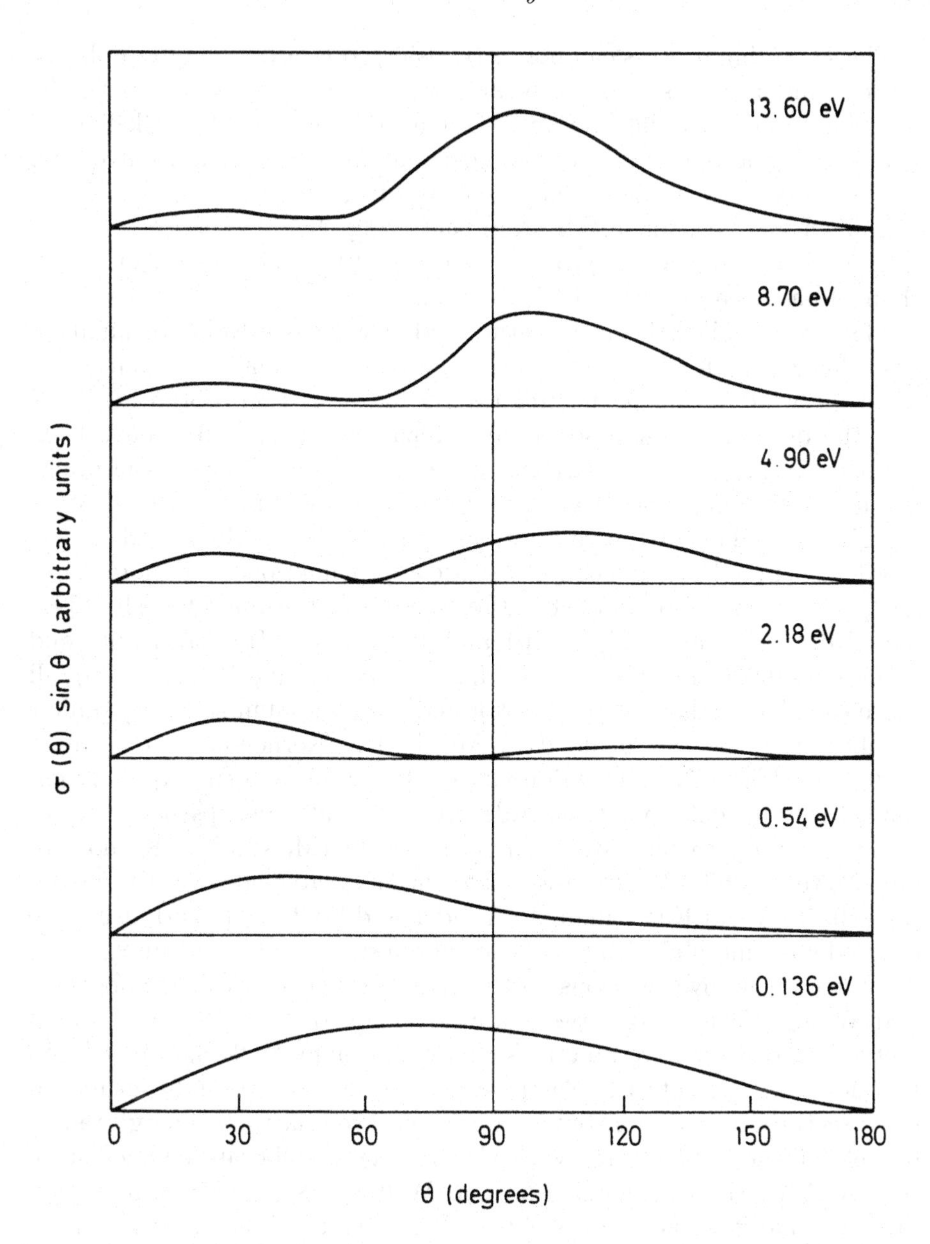

Fig. 3.9. Angular distributions of positrons elastically scattered by helium. The model H5 phase shifts were used to obtain these results.

account of exchange between the valence and core electrons whereas there is, of course, no exchange between the positron and the core electrons. Nevertheless, such a positron–core potential is probably reasonably accurate.

The scattering process can now be considered as a three-body problem, rather similar to positron scattering by atomic hydrogen but with the important difference that, because the ionization energy of an alkali atom is less than the binding energy of positronium, 6.8 eV, the positronium formation channel is open even at zero positron energy.

Low energy positron–alkali atom scattering should therefore be considered as a two-channel process, although the positronium formation channel has been neglected in some calculations.

The most detailed theoretical studies of positron scattering by an alkali atom have been made for lithium, although this remains the only such atom for which no experimental results have yet been obtained. It is also the only alkali atom so far for which scattering results have been obtained using elaborate variational methods in addition to the different forms of coupled-state approximation employed in almost all other calculations. The first investigations were made by Guha and Ghosh (1981) using the Born and coupled-static approximations, and more states have subsequently been added to the expansion. Basu and Ghosh (1991) included three states, Li(2s, 2p) and Ps(1s), and Hewitt, Noble and Bransden (1992b) included seven, Li(2s, 2p, 3s, 3p) and Ps(1s, 2s, 2p), all these calculations having been performed in momentum space. A similar number of states was included by McAlinden, Kernoghan and Walters (1994), working in configuration space. These authors subsequently extended their calculations to include up to 29 states and pseudostates of lithium and up to nine states of positronium (McAlinden, Kernoghan and Walters, 1997). Most of these calculations were made over the energy range 0–50 eV but Kernoghan, McAlinden and Walters (1994a), using the coupled-state method with nine positronium states and five lithium states, restricted their investigations to the energy range < 3 eV. Humberston and Watts (1994) used a two-channel version of the Kohn variational method, with trial wave functions containing many Hylleraas correlation functions, to calculate the elastic scattering and positronium formation cross sections for lithium, but only over an even narrower energy range, 0–2 eV. Their results agree well with the most elaborate coupled-state results of McAlinden, Kernoghan and Walters, as may be seen in Figure 3.10, and it is therefore likely that the results of the latter are quite accurate throughout the energy range they investigated. Further details of these and other calculations in which positronium formation has been included as an open channel are given in section 4.2.

The open positronium formation channel in positron–alkali atom scattering was neglected in the coupled-state calculations of Ward *et al.* (1989) and McEachran, Horbatsch and Stauffer (1991), and only states of the target alkali atom were included in the expansion of their wave function. At low positron energies the elastic scattering cross sections calculated by

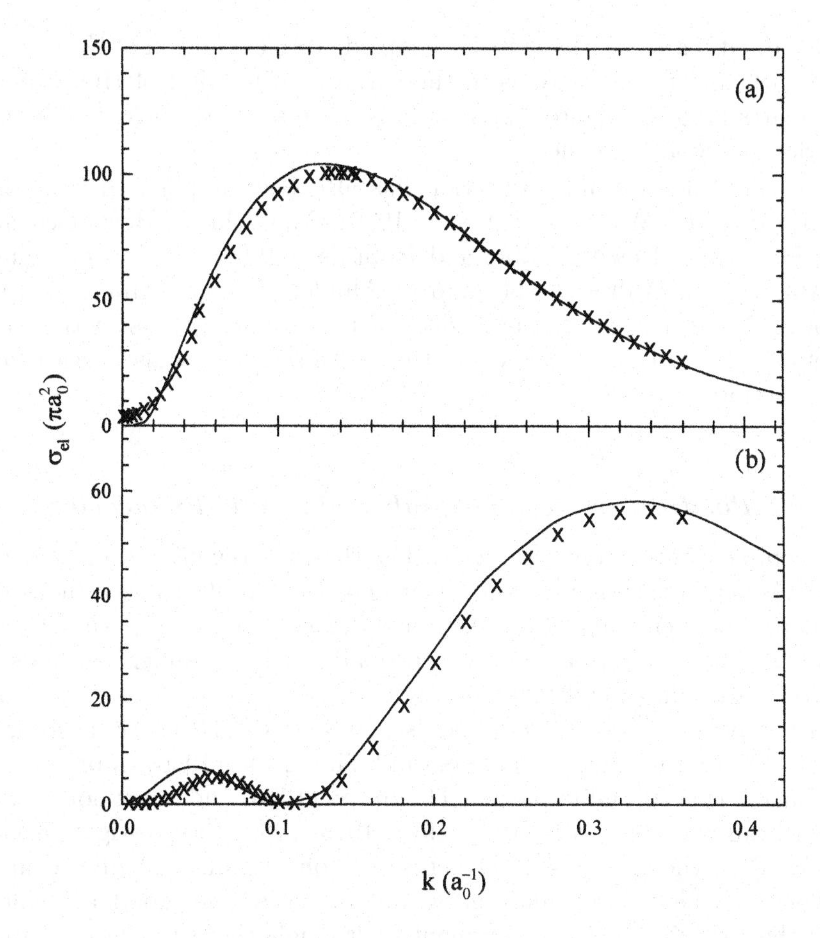

Fig. 3.10. Partial-wave cross sections for positron–lithium elastic scattering: (a) $l = 0$; (b) $l = 1$. The solid curves are the most accurate coupled-state results of Kernoghan, McAlinden and Walters (1994a) and the crosses are the variational results of Humberston and Watts (1994).

these authors differ significantly from those obtained with the inclusion of positronium states, both in overall magnitude and in structure. Furthermore, several resonances were found in the elastic scattering cross sections, but they are probably a consequence of neglect of the open positronium channels, since they are not observed when positronium states are included. At energies beyond 10 eV the effect on the elastic scattering cross section of excluding positronium states from the coupled-state expansion, as was done by Ward *et al.* (1989), is slight, but at lower energies it becomes very pronounced. Without such states, the elastic scattering cross section continues to rise steeply to values of a few hundred πa_0^2 as the positron energy approaches zero, but when positronium states are included, either explicitly as in the coupled-state expansion of Kernoghan *et al.* (1994a) or implicitly as in the variational method of Humberston and Watts (1994), the cross section falls quite abruptly below 1 eV to a very low value at zero energy.

The coupled-state approximation has also been applied to the other alkali atoms, by Ward *et al.* (1988, 1989), McEachran, Horbatsch and Stauffer (1991), Hewitt, Noble and Bransden (1993, 1994), Kernoghan, McAlinden and Walters (1996) and McAlinden, Kernoghan and Walters (1996). As with lithium, the neglect of positronium terms in the wave function has a serious influence on the very-low-energy elastic scattering cross section.

4 Positron scattering by other atoms and molecules

The complexity of atomic targets other than hydrogen, helium and the alkali atoms (considered as equivalent one-electron atoms) precludes the possibility of such elaborate and detailed investigations of positron scattering as those described above. Instead, rather simpler methods of approximation have had to be used. The most satisfactory of these has been the polarized-orbital method (see subsection 3.2.1), some form of which was used by Massey, Lawson and Thompson (1966), Montgomery and LaBahn (1970), Gillespie and Thompson (1975) and, most notably, by McEachran and coworkers (1977, 1978, 1979, 1980) to investigate elastic scattering by the inert gases. The cross sections calculated by these latter authors for the heavier inert gases are in reasonably good agreement with the experimental measurements, although there is evidence from the lifetime spectrum for positrons diffusing in the gas (see Chapter 6) that the differential cross sections are not as accurate as are the integrated elastic scattering cross sections.

The determination of accurate theoretical values of the cross sections for positron scattering by molecules is considerably more complicated than scattering by atoms because in the fixed-nuclei approximation the Hamiltonian of the target molecule no longer has the spherical symmetry of an atomic target. In a diatomic molecule there is only axial symmetry about the internuclear axis, and so the wave function can be expressed only in terms of eigenfunctions of L_z, not those of L^2, resulting in the mixing of spherical partial waves. It is, therefore, not surprising that most theoretical studies of positron–molecule scattering have used rather simple phenomenological model potentials to represent the positron–molecule interaction (see Armour, 1988 for a comprehensive review of the subject).

In the Born–Oppenheimer approximation, in which the nuclear and electronic motions within the molecule are considered separately, the total positron–molecule wave function is written as

$$\Psi_{\nu\nu'}(\boldsymbol{r}_1, \boldsymbol{r}, \boldsymbol{R}) = \Phi_\nu(\boldsymbol{r}_1, \boldsymbol{r}, \boldsymbol{R})\chi_{\nu\nu'}(\boldsymbol{R}), \tag{3.83}$$

where $\boldsymbol{r}_1$ is the position of the positron, $\boldsymbol{r}$ represents the positions of all the electrons in the molecule, $\boldsymbol{R}$ is the internuclear separation and ν and ν' represent all the quantum numbers required to specify the electronic and nuclear wave functions respectively. The function $\Phi_\nu(\boldsymbol{r}_1, \boldsymbol{r}, \boldsymbol{R})$ represents the wave function of the positron and the electrons for a fixed internuclear separation $\boldsymbol{R}$, and $\chi_{\nu\nu'}(\boldsymbol{R})$ is the wave function representing the internuclear motion. For elastic scattering, $\Phi_\nu(\boldsymbol{r}_1, \boldsymbol{r}, \boldsymbol{R})$ has the asymptotic form

$$\Phi_\nu(\boldsymbol{r}_1, \boldsymbol{r}, \boldsymbol{R}) \underset{r_1 \to \infty}{\sim} F(\boldsymbol{r}_1)\Phi_{\rm T}(\boldsymbol{r}, \boldsymbol{R}) \tag{3.84}$$

where $\Phi_{\rm T}(\boldsymbol{r}, \boldsymbol{R})$ is the wave function of the molecular target.

The only molecular target for which very elaborate *ab initio* calculations of elastic scattering have been made is molecular hydrogen, the most accurate results for which were obtained by Armour and his collaborators (Armour, 1988; Armour, Baker and Plummer, 1990; Armour and Humberston, 1991). These authors used the Kohn variational method together with the method of models, in a somewhat similar manner to that described previously for positron–helium scattering (subsection 3.2.2). An accurate Hylleraas-type approximation was used for the H_2 target wave function, with R fixed at the equilibrium separation of $1.4a_0$. Many terms of $\Sigma_{\rm g}^+$, $\Sigma_{\rm u}^+$, $\Pi_{\rm u}$ and $\Pi_{\rm g}$ symmetry, also of Hylleraas form and expressed in terms of prolate spheroidal coordinates, were then included in the total wave function. The most accurate results for $\sigma_{\rm el}$ are given in Figure 2.19, together with accurate experimental measurements of the total positron–H_2 cross section already discussed in Chapter 2. Good agreement is obtained between the theoretical and experimental results up to a positron energy of 5 eV, but thereafter the experimental results exceed the theoretical results, particularly beyond the positronium formation threshold at 8.63 eV, where positronium formation is known to make a substantial contribution to the total cross section. Even below the positronium formation threshold, however, the experimental cross sections include contributions from rotational and vibrational excitations of the molecule, but these processes are not expected to contribute very significantly to the total cross section. Much of the difference between theory and experiment in the energy range 5.0–8.63 eV is probably the result of not having included all possible symmetries in the trial wave function.

The R-matrix method, which has been used extensively in studies of electron–molecule scattering, has also been applied to positron–molecule scattering, notably by Tennyson and his collaborators (Tennyson, 1986;

Tennyson and Danby, 1987), merely by changing the sign of the charge on the projectile and removing the exchange terms. However, such modifications alone cannot provide a very satisfactory representation of electron–positron correlations, because neither real nor virtual states of positronium were included in the wave function. Comparisons with the elastic scattering cross sections of Armour reveal that the R-matrix results for H_2 are indeed significantly less accurate. The R-matrix method has also been applied to other diatomic molecules, including N_2 and CO (Tennyson and Morgan, 1987) and HF (Danby and Tennyson, 1988). These authors found evidence of a very weakly bound state of the positron–HF system.

Earlier studies of positron–molecule elastic scattering did not involve such detailed descriptions of the scattering process as do the variational and R-matrix formulations. Instead, the interaction between the positron and the molecule was represented by a relatively simple model potential, and the positron wave function $F(\boldsymbol{r}_1)$ was assumed to satisfy the equivalent single-particle Schrödinger equation

$$-\tfrac{1}{2}\nabla^2_{r_1}F(\boldsymbol{r}_1) + V(\boldsymbol{r}_1)F(\boldsymbol{r}_1) = \tfrac{1}{2}k^2F(\boldsymbol{r}_1). \qquad (3.85)$$

The potential function was taken to be

$$V(\boldsymbol{r}_1) = V_{\rm stat}(\boldsymbol{r}_1) + V_{\rm pol}(\boldsymbol{r}_1), \qquad (3.86)$$

where the static potential between a positron and a homonuclear diatomic molecule such as H_2 has a short-range spherically symmetric component and a long-range non-spherical component with asymptotic form

$$V_{\rm stat}(\boldsymbol{r}_1) \underset{r_1\to\infty}{\sim} Q\frac{P_2(\cos\theta)}{r_1^3}. \qquad (3.87)$$

Here Q is the quadrupole moment of the molecule, which has the value of 0.49 a.u. for H_2 at the equilibrium internuclear separation, and $P_2(\cos\theta)$ is the second-degree Legendre function, θ being the angle between $\boldsymbol{R}$ and $\boldsymbol{r}_1$. The polarization potential, also with spherical and non-spherical components, has the asymptotic form

$$V_{\rm pol}(\boldsymbol{r}_1) \underset{r_1\to\infty}{\sim} -\frac{\alpha_0 + \alpha_2 P_2(\cos\theta)}{2r_1^4}, \qquad (3.88)$$

where α_0 and α_2 are defined in terms of the polarizabilities parallel and perpendicular to the internuclear axis, $\alpha_\parallel$ and $\alpha_\perp$, as

$$\alpha_0 = (\alpha_\parallel + 2\alpha_\perp)/3 \qquad (3.89)$$

and

$$\alpha_2 = 2(\alpha_\parallel - \alpha_\perp)/3. \qquad (3.90)$$

Instead of expressing the scattering problem in terms of the one-centre polar coordinates r_1 and θ, it is more appropriate to use the two-centre prolate spheroidal coordinates defined by

$$\lambda = (r_A + r_{\rm B})/R, \qquad \mu = (r_A - r_{\rm B})/R, \tag{3.91}$$

where r_A and $r_{\rm B}$ are the distances of the positron from the two nuclei, labelled A and B. Although the use of these variables complicates the formulation in some respects, the two-centre character of the system can thereby be more conveniently represented. In terms of these variables,

$$V_{\rm stat}(\boldsymbol{r}_1) \sim \frac{8QP_2(\mu)}{R^3\lambda^3} \tag{3.92}$$

and

$$V_{\rm pol}(\boldsymbol{r}_1) \sim \frac{-8}{R^4\lambda^4}[\alpha_0 + \alpha_2 P_2(\mu)]. \tag{3.93}$$

These asymptotic forms for $V_{\rm stat}(\boldsymbol{r}_1)$ and $V_{\rm pol}(\boldsymbol{r}_1)$ must be modified at short range, particularly to shield the singularities which would otherwise exist at $r_1 = 0$, and this has been done in several different ways.

In the earliest theoretical study of positron–H_2 scattering, by Massey and Moussa (1958), only the static potential was included, and the polarization potential was first added by Lodge, Darewych and McEachran (1971). Since then, numerous calculations of positron scattering by various molecules have been made using different model potentials. Among the more significant of these investigations for scattering by H_2 are those of Baille, Darewych and Lodge (1974), using a one-centre formalism, Hara (1974) and Morrison *et al.* (1984). Somewhat similar studies with model potentials have also been made for more complicated molecules such as CO_2 (Horbatsch and Darewych, 1983), CH_4 and NH_3 (Jain and Thompson, 1983).

5 Elastic scattering beyond inelastic and rearrangement thresholds

As the positron energy is raised beyond various rearrangement and inelastic thresholds, elastic scattering continues to make a significant contribution to the total scattering cross section $\sigma_{\rm T}$, but it is now just one of several open channels, all of which are coupled together. Even if only the elastic scattering cross section is required, its determination should be considered as part of a more comprehensive multichannel calculation in which the cross sections for all possible transitions between any two open channels are included. Furthermore, the couplings between the open channels give rise to structure in the elastic scattering cross section on either side of the threshold for the opening of each new inelastic channel. This is discussed in section 3.3 in the context of positronium formation.

When considering positron scattering at energies such that several inelastic channels are open, it is usually not feasible to represent every open channel explicitly; however, the neglect of open inelastic channels in the formulation of the collision process may produce extensive pseudoresonant structure in the cross sections for elastic scattering and for those inelastic processes which are explicitly included. Nevertheless, it is still possible to extract meaningful cross sections by a process of T-matrix averaging (Burke, Berrington and Sukumar, 1981). An interesting example of a pseudoresonance was found by Higgins and Burke (1991, 1993) in their calculations of s-wave positron–hydrogen scattering using the coupled-static approximation. These authors found a prominent feature at an energy of 35.6 eV, which revealed itself most conspicuously in the positronium formation cross section. Its existence was confirmed by several other authors (Hewitt, Noble and Bransden, 1991; Sarkar, Basu and Ghosh, 1993) using similar approximations, and it was believed to be authentic, being termed a coupled-channel shape resonance. However, subsequent investigations by Kernoghan, McAlinden and Walters (1994b) and Zhou and Lin (1995a), using more terms in the coupled-state expansion, established that there are no real resonances above the ionization threshold of the target hydrogen atom, a fact consistent with the theorem of Simon (1974, 1978), which states that in a many-body system experiencing only Coulomb interactions there can be no resonances above the total break-up energy.

Among the most accurate values of the positron–hydrogen elastic scattering cross sections in the intermediate energy region up to 80 eV are the 18-coupled-state results of Kernoghan, McAlinden and Walters (1995), which, after being smoothed to remove pseudoresonance structure, are displayed in Figure 3.11. Also given there are the results of the moment T-matrix method of Winick and Reinhardt (1978a, b) and those of the convergent close-coupling method of Bray and Stelbovics (1994), neither of which explicitly includes positronium formation. Nevertheless, the agreement between the results of these three methods is quite good for energies greater than 30 eV.

Elastic scattering cross sections at intermediate energies have been calculated for several other atoms. Important examples are the noble gases, investigated by McEachran and Stauffer (1986) using the polarized-orbital method, and the alkali atoms, investigated by McAlinden, Kernoghan and Walters (1996), Kernoghan, McAlinden and Walters (1996) and Hewitt, Noble and Bransden (1992b, 1993) using various forms of the coupled-state method. The latter technique has also recently been applied by Campbell *et al.* (1998a) to positron–helium scattering.

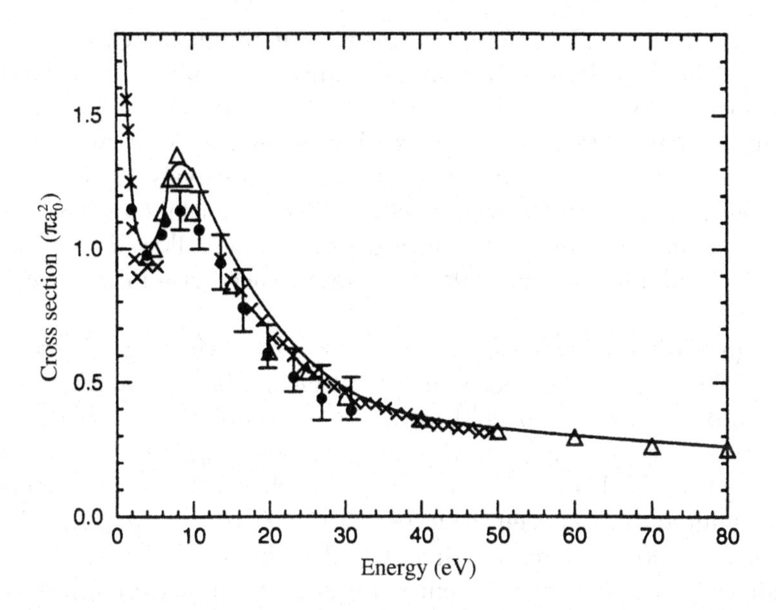

Fig. 3.11. Calculations of the positron–hydrogen elastic scattering cross sections: ——, 18-coupled-state approximation (Kernoghan *et al.*, 1995) (a 33-coupled-state calculation by Kernoghan *et al.*, 1996, yielded very similar results); ×, intermediate energy R-matrix theory (Higgins, Burke and Walters, 1990); •, moment T-matrix method (Winick and Reinhardt, 1978a, b); △, convergent close-coupling method (Bray and Stelbovics, 1994).

Throughout the intermediate and high energy regions, the elastic scattering cross sections for all target atoms decrease steadily with increasing positron energy, eventually merging with the elastic scattering cross sections for electrons. However, for a given atom the energy at which the elastic scattering cross sections for the two projectiles merge is much higher than that at which the two total cross sections merge. For example, in helium at 250 eV, which is comfortably above the energy at which merging of the two total cross sections takes place, the elastic scattering cross section for electrons is still more than three times greater than that for positrons.

In addition to total elastic scattering cross sections, several workers have calculated differential elastic scattering cross sections at intermediate and high energies (Byron and Joachain, 1977a, for hydrogen and helium; Kernoghan, McAlinden and Walters, 1995, for hydrogen; McEachran and Stauffer, 1986, for the rare gases). McEachran and Stauffer (1986) made a detailed study of the energy dependence of the differential cross sections for the rare gases, each of which has the following rather similar structure. From a peak in the forward direction, it falls steeply with increasing scattering angle to a local minimum, followed by an increase to a local maximum, after which it then falls steadily to a much lower value in the backward direction. The angles at which the local minimum and maximum occur decrease with increasing energy, so that at 250 eV in helium, for example, they are at 6° and 22° respectively. This structure is different from that for electrons, where the differential cross section has a somewhat larger value in the forward direction, followed by a steady fall to a minimum at around 90° and a pronounced rise in the backward direction.

At high positron energies, where the speed of the incident positron significantly exceeds the speeds of the electrons in the target, the first Born approximation is expected to yield reasonably accurate elastic scattering cross sections which are the same for positrons and electrons. For atomic hydrogen, this condition is satisfied when the positron energy is several hundred eV. Between the upper limit of 80 eV for the coupled-state calculations of Kernoghan, McAlinden and Walters (1995) and the region of validity of the first Born approximation, various approximation schemes based on higher order terms in the Born series have been employed. The eikonal Born series, which introduces differences between the cross sections for positrons and electrons, was applied to positron–hydrogen scattering by Byron and Joachain (1973, 1977a, b), and an improved unitarized version of this method has been used by Byron, Joachain and Potvliege (1981, 1982, 1985). At an energy of 400 eV, the results of the first Born approximation are already quite accurate, exceeding the two very similar eikonal Born results by less than 10%. At an energy of 100 eV, however, the first Born result exceeds the unitarized eikonel Born result by 30%, and even the results of the two eikonal Born methods differ by more than 5%. These more accurate results match smoothly onto the coupled-state results of Kernoghan, McAlinden and Walters (1995) at lower energies. Somewhat similar results for positron–hydrogen scattering have also been obtained using the third order optical model (Byron and Joachain, 1981) and the second order potential method (Mukherjee and Sural, 1982). Similar techniques were used by Byron and Joachain (1977a) for helium and by Joachain and Potvliege (1987) for argon.

3.3 Threshold effects

The threshold for positronium formation in collisions of positrons with atoms and molecules is an example of a general class of thresholds in collision processes where there is no residual long-range Coulomb interaction between the constituent subsystems in either the initial or final states. Since the original work of Wigner (1948), there has been much discussion of the effect of the opening of a new channel on those already open (Baz 1958; Newton, 1959; Meyerhof, 1963). These studies, though originally motivated by the need to analyse nuclear scattering data, also apply to the onset of positronium formation, which is then in competition with elastic scattering. The requirement of the conservation of flux gives rise to characteristic structures, such as cusps and rounded steps, in the energy dependence of the elastic scattering cross section close to inelastic thresholds. General discussions of these features have been given by Mott and Massey (1965, Chapter 13), McDaniel (1989, Chapter 4) and Newton (1982, Chapter 17). The application to the positronium formation threshold is outlined below.

The first attempt to analyse the behaviour of the elastic scattering cross section close to the positronium formation threshold energy, E_{Ps}, was that of Campeanu *et al.* (1987) who, as described in section 2.7, undertook a detailed partitioning of the total cross section, σ_{T}, for positron–helium scattering. In the energy range between E_{Ps} and the first positron excitation threshold of the target, at energy E_{ex}, these authors determined σ_{el} from the relationship

$$\sigma_{\mathrm{el}} = \sigma_{\mathrm{T}} - \sigma_{\mathrm{Ps}}, \tag{3.94}$$

where they used the total cross sections of Stein *et al.* (1978) and the positronium formation cross sections, σ_{Ps}, of Fromme *et al.* (1986; see also section 4.4). Although these experimental data for σ_{Ps} were sparse in this energy region, Campeanu *et al.* (1987) argued that the feature they deduced in σ_{el}, see the broken curve in Figure 3.12, was genuine. Thus, they argued that according to these measurements the elastic scattering cross section exhibits a cusp-like behaviour centred on E_{Ps}, followed by a steady fall, as the energy increases to E_{ex}, of approximately 20%. A similar analysis was performed by Fromme *et al.* (1988), following their determination of the positronium formation cross section in H_2 gas; here, an even bigger effect was found, the elastic scattering cross section at E_{ex} being lower than its value at E_{Ps} by around 50%. Thus, for both helium and H_2 an important cusp-like feature was deduced in the elastic scattering cross section in the vicinity of E_{Ps}.

In an attempt to explore this interesting energy region further, an experiment was undertaken by Coleman *et al.* (1992) to determine the elastic scattering cross section according to equation (3.94), using measurements of the total and positronium formation cross sections made in the same apparatus. The apparatus and the method used were similar to those developed by the Arlington group (e.g. Fornari, Diana and Coleman 1983; Diana *et al.*, 1986b; see also Figure 4.12 and accompanying discussion). Positrons from a tungsten-mesh moderator, held at a potential V_{m}, were guided by an axial magnetic field of approximately 0.01 T through a system of grids and a localized scattering cell 0.03 m long to a channeltron

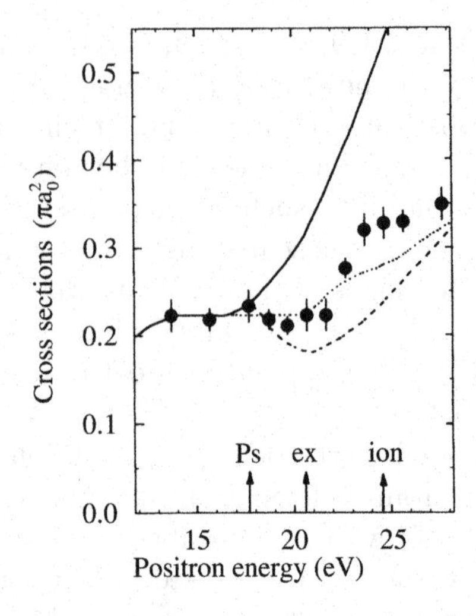

Fig. 3.12. Cross sections for positron–helium scattering in the vicinity of the positronium formation threshold (labelled Ps; 'ex' and 'ion' denote the respective thresholds for excitation and ionization). •, $\sigma_T - \sigma_{Ps}$ from Coleman *et al.* (1992); ——, σ_T from Stein *et al.* (1978); – – –, $\sigma_{el} - \sigma_{ex}$ (see text) deduced by Campeanu *et al.* (1987). The dotted curve is an attempt to account for the $\sigma_T - \sigma_{Ps}$ results, using other known cross sections as the various channels open (see Coleman *et al.*, 1992 for details). Reprinted from *Journal of Physics B* **25**, Coleman *et al.*, Elastic positron–helium scattering near the positronium formation threshold, L585–L588, copyright 1992, with permission from IOP Publishing.

detector located in line of sight with, and 0.6 m distant from, the source and moderator. The grid, located just after the moderator, was biassed to V_m + 1.3 V in order to narrow the energy width of the beam by preventing those positrons with an axial emission energy less than 1.3 eV from traversing the scattering region. This configuration effectively set the energy width of the beam (to approximately 0.8 eV FWHM) and thus the energy resolution with which σ_{el} was determined.

Biassing a grid in front of the channeltron to the voltage V_m prevented all the inelastically scattered, and nearly all the elastically scattered, positrons from being detected, by virtue of their loss of axial kinetic energy. The gas pressure was kept sufficiently low that the target thickness approximated to single-collision conditions, so that the total cross section was then determined according to

$$\sigma_T = f_T/(nL), \tag{3.95}$$

where f_T is the total fractional attenuation of the beam when gas is introduced into the scattering cell, and n and L are the gas number density and the effective path length in the cell.

In contrast, σ_{Ps} was determined by lowering the voltage on the grid in front of the channeltron to zero and so allowing all scattered positrons to reach the detector except those neutralized by forming positronium. To a good approximation, it could be assumed that only those positrons which had formed positronium would then be lost from the beam, and σ_{Ps} could be obtained from

$$\sigma_{Ps} = f_{Ps}/(nL), \tag{3.96}$$

where f_{Ps} is the fractional attenuation of the beam due to positronium formation in the gas. Thus, from the last three equations,

$$\sigma_{el} = (f_T - f_{Ps})/(nL).$$

In order to obtain an absolute scale for the measurements, the elastic scattering cross section was normalized to the total cross sections of Stein *et al.* (1978) at energies below E_{Ps}.

The results obtained by Coleman *et al.* (1992) are shown in Figure 3.12, where the most notable feature is that the elastic scattering cross section remains essentially constant between E_{Ps} and E_{ex}, with no evidence of a cusp as significant as that proposed by Campeanu *et al.* (1987). Recent accurate theoretical results obtained by Van Reeth and Humberston (1999a, b) confirm the constancy of the elastic scattering cross section throughout this energy range. When the positron energy exceeds E_{ex}, $\sigma_T - \sigma_{Ps}$ is no longer equal to σ_{el} but then contains contributions from the various possible inelastic processes, namely excitation and ionization (see Chapter 5). The effects of this are seen in the figure. Also displayed there are the values of σ_T measured by Stein *et al.* (1978) and the cross section $\sigma_{el} + \sigma_{ex}$ as deduced by Campeanu *et al.* (1987), where σ_{ex} is the excitation cross section. Note that this latter cross-section sum corresponds to $\sigma_{\text{no ion}}$, equation (2.19).

Support for the conclusion of Coleman *et al.* (1992) was forthcoming from the work of Moxom, Laricchia and Charlton (1993), who made a detailed study of the positronium formation cross section near to E_{Ps}, but found no evidence of threshold cusps in σ_{el} for helium, argon or molecular hydrogen targets. An attempt to provide elucidation was made by Moxom *et al.* (1994), who were the first to apply Wigner's R-matrix analysis of threshold phenomena to positron collisions. This work followed the theory of Meyerhof (1963), of which a brief account is now given. Our main interest here is to consider how the opening of the positronium formation channel affects the elastic scattering cross section; the energy dependence of the positronium formation cross section close to the threshold is considered in section 4.2.

The theory predicts that, over a small energy range above E_{Ps}, the positronium formation cross section in the lth partial wave increases from zero according to

$$\sigma^l_{\rm Ps} \propto \kappa^{2l'+1}, \tag{3.97}$$

where the superscripts l and l' refer to the orbital angular momenta of the incident positron and the outgoing positronium relative to the residual ion and κ is the wave number of the positronium. This is related to the wave number of the incident positron, k, through energy conservation:

$$\tfrac{1}{2}k^2 = \tfrac{1}{4}\kappa^2 + E_{\rm Ps}. \tag{3.98}$$

If the orbital angular momenta of the ion and the initial target atom are the same, as they are for atomic hydrogen and helium, then $l = l'$. For many other atoms, including all the other rare gases, $l \neq l'$, but for simplicity we shall consider as an example the case where $l = l'$. From equation (3.97), the rate of increase of the partial positronium formation cross section with respect to k near to the positronium formation threshold is

$$\frac{d}{dk}\sigma^l_{\rm Ps} \propto \kappa^{2l-1}. \tag{3.99}$$

The s-wave contribution to $\sigma_{\rm Ps}$ therefore has an infinite gradient at the threshold, where $\kappa = 0$, whereas the gradients of all higher partial-wave contributions are zero at the threshold.

The anomalous energy dependence of the elastic scattering cross section in the vicinity of $E_{\rm Ps}$ may be understood by reference to the properties of the S-matrix (see Mott and Massey, 1965, Chapter 13), in terms of which the elastic scattering and positronium formation cross sections in the lth partial wave are expressed as

$$\sigma^l_{\rm el} = \frac{\pi}{k^2}(2l+1)|1 - S_{11}|^2, \tag{3.100}$$

$$\sigma^l_{\rm Ps} = \frac{\pi}{k^2}(2l+1)|S_{12}|^2, \tag{3.101}$$

where the positron and the positronium channels are represented by the labels 1 and 2 respectively. From the unitarity of the S-matrix, which expresses the conservation of particles,

$$|S_{11}|^2 + |S_{12}|^2 = 1, \tag{3.102}$$

and therefore

$$|S_{11}|^2 = 1 - \frac{k^2\sigma_{\rm Ps}}{\pi(2l+1)}, \tag{3.103}$$

whence

$$S_{11} = \exp(i2\eta_l)\left[1 - \frac{k^2\sigma_{\rm Ps}}{\pi(2l+1)}\right]^{1/2} \tag{3.104}$$

$$\simeq \exp(i2\eta_l)\left[1 - \frac{k^2\sigma_{\rm Ps}}{2\pi(2l+1)}\right] \tag{3.105}$$

at energies just above E_{Ps}, where σ_{Ps} is still very small. At the threshold, $\sigma_{\mathrm{Ps}} = 0$ and $S_{11} = \exp(i2\eta_l)$, where η_l is the elastic scattering phase shift at that energy. If it is assumed that η_l is a background phase which retains a constant value over a small energy range on either side of E_{Ps} then the energy dependence of S_{11} close to the threshold is determined by the energy dependence of σ_{Ps}.

Although $\sigma_{\mathrm{Ps}} = 0$ for energies below E_{Ps}, the expression for S_{11} given by equation (3.105) can nevertheless be analytically continued below the threshold by replacing the positronium wave number κ by $i|\kappa|$ and so replacing $\sigma_{\mathrm{Ps}}(\kappa)$ by $i(-1)^l\sigma_{\mathrm{Ps}}(|\kappa|)$, see equation (3.97). Consequently, just above the positronium formation threshold, S_{11} is of the form given in equation (3.105), but just below this threshold it is given by

$$S_{11} \simeq \exp(i2\eta_l)\left[1 - \frac{k^2(-1)^l}{2\pi(2l+1)}\sigma_{\mathrm{Ps}}(|\kappa|)\right]. \tag{3.106}$$

The corresponding expressions for the elastic scattering cross section above and below the threshold are therefore

$$\sigma_{\mathrm{el}} \simeq \frac{4\pi}{k^2}(2l+1)\sin^2\eta_l - 2\sin^2\eta_l\,\sigma_{\mathrm{Ps}}(\kappa) \qquad \text{(above)} \tag{3.107}$$

$$\sigma_{\mathrm{el}} \simeq \frac{4\pi}{k^2}(2l+1)\sin^2\eta_l - (-1)^{l+1}\sin 2\eta_l\,\sigma_{\mathrm{Ps}}(|\kappa|) \qquad \text{(below).} \tag{3.108}$$

The elastic scattering cross section must fall as the positron energy is increased above the threshold, and it will either rise or fall as the threshold is approached from below, depending on the value of l and on the phase shift at the threshold. Furthermore, because the s-wave contribution to σ_{Ps}, considered as a function of the positron energy, has an infinite slope at the threshold energy E_{Ps}, equation (3.99), so too does σ_{el}^0, and its energy dependence has the shape of either a cusp or a downward rounded step. All other partial-wave contributions to σ_{el}, however, continue through the threshold with no discontinuity of slope.

A detailed study of the behaviour of the elastic scattering and positronium formation cross sections in the vicinity of the positronium formation threshold in positron–hydrogen scattering was made by Humberston *et al.* (1997); see section 4.2. For s-wave scattering, the phase shift just below the positronium formation threshold is -0.055 rad and therefore, according to equations (3.107) and (3.108), the elastic scattering should display a downward rounded step rather than a cusp. This prediction is confirmed by the accurate calculated values of the s-wave contribution to the elastic scattering cross section; this contribution does indeed reveal a very small downward step in passing through the threshold. The insignificant nature of this feature in positron–hydrogen scattering is a consequence of the very small value of the s-wave component of the positronium formation cross section in comparison to the elastic scattering cross section.

The p-wave component of $\sigma_{\rm el}$ continues to rise as the positron energy is increased above $E_{\rm Ps}$, apparently contradicting the claim made above that it must fall. However, a more rigorous analysis of the energy dependence of the various partial-wave contributions to the elastic scattering cross section, based on R-matrix theory (Breit, 1957) introduces a correction term proportional to κ^2 into equations (3.107) and (3.108) for the components of the elastic scattering cross section with $l > 0$; this correction term is positive above the threshold and negative below. Good agreement is then obtained between the predictions of the threshold theory and the results of accurate p-wave variational calculations (Humberston *et al.*, 1997).

The threshold behaviour described above may only apply over a very narrow energy range about the positronium formation threshold, but attempts have been made by Meyerhof *et al.* (1996) and Humberston *et al.* (1997), using R-matrix theory, to investigate the influence of the opening of the positronium formation channel on the elastic scattering cross section over a somewhat wider energy range. A moderately good fit to the accurate variational cross sections for elastic positron–hydrogen scattering, positronium formation and positronium–proton elastic scattering has thereby been achieved over an energy range of several eV on either side of $E_{\rm Ps}$ using only five parameters for each partial wave.

Recent detailed variational calculations of positronium formation in positron–helium scattering, by Van Reeth and Humberston (1995b, 1997, 1999b), reveal a similar threshold structure to that found in positron–hydrogen scattering. Again, the s-wave contribution to the positronium formation cross section is small compared to the elastic scattering cross section (see section 4.2), and the calculated threshold structure in the s-wave component of $\sigma_{\rm el}$ is therefore also small, making it difficult to observe in the experimental measurements of the total elastic scattering cross section. Consequently, the elastic scattering cross section is expected to continue rather smoothly through the threshold without displaying any pronounced threshold feature. This theoretical prediction is consistent with the aforementioned measurements of Coleman *et al.* (1992) and an R-matrix analysis by Moxom *et al.* (1994).

Moxom *et al.* (1994) applied their analysis to all the rare gases, using the polarized-orbital phase shifts of McEachran and his collaborators (McEachran *et al.*, 1977; McEachran, Ryman and Stauffer, 1978, 1979; McEachran, Stauffer and Campbell, 1980) and the experimental measurements of $\sigma_{\rm Ps}$ made by Moxom, Laricchia and Charlton (1993) and Moxom *et al.* (1994). From fits of this data to the theoretical energy dependence given in equation (3.97) (see Figure 4.18), these authors concluded that the positronium formation process was, in each case, dominated by the contribution from one partial wave, this being $l = l' = 1$ for helium and $l = 1$, $l' = 0$ for the other gases, although small admixtures from the partial wave with $l = 0$, $l' = 1$ were also believed to be present for krypton and argon. Thus, in all cases the dominant contribution to the positronium formation cross section seemed to be provided by an incoming positron wave with $l = 1$.

Using this theoretical and experimental information, in conjunction with equations (3.107) and (3.108), Moxom *et al.* (1994) predicted the energy dependence of the elastic and total scattering cross sections on either side of the positronium formation threshold. Comparisons of the predicted and measured values are shown in Figure 3.13. Although the quantitative agreement is not very good, there are important qualitative similarities which show that for helium there is no discernible cusp in the elastic scattering cross section but that a progressively more prominent cusp develops as the atomic number of the target atom increases.

The conclusion of Moxom *et al.* (1994) that the positronium formation process in helium is dominated by the p-wave contribution over an energy range of a few electron volts above the threshold is at variance with accurate theoretical results obtained by Van Reeth and Humberston (1997) and described in detail in Chapter 4. The $l = 0$ contribution to $\sigma_{\rm Ps}$ is indeed small, but the $l = 1$ contribution, although rapidly exceeding the $l = 0$ contribution, is itself quite rapidly overtaken by the $l = 2$ contribution as the positron energy is increased. However, these various partial-wave contributions to $\sigma_{\rm Ps}$ are found to add in such a way as to give an overall energy dependence of $\sigma_{\rm Ps}$ very similar to that which would be expected on the assumption that the p-wave contribution to $\sigma_{\rm Ps}$ is dominant (Van Reeth and Humberston, 1997).

3.4 Angle-resolved elastic scattering

There has been growing experimental activity in the area of angle-resolved elastic scattering since the mid-1980s. It is well known from electron collision physics that the angle-resolved elastic scattering cross section, $d\sigma_{\rm el}/d\Omega$, contains more detailed information than the integrated cross section, and therefore its accurate determination provides a more stringent test of collision theory. As emphasized by McDaniel (1989), the first observations of this quantity for electron–atom scattering provided convincing evidence in favour of the validity of a quantum mechanical description of scattering phenomena. More recently, comparisons between phase shift analyses of experimental differential cross sections and theoretical predictions of these phase shifts have led to a comprehensive understanding of the behaviour of electrons undergoing elastic collisions

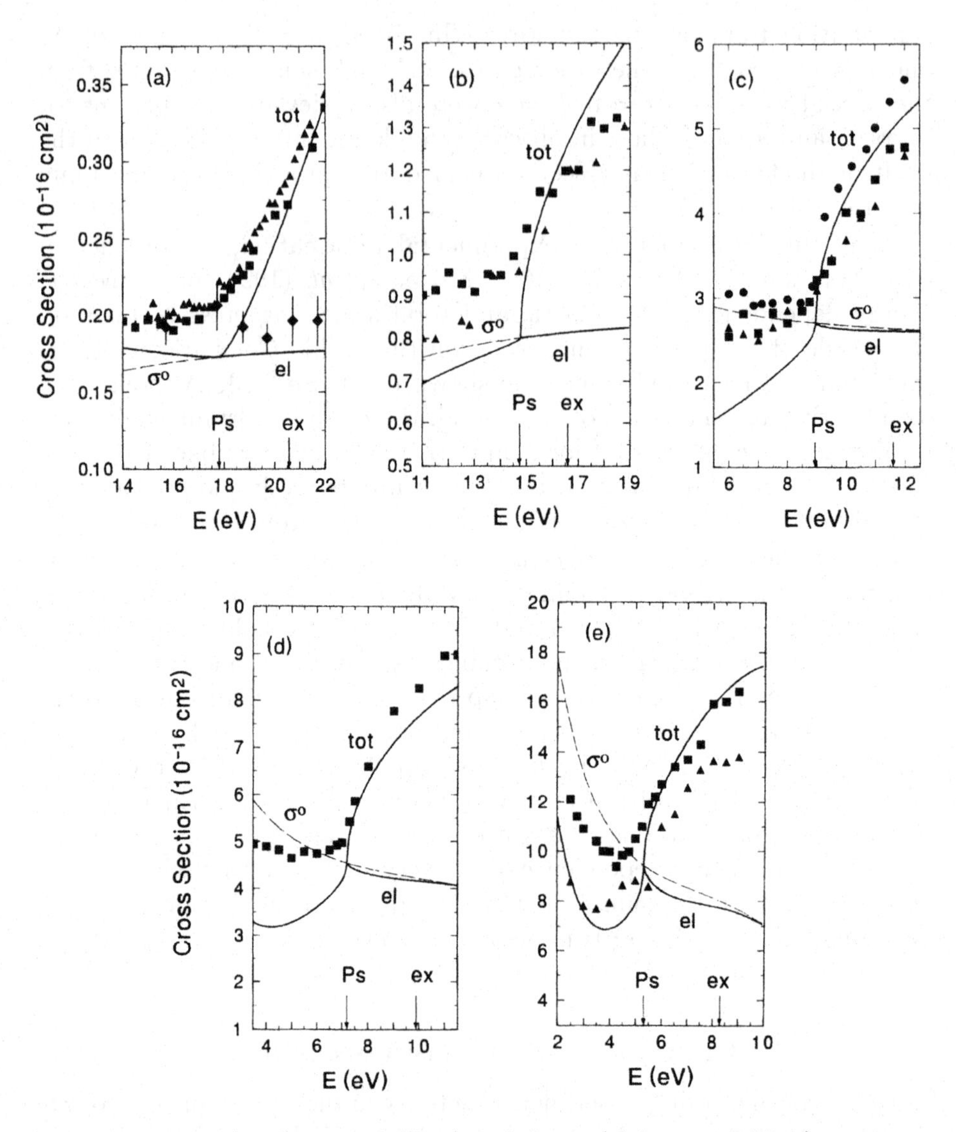

Fig. 3.13. Total (tot, upper solid line) and elastic (el, lower solid line) cross sections for positron–noble gas scattering near the positronium formation threshold from the R-matrix analysis of Moxom *et al.* (1994). Graphs (a)–(e) correspond to helium through to xenon. The data points shown are total cross section measurements from the literature (see Chapter 2 and Moxom *et al.*, 1994, for details) except for the solid diamonds for helium, which are the $\sigma_T - \sigma_{Ps}$ results of Coleman *et al.* (1992) (see Figure 3.12). The curves for σ^0, which is the elastic scattering cross section calculated without the inclusion of positronium formation, are from the work of McEachran and collaborators. Reprinted from *Physical Review* **A50**, Moxom *et al.*, Threshold effects in positron scattering on noble gases, 3129–3133, copyright 1994 by the American Physical Society.

with atoms (see e.g. Newell, Brewer and Smith, 1981; Brewer *et al.*, 1981, and references therein).

For the elastic scattering of a beam of unpolarized projectiles by an unpolarized target, the cross section has axial symmetry about the incident beam direction and therefore no dependence upon the azimuthal angle ϕ, so that the differential elastic cross section is related to its integral counterpart by

$$\sigma_{\rm el} = \int \frac{d\sigma_{\rm el}}{d\Omega} d\Omega = 2\pi \int_0^\pi \frac{d\sigma_{\rm el}}{d\theta} \sin\theta \, d\theta. \qquad (3.109)$$

All the physical information is then contained in $d\sigma_{\rm el}/d\theta$, which is determined by measuring the probability that a particle of fixed kinetic energy is elastically scattered through an angle θ.

Over the years there have been several attempts to measure $d\sigma_{\rm el}/d\theta$ for positron–atom scattering. This work culminated in the experiments of Hyder *et al.* (1986), Floeder *et al.* (1988) and Smith *et al.* (1990), who reported the first measurements obtained using the crossed positron beam–gas beam geometry. Such a configuration is similar to that frequently encountered in electron scattering. These workers investigated positron–noble gas scattering, at first using neon and argon targets rather than more theoretically tractable targets, because their larger elastic scattering cross sections made the measurements practicable using the available fluxes of low energy positrons. Extensions to other targets are now, however, becoming possible.

The very first attempts to derive information concerning the angular distribution of positron–atom scattering were made in the transmission-type experiments of Jaduszliwer and Paul (1973, 1974). These authors found that as the strength of the axial magnetic guiding field in their scattering chamber was increased the number of detected scattered positrons also increased. The extent of this increase depended on the form of the differential cross section, and it proved possible, using a Monte Carlo trajectory analysis, to extract s-, p- and d-wave phase shifts at energies below $E_{\rm Ps}$ for positron scattering by helium (Jaduszliwer and Paul, 1973) or by neon and argon (Jaduszliwer and Paul, 1974). These phase shifts were in reasonable agreement with the results of existing theories, although the p-wave phase shifts for helium were subsequently found to be much larger than the accurate theoretical values of Humberston and Campeanu (1980). Although these data have been superseded by the results of later work, the principle underlying this experiment is sound and the work represents an ingenious attempt to extract angular information using a feeble positron beam.

Similar ingenuity was also displayed by Coleman and coworkers (Coleman and McNutt, 1979; Coleman *et al.*, 1980b). In these experiments a beam of low energy positrons was timed over a flight path 250 mm long in an axial magnetic guiding field of 14 mT, which constrained the positrons scattered through almost 90° to follow a helical trajectory of 1 mm radius at the highest energy investigated (8.7 eV). The scattering was confined to a gas cell 10 mm long located close to the positron moderator, and, owing to their delayed time of arrival at the detector, scattered particles could be distinguished from those which had not been scattered using a time-of-flight (TOF) technique. Coleman and McNutt (1979) showed how to extract absolute values for $d\sigma_{el}/d\theta$ from the observed differences between the TOF spectra obtained with and without the gas present. A full discussion of the possible systematic effects in this experiment was given by Coleman *et al.* (1980b).

Results for positron–argon collisions are shown in Figure 3.14 at various mean positron impact energies and over the angular range 20°–60°. Also shown in Figure 3.14 are the polarized-orbital results of McEachran *et al.* (1979) and those of Schrader (1979), who used a semi-empirical polarization-potential method. There is good agreement between the shapes of the theoretical and experimental data, although the results of McEachran *et al.* (1979) have been scaled down by an energy-dependent factor which reflects the deviations of their calculated total cross sections from the experimental values. The validity of this procedure is open to question.

Coleman *et al.* (1980b) also performed electron–argon scattering experiments using the same apparatus and method by utilizing the secondary electrons emitted from their moderator after high energy β^+ bombardment. They reported results over the same 40° angular range, at energies up to 20 eV, which were in good agreement with those of Williams (1979). This test established the reliability of their procedure and measurements.

Following this early work, groups in Detroit and Bielefeld initiated measurements of $d\sigma_{el}/d\theta$ using a crossed-beam geometry. The two groups used rather similar apparatus, and in Figure 3.15 we show the layout of Hyder *et al.* (1986), which, with some small modifications, has been used for many subsequent measurements.

The atomic beam was formed by a multichannel capillary array, placed perpendicular to the positron beam, with a 2.5 mm^2 effusing area and a length-to-diameter ratio of 25 : 1. The head pressure behind the array was kept at 9 torr ($\approx 10^3$ Pa) in the initial measurements. An annealed tungsten moderator was used to provide a beam of more than 10^5 positrons per second at 200 eV. A much more intense beam of electrons could also be obtained by reversing the electrostatic potentials on the various elements which made up the transport system. Channel electron multipliers (CEM1 and CEM2 respectively) were used to monitor the incident and scattered beams. In later versions of the apparatus, a third

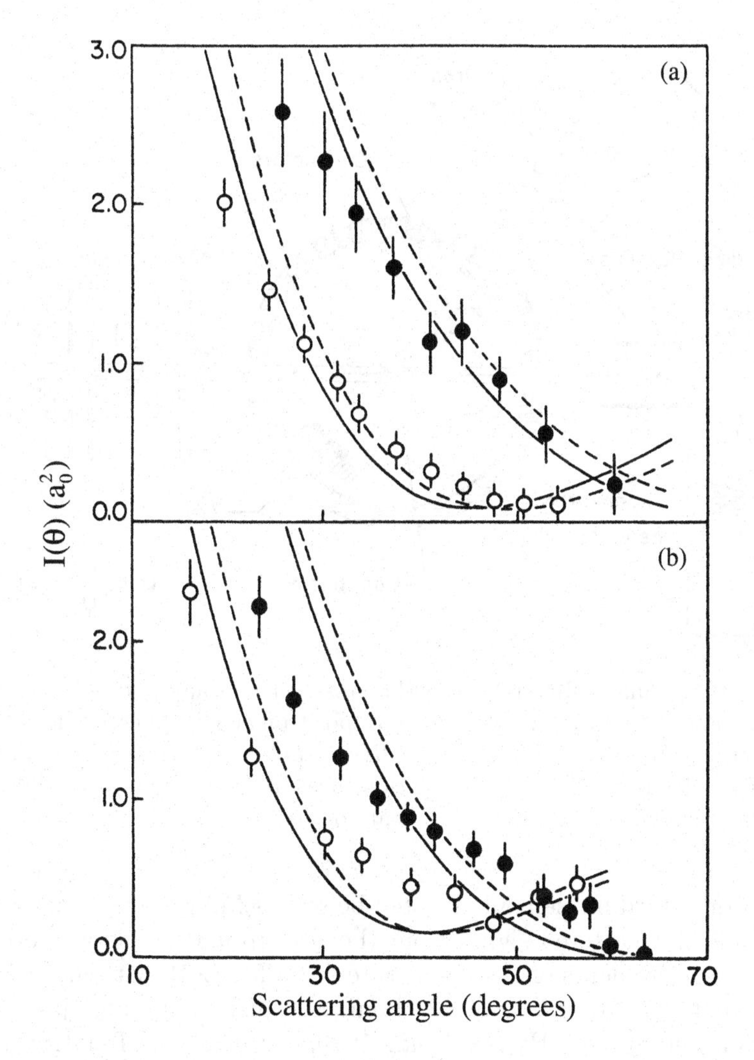

Fig. 3.14. Differential cross sections for positron–argon elastic scattering at the following energies: (a) •, 2.2 eV; (b) •, 3.4 eV; (a) ∘, 6.7 eV; (b) ∘, 8.7 eV. (The corresponding k-values are respectively 0.4, 0.5, 0.7 and 0.8 a.u.) The solid curves give the theory of Schrader (1979) whilst the broken curves are the scaled results of McEachran, Ryman and Stauffer (1979).

CEM was added to monitor the scattered beam at a different angle. Note that CEM1 was offset from the direct line of sight with the ^{22}Na source, in order to reduce background counts from β^+ particles and γ-rays. The angular acceptance of CEM2, which was defined by the geometry and the collimators, was estimated to be $\pm 8^\circ$. In order to reduce the noise counts in this detector, a non-reflective surface composed of a stack of knife-edge plates was located directly opposite it in the chamber.

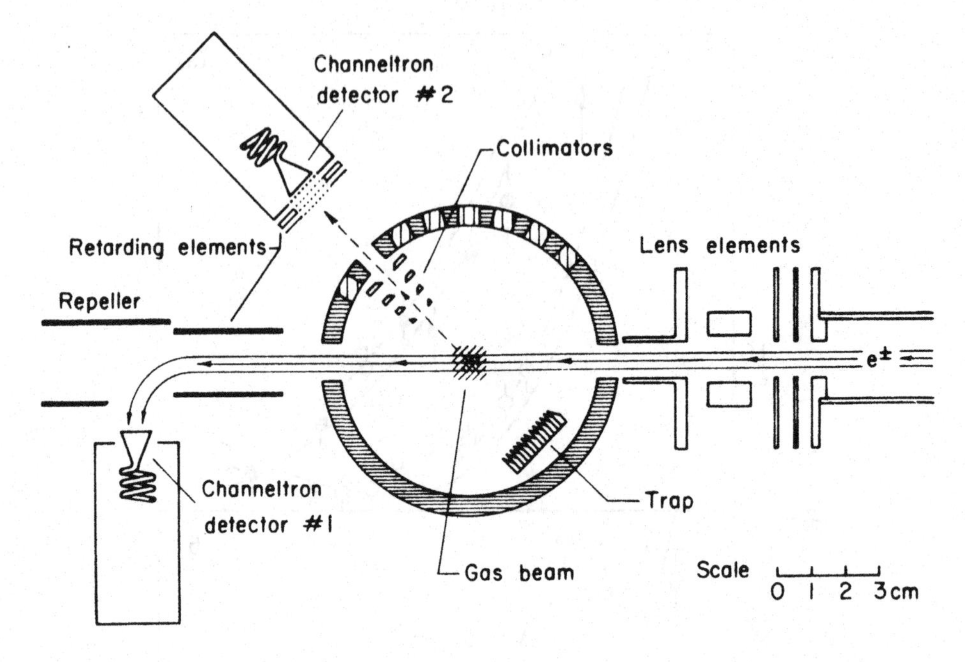

Fig. 3.15. Schematic illustration of the crossed-beam apparatus developed by Hyder *et al.* (1986) for the measurement of positron elastic differential scattering cross sections. Reprinted from *Physical Review Letters* **57**, Hyder *et al.*, Positron differential elastic scattering cross section measurements for argon, 2252–2255, copyright 1986 by the American Physical Society.

The grid retarding elements located before CEM2 were used to separate the elastically scattered signal from the background. This was done by making measurements of the count rate of CEM2 both with and without gas present, the retarder voltage being set just above and just below the beam energy (see Hyder *et al.*, 1986, for details). Relative values of $d\sigma_{\rm el}/d\theta$ at different scattering angles were obtained by dividing the scattered signal by that for the incident beam measured at CEM1. This ratio was found to be in the range 10^{-5}–10^{-7}, the signal-to-noise ratio for positrons in CEM2 being typically 10^{-1}–10^{-2} for angles greater than 45°. This entailed the use of long run times and computer-controlled automated data taking. The demanding nature of such experiments also explains why a large acceptance angle was tolerated for CEM2 and a large drive pressure was used for the capillary-array gas source.

Some of the data of Hyder *et al.* (1986) for positron and electron impact on argon gas at 100 eV and 200 eV are shown in Figures 3.16(a)–(d). The data were independently normalized, at 90° for both projectiles, to the existing experimental data of Srivastava *et al.* (1981) and DuBois and Rudd (1975) for electrons and to the theoretical results of McEachran

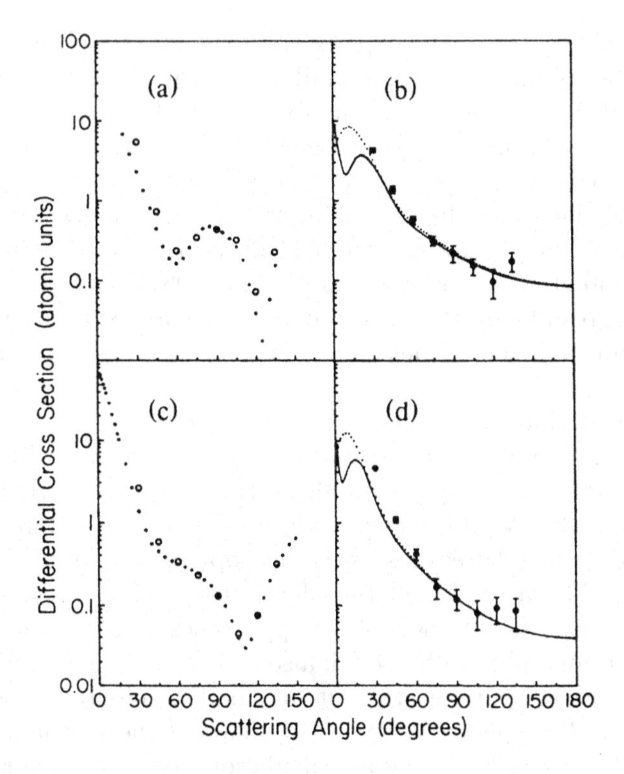

Fig. 3.16. Elastic differential cross sections for positrons and electrons scattering from argon. (a) Electrons at 100 eV: ∘, Hyder *et al.* (1986); •, Srivastava *et al.* (1981). (b) Positrons at 100 eV: Hyder *et al.* (1986); ——, theory of McEachran and Stauffer (1986); ·····, theory of Nahar and Wadehra (1987). (c) Electrons at 200 eV: ∘, Hyder *et al.* (1986); •, DuBois and Rudd (1975). (d) Positrons at 200 eV, same key as (b). Reprinted from *Physical Review Letters* **57**, Hyder *et al.*, Positron differential elastic scattering cross section measurements for argon, 2252–2255, copyright 1986 by the American Physical Society.

and Stauffer (1986) and Nahar and Wadehra (1987) (which are in good agreement at this angle) for positrons. As stated by Hyder *et al.* (1986), their electron measurements compare favourably with those of other workers except at small angles; the latter effect may be attributed to the geometry of their gas beam. A similar situation arises when their positron data are compared with theory. Joachain and Potvliege (1987) also performed calculations for this system; their results are somewhat lower than the other theoretical values and exhibit markedly different behaviour at small angles, lacking the structures found by McEachran and Stauffer (1986) and Nahar and Wadehra (1987). It should be noted that if the data of Hyder *et al.* (1986) are normalized to the theoretical results of Joachain and Potvliege (1987) at 90°, for both

100 eV and 200 eV, then the experimental results still exceed the calculated values at small angles in a similar way to that seen in the figure. Joachain and Potvliege (1987), in extending the theory of Joachain *et al.* (1977) to lower energies, discussed the differing trends of the theories at small angles ($< 30^\circ$ at 100 eV). They showed that similar behaviour to that found by McEachran and Stauffer (1986) and Nahar and Wadehra (1987) can also be obtained from their calculations if the absorption potential is neglected. Joachain (1987) argued that the full optical model results are the most reliable since, as required by unitarity, proper account is taken of inelastic channels by means of the absorption potential.

Measurements have also been made at lower energies, where the predicted structure in $d\sigma_{\rm el}/d\theta$ moves to larger, experimentally amenable angles. Theoretical and experimental results for positron–argon collisions at 8.7 eV and 30 eV are shown in Figure 3.17. At small angles the experimental results agree best with the optical potential calculations of Bartschat, McEachran and Stauffer (1988), which take account of absorption, whilst other theoretical approaches (Nahar and Wadehra, 1987; McEachran and Stauffer, 1986; also Nakanishi and Schrader, 1986b) apparently produce spurious structure at small angles. It should be noted that there is a discrepancy between the shape of the normalized data of Smith *et al.* (1990) at 30 eV and all calculations, even when normalization at different angles is attempted; see Figure 3.17.

At lower energies, below the threshold for positronium formation in argon (8.9 eV), the results of both polarized-orbital calculations (McEachran and Stauffer, 1986) and model-potential calculations (Nakanishi and Schrader, 1986a) are in better agreement with the shape of the experimental data. The results shown in Figure 3.17 are at 8.7 eV, an energy also investigated earlier by Coleman and McNutt (1979), as shown in Figure 3.14. It should be noted that the data of Floeder *et al.* (1988) (which were actually taken at a beam energy of 8.5 eV) were normalized to those of Coleman and McNutt (1979) in a manner which produced the best overall comparison of the two sets of values. The shape of these data are in good agreement with one another and with the theories of McEachran *et al.* (1979) and Montgomery and LaBahn (1970) and also with that of Nakanishi and Schrader (1986a) (not shown). However, the data of Smith *et al.* (1990) display a significantly shallower minimum than all the other results in the vicinity of a scattering angle of 40°. It is unlikely that this discrepancy can be explained by the differing angular resolutions of the experiments.

The above discussion shows that some useful information has been obtained on the need to include absorption effects into the theory. Indeed, as predicted by Joachain and Potvliege (1987), it appears that any

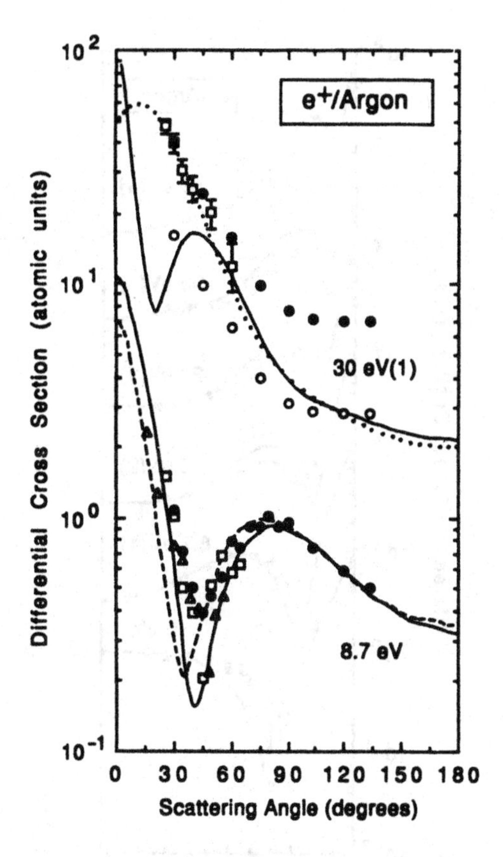

Fig. 3.17. Positron–argon elastic differential scattering cross sections. Experiment: •, ∘, Smith *et al.* (1990); △, Coleman and McNutt (1979); □, Floeder *et al.* (1988). Theory: ——, McEachran, Ryman and Stauffer (1979) and McEachran and Stauffer (1986); · · · · ·, Bartschat, McEachran and Stauffer (1988); – – –, Montgomery and LaBahn (1970). The number in parentheses following 30 eV indicates the power of ten by which the cross sections have been multiplied. Normalization to theory was done at 30° and 120° by Smith *et al.* (1990) and at 30° and 60° by Floeder *et al.* (1988). Reprinted from *Physical Review Letters* **64**, Smith *et al.*, Evidence for absorption effects in positron elastic scattering by argon, 1227–1230, copyright 1990 by the American Physical Society.

structure which could appear in the calculated positron differential elastic scattering cross sections, at small angles and in the intermediate energy region, is washed out when these effects are properly taken into account. At lower energies the experimental data begin to show features similar to those present in the polarized-orbital (without absorption) results of McEachran *et al.* (1979), and theory and experiment are in reasonable agreement below the positronium formation threshold. The trend of the

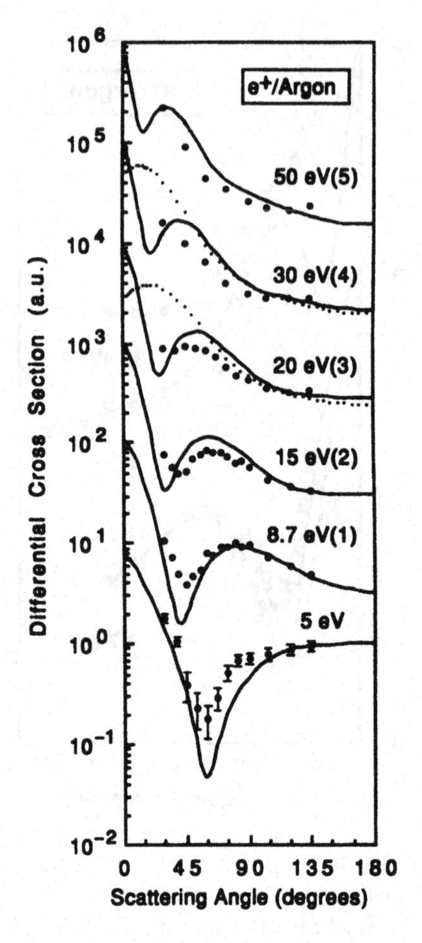

Fig. 3.18. Experimental elastic positron–argon differential scattering cross sections in the range 5–50 eV from Smith *et al.* (1990). The theoretical data are from McEachran and Stauffer (1986) and McEachran *et al.* (1979) (——) and from Bartschat, McEachran and Stauffer (1988) (· · · · ·). The numbers in parentheses following the energy values indicate the power of ten by which the cross sections have been multiplied for clarity of display. Reprinted from *Physical Review Letters* **64**, Smith *et al.*, Evidence for absorption effects in positron elastic scattering by argon, 1227–1230, copyright 1990 by the American Physical Society.

data as the positron energy is lowered is nicely illustrated in Figure 3.18, which shows a compendium of data by Smith *et al.* (1990) for positrons at 5–50 eV scattering from argon. The data are all normalized to theory at 120°. As summarized by Kauppila *et al.* (1996), similar conclusions can be drawn from studies of the other inert gases at energies below their respective positronium formation thresholds.

All the values of $d\sigma_{el}/d\theta$ described here so far have been relative, the absolute scale usually having been obtained by normalization to theory. Efforts have been made by the Detroit group to make direct absolute differential cross section measurements for positrons, where the only comparison was between the positron data and their own normalized electron data. Absolute values of $d\sigma_{el}/d\theta$ for positrons were reported by Dou *et al.* (1992a, b), but Kauppila *et al.* (1996) were subsequently unable to reproduce these data. Clearly, the unambiguous determination of absolute positron differential cross sections remains a task for the future.

One innovation introduced by Dou *et al.* (1992a, b) was to measure $d\sigma_{el}/d\theta$ not at a fixed energy and varying angle but at a fixed angle and varying energy. In so doing, they discovered apparent structure, most notably in the energy range 55–60 eV, at various angles for positron scattering in argon gas. In particular, they found that at 60° the differential cross section appeared to fall abruptly by a factor of approximately two in this energy region. They tentatively raised the possibility that this effect was linked to channel coupling but, as mentioned above, subsequent work reported by Kauppila *et al.* (1996) failed to reproduce the observed structure as also, independently, did that of Finch *et al.* (1996a, b) in a study of single differential cross sections for elastic scattering, positronium formation and ionization at an angle of 60° in argon gas.

4
Positronium formation

4.1 Introduction

Positronium formation involves the capture by an incident positron of one of the target electrons, to form the bound state Ps. This is one of the simplest examples of a rearrangement collision and accordingly it has attracted considerable attention, both experimental and theoretical.

Although, as described in section 1.5, positronium can be formed when positrons interact with many different media, in this chapter we are mainly concerned with the reaction

$$e^+ + X \to \mathrm{Ps} + X^+, \tag{4.1}$$

where the target atom or molecule, X, is in its ground state. Reaction (4.1) can be generalized to

$$e^+ + X \to \mathrm{Ps}(n_{\mathrm{Ps}}, l_{\mathrm{Ps}}) + X^+(n_{X^+}, l_{X^+}), \tag{4.2}$$

where the quantities in parentheses are the relevant principal and orbital angular momentum quantum numbers of the positronium and of the residual ionic state. In a few instances we shall present data in which the n_{Ps} and n_{X^+} values are not those of the appropriate ground states. The total cross section for positronium formation, which includes all possible states of the positronium and of the residual ion, is denoted by σ_{Ps}.

Whilst reactions (4.1) and (4.2) are unique to positrons, they do have close counterparts in the charge-transfer reactions of heavy positively charged projectiles. Most notable are those involving protons, p^+, namely,

$$p^+ + X \to \mathrm{H} + X^+, \tag{4.3}$$

and it is instructive in some cases to compare results for the two projectiles. This is attempted in section 4.6.

Positronium formation in gases has been the subject of investigation since the pioneering works of Ore (1949) and Deutsch (1951) (see section 1.5), who studied the slowing down of β^+ particles, and the consequential positronium formation, in dense gases. We describe in section 4.8 the current situation in this field.

After discussing the theory of positronium formation, with applications to several relatively simple systems, we shall describe various techniques used to measure positronium formation cross sections and present the results so obtained, comparing them with theoretical predictions wherever applicable. The chapter also includes a discussion of the angular dependence of the positronium formation cross section. As well as being of intrinsic interest as a test of theory, the differential formation cross section, $d\sigma_{\rm Ps}/d\Omega$, is also relevant for the production of energy-tunable beams of positronium atoms. This topic is treated more fully in section 7.6.

4.2 Theory

Positronium formed in a positron–atom collision can be in a state with principal quantum number up to $n_{\rm Ps}$ provided that the kinetic energy of the incident positron, E, exceeds the difference between the ionization energy of the target, $E_{\rm i}$, and the binding energy of the positronium in that state, i.e.

$$E \geq E_{\rm i} - 6.8/n_{\rm Ps}^2, \tag{4.4}$$

where energies are measured in eV. This inequality follows from the definition of the threshold $E_{\rm Ps}$ given in equation (1.13). All states of positronium with principal quantum numbers $< n_{\rm Ps}$ can also be formed in such a collision. If the ionization energy of the target atom is less than 6.8 eV, as it is for all the alkali metal atoms, then positronium formation into the ground state is possible even at zero incident positron energy. More commonly, however, the positronium formation threshold is at some positive energy, for example, it is at 6.8 eV for atomic hydrogen and 17.8 eV for helium, the highest value for any atom.

If the first inelastic threshold of the target is at a higher energy than that of the ground state positronium formation threshold, there is an energy interval between these two thresholds, in which the only two scattering processes are elastic scattering and ground state positronium formation. It is in this energy interval, the so-called Ore gap, that the most detailed theoretical investigations of positronium formation have been made, as will now be described.

1 Positron–hydrogen scattering

Because of its relative simplicity, particular attention has been devoted to positronium formation in positron collisions with atomic hydrogen. Within the Ore gap the two open channels (other than direct annihilation) are

$$\begin{aligned} e^+ + \mathrm{H} &\rightarrow e^+ + \mathrm{H} \\ &\rightarrow \mathrm{Ps} + \mathrm{p}. \end{aligned} \tag{4.5}$$

The total wave function therefore has two components:

$$\Psi = \begin{pmatrix} \Psi_1 \\ \Psi_2 \end{pmatrix}, \tag{4.6}$$

where Ψ_1 represents positron–hydrogen elastic scattering plus positronium formation, and Ψ_2 represents positronium–proton elastic scattering plus hydrogen formation. Using real wave functions, the asymptotic forms of the two components for the lth partial wave are, using the nomenclature of Figure 3.1,

$$\begin{aligned} \Psi_1 &\underset{r_1\to\infty}{\sim} Y_{l,0}(\theta_1,\phi_1)\sqrt{k}\,[j_l(kr_1) - K_{11}n_l(kr_1)]\,\Phi_{\mathrm{H}}(r_2) \\ &\underset{\rho\to\infty}{\sim} -Y_{l,0}(\theta_\rho,\phi_\rho)\sqrt{2\kappa}K_{21}n_l(\kappa\rho)\Phi_{\mathrm{Ps}}(r_{12}) \end{aligned} \tag{4.7}$$

$$\begin{aligned} \Psi_2 &\underset{\rho\to\infty}{\sim} Y_{l,0}(\theta_\rho,\phi_\rho)\sqrt{2\kappa}\,[j_l(\kappa\rho) - K_{22}n_l(\kappa\rho)]\,\Phi_{\mathrm{Ps}}(r_{12}) \\ &\underset{r_1\to\infty}{\sim} -Y_{l,0}(\theta_1,\phi_1)\sqrt{k}K_{12}n_l(kr_1)\Phi_{\mathrm{H}}(r_2), \end{aligned} \tag{4.8}$$

where K_{11}, K_{12}, K_{21} and K_{22} are the elements of the K-matrix, $\phi_{\mathrm{H}}(r_2)$ and $\phi_{\mathrm{Ps}}(r_{12})$ are the ground state wave functions of hydrogen and positronium respectively and $\boldsymbol{\rho} = (\boldsymbol{r}_1 + \boldsymbol{r}_2)/2$ is the position of the centre of mass of the positronium relative to the proton. Conservation of the total energy of the entire system of positron plus target atom, E_{T}, gives the relationship between the wave numbers of the positron and positronium, k and κ respectively, as

$$E_{\mathrm{T}} = \tfrac{1}{2}k^2 - \tfrac{1}{2} = \tfrac{1}{4}\kappa^2 - \tfrac{1}{4}. \tag{4.9}$$

The relationships between the K-, S- and T-matrices (Mott and Massey, 1965, Chapter 13) are

$$\mathbf{S} = (\mathbf{1} + i\mathbf{K})/(\mathbf{1} - i\mathbf{K}), \tag{4.10}$$

$$\mathbf{T} = \mathbf{1} - \mathbf{S} = -2i\mathbf{K}/(\mathbf{1} - i\mathbf{K}). \tag{4.11}$$

In terms of the K-matrix, the cross section for scattering between an initial channel i and a final channel f is expressed as

$$\sigma_{if} = \frac{4\pi(2l+1)}{k_i^2}\left|\frac{\mathbf{K}}{\mathbf{1} - i\mathbf{K}}\right|^2_{if}, \tag{4.12}$$

where the subscripts i and f, with values 1 or 2, refer to the positron–hydrogen and the positronium–proton channels respectively; also, $k_1 = k$ and $k_2 = \kappa$. Thus, σ_{11} is the positron–hydrogen elastic scattering cross section $\sigma_{\rm el}$ and σ_{12} is the positronium formation cross section $\sigma_{\rm Ps}$.

Among the most detailed and accurate investigations of positronium formation in the Ore gap are those of Humberston (1982, 1984) and Brown and Humberston (1984, 1985), who used an extension of the Kohn variational method described previously, see section 3.2, to two open channels. The single-channel Kohn functional, equation (3.37), is now replaced by the following stationary functional for the K-matrix:

$$\begin{vmatrix} K^{\rm v}_{11} & K^{\rm v}_{12} \\ K^{\rm v}_{21} & K^{\rm v}_{22} \end{vmatrix} = \begin{vmatrix} K^{\rm t}_{11} & K^{\rm t}_{12} \\ K^{\rm t}_{21} & K^{\rm t}_{22} \end{vmatrix} - \begin{vmatrix} \langle\Psi_1|L|\Psi_1\rangle & \langle\Psi_1|L|\Psi_2\rangle \\ \langle\Psi_2|L|\Psi_1\rangle & \langle\Psi_2|L|\Psi_2\rangle \end{vmatrix}, \tag{4.13}$$

where Ψ_1 and Ψ_2 are trial functions representing the two components of the total wave function, as in equation (4.6). These trial functions must satisfy the asymptotic forms given in equations (4.7) and (4.8), and they should also provide adequate representations of the short-range correlations between all the particles in the system. Suitable trial functions for s-wave scattering, similar in form to those used in the elaborate variational calculations of the s-wave positron–hydrogen elastic scattering phase shift, equation (3.42), are as follows:

$$\begin{aligned}\Psi_1 = {} & \sqrt{\frac{k}{4\pi}}\,\{j_0(kr_1) - K^{\rm t}_{11}n_0(kr_1)[1-\exp(-\lambda r_1)]\}\,\Phi_{\rm H}(r_2) \\ & - \sqrt{\frac{2\kappa}{4\pi}}K^{\rm t}_{21}n_0(\kappa\rho)\left[1-\exp(-\mu\rho)\left(1+\frac{\mu\rho}{2}\right)\right]\Phi_{\rm Ps}(r_{12}) \\ & + \sum_{i=1}^{N} c_i \exp[-(\alpha r_1 + \beta r_2 + \gamma r_{12})]r_1^{k_i}r_2^{l_i}r_{12}^{m_i}\end{aligned} \tag{4.14}$$

$$\begin{aligned}\Psi_2 = {} & \sqrt{\frac{2\kappa}{4\pi}}\left\{j_0(\kappa\rho) - K^{\rm t}_{22}n_0(\kappa\rho)\left[1-\exp(-\mu\rho)\left(1+\frac{\mu\rho}{2}\right)\right]\right\}\Phi_{\rm Ps}(r_{12}) \\ & - \sqrt{\frac{k}{4\pi}}\,kK^{\rm t}_{12}n_0(kr_1)[1-\exp(-\lambda r_1)]\Phi_{\rm H}(r_2) \\ & + \sum_{j=1}^{N} d_i \exp[-(\alpha r_1 + \beta r_2 + \gamma r_{12})]r_1^{k_j}r_2^{l_j}r_{12}^{m_j}.\end{aligned} \tag{4.15}$$

These trial functions may be written in a more concise form as

$$\Psi_1 = S_1 + K^{\mathrm{t}}_{11}C_1 + K^{\mathrm{t}}_{21}C_2 + \sum_{i=1}^{N} c_i\phi_i \tag{4.16}$$

$$\Psi_2 = S_2 + K^{\mathrm{t}}_{22}C_2 + K^{\mathrm{t}}_{12}C_1 + \sum_{i=1}^{N} d_j\phi_j, \tag{4.17}$$

where

$$S_1 = \sqrt{\frac{k}{4\pi}}\, kj_0(kr_1)\Phi_{\mathrm{H}}(r_2) \tag{4.18}$$

$$C_1 = \sqrt{\frac{k}{4\pi}}\, n_0(kr_1)[1 - \exp(\lambda r_1)]\Phi_{\mathrm{H}}(r_2) \tag{4.19}$$

$$S_2 = \sqrt{\frac{2\kappa}{4\pi}}\, j_0(\kappa\rho)\Phi_{\mathrm{Ps}}(r_{12}) \tag{4.20}$$

$$C_2 = \sqrt{\frac{2\kappa}{4\pi}}\, n_0(\kappa\rho)\left[1 - \exp(-\mu\rho)\left(1 + \frac{\mu\rho}{2}\right)\right]\Phi_{\mathrm{Ps}}(r_{12}). \tag{4.21}$$

Now, with two types of open-channel function, two forms of the kinetic energy operator T are required:

$$T = -\tfrac{1}{2}\left(\nabla^2_{r_1} + \nabla^2_{r_2}\right) \tag{4.22}$$

and

$$T = -\left(\tfrac{1}{4}\nabla^2_{\rho} + \nabla^2_{r_{12}}\right), \tag{4.23}$$

the latter form being appropriate when operating on the positronium open-channel functions S_2 and C_2. If T operates on a function with spherical symmetry, representing a system with zero total orbital angular momentum, it can be expressed in terms of ρ, r_{12} and α (the angle between ρ and r_{12}) as

$$T = -\left[\frac{1}{4\rho^2}\frac{\partial}{\partial\rho}\left(\rho^2\frac{\partial}{\partial\rho}\right) + \frac{1}{r_{12}^2}\frac{\partial}{\partial r_{12}}\left(r_{12}^2\frac{\partial}{\partial r_{12}}\right) + \left(\frac{1}{4\rho^2} + \frac{1}{r_{12}^2}\right)\frac{1}{\sin\alpha}\frac{\partial}{\partial\alpha}\left(\sin\alpha\frac{\partial}{\partial\alpha}\right)\right]. \tag{4.24}$$

Its form in terms of the interparticle coordinates r_1, r_2 and r_{12} has been given in equation (3.46).

The variational K-matrix is symmetric, as is required by unitarity, but the trial K-matrix is not: $K^{\mathrm{v}}_{12} = K^{\mathrm{v}}_{21}$ but $K^{\mathrm{t}}_{12} \neq K^{\mathrm{t}}_{21}$. This identity may be readily proved by applying Green's theorem to

$$K^{\mathrm{v}}_{12} - K^{\mathrm{v}}_{21} = K^{\mathrm{t}}_{12} - K^{\mathrm{t}}_{21} - [\langle\Psi_1|L|\Psi_2\rangle - \langle\Psi_2|L|\Psi_1\rangle] \tag{4.25}$$

and noting that $S_2 \to 0$ and $C_2 \to 0$ sufficiently rapidly on the surface $r_1 \to \infty$, and that $S_1 \to 0$ and $C_1 \to 0$ sufficiently rapidly on the surface $\rho \to \infty$.

The stationary property of the Kohn-matrix functional (3.37) requires that the partial derivatives of all the elements of the variational K-matrix, K_{11}^{v}, K_{12}^{v} $(= K_{21}^{\mathrm{v}})$ and K_{22}^{v}, with respect to the linear parameters in Ψ_1 and Ψ_2, i.e. K_{11}^{t}, K_{12}^{t}, K_{21}^{t}, K_{22}^{t}, c_j $(j = 1, \ldots, N)$ and d_j $(j = 1, \ldots, N)$, be zero.

As with single-channel scattering (see section 3.2), the resulting set of linear simultaneous equations can be written as

$$\mathbf{AX} = -\mathbf{B}, \tag{4.26}$$

but now

$$\mathbf{A} = \begin{bmatrix} \langle C_1|L|C_1\rangle & \langle C_1|L|C_2\rangle & \ldots & \langle C_1|L|\phi_j\rangle & \ldots \\ \langle C_2|L|C_1\rangle & \langle C_2|L|C_2\rangle & \ldots & \langle C_2|L|\phi_j\rangle & \ldots \\ \vdots & \vdots & & \vdots & \\ \langle \phi_i|L|C_1\rangle & \langle \phi_i|L|C_2\rangle & \ldots & \langle \phi_i|L|\phi_j\rangle & \ldots \\ \vdots & \vdots & & \vdots & \end{bmatrix},$$

$$\mathbf{B} = \begin{bmatrix} \langle C_1|L|S_1\rangle & \langle C_1|L|S_2\rangle \\ \langle C_2|L|S_1\rangle & \langle C_2|L|S_2\rangle \\ \vdots & \vdots \\ \langle \phi_i|L|S_1\rangle & \langle \phi_i|L|S_2\rangle \\ \vdots & \vdots \end{bmatrix}, \qquad \mathbf{X} = \begin{bmatrix} K_{11}^{\mathrm{t}} & K_{12}^{\mathrm{t}} \\ K_{21}^{\mathrm{t}} & K_{22}^{\mathrm{t}} \\ \vdots & \vdots \\ c_i & d_i \\ \vdots & \vdots \end{bmatrix}.$$

This is a rather obvious and natural extension, to two open channels, of the set of linear simultaneous equations for a single open channel, equation (3.53). The formulation can be readily extended to accommodate more open channels. Once all the matrix elements have been calculated, the determination of the variational K-matrix proceeds in a similar manner to that described previously for single-channel scattering. Further details were given by Armour and Humberston (1991).

The multichannel Kohn variational method does not yield rigorous lower bounds on any of the scattering parameters. However, the addition of more short-range correlation terms to the trial wave function usually produces an increase in the values of the diagonal K-matrix elements and the eigenphase shifts; there is a similar pattern of convergence with respect to increasing the number of terms in the trial wave function, as described previously for single-channel scattering, see equation (3.54). The values of the diagonal K-matrix elements obtained in this way are

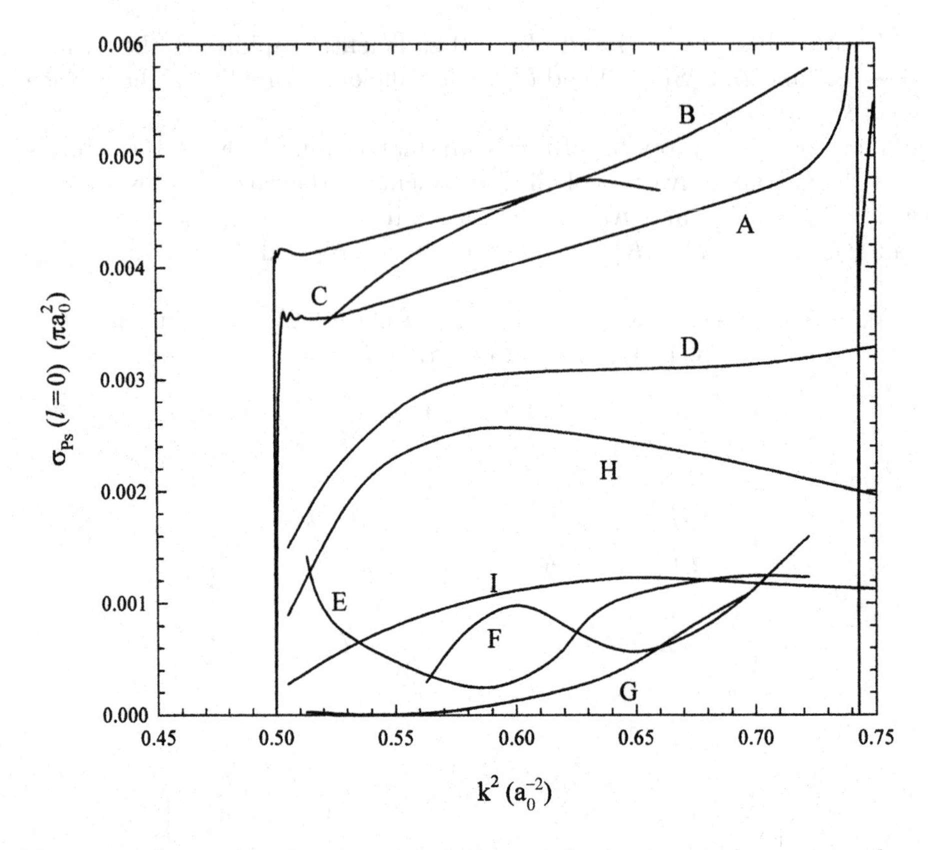

Fig. 4.1. The results of various calculations of the $l = 0$ partial-wave contribution to the positronium formation cross section in positron–hydrogen scattering in the Ore gap: A, Archer, Parker and Pack (1990); B, Humberston (1982); C, Stein and Sternlicht (1972); D, Chan and Fraser (1973); E, Wakid (1973); F, Dirks and Hahn ($\times 10$) (1971); G, Wakid and LaBahn (1972); H, Khan and Ghosh ($\times 10^{-1}$) (1983); I, Born approximation ($\times 10^{-3}$).

therefore probably lower bounds on the exact values, but they do not translate into bounds on σ_{Ps}.

The most accurate values of the s-wave positronium formation cross section calculated by Humberston (1982) and Humberston *et al.* (1997) are shown in Figure 4.1. (The latter results are more accurate but there is no difference between the two sets of results on the scale of this figure.) This cross section is much smaller than the s-wave elastic scattering cross section and also, as we shall see, much smaller than other contributions to σ_{Ps} of low orbital angular momentum. It has recently been shown by Ward, Macek and Ovchinnikov (1998), using hidden crossing theory, that the small magnitude of the s-wave contribution to σ_{Ps} is a consequence of the near-total destructive interference of two terms in the S-matrix element for positronium formation.

The results of several other calculations of the s-wave contribution to $\sigma_{\rm Ps}$ are also shown in Figure 4.1, and they reveal extreme sensitivity to the method of approximation being used. The range of results spans several orders of magnitude, with the values obtained using the first Born approximation being a factor 200 times too large and those obtained using the coupled-static approximation (see Cody *et al.*, 1964, for a description of this approximation) being approximately a factor of ten too small. However, several sets of results obtained using variational methods with flexible trial functions are in reasonable agreement with each other. Stein and Sternlicht (1972) used a similar technique to that of Humberston, with a somewhat similar but rather less flexible trial function, and the two sets of results are indeed in rather good agreement except close to the positronium formation threshold. The only other calculation dating from that time which yielded moderately accurate results was that of Chan and Fraser (1973). These authors used a formulation based on the coupled-static approximation, with the addition of several short-range Hylleraas correlation terms, and they obtained rigorous lower bounds on the diagonal elements of the K-matrix and the eigenphases, but their values of $\sigma_{\rm Ps}$ are little more than half those of Humberston. Winick and Reinhardt (1978b) used a moment T-matrix method to determine the elastic scattering amplitude, from which they then calculated the elastic scattering cross section $\sigma_{\rm el}$ and also, using the optical theorem, equation (2.3), the total scattering cross section $\sigma_{\rm T}$. In the Ore gap the difference between these two cross sections is then $\sigma_{\rm Ps}$. It is perhaps surprising that, although their wave functions contained many Hylleraas correlation terms, their results, not shown in Figure 4.1, are approximately five times larger than those of Humberston. The probable reason for this discrepancy is that the subtraction procedure involved in obtaining $\sigma_{\rm Ps}$ is rather inaccurate, the magnitudes of $\sigma_{\rm el}$ and $\sigma_{\rm T}$ being very similar.

More recent, detailed investigations by Archer, Parker and Pack (1990), who used the reactive scattering method of Pack and Parker (1987), yielded results, shown in Figure 4.1, which, within the Ore gap, are approximately 15% lower than those of Humberston. These authors also found two resonances just below the $n_{\rm H} = 2$ excitation threshold of hydrogen, as well as other resonances just below higher excitation thresholds, which Humberston failed to find. Similar results to those of Archer *et al.* were obtained by McAlinden, Kernoghan and Walters (1994) and Kernoghan, McAlinden and Walters (1995) using the coupled-state approximation with the following 18 states: H(1s, 2s, $\overline{3\rm s}$, $\overline{4\rm s}$, 2p, $\overline{3\rm p}$, $\overline{4\rm p}$, $\overline{3\rm d}$, $\overline{4\rm d}$), Ps(1s, 2s, $\overline{3\rm s}$, $\overline{4\rm s}$, 2p, $\overline{3\rm p}$, $\overline{4\rm p}$, $\overline{3\rm d}$, $\overline{4\rm d}$), where a bar implies a pseudostate. However, Kvitsinsky, Carbonell and Gignoux (1995) and

Kvitsinsky, Wu and Hu (1995), using two different methods to solve the Fadeev equations in configuration space, obtained well-converged results in excellent agreement with Humberston's, as also did Mitroy, Berge and Stelbovics (1994), Zhou and Lin (1994, 1995b) and Gien (1997).

Several other recent studies of positronium formation have been based on some form of the coupled-state approximation. Basu, Mukherjee and Ghosh (1990) used an integral form of the coupled equations in momentum space with the expansion H(1s, 2s, $\overline{2\text{p}}$, $\overline{3\text{d}}$), Ps(1s) and obtained s-wave positronium formation cross sections in moderately good agreement with those of Humberston, although their elastic scattering cross sections are rather less accurate. Higgins and Burke (1993) used the R-matrix method with a more elaborate six-state approximation of the form H(1s, $\overline{2\text{s}}$, $\overline{2\text{p}}$), Ps(1s, $\overline{2\text{s}}$, $\overline{2\text{p}}$), but their s-wave contribution to σ_{Ps} is more than twice as large as Humberston's, and with a rather different energy dependence.

An interesting feature of the accurate s-wave positronium formation cross section (curve B of Figure 4.1) is the very rapid rise just above threshold, from zero at the threshold, $k = 1/\sqrt{2} = 0.707\,12a_0^{-1}$, to a value of $0.004\pi a_0^2$ at $k = 0.71a_0^{-1}$ and increasing slowly thereafter up to the $n_{\text{H}} = 2$ threshold. As mentioned in section 3.3, when discussing the effect of the opening of the positronium formation channel on the elastic scattering cross section, Wigner's threshold law predicts that the positronium formation cross section for a given partial wave l should increase just above threshold according to $\sigma_{\text{Ps}}^l \propto E_1^{l+1/2}$, where $E_1 = E - E_{\text{Ps}} = \frac{1}{4}\kappa^2$ is the kinetic energy of the positronium. Consequently σ_{Ps}^0 should increase linearly with κ and have an infinite derivative with respect to the positron energy at the positronium formation threshold. The results of Humberston (1982), and more recently those of Archer, Parker and Pack (1990), Kvitsinsky and coworkers (1995), Mitroy, Berge and Stelbovics (1994), Zhou and Lin (1995b) and Humberston *et al.* (1997) all confirm this law, but the linear increase with κ is restricted to a very narrow range of positron energies just above the threshold.

Brown and Humberston (1984, 1985) used a similar variational method to that of Humberston (1982) to calculate the p- and d-wave contributions to σ_{Ps} in the Ore gap. Their trial functions were obvious modifications to higher partial waves of the s-wave trial functions of Humberston, equations (4.14) and (4.15), but, as with the trial functions for elastic scattering, there were now $l + 1$ groups of short-range correlation functions, each group being associated with a different rotational harmonic; see equation (3.65). The results of these and some other calculations are given in Figures 4.2 and 4.3. There is still significant sensitivity to the method of calculation, and to the form of the trial function, in the p-wave results, although much less so than with the s-wave. The results of the

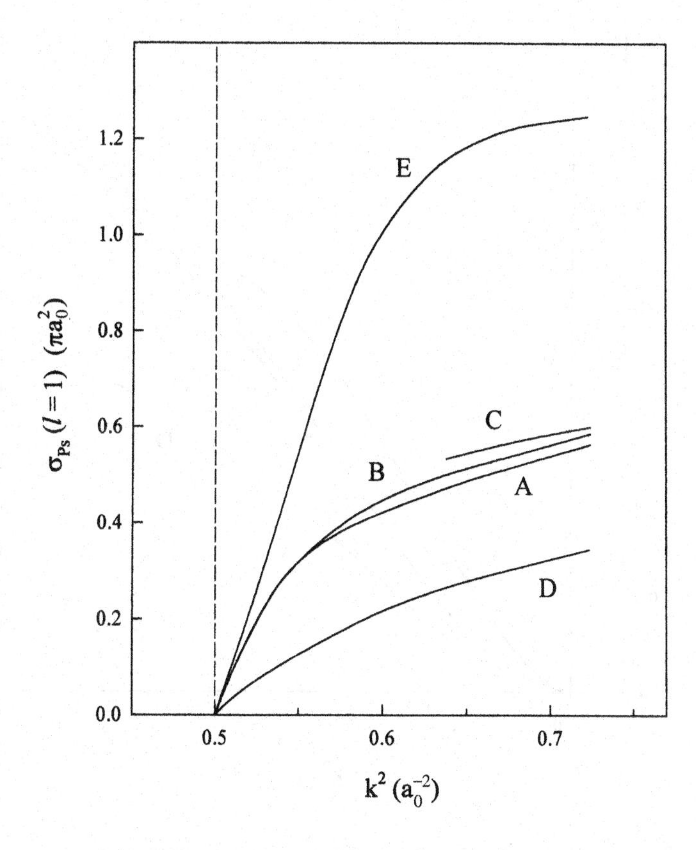

Fig. 4.2. The results of various calculations of the $l = 1$ partial-wave contribution to the positronium formation cross section in positron–hydrogen scattering in the Ore gap: A, Brown and Humberston (1985); B, Chan and McEachran (1976); C, Winick and Reinhardt (1978b); D, coupled-static approximation; E, Born approximation.

Born approximation are now only a factor of three too large, and those of the coupled-static approximation are a little less than half of the accurate values. Furthermore, because the difference between $\sigma_{\rm el}$ and $\sigma_{\rm T}$ is no longer very small, the subtraction procedure of Winick and Reinhardt (1978b) yields quite accurate values for the p-wave contribution to $\sigma_{\rm Ps}$. The R-matrix results of Higgins and Burke (1993) and the coupled-state results of McAlinden, Kernoghan and Walters (1994) and Kernoghan and coworkers (1995, 1996) are in reasonable agreement with those of Brown and Humberston, and also with those of Basu, Mukherjee and Ghosh (1990), although the agreement between the latter results and those of Brown and Humberston is worse than for the s-wave.

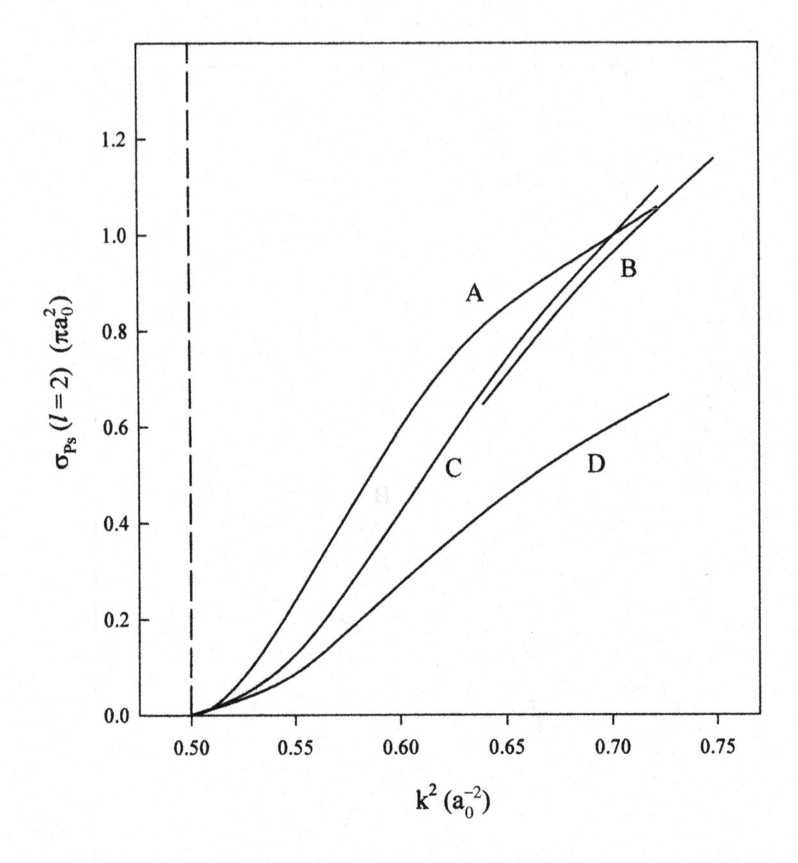

Fig. 4.3. The results of various calculations of the $l = 2$ partial-wave contribution to the positronium formation cross section in positron–hydrogen scattering in the Ore gap: A, Brown and Humberston (1985); B, Winick and Reinhardt (1978b); C, Born approximation; D, coupled-static approximation.

The d-wave contribution to σ_{Ps} is relatively insensitive to the method of approximation, and even the results of the Born approximation agree quite well with the accurate variational calculations. It is therefore to be expected that reasonably accurate values can be obtained for all higher-partial-wave contributions to σ_{Ps} by using the Born approximation, and this has been confirmed by the results of Gien (1997).

The total positronium formation cross section in the Ore gap, constructed from the addition of accurate variational results for the first three partial waves and the values given by the Born approximation for all partial waves with $l > 2$, is plotted in Figure 4.4. On the scale of the ordinate, the s-wave contribution is too small to be visible. A very small s-wave contribution is found to be a feature of the positronium formation cross section for several other atoms.

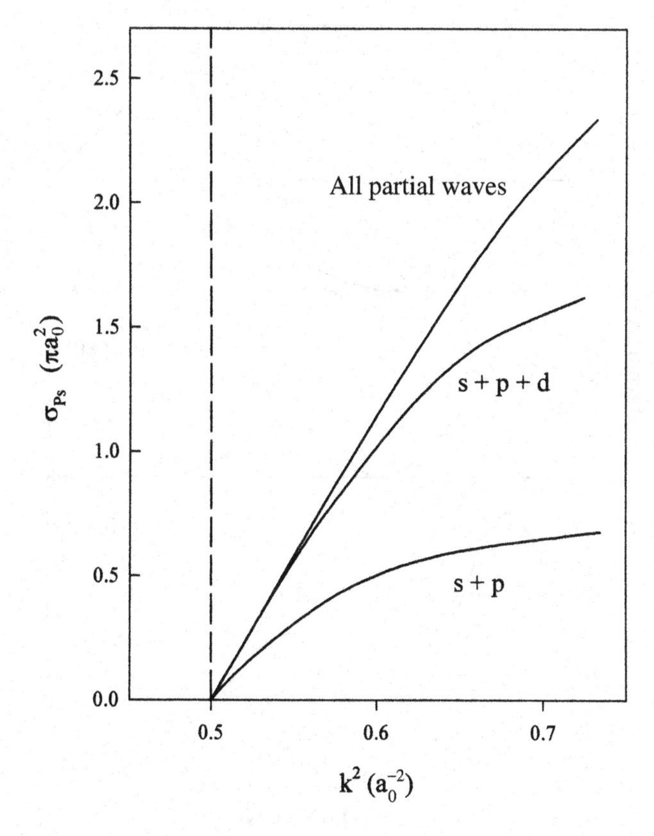

Fig. 4.4. The total positronium formation cross section in positron–hydrogen scattering in the Ore gap as calculated using the results of Brown and Humberston (1985) for $l \leq 2$ and the Born approximation for $l > 2$.

The differential cross section for positronium formation may be expressed in terms of the partial-wave K-matrix elements as

$$\frac{d\sigma_{\rm Ps}}{d\Omega}(\theta) = \frac{\pi}{k^2}\left|\sum_l (2l+1)\left(\frac{\mathbf{K}}{\mathbf{1} - i\mathbf{K}}\right)_{12} P_l(\cos\theta)\right|^2 , \qquad (4.27)$$

and the angular distributions obtained from the accurate data for the first three partial waves are plotted in Figure 4.5 for several energies in the Ore gap. At energies just above the positronium formation threshold, rather more than half the positronium is produced at angles greater than 90° relative to the incident positron beam direction but, as the positron energy is increased, the angular distribution of the positronium becomes more peaked in the forward direction. It is this feature of positronium formation which is exploited in the production of positronium beams (see subsection 7.6.1).

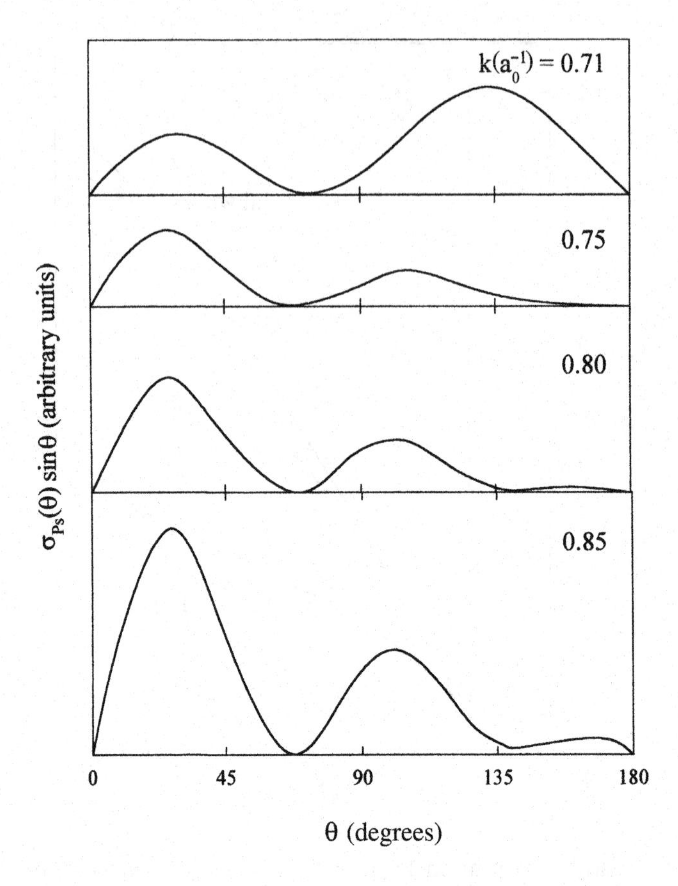

Fig. 4.5. The angular distribution of positronium formation in positron–hydrogen scattering at various incident positron wavenumbers in the Ore gap.

Most investigations of positronium formation in positron–hydrogen scattering have been made over a wider energy range than just the Ore gap. One of the most detailed studies of s-wave scattering over the energy range up to the $n_{\rm H} = 4$ excitation threshold of hydrogen is that of Archer, Parker and Pack (1990), who included all energetically possible reaction channels in the formulation. Their results reveal interesting resonance phenomena associated with various excitation thresholds of hydrogen and positronium. Rather less detailed studies, but extending over an even wider energy range, up to 60 eV, and including more partial waves, have been made by Higgins and Burke (1993), Hewitt, Noble and Bransden (1990), McAlinden, Kernoghan and Walters (1994), Kernoghan and coworkers (1995, 1996) and Mitroy (1996), all using various forms of the coupled-state approximation. Higgins and Burke (1991, 1993) found resonance structure in the s- and p-wave cross sections for both positronium formation and elastic scattering at incident positron energies

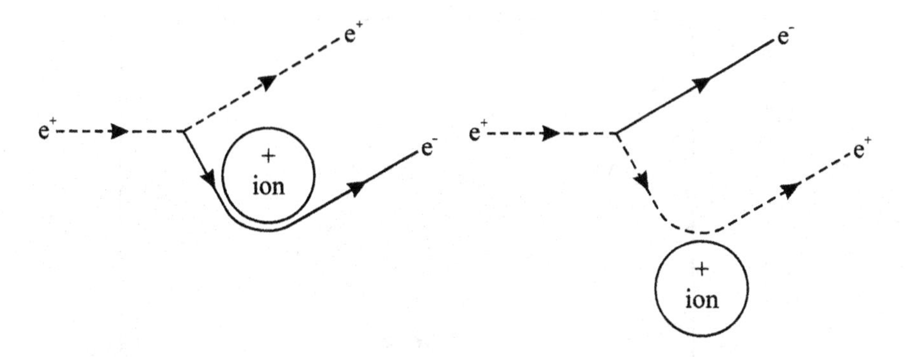

Fig. 4.6. The two double-binary collision processes resulting in positronium formation following positron impact at high energies. The positron collides with an atomic electron: on the left the electron then scatters off the residual ion into the same direction as the positron, whilst on the right the process is shown in which the positron scatters off the residual ion.

between 2.7 ryd and 3.5 ryd (37–48 eV), well above the break-up energy of the target atom, and they cannot therefore be Feshbach resonances associated with degenerate excitation thresholds of the hydrogen or the positronium atoms. Originally they were believed to be so-called 'coupled-channel shape resonances', arising from the coupling between the hydrogen and positronium channels, but they have since been shown by Kernoghan, McAlinden and Walters (1994b) to be artifacts of neglect of the other open channels.

Shakeshaft and Wadehra (1980) used the distorted-wave Born approximation to calculate positronium formation cross sections over the positron energy range 13.6–200 eV. Their results for ground state positronium formation agree reasonably well with more accurate values at the lower end of this range, and so they are probably quite accurate at all energies. Another variant of the distorted-wave Born approximation was used by Mandal, Guha and Sil (1979), but their results are probably less accurate than those of Shakeshaft and Wadehra, particularly at the higher energies.

The first Born approximation is known to provide a rather inaccurate description of positronium formation, even at high energies, because the process then becomes essentially two-stage; this can be understood as follows. In order to form positronium, the positron and an electron must emerge from the target with very similar velocities, and the simplest way in which this can be achieved is via one or other of the processes represented in Figure 4.6. In both cases, first the positron scatters from the electron and then either the electron or the positron is scattered into the required final direction by the nucleus. It is therefore to be expected that the second Born approximation, with its quadratic

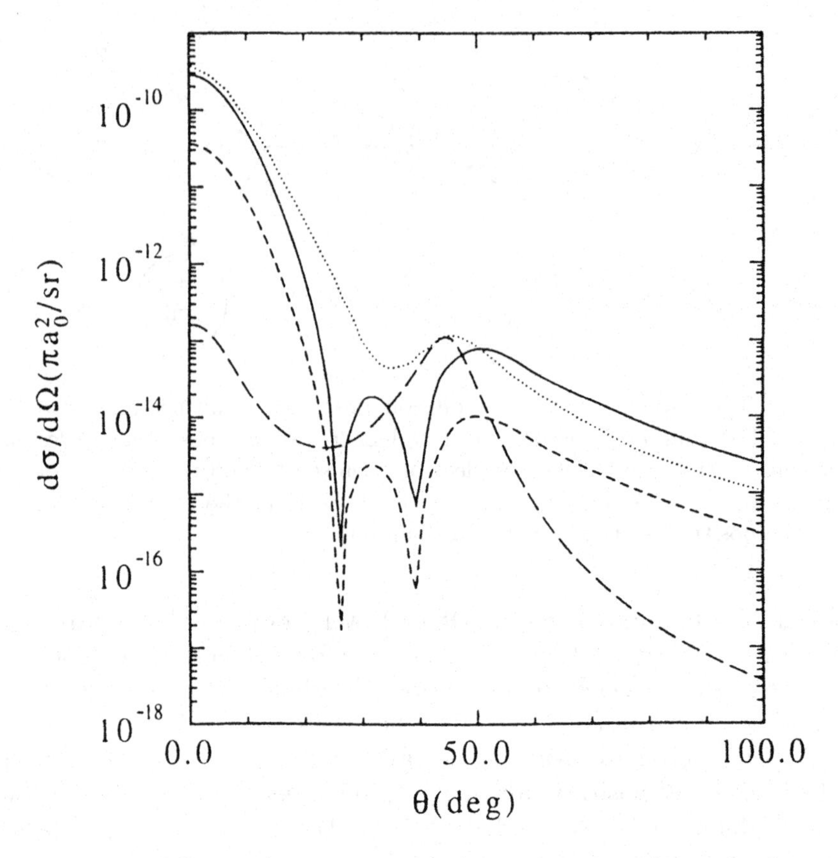

Fig. 4.7. The differential cross sections for positronium formation into the ground state and the $n_{\mathrm{Ps}} = 2$ excited states in positron–hydrogen collisions at an incident positron energy of 5.44 keV: ——, ground state formation; - - - -, formation into the 2S state; — —, formation into the 2P state (Igarashi and Toshima, 1992); · · · · ·, ground state formation (Deb, McGuire and Sil, 1987). Reprinted from *Physical Review* **A46**, Igarashi and Toshima, Destructive and constructive interferences of the second Born amplitudes for positronium formation, R1159–R1162, copyright 1992 by the American Physical Society.

term in the interaction potential, provides a more appropriate representation of this process than does the first Born approximation, although amplitudes of all orders do of course contribute to some extent. The distorted-wave Born approximation implicitly includes second and higher order terms, which explains why the results of Shakeshaft and Wadehra (1980) and Mandal, Guha and Sil (1979) are rather similar to those obtained by Deb, McGuire and Sil (1987) and Basu and Ghosh (1988), who used the second Born approximation directly. These second order results are strongly forward peaked at high energies, as shown in Figure 4.7, which relates to a positron energy of 5.44 keV, but they

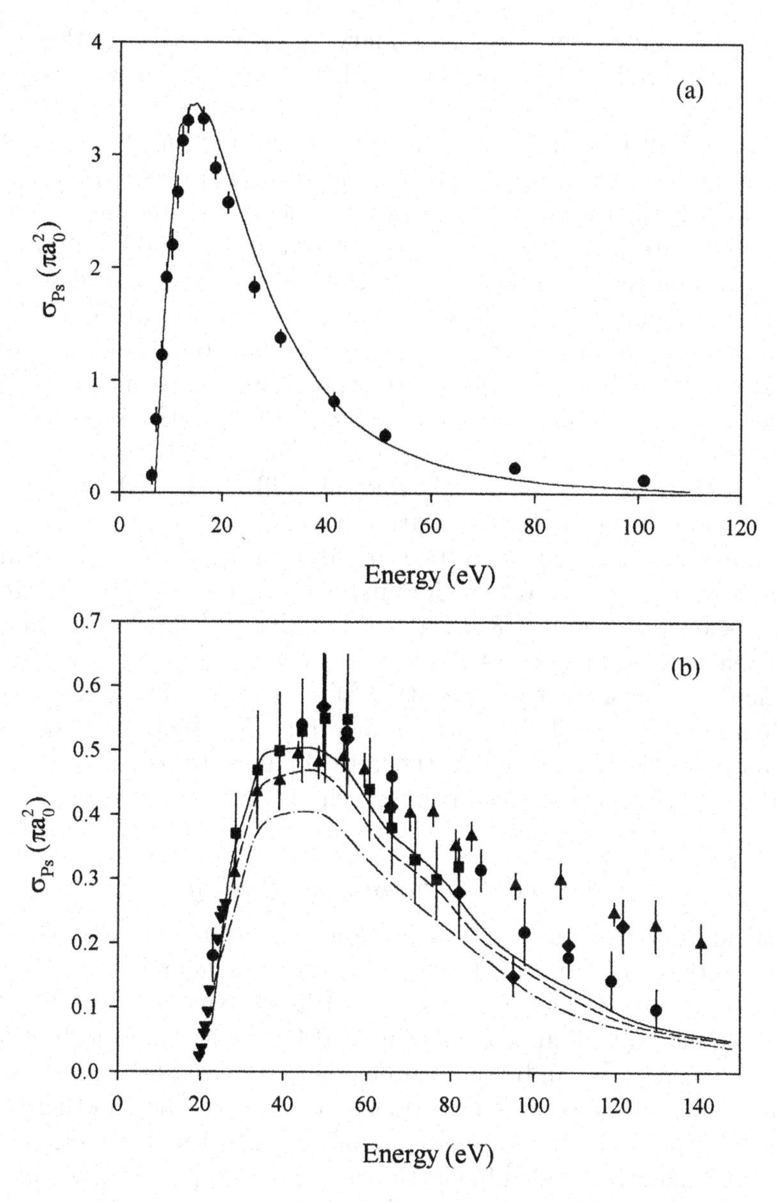

Fig. 4.8. Cross sections for positronium formation in positron collisions with atomic hydrogen and helium. (a) Hydrogen: ——, 33-coupled-state results of Kernoghan, Robinson, McAlinden and Walters (1996) for the total positronium formation cross section together with the experimental points of Zhou *et al.* (1997). (b) Helium: the theoretical results were obtained by Campbell *et al.* (1998a) using a 27-coupled-state approximation; — · —, ground state formation; – – –, sum of the the formation cross sections for $n_{Ps} = 1$ and $n_{Ps} = 2$; ——, total positronium formation cross section. Experimental results: ■, Fornari, Diana and Coleman (1983); ◆, Diana *et al.* (1986b); ▲, Fromme *et al.* (1986); ▼, Moxom *et al.* (1993); ●, Overton *et al.* (1993). See also Figure 4.17 for more data on this well-studied system.

also reveal a quite prominent secondary peak around 45°; this is the Thomas peak predicted by the classical kinematics of the two-stage capture process.

The results discussed above relate to the formation of ground state positronium in collisions of positrons with hydrogen atoms in their ground state, but investigations have also been made of positronium formation into excited states, notably by McAlinden, Kernoghan and Walters (1994) and Kernoghan and coworkers (1995, 1996) using the coupled-state approximation mentioned previously. These authors obtained results for positronium formation into states with $n_{\rm Ps} \leq 4$ but, because their positronium states with $n_{\rm Ps} = 3$ and 4 were represented by pseudostates, they considered it to be more accurate to estimate the positronium formation cross sections for $n_{\rm Ps} > 2$ using a scaling rule derived from the Born approximation. According to this rule, at sufficiently high energies the cross section for positronium formation into a state with principal quantum number $n_{\rm Ps}$ is proportional to $1/n_{\rm Ps}^3$ (Omidvar, 1975). The resulting total $\sigma_{\rm Ps}$, which for hydrogen is dominated by formation into the ground state, is shown in Figure 4.8(a). Its peak value of $3.5\pi a_0^2$ is attained at an incident positron energy of 15 eV, after which it declines fairly rapidly with increasing positron energy, so that at an energy of 80 eV it is already less than one quarter of the elastic scattering cross section, whose value is $0.25\pi a_0^2$ at this energy. These theoretical values for $\sigma_{\rm Ps}$ are compared with the experimental results in subsection 4.4.2.

2 Positron–helium scattering

The calculation of accurate cross sections for positronium formation is a particularly challenging task when the target is helium or some other complex atom. As we have already seen with hydrogen (subsection 4.2.1), the simple methods of approximation used for positronium formation at low positron energies can be very unreliable for the first few partial waves, and the results obtained may be seriously in error. For helium there is the additional problem of having to use an inexact target wave function. This can be conveniently avoided in elastic scattering by the use of the method of models (subsection 3.2.2), but no such self-consistent formulation is possible for a rearrangement collision: the model potential describing the interaction between each electron and the nucleus is inconsistent with the Coulomb interaction between the electron and the nucleus in the residual ion. Consequently, the exact Hamiltonian should be used throughout the formulation.

Because of its complexity, most calculations of positronium formation in positron–helium scattering have been made using relatively crude methods of approximation with rather simple uncorrelated helium wave

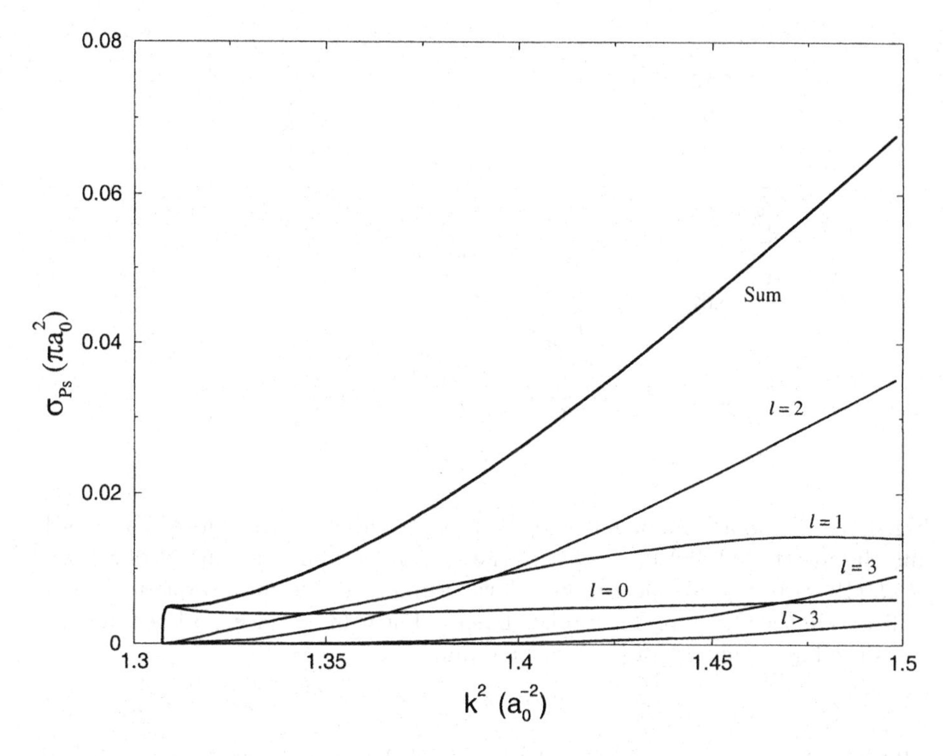

Fig. 4.9. The total positronium formation cross section, and the various partial-wave contributions to it, for positron–helium scattering in the Ore gap. The contributions with $l \leq 3$ are determined variationally whilst the sum of all higher partial waves is calculated in the Born approximation.

functions, and the results are therefore of somewhat uncertain accuracy. Massey and Moussa (1961) used the first Born approximation but their results were subsequently found to be in error, and were corrected by Mandal, Ghosh and Sil (1975), who also applied the coupled-static approximation to the problem. The first use of the coupled-static approximation was by Kraidy and Fraser (1967), who also introduced extra potential terms to take some account of the polarization of both the positronium and the helium atom. Their results, which were obtained in the energy range of 10 eV above the positronium formation threshold, suggested that the d-wave contribution to σ_{Ps} is dominant, as has been confirmed by Van Reeth and Humberston (1999b).

The only detailed investigations of low energy positronium formation in positron–helium scattering in the Ore gap have been made by Van Reeth and Humberston (1995b, 1997, 1999b); see also Humberston and Van Reeth (1996) and Van Reeth and Humberston (1999a). These authors used the Kohn variational method in a similar manner to that described in

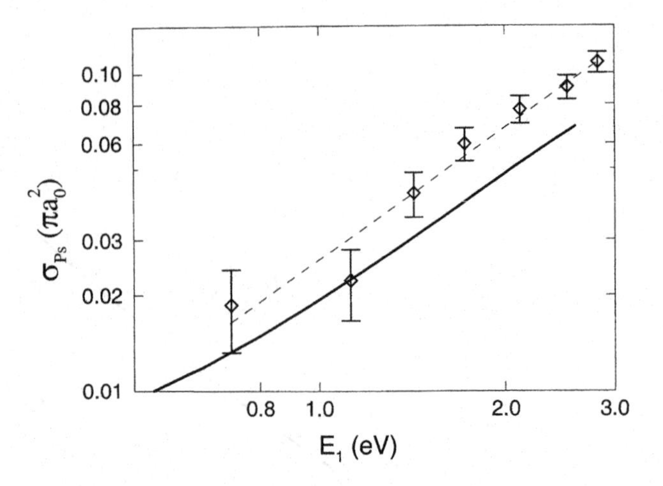

Fig. 4.10. Comparison, on a log–log plot, of the theoretical results of Van Reeth and Humberston (1999b) (——) with the experimental measurements of Moxom *et al.* (1994) (◇) for the positronium formation cross section in positron–helium collisions in the Ore gap. The broken line is a linear fit to the experimental data; E_1 is the kinetic energy of the positronium.

subsection 4.2.1 for positronium formation in positron–hydrogen scattering, but with the short-range correlation terms in the trial wave function having a form like those used in the detailed investigations of elastic positron–helium scattering, described in subsection 3.2.2. A very accurate correlated Hylleraas-type helium wave function was used, similar in form to that described by equation (3.79) but containing 22 terms, so as to minimize any problems that might be associated with the use of an inexact target wave function.

The most accurate results obtained by these authors for the s-, p-, d- and f-wave contributions to σ_{Ps} are displayed in Figure 4.9. Also given there is the total σ_{Ps} obtained by summing the first four partial-wave contributions and adding the first Born results for all higher partial waves. The s-wave contribution exhibits a very steep rise from zero at the threshold to a gently rising plateau of very similar magnitude to that obtained for hydrogen, although in helium it constitutes a larger fraction of what is a much smaller total positronium formation cross section. The p-wave contribution rapidly exceeds that of the s-wave, and this in turn is exceeded by that of the d-wave, which forms the largest single partial-wave contribution to σ_{Ps} throughout the upper half of the Ore gap. A comparison of the accurate theoretical results of Van Reeth and Humberston (1999b) with the low energy experimental measurements of Moxom *et al.* (1994) is shown in Figure 4.10. The differential positronium formation

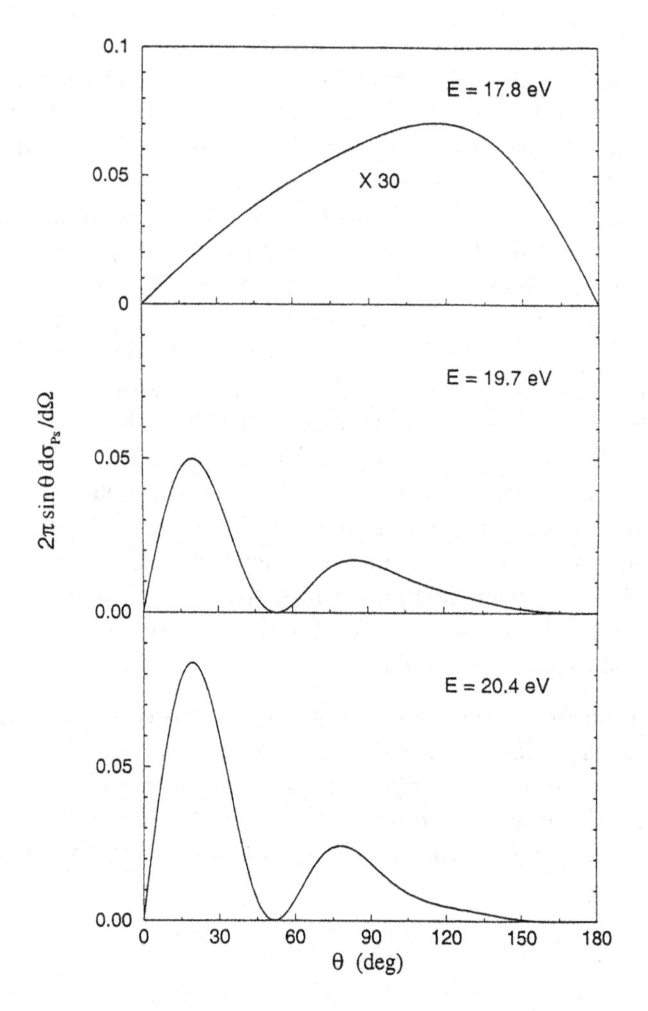

Fig. 4.11. The angular distribution of positronium formation in positron–helium collisions at various energies in the Ore gap (Van Reeth and Humberston (1999b). The results for $E = 17.8$ eV, just above the positronium formation threshold, have been multiplied by a factor of 30.

cross sections obtained by Van Reeth and Humberston (1999b) at three energies in the Ore gap are displayed in Figure 4.11.

Most other calculations of positronium formation in positron–helium scattering have employed much simpler methods of approximation, but results have usually been obtained over energy ranges extending well beyond the Ore gap. It must therefore be borne in mind that the experimental results include contributions from positronium formation into excited states as well as into the ground state. The Born approximation, used first by Massey and Moussa (1961) and subsequently by Mandal, Ghosh and Sil (1975), gives a more sharply peaked maximum, at a rather lower energy, but more than three times larger, than that in the experimental measurements of Fornari, Diana and Coleman (1983) and Fromme *et al.* (1986), which have a maximum at approximately 50 eV. Only beyond an energy of 80 eV does the Born approximation give reasonably good agreement with experiment. The distorted-wave approximation, used by Mandal, Guha and Sil (1975), and the coupled-static approximation, used by McAlinden and Walters (1992), both give results in reasonably good agreement with experiment in the vicinity of the maximum and up to a positron energy of 80 eV; however, they fall below the experimental values as the energy is increased further. A more elaborate calculation was performed by Hewitt, Noble and Bransden (1992a) using the coupled-state method with five states of helium and three states of positronium, but an equivalent one-electron model was used for the helium atom. Their results for ground state positronium formation fall significantly below the experimental cross sections, although rather better agreement is obtained when contributions from formation into $n_{\rm Ps} = 2$ states, and estimates based on the $1/n_{\rm Ps}^3$ scaling law for formation into even higher excited positronium states, are included.

The most accurate theoretical results for positronium formation in positron–helium collisions in the energy range 20–150 eV are probably those of Campbell *et al.* (1998a), who used the coupled-state method with the lowest three positronium states and 24 helium states, each of which was represented by an uncorrelated frozen orbital wave function

$$\Psi^n_{\rm He}(\boldsymbol{r}_2, \boldsymbol{r}_3) = N_n(1 \pm {\rm P}_{23})\phi_n(r_2)\bar{\phi}(r_3), \tag{4.28}$$

where $\bar{\phi}$ is the frozen orbital and $\rm P_{23}$ is the exchange operator for the two electrons. This form provides a reasonably good representation of the various excited states of helium, but the ground state is much less accurate than that used by Van Reeth and Humberston (1999b). Cross sections for positronium formation into states with $n_{\rm Ps} = 1$ and 2 were evaluated explicitly and the $1/n_{\rm Ps}^3$ scaling formula was used to estimate the contributions from all higher states. As in hydrogen, the total positronium formation cross section is dominated by formation into the ground state. The results obtained by Campbell *et al.* (1998a) agree reasonably well with several sets of experimental data up to 60 eV, and with the data of Fornari, Diana and Coleman (1983) and Diana *et al.* (1986b) up to 90 eV, but thereafter they fall some way below all sets of experimental data; see Figure 4.8(b). Further discussion and comparison with experiment can be found in subsection 4.4.1.

3 Positron–alkali atom scattering

As already mentioned in subsection 3.2.3, in discussing elastic scattering, the positronium formation channel is open even at zero incident positron energy for all the alkali atoms. Furthermore, because this reaction is exothermic, the positronium formation cross section is infinite at zero incident positron energy. It is a reasonably good approximation to consider an alkali atom as an equivalent one-electron atom; consequently the formulation of positronium formation in positron scattering by alkali atoms is rather similar to that in positron–hydrogen scattering. All such theoretical investigations of positron-alkali atom scattering have assumed this three-body structure.

Lithium is the simplest alkali atom; in this case representation of the atom as an electron in the central field of the core is particularly good, and more theoretical attention has been given to positron scattering by lithium than by any other alkali atom, despite the complete lack of experimental data with which to compare. Detailed investigations of low energy positron–lithium scattering in the energy region from zero up to the first excitation threshold have been made by Humberston and Watts (1994), using the Kohn variational method with similar forms of trial wave function to those used by Humberston (1982), and Brown and Humberston (1985) for positron–hydrogen scattering (see subsection 4.2.1). The representation of the lithium atom was based on an electron–core potential devised by Peach, Saraph and Seaton (1988) and of the form

$$V_{-}(r) = -\frac{1}{r} - 2\frac{e^{-\gamma r}}{r}(1 + \delta_1 r + \delta_2 r^2) - \frac{\alpha}{2r^4}\omega(r), \qquad (4.29)$$

where the first two terms represent the static interaction and the last represents the polarization of the core, which has a dipole polarizability α. The values of the parameters δ_1, δ_2 and γ were obtained by fitting to the energy spectrum of lithium. Changing the sign of the static component of the potential, but retaining the sign for the core polarization term, yields the positron–core potential V_{+}. The wave function of the valence electron in the lithium atom should have the radial character of a 2s orbital, which means that the electron in this equivalent one-electron model is not in its ground state but in the first excited state. The 1s ground state, however, is merely an artifact of the model with no real physical significance, and its energy, ≈ -50 eV, is so far below the energy range under consideration in the scattering calculation that it can probably be ignored. Nevertheless, in order to test the validity of this assumption, another model of the electron–core potential was also used by Humberston and Watts (1994) in which the required valence electron is in its ground state. The corresponding wave function is then of 1s character and nodeless, but the differences between the two wave functions are confined to small radial distances of little more than a_0 from the nucleus. Accurate approximations to both lithium wave functions of the form

$$\Phi_{\rm Li}(r) = \exp(-Ar) \sum_{j=0}^{N_0} e_j r^j, \qquad (4.30)$$

where A and e_j ($j = 0, \ldots, N_0$) are variational parameters, were generated using the Rayleigh–Ritz variational method.

Elastic and positronium formation cross sections for $l = 0$, 1 and 2 were calculated in a similar manner to that previously described in subsection 4.2.1 for positron–hydrogen scattering. However, the convergence of the results with respect to increasing the number of correlation terms in the trial functions is now worse, particularly at very low positron energies; the reason is probably that the short-range character of the correlation terms used in these trial wave functions makes them rather less suitable for representing the long-range distortions associated with the high dipole polarizability of the lithium atom. Better convergence could be achieved by adding longer-range terms to the trial wave function, as was done to improve the convergence of the positron–hydrogen scattering length (see subsection 3.2.1). Despite these reservations, these contributions to $\sigma_{\rm Ps}$ are believed to be accurate to within 20%. From its infinite value at $k = 0$, the s-wave contribution $\sigma^0_{\rm Ps}$ rapidly falls to be several orders of magnitude smaller than $\sigma^0_{\rm el}$ so that again, as with hydrogen and helium, the s-wave only makes a small contribution to $\sigma_{\rm Ps}$.

All other calculations for lithium, and all those for the other alkali atoms, have used some form of the coupled-state approximation, and results have usually been obtained over a much wider energy range, typically 0–60 eV. The simplest approximation of this type, the coupled-static approximation, was used by Guha and Ghosh (1981) and Abdel-Raouf (1988), but more recent calculations have included several states of both lithium and positronium. In such calculations, cross sections have been obtained not only for elastic scattering and ground state positronium formation but also for positronium formation into various excited states and for excitation, and sometimes ionization, of the target atom. Among the most important of these calculations are those of Hewitt, Noble and Bransden (1992b), who included the states Li (2s, 3s, 2p, 3p) and Ps (1s, 2s, 2p), Kernoghan, McAlinden and Walters (1994a), who included Li (2s, 2p, 3p, 3d) and Ps (1s, 2s, $\overline{\rm 3s}$, $\overline{\rm 4s}$, 2p, $\overline{\rm 3p}$, $\overline{\rm 4p}$, $\overline{\rm 3d}$, $\overline{\rm 4d}$), and McAlinden, Kernoghan and Walters (1997), who included Li (39 states and pseudostates) and Ps (1s, 2s, 2p). The low energy results of Kernoghan, McAlinden and Walters (1994a) are in good agreement with the variational results of Humberston and Watts (1994), whereas those of Hewitt,

Noble and Bransden (1992b) are significantly larger. The most extensive investigations were made by McAlinden, Kernoghan and Walters (1997), who determined the total cross section for positronium formation into all possible states. They calculated directly the cross sections for positronium formation into its ground state and the $n_{\rm Ps} = 2$ excited states and estimated the contributions from all other excited positronium states using the $1/n_{\rm Ps}^3$ scaling law. Thus

$$\sigma_{\rm Ps} = \sigma_{\rm Ps}(n_{\rm Ps} = 1) + \sigma_{\rm Ps}(n_{\rm Ps} = 2) + \sigma_{\rm Ps}(n_{\rm Ps} \geq 3), \tag{4.31}$$

where

$$\begin{aligned} \sigma_{\rm Ps}(n_{\rm Ps} \geq 3) &= 8\sigma_{\rm Ps}(n_{\rm Ps} = 2) \sum_{n_{\rm Ps}=3}^{\infty} 1/n_{\rm Ps}^3 \\ &= 0.6165\sigma_{\rm Ps}(n_{\rm Ps} = 2). \end{aligned} \tag{4.32}$$

The total positronium formation cross section obtained in this way falls steadily with increasing positron energy, so that at an energy of 25 eV it is already very small compared with the cross sections for elastic scattering and excitation of the target atom. Formation is mainly into the ground state at energies below 10 eV, but formation into excited states becomes more probable at higher energies.

McAlinden, Kernoghan and Walters (1994) also calculated the differential cross sections for the formation of positronium into its various states. The cross section for ground state formation, Ps(1S), is particularly strongly peaked in the forward direction (angles $< 20^\circ$) as also, to a slightly less extent, is that for Ps(2S) formation. The cross section for Ps(2P) formation has a rather broader angular spread, being significant out to 40°, although the forward peaking increases with increasing incident positron energy.

Similar coupled-state methods have been applied to the other alkali atoms by Hewitt, Noble and Bransden (1993), McAlinden, Kernoghan and Walters (1994, 1996) and Kernoghan, McAlinden and Walters (1996). These results reveal that, except at those low energies where only ground state positronium formation is energetically possible, the ground state formation cross section $\sigma_{\rm Ps}(n_{\rm Ps} = 1)$ becomes a progressively smaller fraction of the total positronium formation cross section as the atomic number of the target alkali atom is increased. The contributions to $\sigma_{\rm Ps}$ from positronium formation into excited states other than those explicitly represented in the coupled-state expansion were again estimated using the $1/n_{\rm Ps}^3$ scaling law, in a somewhat similar manner to that described above for lithium.

The total positronium formation cross sections obtained in this way are, as will be described in section 4.4, in reasonably good agreement with the lower-bound measurements of Kwan *et al.* (1994), Zhou *et al.* (1994b) and Surdutovich *et al.* (1996) for energies below 30 eV. Beyond this energy, the total positronium formation cross sections for all the alkali atoms become very small.

4 Other atoms and molecules

Theoretical studies of positronium formation have been made for several other atoms, particularly the rare gases and the alkaline earth atoms, magnesium, calcium and zinc, although in most cases only simple approximation methods have been used. In the first such investigation for any atom other than hydrogen and helium, Gillespie and Thompson (1977) applied the first Born approximation and two forms of distorted-wave approximation to positronium formation in neon and argon, but they restricted the energy range of the calculations to within 10 eV of the positronium formation threshold. For both atoms, the results of the Born approximation exceed the experimental measurements, although by a smaller factor for argon than for neon. The results of the distorted-wave approximation are much lower than the experimental values. McAlinden and Walters (1992) used the coupled-static approximation to calculate the elastic scattering and positronium formation cross sections for all the rare gases, and these same authors also calculated the differential positronium formation cross sections in the same approximation for all the rare gases except helium (McAlinden and Walters, 1994). The results in this approximation compare reasonably well with the experimental measurements, particularly in the vicinity of the maximum in the positronium formation cross section. At higher energies these theoretical results fall significantly below the experimental data, although the latter also include positronium formation into excited states. However, McAlinden and Walters (1992) expressed doubt that the sum of the partial cross sections for positronium formation into all excited states is sufficiently large to account fully for the discrepancy.

Very few theoretical investigations of positronium formation have been made for molecular targets, and virtually all these relate to hydrogen. The first such study was made by Sural and Mukherjee (1970), who used the first Born approximation in the energy range 50–544 eV. A rather simple hydrogen molecular wave function was used, consisting of a sum of products of single electron wave functions of the form $\exp(-1.193r)$, centred on one or other of the two protons. Similarly, the wave function of the H_2^+ molecular ion was taken to be the sum of two terms, each one being the same function as that given above and centred on one or other of the protons. Despite the simplicity of these various approximations, the results obtained are in moderately good agreement with the experimental measurements, bearing in mind that the theoretical results relate only to ground state positronium formation. Rather similar results for the same system were also obtained by Biswas, Mukherjee and Ghosh (1991) using essentially the same method of approximation, but these latter authors performed their calculations over a wider energy range and found that their results began to exceed the experimental measurements by an increasingly large factor as the incident positron energy was reduced below 50 eV. Both groups of authors only considered positronium formation with the H_2^+ ion left in the gerade state. Ray, Ray and Saha (1980), however, using a simplified form of the full Born approximation known as the molecular Jackson–Schiff approximation, also investigated positronium formation with the ion left in the ungerade state but found that this made only a small contribution to the overall positronium formation cross section, particularly at low positron energies.

If the hydrogen molecule is considered as two hydrogen atoms it might naively be assumed that the positronium formation cross section for the molecule should be approximately double that for the atom. Sural and Mukherjee (1970) tested the validity of this assumption by comparing their Born results for the molecule with the Born results for atomic hydrogen and they found that, instead of the results for the molecule being approximately double those for the atom, they were approximately three times as large. However, a comparison of the experimental data reveals that the positronium formation cross sections for the two target systems are in fact rather similar in magnitude throughout the energy range in which this process makes a significant contribution to the total scattering cross section.

Positronium formation into $n_{Ps} = 2$ excited states in positron–H_2 scattering was investigated in the first Born approximation by Ray, Ray and Saha (1980) and also by Biswas *et al.* (1991b).

4.3 Experimental techniques

Experimentally, positronium formation may be identified by a number of different signals, and these can be summarized as follows (Raith, 1987):

(i) Positrons are lost from the beam: positronium formation is the only channel which effectively removes positrons, apart from annihilation in flight, which usually has a very small cross section.

(ii) After formation, ortho-positronium will annihilate either via three gamma-rays in vacuum or, perhaps, with the emission of two gamma-rays upon striking part of the experimental apparatus, whereas para-positronium, with its characteristic lifetime of around 125 ps, will annihilate in flight into two gamma-rays.

(iii) The formation of positronium can be monitored because the residual ion can be extracted from the scattering region and detected. Positronium formation can be distinguished from other processes which produce ions, e.g. impact ionization (see Chapter 5), since in these cases the positron remains as an isolated particle in the final state, in addition to the ion.

As will be shown, all these methods have been used in investigations of positronium formation.

The first experiment to identify positronium production directly in positron–gas collisions using a low energy beam was that of Charlton *et al.* (1980c). These workers used a magnetically confined beam similar to that described in subsection 1.4.2 except that a curved solenoid was used to remove the scattering cell from the line of sight with the ^{22}Na source. The collision gas cell was surrounded by three large NaI(Tl) gamma-ray detectors set to monitor triple coincidences. Thus, the signal used was the three-gamma-ray annihilation of ortho-positronium described above under heading (ii). Several gases were investigated, and in each case, as the positron energy was increased beyond the positronium formation threshold the three-gamma-ray signal was found to rise sharply before starting to fall after a few eV.

Later experiments by other workers, described below, showed that this energy dependence was not that of σ_{Ps} and that the cross sections which were published using this technique (Charlton *et al.*, 1983b) were in error. This was caused by the combination of the following effects. First, most of the fast ortho-positronium was quenched on the walls of the scattering cell, with a resultant loss in the three-gamma-ray signal. Second, the natural forward collimation of the differential cross section $d\sigma_{Ps}/d\Omega$, discussed in sections 4.2 and 4.7, meant that a significant fraction of the ortho-positronium formed at kinetic energies greater than a few eV either escaped from the gas cell or moved to a region of lower three-gamma-ray detection efficiency. In any event, the true σ_{Ps} was much underestimated. Later, Clark (1984) investigated a two-gamma-ray arrangement for measuring σ_{Ps} and, although this gave results much closer to others obtained at that time using different methods, there was still the suspicion that not all systematic errors had been eliminated. A similar two-gamma-ray method has since been employed by the Detroit group (Zhou *et al.*, 1994b; Surdutovich *et al.*, 1996) in the first attempts to measure σ_{Ps} for the alkali metals.

Reliable values of σ_{Ps} were first obtained by Fornari, Diana and Coleman (1983), using a technique based upon the detection method (i). The apparatus, shown in Figure 4.12, consisted of a flight path 2.3 m long immersed in a 10^{-2} T magnetic field, the entire beamline having been

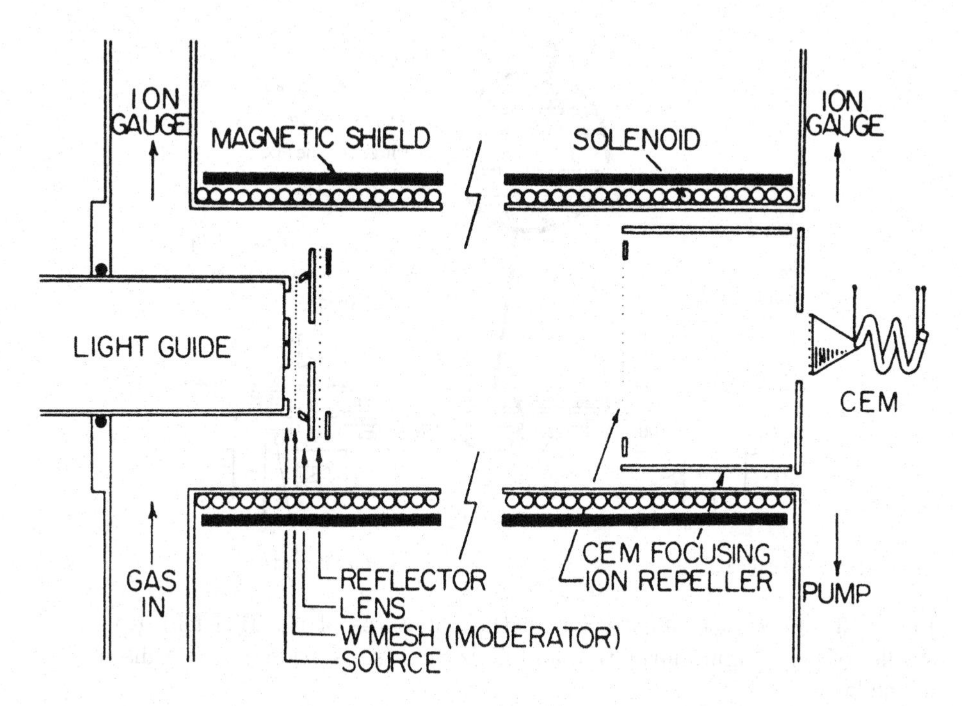

Fig. 4.12. Schematic of the apparatus developed by Fornari, Diana and Coleman (1983) for studies of positronium formation. Reprinted from *Physical Review Letters* **51**, Fornari *et al.*, Positronium formation in collisions of positrons with He, Ar and H_2, 2276–2279, copyright by the American Physical Society.

shielded from the Earth's magnetic field. The positron beam originated from an annealed tungsten mesh moderator which was held at a potential V_W to fix the kinetic energy of the positrons (found to be at a mean energy of 1.3 eV above eV_W). In the first experiments of this type carried out at Texas, the slow positrons were timed using the thin-scintillator technique, as shown in Figure 4.12; see section 2.3. This arrangement also allowed the total attenuation of the beam to be measured. At the end of the flight path the beam was detected using a channel electron multiplier.

For cross section measurements, gas could be admitted to the chamber, its pressure being recorded at each end of the flight path using an ionization gauge. A correction for the pressure gradient along the flight path was necessary owing to the pumping arrangement, which dispensed with the need for gas-confining apertures which might have absorbed some of the scattered beam. This was essential to the technique, which relied upon detecting all scattered positrons except those lost through positronium formation. Full detection was aided by reducing the initial beam spot to 5 mm diameter, which may be compared with the 10 mm opening of the detector, and using a high transmission grid placed near the

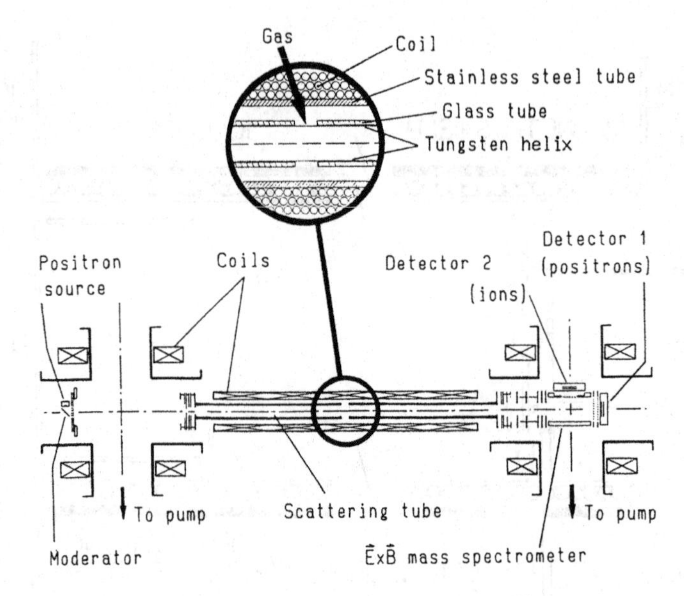

Fig. 4.13. Schematic illustration of the apparatus of the Bielefeld group for studies of positronium formation and ionization (see section 5.3 for a discussion of the latter).

source/moderator and held at a potential V_W. This reflected practically all back-scattered positrons which, along with those scattered in the forward direction, were confined by the axial magnetic field to reach the detector.

The channel electron multiplier count rate and the time-of-flight spectra were recorded simultaneously. When gas was admitted into the system the decrease in the count rate gave the fractional attenuation, f_{Ps}, due to incident positrons which formed positronium and the time-of-flight spectrum yielded the total fraction, f_T, of scattered positrons, some corrections having been made at higher energies to compensate for the effects of elastic scattering at small angles. The positronium formation cross section was then given simply by $\sigma_{Ps} = f_{Ps}\sigma_T/f_T$, where the values of σ_T were taken from the literature.

Fornari, Diana and Coleman (1983) investigated positronium formation in helium, H_2 and argon gases over the energy ranges from their respective formation thresholds up to 76 eV. The number of targets and the energy range were extended by Diana and his collaborators. In a later variant of their apparatus, Diana *et al.* (1986b) dispensed with the time-of-flight arrangement and instead deduced f_T using a retarding tube located just before the detector; this allowed the use of stronger radioactive sources and higher positron fluxes. This instrument is similar to that used by Coleman *et al.* (1992) in their investigations of the behaviour of σ_{el} at impact energies near E_{Ps} (see section 3.3).

The results obtained by the Texas group (Fornari, Diana and Coleman, 1983; Diana *et al.*, 1986b) are discussed below along with those obtained by the Bielefeld group, whose apparatus is shown schematically in Figure 4.13. In this case the manner of detecting positronium depended upon method (iii) outlined above, i.e. detection of the ions formed without an accompanying free positron in the final state.

Figure 4.13 shows the differentially pumped gas cell, of length 50 cm, used by Fromme *et al.* (1986) to form their scattering target; the inset illustrates the inner construction of the cell, which consisted of a glass tube with a tungsten helix lining. The latter was used to define the electrical potential of the wall and, by having a potential difference along the wire of approximately 10 V, to provide an extraction field for the ions created along the axis of the tube. However, this electric field also had the detrimental effect of accelerating the positrons as they passed through the scattering region. The beam was transported in an axial magnetic field with a strength in the scattering region of approximately 35 mT, which also helped to confine the ions and facilitate their detection. Later (e.g. Kruse *et al.*, 1991) a field of 0.12 T was used to increase the efficiency of ion transport.

The beam was formed using a 70 MBq ^{22}Na source, held off the axis of the beamline to remove it from the line of sight of the detectors, together with a tungsten plate moderator held at 45° with respect to the same axis. The beam diameter was set at 4 mm by the scattering-cell entrance aperture, and its energy was varied by varying the voltage applied to the moderator. Positrons were confined by the magnetic field to reach detector 1, a channel electron multiplier array. An $\boldsymbol{E} \times \boldsymbol{B}$ mass separator was located after the scattering region, and the ions and positrons were both accelerated into this region, the ions being deflected by the field combination to detector 2. After appropriate allowance for background counts in each detector and for the subtraction of any ions originating outside the scattering region, a measurement consisted of recording the count rates from detectors 1 and 2, their ratio being proportional to the combined cross section $\sigma_{\rm Ps} + \sigma_{\rm i}^{+}$, where $\sigma_{\rm i}^{+}$ is the total single-ionization cross section. By recording also the (background-subtracted) coincidences between detectors 1 and 2 and selecting the appropriate time of flight for the ions required, a relative measurement of $\sigma_{\rm i}^{+}$ was obtained.

The absolute scale of the cross sections was obtained by making measurements with the secondary electron beam produced by β^{+} bombardment of the moderator. Comparing the ion count rates measured at detector 2 obtained from both electron and positron impact gave the ratio $[\sigma_{\rm Ps} + \sigma_{\rm i}^{+}({\rm e}^{+})]/\sigma_{\rm i}^{+}({\rm e}^{-})$, which converged to $\sigma_{\rm i}^{+}({\rm e}^{+})/\sigma_{\rm i}^{+}({\rm e}^{-})$ above a positron energy of 300 eV. Above 600 eV this ratio remained constant,

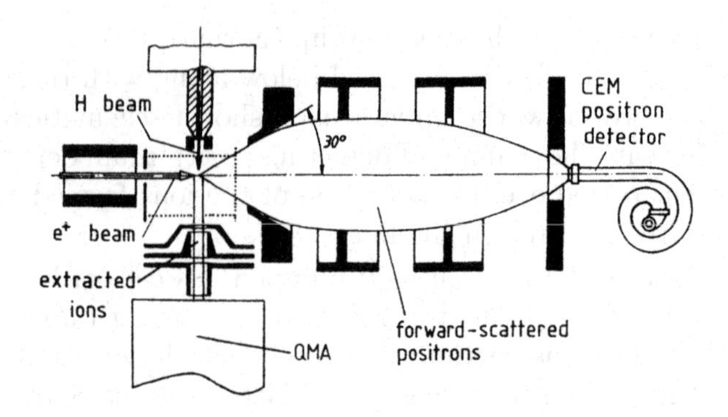

Fig. 4.14. Layout of the positron–atomic hydrogen scattering experiment developed by the Bielefeld–Brookhaven collaboration. Reprinted from *Physical Review Letters* **68**, Sperber *et al.*, Measurement of positronium formation in positron collisions with hydrogen atoms, 3690–3693, copyright 1992 by the American Physical Society.

indicating that the cross sections for the two projectiles had merged (see section 5.4 for further discussion). Values above 750 eV were used for normalization, which introduced an estimated ±8% systematic error into all the final results. Finally, $\sigma_{\rm Ps}$ was obtained by subtracting $\sigma_{\rm i}^{+}$ from $\sigma_{\rm Ps} + \sigma_{\rm i}^{+}$.

Although there have been many theoretical studies of positronium formation in positron scattering by atomic hydrogen (see subsection 4.2.1), the only experimental studies have been those of Sperber *et al.* (1992), Weber *et al.* (1994), Zhou *et al.* (1997) and Kara (1999). The apparatus, shown in Figure 4.14, is a variant of that used by Spicher *et al.* (1990), see section 5.3, to measure the positron impact-ionization cross section for atomic hydrogen. One of the main difficulties in that work was the low ion signal rate, caused by a combination of the low density of atomic hydrogen in the gas beam and a relatively weak positron beam. In order to overcome the latter restriction, the apparatus was moved to the high intensity positron beam facility at Brookhaven National Laboratory, USA, where the positronium formation experiments were performed.

The principle of the experiment was similar to that used by the Bielefeld group in their earlier measurements of positronium formation in helium and H_2 in that the positronium signal was identified by the production of an ion (proton) without an accompanying free positron in the final state. The apparatus consisted of a target gas beam, which was a mixture of atomic and molecular hydrogen, crossed by an electrostatically confined positron beam. Ions were extracted using a d.c. electric field of 8 V cm^{-1} and passed through a quadrupole mass analyser (QMA), set to transmit protons, before being detected.

The atomic hydrogen was produced in a radio-frequency Slevin-type discharge, whose operating conditions were held stable to facilitate data-taking over long periods, and the degree of dissociation of the hydrogen was found to be $\approx 55\%$ in the region where the scattering took place. The CEM detector shown in Figure 4.14 could only detect positrons scattered into a cone with a half-angle of 30°, as well as the unscattered beam, the effect of which will be described below.

The measurement consisted of recording the background-corrected QMA count rate for particles with unit mass, which, since both positronium formation and ionization produce protons, is proportional to $\sigma_{\mathrm{Ps}} + \sigma_{\mathrm{i}}^{+}$, and also monitoring the QMA–CEM coincidence rate, which should be proportional to the ionization cross section alone. The positronium formation cross section can then be determined by subtraction, once appropriate normalization has taken place at an energy where $\sigma_{\mathrm{Ps}} \ll \sigma_{\mathrm{i}}^{+}$.

Several systematic effects were considered by Sperber *et al.* (1992) and Weber *et al.* (1994), but the most serious potential source of error arose from the limited angular acceptance of the CEM for scattered positrons. In general terms it is expected that at higher positron energies the ionization cross section will be predominantly forward peaked, so that the scattered positrons can be detected by the CEM. At lower energies, however, some positrons may be scattered through angles $> 30°$, leading to loss of the positron and the signal thus being wrongly classified as due to positronium formation. Although this effect becomes potentially serious as the positron energy is lowered, it is offset by the fact that σ_{i}^{+} decreases relative to σ_{Ps} in this region. In an attempt to correct for this effect, angular distributions obtained from the first Born approximation were used to estimate the fraction of ionizing positrons scattered through angles $> 30°$. Sperber *et al.* (1992) found that their measured values of σ_{Ps} had to be lowered by an amount which varied between 4% and 20% in the energy range 15–65 eV.

Investigations of the relative behaviour of the QMA and QMA–CEM coincidence signals showed that their ratio remained constant within the experimental errors at energies greater than approximately 70 eV. This was taken to imply that in this energy region $\sigma_{\mathrm{Ps}} \approx 0$ and as a result both signals were proportional to σ_{i}^{+}. Thus, relative efficiencies of the two signals could be determined, so that a relative measure of σ_{i}^{+} could be subtracted from the QMA signal, the residue being directly proportional to σ_{Ps}. Weber *et al.* (1994) described how this was done in detail and presented some new data and corrections to the original values of Sperber *et al.* (1992). They also included a discussion of the normalization procedure and tabulated the final cross sections. We note, however, that in correcting the ionization work of Spicher *et al.* (1990), Hofmann *et al.*

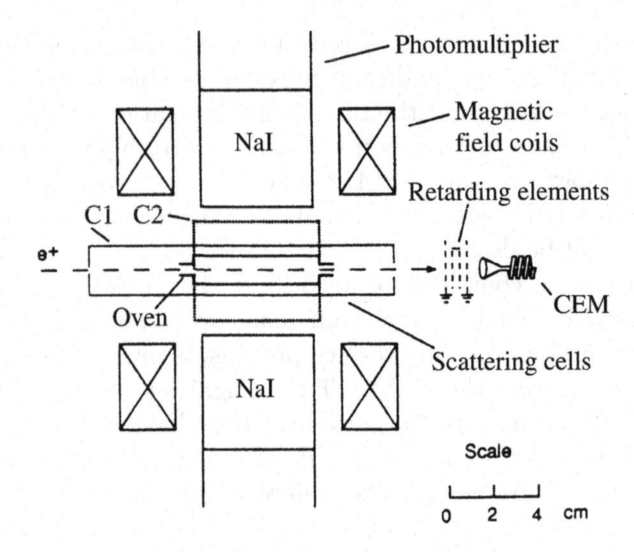

Fig. 4.15. Illustration of the apparatus developed by Zhou *et al.* (1994b) for studies of positronium formation in positron–alkali metal collisions.

(1997) pointed out that there are large changes to the values of σ_{Ps} determined by Weber *et al.* (1994).

Positronium formation cross sections for atomic hydrogen have also been reported by Zhou *et al.* (1997), who used the special hydrogen scattering cell described in subsection 2.5.4 with reference to their measurements of total scattering cross sections. Details were given by Zhou *et al.* (1997) and descriptions can be found below, since the same methods were used by the Detroit group (e.g. Zhou *et al.*, 1994b) in their alkali metal work.

The apparatus employed by Zhou *et al.* (1994b) and Surdutovich *et al.* (1996) for measurements of positronium formation in the alkali metals is shown in Figure 4.15. The experiment used more than one method to estimate σ_{Ps}, which resulted in the setting of both upper and lower limits on the cross section. Comparisons with the results of the Arlington group (Fornari, Diana and Coleman, 1983) for argon gas were made, to test the reliability of the methods. In order to investigate the effects of different geometries on the apparent value of σ_{Ps} derived by a gamma-ray technique, two gas cells with different length-to-diameter ratios were employed for the argon studies, and both are shown in Figure 4.15 along with the oven used for the alkalis.

After the scattering region, the beam passed a retarding grid assembly, which could be used to monitor the longitudinal energy spread of the beam and also to provide angular discrimination against forward elastically scattered projectiles when the apparatus was used to determine total cross sections. These measurements were undertaken to check the product of the gas density and the path length as seen by the beam, which was monitored at the end of the flight path using a CEM detector. Annihilations occurring in the gas-cell region could be monitored using the two NaI(Tl) detectors set to count coincident pairs of 511 keV gamma-rays. These could arise either from annihilations of para-positronium or scattered positrons, as they strike one of the apertures, or from ortho-positronium which quenched on collision with the walls of the cell. Full details of the procedure were given by Zhou *et al.* (1994b). In short, they were able to set upper limits to σ_{Ps} by using the Arlington technique, described above, to confine the scattered positrons and ascribing any lost positrons to positronium formation. Care was taken to ensure that all forward-scattered positrons passed through the relevant apertures to reach the CEM; the retarder in front of the CEM was, of course, set to zero bias. However, positrons scattered into the backward hemisphere could also fail to reach the detector and so contribute to the apparent positronium formation cross section, thus rendering these results upper limits to the true values.

The apparent para-positronium formation cross section, obtained using the 511 keV coincidence signal, was used as a lower limit on σ_{Ps}. If significant ortho-positronium quenching takes place at the chamber walls within sight of the gamma-ray detectors, then this cross section may be close to the true value. In practice, experiments with the two different gas-cell geometries found this to be the case, the quenching effect being highest in the scattering cell having the smallest inner diameter. In a study of argon, Zhou *et al.* (1994b) found that most of the experimental results of Fornari *et al.* (1983), which were considered to be the most reliable, lay between their measured upper and lower limits.

The final system we describe here is that used by Moxom and coworkers (1993, 1994) for studies of the near-threshold behaviour of the positronium formation cross section. Their apparatus, which is described in detail by Moxom, Laricchia and Charlton (1995a), actually measured the total ion yield and thus gave values of $\sigma_{Ps} + \sigma_i^+$. The scattering cell they employed is shown in Figure 4.16. The positron beam was guided by an axial magnetic field through various electrostatic elements located prior to the cell, which enabled unwanted secondary electrons to be eliminated and the energy spread of the beam to be reduced for the near-threshold work. This property of the beam could be analysed using the grid arrangement in front of the ceratron detector. A Wien filter was also used to help prevent unwanted particles from reaching the scattering cell and also to chop the beam.

The scattering took place in the hemispherical chamber. The base of the hemisphere consisted of a series of electrodes which, when appropriately biassed, created a radial electric field throughout the interaction region.

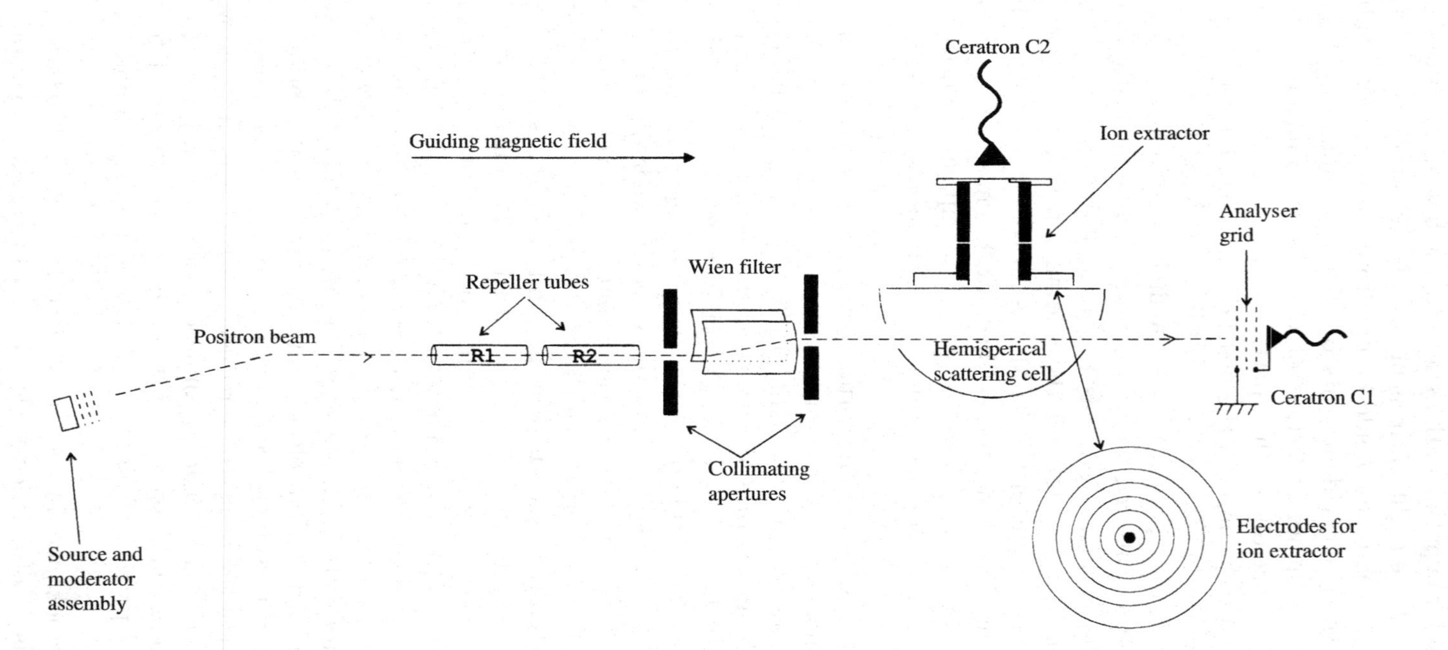

Fig. 4.16. Layout of the apparatus of Moxom, Laricchia and Charlton (1995a) for studies of positronium formation and ionization (not to scale). Reprinted from *Applied Surface Science* **85**, Moxom *et al.*, A gated positron beam incorporating a scattering cell and novel ion extractor, 118–123, copyright 1995, with permission from Elsevier Science.

Thus, ions from all points in the cell were focussed onto a hole in the centre of the extractor, through which they could pass to be detected by a second ceratron. The resultant efficient ion extraction and detection facilitated measurements near threshold, where the cross sections are small. The measurement cycle was as follows: the beam was allowed to pass through the scattering chamber for a period of around 40 μs, before being chopped by grounding the Wien filter; after a 1 μs delay to allow all the beam to leave the cell, a voltage of −180 V was applied to the extraction electrodes for a duration of 5 μs and the total number of ions created was counted. This cycle was repeated at each desired beam energy until the required statistical accuracy on the ion counts was achieved. It was found that the lifetime of the ions in the scattering cell was sufficiently long that the majority were extracted before they were lost by collision with the cell wall.

4.4 Results

1 Noble gases and molecules

In Figure 4.17 experimental and theoretical data for σ_{Ps} for helium gas are shown for the energy range from threshold up to around 300 eV. (A closer look at the data near threshold is given later in this section.) Whereas the experimental results of Fornari, Diana and Coleman (1983) are in good agreement with those of Fromme *et al.* (1986), the higher energy extensions of the Texas work by Diana *et al.* (1986b) are not in such good accord. In particular, the oscillatory behaviour observed by Diana *et al.* (1986b) is not found in the results of Fromme *et al.*, which appear to vary smoothly up to the highest energy of approximately 300 eV. This is also true of the data of Overton, Mills and Coleman (1993), who tentatively attributed the phenomenon observed by the Texas group to beam optic effects (see below).

Among the theoretical data shown in Figure 4.17 are the polarized-orbital approximation results of Khan and Ghosh (1983) and Khan, Mazumdar and Ghosh (1985), which together give the contributions from positronium formation into the $n_{Ps} = 1$ and $n_{Ps} = 2$ states. Also shown are the results of the distorted-wave method used by Mandal, Guha and Sil (1980), the coupled-static (plus second order optical potential) approximation of McAlinden and Walters (1992) and the coupled-state calculations of Hewitt, Noble and Bransden (1992a). Although some of these theories are rather crude, nonetheless inspection of Figure 4.17 reveals that the magnitude and position of the observed maximum are reasonably well reproduced. A comparison up to 140 eV of the experi-

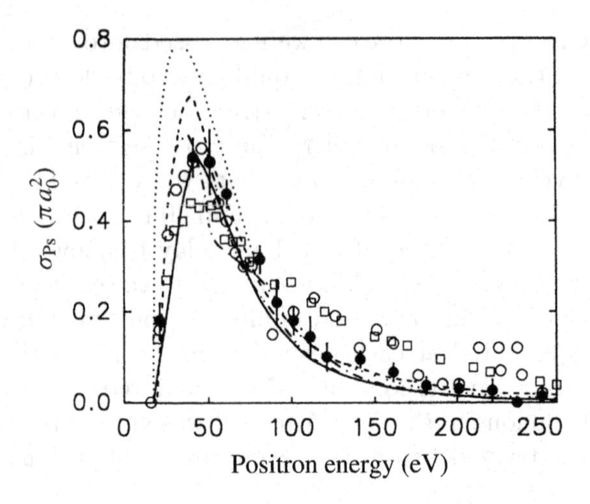

Fig. 4.17. Total positronium formation cross sections for positron–helium scattering. Experiment: •, Overton, Mills and Coleman (1993); ○, Fornari, Diana and Coleman (1983) and Diana *et al.* (1986b); □, Fromme *et al.* (1986). The error bars have been omitted from the latter two sets of data for clarity. Theory: ——, McAlinden and Walters (1992); · · · · ·, Schultz and Olson (1988); – – – –, Mandal, Guha and Sil (1980); — · —, Hewitt, Noble and Bransden (1992a); — · · —, Igarashi and Toshima (1992); — —, Deb, Crothers and Fromme (1990). See also Figure 4.8(b).

mental data with the 27-coupled-state approximation of Campbell *et al.* (1998a) was shown in Figure 4.8(b).

It is notable that the experimental results of Diana *et al.* (1986b) and Fromme *et al.* (1986) exceed the theoretical results at energies above approximately 70–80 eV. Indeed, as Schultz and Olson (1988) pointed out, the energy dependences of the measured and calculated cross sections do not agree, even allowing for the size of the experimental errors. Schultz and Olson (1988) noted that above 100 eV the energy dependence of the experimental data is intermediate between E^{-1} and $E^{-1.5}$, whilst that of the theoretical results is closer to $E^{-3.5}$. These authors have argued further that the discrepancy may be an experimental error caused by an incomplete collection of positrons scattered at large angles in the process of direct ionization. If this were so, it would lead to an overestimate of σ_{Ps} in both experiments. This error would be most pronounced where σ_{Ps} is small, but the resulting difference in the ionization cross section would be negligible in view of the size of σ_{Ps} at these energies. This argument was taken further by direct calculations of Schultz, Reinhold and Olson (1989), although the basis of their argument was criticized by Deb *et al.* (1990) from both experimental and theoretical standpoints.

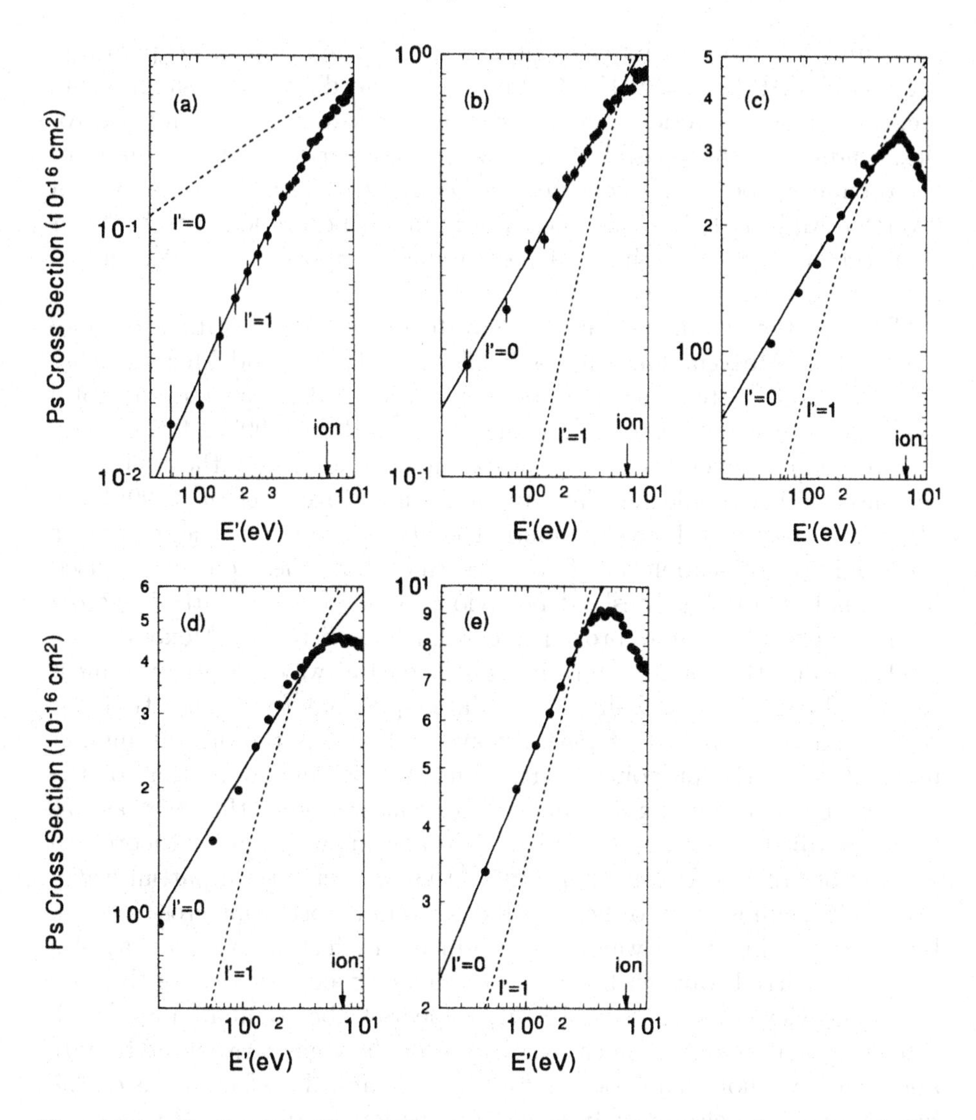

Fig. 4.18. Positronium formation cross sections plotted as functions of the positronium kinetic energy for the following gases: (a) helium, (b) neon, (c) argon, (d) krypton, (e) xenon. The ionization threshold in each case is indicated by 'ion'. Reprinted from *Physical Review* **A50**, Moxom *et al.*, Threshold effects in positron scattering on noble gases, 3129–3133, copyright 1994 by the American Physical Society.

More recent experiments by Overton, Mills and Coleman (1993) may have gone some way towards resolving this dispute in favour of theory. As mentioned above, this group used the Texas technique to measure σ_{Ps} but paid particular attention to obtaining a well-controlled positron beam, particularly at intermediate energies. Based on their previous experience, they concluded that small $\boldsymbol{E} \times \boldsymbol{B}$ effects introduced by grid misalignments etc. could have deleterious energy-dependent effects on positron trajectories, leading to the possible loss of scattered particles and perhaps to the oscillatory behaviour reported by Diana *et al.* (1986b). The data of Overton, Mills and Coleman, shown in Figure 4.17, are found to be lower than the previous experimental measurements above ≈ 100 eV, and in better accord with theory.

Close to threshold, the most accurate experimental data are those obtained by Moxom, Laricchia and Charlton (1993) and Moxom *et al.* (1994), whose system, as described in section 4.3, measured the total ion yield (see e.g. Moxom, Laricchia and Charlton, 1995b, for a study at intermediate energies) but gave σ_{Ps} at energies below the ionization threshold. The noble-gas data reported by Moxom *et al.* (1994) are plotted in Figure 4.18 as a function of E', the kinetic energy of the positronium. In section 3.3 it was described how these data were used in an analysis of the threshold behaviour of the elastic scattering cross section. The plots show broken and solid lines fitted by Moxom *et al.* (1994) in an attempt to determine which partial waves contribute most to σ_{Ps}. In the case of helium, the data appeared to suggest that σ_{Ps} is dominated by the $l' = 1$ partial wave, where l' is the orbital angular momentum of the outgoing positronium and is related to that of the incident positron by angular momentum conservation. However, as has been described in sections 3.3 and 4.2, it is now known from the theoretical calculations of Van Reeth *et al.* (1997) that in fact several partial waves contribute significantly to the overall positronium formation cross section, but they add in such a way as to produce an effect similar to that of a dominant partial wave with $l' = 1$. A similar effect may also be the case for the heavier noble gases, where σ_{Ps} appears to be dominated by $l' = 0$. The positronium formation cross sections for these gases rise more rapidly above the threshold than does that for helium, and the effect of this on the behaviour of the elastic scattering cross sections is discussed by Moxom *et al.* (1994).

Results for the heavier noble gases were obtained by the Arlington group (Fornari, Diana and Coleman, 1983; Diana and coworkers, 1986a, c). As in the case of helium, the oscillations observed at higher energies (shown most clearly for argon in Figure 4.19) have yet to find a plausible physical explanation. The data of Zhou *et al.* (1994b) for argon, obtained, as described in section 4.3, in an investigation into positronium formation from the alkali metals, do not agree with the data of Diana *et al.* (1986c) and reveal no structure. The only calculations of positronium formation for argon have been, as described in subsection 4.2.4, those of McAlinden and Walters (1992), who used a truncated coupled-static

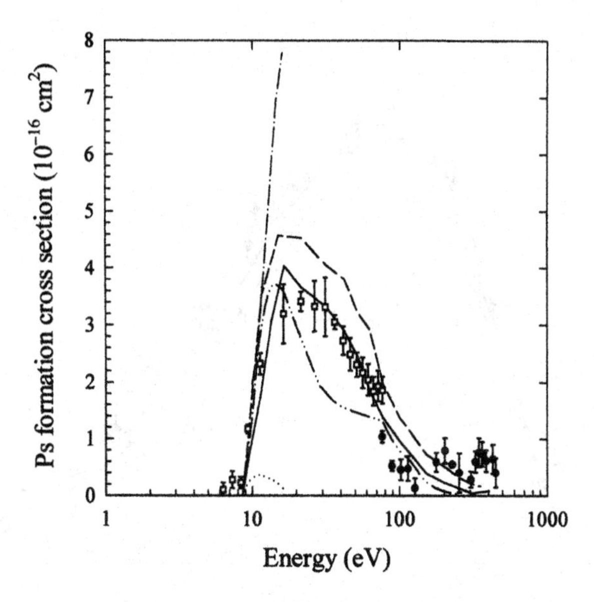

Fig. 4.19. Positronium formation cross sections for positron–argon scattering. Experiment: □, Fornari, Diana and Coleman (1983); •, Diana *et al.* (1986c); the solid line and the broken line are drawn through the upper and lower best estimates of Zhou *et al.* (1994b). Theory: · · · · ·, distorted-wave calculation, — · —, first Born approximation, both from Gillespie and Thompson (1977); — · · —, McAlinden and Walters (1992).

approximation, and those of Gillespie and Thompson (1977), who used a distorted-wave approximation up to 8 eV above threshold. The results of the latter are smaller than the experimental measurements by more than an order of magnitude. Given the nature of the approximation and the complexity of the target, the data of McAlinden and Walters (1992) are in fair accord with experiment. The cross sections shown comprise the capture of electrons from both the 3p and 3s shells and the feature observed around 75 eV is caused by capture from the 3s shell.

Data also exist for H_2 and in this case reasonable agreement has been found between the theoretical estimates of Bussard, Ramaty and Drachman (1979) and the experimental results of the Arlington and Bielefeld groups at energies up to ≈ 20 eV. Between 20 eV and 35 eV, however, theory exceeds experiment by up to 30%. At higher energies the calculations of Ray, Ray and Saha (1980), using the molecular Jackson–Schiff approximation, and those of Sural and Mukherjee (1970), using the Born approximation, are consistent with experiment, the latter only slightly exceeding the theoretical values.

The near-threshold behaviour of the positronium formation cross section has also been investigated for O_2 (Laricchia, Moxom and Charlton,

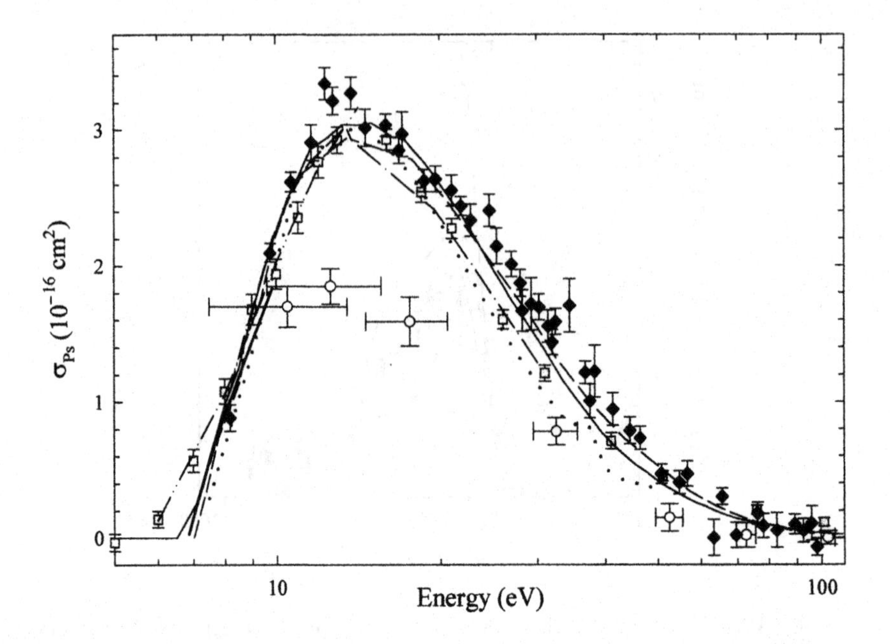

Fig. 4.20. Positronium formation in positron–hydrogen scattering. Experiment: □, Zhou *et al.* (1997); ○, Hofmann *et al.* (1997); •, Kara (1999). Theory: ——, Brown and Humberston (1985); ——, Kernoghan *et al.* (1996); — —, Mitroy (1996); — · —, Igarishi and Toshima (1994); · · · · ·, Higgins and Burke (1993); — · · —, Janev and Solov'ev (1998).

1993) and CO_2 (Laricchia and Moxom, 1993). In the former case the cross section was found to have a distinct peak a few eV above threshold, which was linked by Laricchia and coworkers to the behaviour of the cross section for positron impact excitation of the molecule, as will be described in subsection 5.1.2. In the case of CO_2 a detailed investigation of the energy dependence of the ion yield supported the interpretation of Laricchia, Charlton and Griffith (1988) that simultaneous positronium formation and excitation of the residual ion is an important process in this gas. This phenomenon is discussed in more detail in section 4.5.

2 Atomic hydrogen

The results obtained by Hofmann *et al.* (1997), who corrected those of Sperber *et al.* (1992), are shown in Figure 4.20 along with the data of Zhou *et al.* (1997), Kara (1999) and the results obtained from several recent theoretical studies. A comprehensive discussion of the theoretical data was given in subsection 4.2.1. The observed maximum in σ_{Ps} occurs at around 15 eV and has a value of approximately 3.0×10^{-16} cm^2.

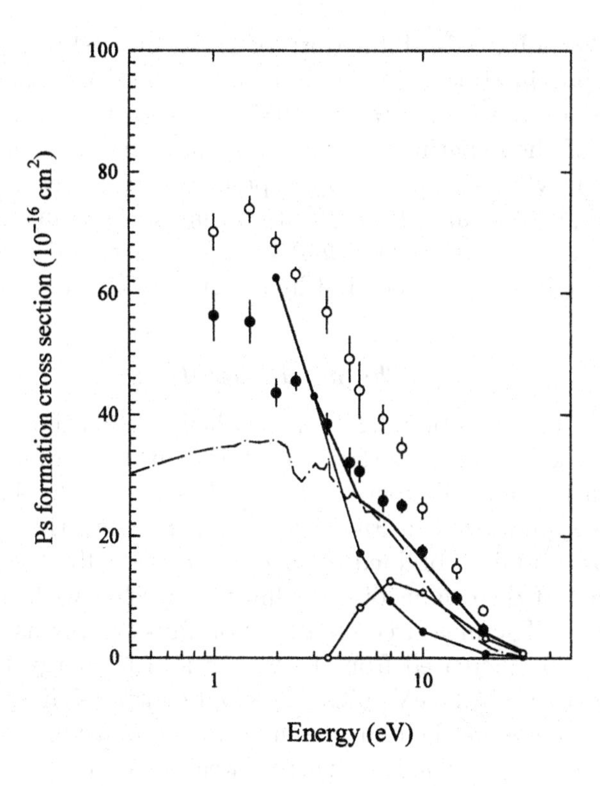

Fig. 4.21. Positronium formation in positron–sodium collisions. The experimental points are from Zhou *et al.* (1994b): the open and solid circles give the upper and lower limits respectively. Theory: upper solid line, the cross section summed for $n_{\rm Ps} = 1$ and $n_{\rm Ps} = 2$, lower solid line with solid circles, cross section for $n_{\rm Ps} = 1$, lower solid line with open circles, cross section for $n_{\rm Ps} = 2$, all from the calculations of Hewitt, Noble and Bransden (1993); — · —, total positronium formation cross section from Campbell *et al.* (1998a).

Thereafter, the cross section appears to fall monotonically to a value close to zero at 100 eV. As explained by Hofmann *et al.* (1997) the re-evaluated data of Sperber *et al.* (1992), which are lower than both theory and other experiments over the entire low energy region, are now thought to be incorrect.

The best agreement with experiment has been obtained by Kernoghan *et al.* (1996) and Mitroy (1996), using the coupled-state approximation with large basis sets (see subsection 4.2.1), and by Igarashi and Toshima (1994), Janev and Solov'ev (1998) and Higgins and Burke (1993). The results of several other calculations (not shown in Figure 4.20) fall below the experimental values, particularly near the cross section maximum, e.g. the R-matrix method as used by Higgins and Burke (1991), the classical calculations of Ohsaki *et al.* (1985) and Wetmore and Olson

(1986) and the polarized-orbital work of Khan and Ghosh (1983). The first Born approximation (Massey and Mohr, 1954) and the Fock–Tani field-theoretic approach of Straton (1987) both yield cross sections that are in excess of the experimental values in the vicinity of the maximum and have a somewhat sharper energy dependence at higher energies. Also shown in Figure 4.20 are the accurate variational results obtained by Brown and Humberston (1984, 1985) for positron energies between $E_{\rm Ps}$ and the first excitation threshold of hydrogen at 10.2 eV.

3 The alkali metals

As emphasized in subsection 4.2.3, of the alkali metals the most detailed theoretical work has been performed on the lithium atom, though to date there have been no experimental studies reported. The situation for sodium is summarized in Figure 4.21, which shows the experimental upper and lower limits obtained by Zhou *et al.* (1994b) (see section 4.3 for a discussion of the origins of these limits) together with the results of various theories. The measured cross section falls steeply as the incident positron energy is increased from 1 eV, the lowest energy investigated, reaching almost zero by 25 eV. This behaviour contrasts sharply with that for atomic hydrogen and helium presented in the preceding sections. For the alkalis the outer electron is sufficiently weakly bound that positronium formation is an exothermic reaction, and this fact governs the shape of the cross section. In particular, as mentioned previously and in accord with the measurements, the cross section is expected to diverge as the incident positron energy is lowered to zero.

The theoretical work of Hewitt, Noble and Bransden (1993) used a seven-coupled-state approximation (four states of sodium and three of positronium) to calculate cross sections for positronium formation into states with $n_{\rm Ps} = 1$ and 2. Both results are shown in Figure 4.21, together with the sum. Reasonable accord is found with the lower-limit measurements of Zhou *et al.* (1994b). As expected, the calculations show that positronium formation into the ground state dominates at low energies and that the $n_{\rm Ps} = 2$ contribution, with its threshold at 3.4 eV, is more important above approximately 7 eV. Only limited agreement is found between the experimental measurements and the results of the distorted-wave and first Born approximation calculations of Guha and Mandal (1980). In particular, the first Born approximation gives completely the wrong energy dependence. Coupled-state calculations on this system have also been reported by McAlinden, Kernoghan and Walters (1994), though using a smaller set of states. The agreement with experiment is poorer than that found by Hewitt, Noble and Bransden (1993) and the results are not shown in Figure 4.21. The most elaborate

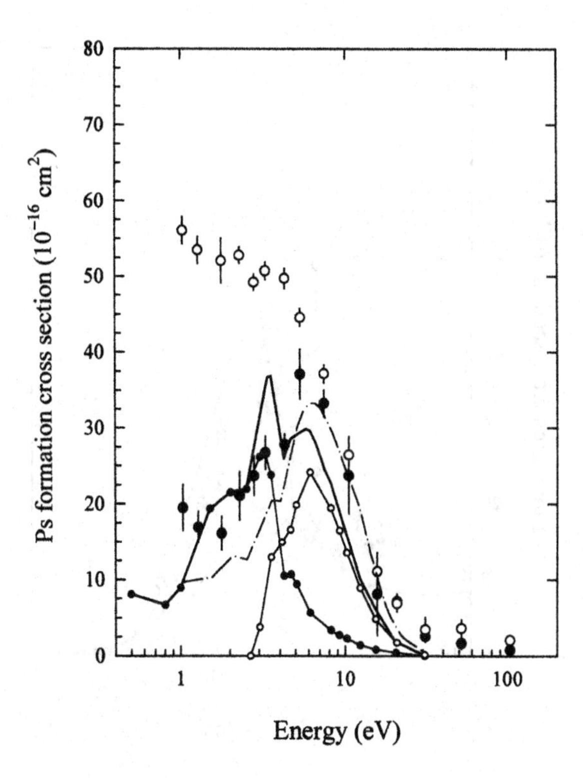

Fig. 4.22. Positronium formation in positron–potassium collisions. The experimental points are from Zhou *et al.* (1994b); the open and solid circles are the upper and lower limits respectively. Theory: thick solid line, cross section summed for $n_{\rm Ps} = 1$ and $n_{\rm Ps} = 2$, thin solid line with solid circles, cross section for $n_{\rm Ps} = 1$, thin solid line with open circles, cross section for $n_{\rm Ps} = 2$, all from calculations of Hewitt, Noble and Bransden (1993); — · —, total positronium formation cross section from McAlinden, Kernoghan and Walters (1996).

coupled-state calculations are those of Campbell *et al.* (1998a), who used six states of positronium and 27 of sodium. Their results are in fair agreement with experiment beyond 4 eV but are too low at lower energies.

Results for potassium and rubidium are given in Figures 4.22 and 4.23. Note that the simpler approximations of Guha and Mandal (1980), as shown in the summary of Stein *et al.* (1996), are in even poorer agreement with experiment than was the case with sodium and are therefore not presented here. The experimental data for potassium show a larger difference between the upper and lower limits than was the case for sodium. Indeed the two sets of measurements exhibit opposite trends below approximately 5 eV, the lower-limit results showing a dramatic decrease whilst the upper-limit values continue to rise. Although at first sight this might suggest that little of significance could be said about $\sigma_{\rm Ps}$

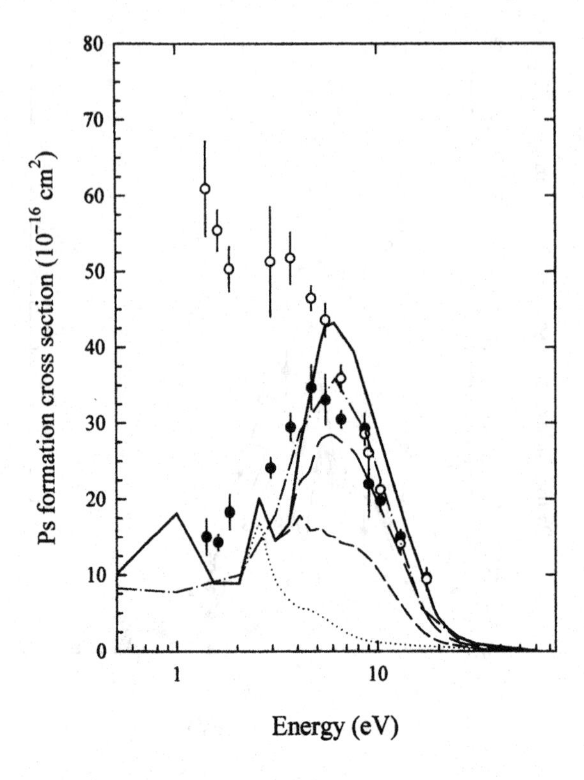

Fig. 4.23. Positronium formation in positron–rubidium collisions. The experimental points are from Surdutovitch *et al.* (1996); the open and solid circles are the upper and lower limits respectively. Theory: — · —, the total positronium formation cross section from Hewitt, Noble and Bransden (1993): ——, total positronium formation cross section; — —, the contributions from $n_{\rm Ps} = 1$, 2 and 3; - - - -, the contributions from $n_{\rm Ps} = 1$ and 2; · · · · ·, the contribution from $n_{\rm Ps} = 1$. All these theoretical results are taken from the work of Kernoghan *et al.* (1996).

for this target, theory suggests that the lower-limit results are closer to the true cross sections and that the trend discovered is a genuine effect. It is notable that the lower limits on the cross sections deduced by Zhou *et al.* (1994b) for argon gas are closest to the data of Fornari, Diana and Coleman (1983), which also lends support to their alkali-metal measurements.

The theoretical results shown are from the coupled-state calculations of Hewitt, Noble and Bransden (1993) and McAlinden, Kernoghan and Walters (1996) (for potassium), and Kernoghan *et al.* (1996) (for rubidium). These authors have also studied caesium and have obtained similar conclusions to those for rubidium. Hewitt and coworkers again used a seven-state approximation, whereas McAlinden, Kernoghan and

Walters (1996) and Kernoghan *et al.* (1996) both employed 11 states (five atomic and six positronium). A common feature of all the calculations is that, although they differ in the details of the magnitudes, they do show that the formation of positronium into excited states is a very important process for these targets and is responsible for the peak-shaped cross sections measured by Zhou *et al.* (1994b). A detailed study by McAlinden, Kernoghan and Walters (1996) shows how the breakdown into the various positronium states occurs. These authors obtained cross sections which, at least for positronium formed in the 2s and 2p states, were a factor 2–3 smaller than those calculated by Hewitt, Noble and Bransden (1993) using the seven-state approximation.

Reasonable agreement is obtained by McAlinden, Kernoghan and Walters (1996) and Kernoghan *et al.* (1996) with the experimental measurements of Zhou *et al.* (1994b) and Surdutovich *et al.* (1996), particularly given the complex nature of the targets and the approximations inherent in both theory and experiment. In calling for a tightening up on the cross sections available from both sources, McAlinden, Kernoghan and Walters (1996) issued some warnings. They highlighted the fact that yet more atomic and positronium states may be needed in the calculation, together with some pseudostates which could help take account of coupling to the continuum (these were found to be of great value in the positron–atomic hydrogen case; see subsection 4.2.1). A better representation of exchange and inner shell effects may also need to be incorporated into the theory.

4.5 Other processes involving positronium formation

The formation of excited states of positronium, usually termed Ps*, through reaction 4.2, can make significant contributions to σ_{Ps}. This has already been discussed briefly in the theoretical section 4.2 and in subsection 4.4.3, where we saw that excited state positronium is thought to dominate σ_{Ps} at certain energies in some of the alkali metals. Here we describe the only experiment to date which has directly detected excited state positronium formation in gases, namely that of Laricchia *et al.* (1985).

The principle of the experiment is similar to that of Canter, Mills and Berko (1975), in that Ps* was detected by monitoring coincidences between an ultraviolet photon (the positronium Lyman-α line at 243 nm) and a gamma-ray from the subsequent annihilation of ground state ortho-positronium. The apparatus employed by Laricchia *et al.* (1985) is shown in Figure 4.24. A low energy positron beam passed through a hemispherical aluminium scattering cell, which was coupled, via a light pipe made from (Al + MgF_2)-coated glass tubes, to an ultraviolet-sensitive phototube; the coated glass enhanced the transmission of photons to the

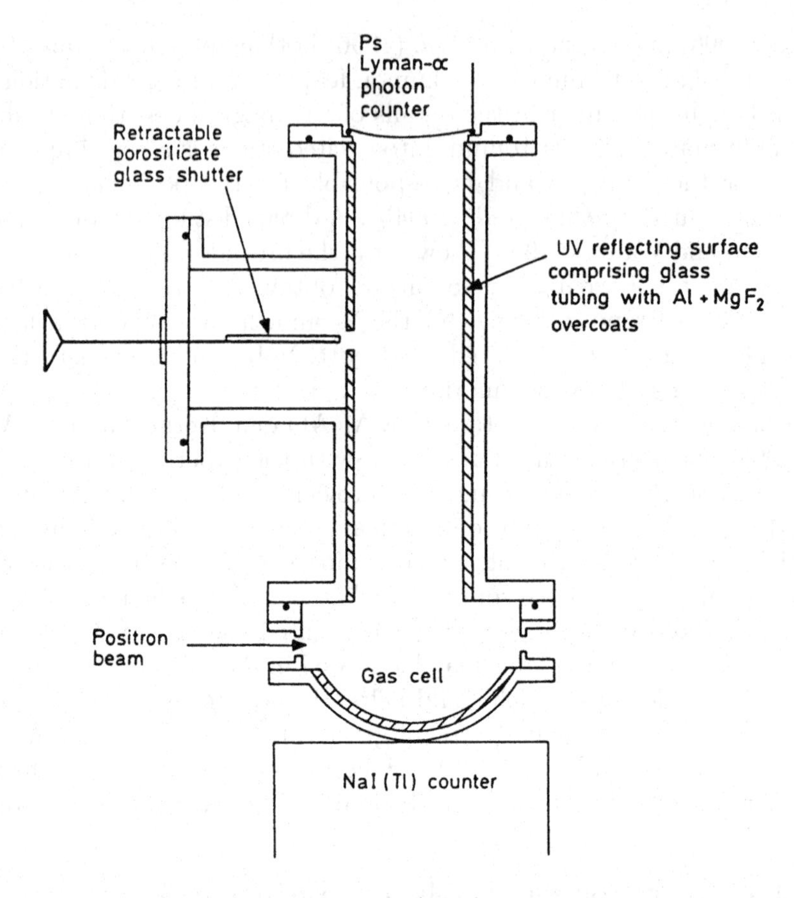

Fig. 4.24. Apparatus of Laricchia *et al.* (1985) for the detection of excited state positronium formed in positron–gas collisions.

detector along the light guide. This arrangement allowed the phototube to be removed from the magnetic field used to guide the beam. A borosilicate glass disc was used as a crude filter to prevent photons with wavelengths $\leq$ 285 nm from reaching the phototube. This disc could be inserted into the light path using a retractable shaft. The phototube signals were used to start a conventional delayed-coincidence timing arrangement; timing was stopped by the arrival of a signal from a large NaI(Tl) gamma-ray detector placed close to the scattering cell.

At each of the energies investigated, two coincidence spectra were accumulated, one with, and one without, the borosilicate glass shutter in position. Examples are given in Figure 4.25(a), which shows total (shutter out) and partial (shutter in) spectra for 17 eV positrons incident upon argon gas. The total spectrum, offset from zero by 200 ns, contains a peak around the time $t = 0$, which is due to gamma-ray–gamma-ray

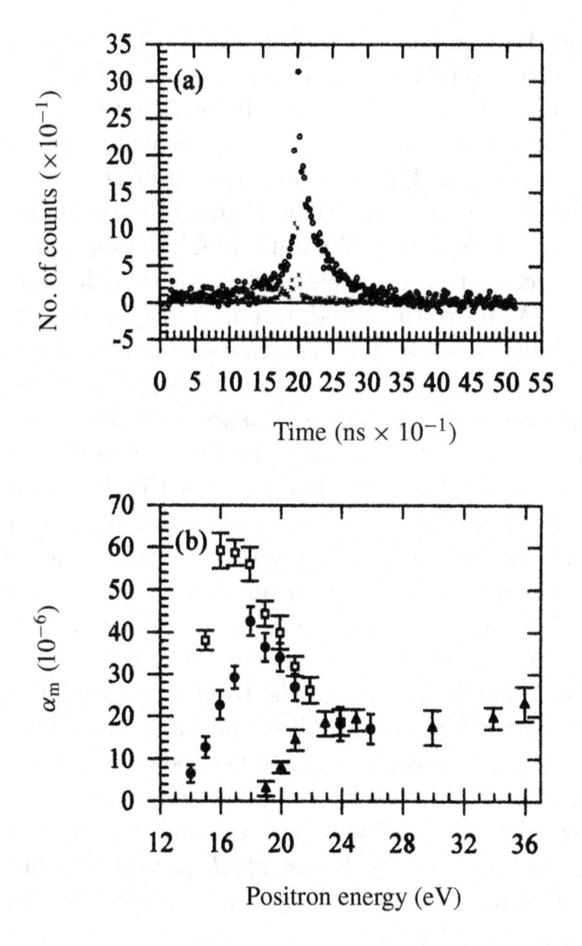

Fig. 4.25. (a) Timing spectra for Lyman-α and gamma-ray coincidence, for 17 eV positrons on argon gas: ∘, total spectrum; ×, partial spectrum. (b) The measured excited state positronium formation efficiency versus positron impact energy for H_2 (□), Ne (▲) and Ar (•) gases.

coincidences plus events on both sides of $t = 0$. The partial spectrum contains no events at $t > 0$ which, because of the constraints thus placed upon the energies of the photons causing these events, could be attributed to Ps* formation into the 2^3P state. The results of Laricchia *et al.* (1985) for the measured Ps* formation efficiency, α_m, deduced from the $t > 0$ events per scattered positron, are presented in Figure 4.25(b) for the targets argon, neon and H_2. The measured yields exhibit definite thresholds at energies which are in accord with the known values of 19.9 eV for neon, 14.1 eV for argon and 13.8 eV for H_2.

By estimating the detection efficiency of the Lyman-α–gamma-ray coincidence arrangement, Laricchia *et al.* (1985) were able to estimate the absolute Ps* production efficiency. Using a correction for the solid angles of the two detectors, plus the ultraviolet-photon transport efficiency and the photocathode quantum efficiency, it was found that the true Ps* formation efficiency was $\alpha = (950 \pm 380)\alpha_m$, resulting in a maximum yield of around 5% for H_2 gas. This is somewhat larger than the efficiencies of 10^{-2}–10^{-3} quoted by workers who have studied Ps* production at surfaces (Schoepf *et al.*, 1992; Steiger and Conti, 1992). Unfortunately, it has not been possible to convert these measured yields, which were found to fall or level off a few eV above threshold, into absolute excited state formation cross sections since the detection efficiencies would be strong functions of energy, owing to the rapid motion of the ortho-positronium. To date, no further attempts have been made to measure Ps* formation cross sections using such arrangements as ultraviolet photon–ion coincidence.

In an experiment similar to the one just described, Laricchia, Charlton and Griffith (1988) reported the observation of simultaneous excitation and positronium formation, whereby positronium is produced in the collision and the residual ion is left in an excited state. This study was performed in CO_2 and was prompted by the suggestion of Kwan *et al.* (1984) that the rise in σ_T observed above 12 eV may be due to the onset of Ps* formation. The behaviour of the total cross section for low energy positron–CO_2 scattering was described in subsection 2.6.3, and it was noted there that good agreement exists between the data of Kwan *et al.* (1984) and Charlton *et al.* (1983a): both measurements show the apparent onset of a new channel, although the threshold energy is not easy to determine. As pointed out by Kwan *et al.* (1984), similar behaviour is found in the total cross section for positron scattering by N_2O, a molecule structurally similar to CO_2.

The apparatus used was that shown in Figure 4.24. Copious coincidences were found between ultraviolet photons and annihilation gamma-rays and, as seen also in the spectrum shown in Figure 4.25(a), although now much more pronounced, they occurred on both sides of $t = 0$. It was also demonstrated that the signal at $t > 0$ did not disappear when the borosilicate slide was inserted into the light path. Thus, this component did not originate from photons with energies > 4.3 eV and was not due to excited state positronium. Laricchia, Charlton and Griffith (1988) fitted exponentials to the data on both sides of $t = 0$ to obtain yields over the energy range 12–20 eV. These were an order of magnitude greater than those found in their Ps* studies and consistent with cross sections of approximately 10^{-16} cm^2.

Details of the interpretation of the features of the spectra are given in the original work; overall the data were found to be consistent with photons having energies around 3.5 eV, emitted from a slowly moving source. It was proposed that the events at $t < 0$ (i.e. gamma-rays followed in time by an ultraviolet photon) could be explained by the single-collision reaction

$$
\begin{array}{lcccc}
e^+ + CO_2 & \longrightarrow & \text{para-Ps} & + & CO_2^{+*} \\
 & & 0.125\text{ ns} \downarrow \text{prompt} & & \downarrow \text{delayed} \\
 & & \text{two } \gamma\text{-rays} & & h\nu + CO_2^+
\end{array}
\qquad (4.33)
$$

This reaction is consistent with the known energy levels of CO_2^+; in particular, a 3.5 eV photon could arise from the ${}^2\Pi_u$–${}^2\Pi_g$ transition (Turner, 1969), whose rate, $9.23 \pm 0.71\ \mu s^{-1}$ (Smith, Read and Imhof, 1975) is close to that found by Laricchia, Charlton and Griffith (1988). It was also proposed that the events at $t > 0$ were due to the related reaction

$$
\begin{array}{lcccc}
e^+ + CO_2 & \longrightarrow & \text{ortho-Ps} & + & CO_2^{+*} \\
 & & 142\text{ ns} \downarrow \text{delayed} & & \downarrow \text{prompt} \\
 & & \text{three } \gamma\text{-rays} & & h\nu + CO_2^+
\end{array}
\qquad (4.34)
$$

but now complications arise due to the fast-moving, long-lived ortho-positronium and the possibility that events are distributed on both sides of $t = 0$. In general, at low positronium speeds the events at $t < 0$ will have a decay rate equal to that of CO_2^{+*}, whilst on the opposite side of $t = 0$ the rate is that due to the annihilation of ortho-positronium. The measured yields, accounting also for the reaction (4.33) at $t < 0$, were in broad agreement with these expectations.

Laricchia, Charlton and Griffith (1988) also reported similar effects in N_2O gas, with an energy threshold consistent with a rise in the total cross section around 11 eV and with the energetics of N_2O and N_2O^+. As reported by Charlton and Laricchia (1986), a variety of other molecular targets were also investigated in the same study, but none displayed the same features as found for CO_2 and N_2O. In a later investigation of the total ion yield in CO_2, Laricchia and Moxom (1993) found that the threshold energies for the reactions (4.33) and (4.34) were located in the range 10–11 eV; they postulated that the relatively high cross section for these processes arose from an accidental quasi-resonant mechanism. Here, the energy of an excited state of the parent molecule (i.e. CO_2^*) was very close to that required to form positronium and leave the ion in an excited state. Thus, the positron could be pictured as virtually exciting the molecule but then emerging from the collision as positronium, leaving a residual excited ion.

Finally, Bluhme *et al.* (1998) have reported a study of near-threshold transfer ionization (in which the target is twice ionized and an electron and positronium are liberated) in collisions of positrons with helium and neon atoms. In contrast to the case for simple positronium formation, process (4.1), which usually has a rapid onset near threshold, the transfer-ionization channel was found to be strongly suppressed. These data are described further in section 5.5.

4.6 Comparisons with protons

Once reliable data for electron capture by positrons became available it was natural to compare the behaviour of the cross sections for this process with those for the analogous capture process in heavy positive particle impact, in particular, the cross section for the formation of atomic hydrogen in collisions of protons with various atoms and molecules, reaction (4.3). It is also pertinent to note that comparisons between the behaviour of protons and positrons are usually made at equal projectile speeds v rather than at equal energies.

McGuire (1986) first pointed out, on the basis of a calculation for an atomic hydrogen target, that the cross section for positronium formation, $\sigma_{\rm Ps}$, is much greater than the cross section for atomic hydrogen formation, $\sigma_{\rm H}$, at intermediate and high speeds ($v \geq 5$ a.u.). He attributed this to the fact that the positron must share its kinetic energy with an electron in order to capture it and must therefore slow down during the collision. Far above the formation threshold, the positronium emerges with a speed of approximately $1/\sqrt{2}$ times that of the incident positron. In contrast, the initial speed of the proton and the final speed of the hydrogen atom are almost identical. Since the capture cross section declines rapidly with increasing speed (see section 4.2 and the discussion in subsection 4.4.1), $\sigma_{\rm Ps}$ is expected to be larger than $\sigma_{\rm H}$ at a given initial projectile speed. We note in passing that, at the speeds under consideration, for either projectile charge transfer is a relatively minor constituent of the total scattering cross section.

The ratio $\sigma_{\rm Ps}/\sigma_{\rm H}$ for helium, covering the entire range of speeds for which experimental positronium formation data exists, is shown in Figure 4.26 (Schultz and Olson, 1988). The data are those of the Bielefeld and Arlington groups, divided by the accepted proton results; also shown are the classical trajectory Monte Carlo calculations. At low speeds, less than 2 a.u., electron capture is less likely by positrons than by protons. This can be attributed to a 'threshold effect', whereby the cross section is lowered by virtue of the low kinetic energy of the positron and the fact that it must expend a significant fraction of this in capturing the electron. A similar effect has been observed in scattering involving single and double ionization (see sections 5.4 and 5.5).

As the speed of the projectile is raised, experiment and theory both predict that $\sigma_{\rm Ps}/\sigma_{\rm H}$ becomes greater than unity, although, as outlined above, there is a discrepancy between the results of the Arlington and Bielefeld

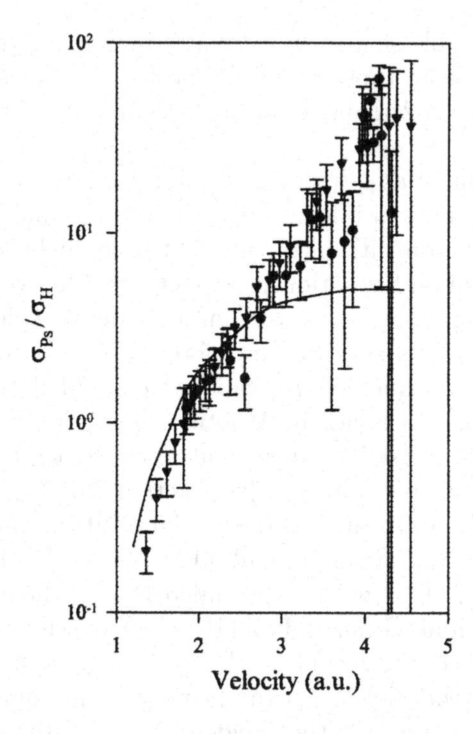

Fig. 4.26. Plot of the ratio σ_{Ps}/σ_H for positron and proton collisions with helium, versus the velocity of the projectiles (following Schultz and Olson, 1988). The curve is the result of their classical trajectory Monte Carlo calculation. The triangles are based on the positron data of Fromme *et al.* (1986) whilst the circles are from Diana *et al.* (1986b).

groups and the more recent data of Overton, Mills and Coleman (1993). Overall, theory tends to predict a lower value of σ_{Ps}/σ_H and, since the values of σ_H are well established from both theory and experiment, Schultz and Olson (1988) argued that the divergence is due to discrepancies in σ_{Ps}. Schultz and coworkers have proposed, as was outlined in subsection 4.4.1, that the cause of this discrepancy is the difficulty in making accurate measurements of the small values of σ_{Ps} at these speeds. This situation was somewhat remedied by the work of Overton *et al.* (1993).

4.7 Differential cross sections

In common with all angle-resolved cross sections, the differential positronium formation cross section, $d\sigma_{Ps}/d\Omega$, contains more information than its integrated counterpart. In particular, it sheds light on the dynamics and detailed mechanisms involved in this unique capture process. This is apparent at the higher kinetic energies, where striking effects related to the Thomas double-scattering mechanism are predicted, which are markedly different from those found for proton collisions. The latter are described in section 4.2 along with other examples of the behaviour of $d\sigma_{Ps}/d\Omega$.

The measurement of $d\sigma_{Ps}/d\Omega$ is not an easy task and, as we shall see, there is little experimental information presently available. We note here that many experimental arrangements which could be used to measure differential positronium formation cross sections sum over all the possible quantum states (n_{Ps}, l_{Ps}) of the positronium, whereas calculations usually refer to one particular state. Some differentiation between positronium states with different values of n_{Ps} can be achieved if the time of flight of the positronium, and hence its kinetic energy, is measured, and such a technique has been used to investigate beams of positronium atoms produced in positron–gas collisions (see section 7.6).

The first experimental study of differential positronium formation cross sections was made by Laricchia *et al.* (1987b) during their positronium beam development. This work was guided by the theoretical distorted-wave results of Mandal, Guha and Sil (1979), which show the cross section to be forward peaked. Laricchia *et al.* (1987b), using a simple detection system, measured the yield of ortho-positronium atoms produced in a narrow angular range about the incident positron direction for several impact energies. Their data were in reasonable agreement with the small-angle behaviour predicted by Mandal, Guha and Sil (1979).

Further information was forthcoming at small angles from the studies of Tang and Surko (1993) on molecular hydrogen. A schematic view of their apparatus is shown in Figure 4.27(a). The various grids shown were used to prevent all charged particles from reaching the channel plate detector. The positronium signal was identified as a channel plate count in coincidence with a gamma-ray detected by the photodiode detector, which incorporated a CsI scintillator. The entire detector assembly, which was located approximately 175 mm from the exit of the gas cell, with a 25 mm diameter aperture in front to define the solid angle, could be moved vertically through the positronium flux as indicated by the large solid arrows on the figure.

The fraction of the positron beam forming positronium at detector angles in the range $\pm 14°$ at impact energies of 50 eV, 80 eV and 100 eV is shown in Figure 4.27(b), which also gives the theoretical results of Biswas, Mukherjee and Ghosh (1991) and Biswas *et al.* (1991); these yielded values for $d\sigma_{Ps}/d\Omega$ for positronium formation into the 1S, 2S and 2P states using the first Born approximation. Given the crude nature of this approximation, the accord between theory and experiment is reasonable and confirms the forward-peaked nature of the positronium formation process.

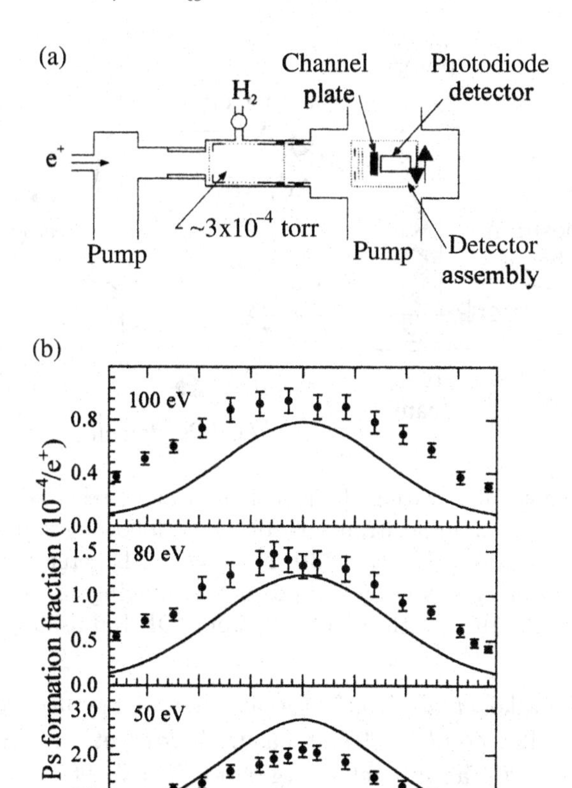

Fig. 4.27. (a) Side view of the apparatus of Tang and Surko (1993), which was used for measurements of the angular dependence of positronium formation. (b) The positronium formation fraction versus the angle between the direction of the incident positron beam (at the kinetic energies shown) and the line joining the centre of the gas cell to the centre of the detector aperture. The solid lines are predictions of this quantity derived from the theoretical work of Biswas, Mukherjee and Ghosh (1991) and Biswas *et al.* (1991).

Measurements of $d\sigma_{\rm Ps}/d\Omega$ for positron–argon scattering have also been made by Finch *et al.* (1996a, b) and Falke *et al.* (1995, 1997) and for positron–krypton scattering by Falke *et al.* (1997). The principle of the experiments, involving the detection of the positronium in coincidence with an atomic ion, is illustrated schematically in Figure 4.28. More details of the system used by Finch *et al.* (1996a), which has also been used to study the differential ionization cross section, can be found in section 5.6.

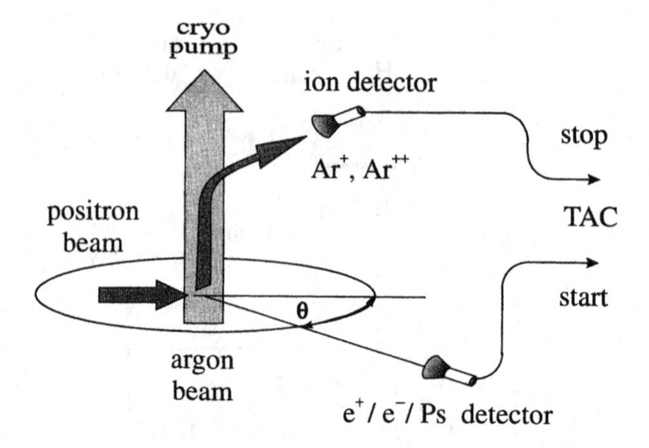

Fig. 4.28. Schematic illustrating the principle of angle-resolved measurements of positronium formation in positron–gas collisions (following Falke *et al.*, 1997). Reprinted from *Journal of Physics* **B30**, Falke *et al.*, Differential Ps-formation and impact-ionization cross sections for positron scattering on Ar and Kr atoms, 3247–3256, copyright 1997, with permission from IOP Publishing.

The data of Falke *et al.* (1997) for argon and krypton are shown in Figure 4.29. Falke *et al.* (1997) measured $d\sigma_{Ps}/d\Omega$ at various angles between 0° and 120° at impact energies of 75 eV, 90 eV and 120 eV. Their data are plotted as the ratio of the cross section at a particular angle to that at 0°; also shown are the calculations of McAlinden and Walters (1994). Good agreement is found between theory and experiment at 75 eV, particularly when allowance is made for the angular resolution of the experiment (stated as ±2°). Falke *et al.* (1995) investigated positronium formation at 30 eV impact energy at angles in the range 0°–50°. In this case the theoretical prediction by McAlinden and Walters (1994) of a minimum in $d\sigma_{Ps}/d\Omega$ at 0° was found to be in marked disagreement with the experimental measurements, which revealed a distinct maximum at this angle.

Finch *et al.* (1996a) measured $d\sigma_{Ps}/d\Omega$ at a fixed scattering angle of 60°, but at energies in the range 40–150 eV. Their results show a steady decline as the impact energy is increased, in contrast to the calculations of McAlinden and Walters (1994), to which they were normalized at 60 eV. It is notable that the detection efficiency of the channeltron used in this work was not known for positronium, but it can be presumed to have been energy dependent and would most likely increase with the kinetic energy of the positronium in the range investigated. Thus, this effect cannot be responsible for the discrepancy between theory and experiment.

The final experimental study of differential positronium formation summarized here is that of Falke *et al.* (1995, 1997), who reported

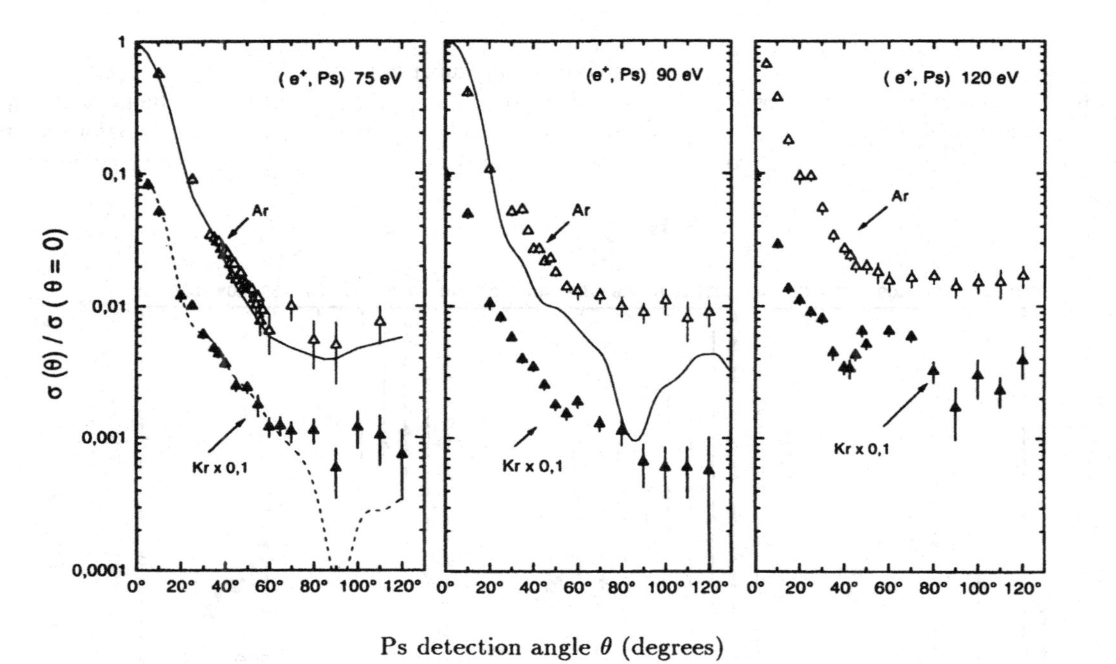

Fig. 4.29. Angle-resolved positronium formation cross sections, relative to those at 0°, for positron–argon and positron–krypton collisions at impact energies of 75 eV, 90 eV and 120 eV (Falke *et al.*, 1997). The curves are from the theoretical work of McAlinden and Walters (1994). Reprinted from *Journal of Physics*, **B30**, Falke *et al.*, Differential Ps-formation and impact-ionization cross sections for positron scattering on Ar and Kr atoms, 3247–3256, copyright 1997, with permission from IOP Publishing.

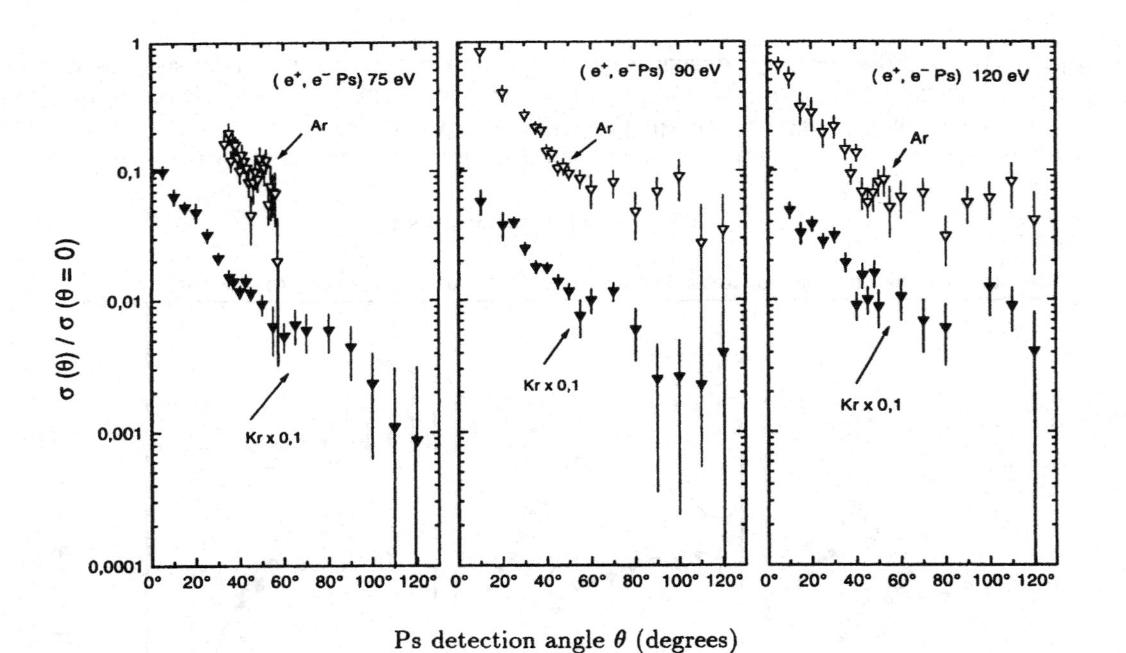

Fig. 4.30. Angle-resolved transfer-ionization cross sections, reaction (4.35), relative to those at 0°, for positron–argon and positron–krypton collisions at impact energies of 75 eV, 90 eV and 120 eV (Falke *et al.*, 1995, 1997). Reprinted from *Journal of Physics* **B30**, Falke *et al.*, Differential Ps-formation and impact-ionization cross sections for positron scattering on Ar and Kr atoms, 3247–3256, copyright 1997, with permission from IOP Publishing.

measurements on transfer ionization in positron–argon and positron–krypton collisions. The process can be written as

$$e^+ + Ar(Kr) \rightarrow Ps + e^- + Ar^{2+}(Kr^{2+}), \tag{4.35}$$

and again the positronium was detected at each angle (with an angular definition of around 2.5°) in coincidence with the Ar^{2+} ion. The results are shown in Figure 4.30. It is notable that the minimum in the cross section observed by Falke *et al.* (1995) at 75 eV in the positron–argon system and tentatively linked to a Thomas-scattering-like process (see sections 4.2 and 5.2 for discussion relevant to this topic), does not seem to persist at higher energies and is not present in the krypton data. This would seem to rule out a connection to Thomas scattering, though the data are highly suggestive and further work in this area is to be expected.

4.8 Dense gases

1 Models of the positronium fraction

Before the advent of low energy positron beams the only method of studying positronium formation in gases was by allowing β^+ particles to stop in dense samples. (Typically gas densities were $\geq 10^{25}$ m^{-3}.) Such experiments were useful in elucidating the basic mechanisms by which positronium can be formed, and in this section we briefly review this body of work.

As will be described in section 6.2, one of the parameters which can be derived from traditional positron lifetime experiments in dense gases is the positronium fraction F, the fraction of positrons which stop in the medium and form positronium. The simplest approach aimed at deriving values of F to be expected in such experiments was proposed by Ore (1949), and the so-called Ore model has since been expounded several times in essentially its original form (e.g. Massey, 1975; Griffith and Heyland, 1978; Schrader and Svetic, 1982; Charlton, 1985a). It involves no other physical input than the threshold energies for positronium formation (E_{Ps}), excitation (E_{ex}) and ionization (E_i) and assumptions concerning the touch-down energy distribution, $TD(E)$ (Schrader and Svetic, 1982), of the positrons as their kinetic energies E are first moderated below E_i. The simplest cases to consider are atomic species in which there are no low-lying excited states, so that positronium formation via reaction (4.1) is the first inelastic channel to open. The positronium fraction can then be written as

$$F = \frac{\int_0^{E_i} TD(E)\{\sigma_{Ps}(E)/[\sigma_{Ps}(E) + \sigma_{ex}(E)]\}\, dE}{\int_0^{E_i} TD(E)\, dE}, \tag{4.36}$$

where $\sigma_{Ps}(E)$ and $\sigma_{ex}(E)$ are the cross sections for positronium formation and excitation. The touch-down distribution is determined largely by the details of how positrons slow down as they lose the last hundred eV or so of their kinetic energy. Whilst some of the relevant processes are now relatively well understood, other factors, for example the role played by the formation and break-up of energetic positronium, remain unclear.

It should be noted that the upper positron energy limit considered in this model is E_i. Above this energy, neglecting small target-recoil effects, any positronium that is formed has a kinetic energy in excess of its binding energy and is assumed to break up in a subsequent collision. Thus, positronium formed in the ground state with a kinetic energy above 6.8 eV is regarded as not contributing to the measured positronium fraction. This assumption is only valid if the kinetic energy of the positronium is not reduced to below its binding energy in subsequent collisions and at sufficiently high gas densities that it does not self-annihilate before break-up can occur. This latter restriction is most serious for para-positronium because of its high (8 ns^{-1}) decay rate. Assuming a break-up cross section of 10^{-16} cm^2 leads to a gas density of approximately 10^{18} cm^{-3}, above which break-up will dominate.

It has been commonly assumed that the TD distribution is uniform in energy, in which case equation (4.36) reduces to

$$F = \frac{E_{ex} - E_{Ps}}{E_i} + \frac{1}{E_i}\int_{E_{ex}}^{E_i} \frac{\sigma_{Ps}(E)}{\sigma_{Ps}(E) + \sigma_{ex}(E)}\, dE, \tag{4.37}$$

where the first term arises because, for all atomic systems undergoing positron impact, $E_{ex} > E_{Ps}$. This equation, and the two following, are often used for both atoms and molecules, though for the latter there are also low-lying vibrational and rotational levels and, in some cases (e.g. O_2), electronic levels to consider. By setting the contribution to F from positrons with kinetic energies above E_{ex} to be zero, the minimum fraction F_{min} predicted by the Ore model is just

$$F_{min} = (E_{ex} - E_{Ps})/E_i. \tag{4.38}$$

However, if electronic excitation is unimportant compared to positronium formation, $\sigma_{Ps}(E) \gg \sigma_{ex}(E)$, the maximum Ore model prediction is obtained from equation (4.36) as

$$F_{max} = (E_i - E_{Ps})/E_i = 6.8 \text{ eV}/E_i. \tag{4.39}$$

Without extra input from either theory or experiment for the relevant cross sections, we are left with predictions for F_{max} and F_{min} which define a band within which the experimental value is expected to fall.

The simple Ore model described above has been used extensively in discussing positronium formation in dense gases, though there is no *a priori* reason to assume that the uniform-energy TD distribution is valid. Indeed, Paul and coworkers (e.g. Paul and Böse, 1982) found, in an investigation of positron drift in dilute gases, see section 6.4, that their data could best be fitted using a phase-space Ore model in which the TD distribution below E_i is uniform in positron momentum, not energy. The modified Ore predictions for F_{max} and F_{min} are similar to those given in equations (4.38) and (4.39), except with the energies raised to the power 3/2. The band of allowed F-values thus differs substantially from those given by equations (4.38) and (4.39). Comparisons with experiment can be found below.

Another model of positronium formation, the so-called spur model, was originally developed by Mogensen (1974) to describe positronium formation in liquids, but it has found some applications to dense gases. The basic premise of this model is that when the positron loses its last few hundred eV of kinetic energy, it creates a track, or so-called spur, in which it resides along with atoms and molecules (excited or otherwise), ions and electrons. The size of the spur is governed by the density and nature of the medium since these, loosely speaking, control the thermalization distances of the positron and the secondary electrons. It is clear that electrostatic attraction between the positron and electron(s) in the spur can result in positronium formation, which will be in competition with other processes such as ion–electron recombination, diffusion out of the spur and annihilation.

A semi-quantitative picture of positronium formation in a spur in a dense gas was developed by Mogensen (1982) and Jacobsen (1984, 1986). If the separation of the positron from an electron is r, and there is assumed to be only one electron in the spur (a so-called single-pair spur), then the probability of positronium formation in the spur, in the absence of other competing processes, can be written as $[1 - \exp(-r_c/r)]$; here r_c is the critical, or Onsager, radius (Onsager, 1938), given for a medium of dielectric constant ϵ by

$$r_c = e^2/(4\pi\epsilon k_B T), \tag{4.40}$$

which is the separation at which the attractive Coulomb energy is equal to $k_B T$. Thus, if the pair separation upon thermalization, R, is $\gg r_c$, positronium formation is unlikely.

Jacobsen (1984) gave a full discussion of the effect of thermalization and concluded that positronium formation by the spur mechanism is unlikely in atomic gases, since $R \gg r_c$ irrespective of density. This is not the case for molecular gases, where R can be of a similar order of magnitude to r_c at high densities. Thus, the positronium formation fraction in molecular gases is expected to be both density and temperature dependent, and it has been semi-quantitatively expressed by Jacobsen (1986), in a combined Ore-plus-spur approach, as

$$F = F_0 + (1 - F_0)[1 - \exp(-r_c/R)] \exp(-\lambda_f \tau_{Ps}), \qquad (4.41)$$

where F_0 is the low density, or Ore, value and is assumed to be constant. The factor $\exp(-\lambda_f \tau_{Ps})$ is the probability that the positron has not annihilated before positronium has formed; τ_{Ps} is the time taken for positronium to form in the spur. In many cases this factor appears to be close to unity, though it can be important at high gas densities and if the particles are immobile (see subsection 6.3.3). Jacobsen (1986) provided estimates of τ_{Ps}.

Finally, Zhang and Ito (1990) proposed the resonant model, which they claim is more successful than the Ore and spur approaches in explaining data for a wide variety of systems. This model seems to incorporate various aspects of these other models, although one crucial difference is the proposal that positronium may be created via the resonant formation of intermediate excited complexes. However, at the time of writing, there is no evidence from positron–gas scattering experiments that the Ore model needs to be modified for atomic species by the addition of an intermediate stage involving positronic complexes. For dense molecular gases, only the conventional spur model has been applied so far and it can offer qualitative explanations of the data, although an extension of the type suggested by Zhang and Ito (1990) cannot be ruled out.

2 *Positronium formation fraction – results*

This section is devoted to experimental values of F and comparisons with predictions from the models described above. The data discussed have mainly been obtained since 1975 because, as outlined by Coleman *et al.* (1975a), earlier results, especially for the noble gases, are thought to be prone to systematic errors.

We first consider the noble gases, together with various mixtures containing them. Our discussion is based largely around the work of Coleman *et al.* (1975a), Griffith and Heyland (1978) and Wright *et al.* (1985), and the Ore model is used to interpret the results since, as described above, there is no contribution to F from spur processes. A selection of values of F_{min} and F_{max}, see equations (4.37)–(4.39), is shown in table 4.1 along with the observed fractions. The values for helium, neon and argon, which are found experimentally to be independent of gas density, lie between F_{min} and F_{max}. Thus, positronium formation in these gases is usually regarded as being in accord with the Ore model. Surprisingly, though,

Table 4.1. Ore model parameters for both standard and modified Ore approaches ($F_{\rm min}$, $F_{\rm max}$, $F^{\rm mod}_{\rm min}$ and $F^{\rm mod}_{\rm max}$), and the experimental fractions F for the noble gases and a variety of molecules. Note that when $E_{\rm ex} < E_{\rm Ps}$, the minimum predictions have been set to zero; see equation (4.38). See Charlton (1985a) for the origin of the measurements. In general, the fractions for the molecular gases have been found to be both density and temperature dependent. The value quoted here is for low densities and is thus expected to be the Ore contribution to the overall positronium fraction in these gases at higher densities

Gas	$F_{\rm min}$	$F^{\rm mod}_{\rm min}$	$F_{\rm max}$	$F^{\rm mod}_{\rm max}$	F
He	0.14	0.19	0.28	0.38	0.23
Ne	0.09	0.11	0.32	0.43	0.26
Ar	0.17	0.20	0.43	0.57	0.33
Kr	0.20	0.24	0.49	0.63	0.11
Xe	0.26	0.29	0.56	0.71	0.03
H_2	0.18	0.21	0.44	0.58	0.32
N_2	0	0	0.44	0.58	0.19
CO_2	0	0	0.49	0.64	0.40
O_2	0	0	0.56	0.71	0.32
CO	0.06	0.06	0.49	0.63	0.28
CH_4	0.18	0.20	0.52	0.67	0.40

the values of F for krypton and xenon, based upon the measured intensity of the ortho-positronium component, fall below $F_{\rm min}$.

Experiments on these two gases, reported by Griffith and Heyland (1978), showed that a fast component, with a density-dependent decay rate, was present in the lifetime spectra, and this was tentatively linked to the dearth of long-lived ortho-positronium. Furthermore, it was found for mixtures of krypton with helium that the maximum value of F, which was observed at a concentration of around 0.01% of krypton, was in excess of the sum of the individual F-values for the two gases when pure.

Similar features were later found by Wright *et al.* (1985), whose experiments were an attempt to shed further light on the low measured values of F in krypton and xenon by a detailed study of the lifetime spectra of the pure gases and mixtures. The spectrum obtained for xenon at a density of 9.64 amagat at room temperature is shown in Figure 6.5, where the paucity of long-lived ortho-positronium is apparent, particularly when compared with the accompanying spectrum for argon. Also, clearly resolved from the prompt peak are the so-called fast components, which Wright *et al.* (1985) were able to fit to two overlapping exponentials.

These authors also found that adding gases to krypton and xenon markedly speeded up the fast components and also raised the measured value of F. Their conclusion was that the fast components were indeed due to the formation of positronium but that the subsequent interactions of the positronium led to rapid annihilation. When this contribution was included in the evaluation of F, its value was found to be in accord with the Ore model. The model proposed by Wright *et al.* (1985) and later supported by the work of Kakimoto and Hyodo (1988), who used the ACAR technique (see sections 7.3 and 7.4), is that positronium moderation plays an important role, i.e. once formed, with a distribution of kinetic energies, the ortho-positronium moderates only slowly in these heavy gases. During this time the ortho-positronium can undergo collisions with the krypton and xenon atoms; this leads to quenching and the appearance of fast components. The addition of small quantities of impurities, particularly of low mass, speeds up the moderating process, making quenching collisions less likely and increasing the apparent intensity of long-lived ortho-positronium. Wright *et al.* (1985) discussed their data in terms of some kind of temporary attachment of the positronium to krypton and xenon, though the physical nature of this state remains unclear. Later, an alternative interpretation was offered (Tuomisaari, Rytsölä and Hautojärvi, 1988), based on a suggestion of Manninen (1987), in which the fast components were a natural consequence of the time dependence of the pick-off collision rate. This rate varies as the product of the quenching cross section and the speed of the positronium. At higher energies the increased speed clearly enhances the quenching rate, but the cross section may also be higher since the positronium is then more likely to overcome the exchange repulsion and penetrate further into the the electron cloud of the atom. Some quantitative support for this view comes from early theoretical work of Barker and Bransden (1968) on the positronium–helium system. This, and other work concerning positronium interactions, is reviewed in Chapter 7.

The work of Jacobsen (1984, 1986) and Mogensen (1982), described in subsection 4.8.1 above, pointed to the potential importance of positronium formation as a consequence of spur processes in dense molecular species. Table 4.1 drew attention to the fact that the positronium fractions for many molecular gases have been found to be both density and temperature dependent. We will not attempt a detailed compilation of these data here, but examples of the density and temperature variations observed are shown in Figure 4.31 for SF_6 and CO_2 gases. The lines are fits to equation (4.41), and the reader is referred to the work of Jacobsen (1986) for a full discussion of the fitting procedures and assumptions. The fact that a reasonable fit to the data can be produced is strong supporting evidence for positronium formation in spurs.

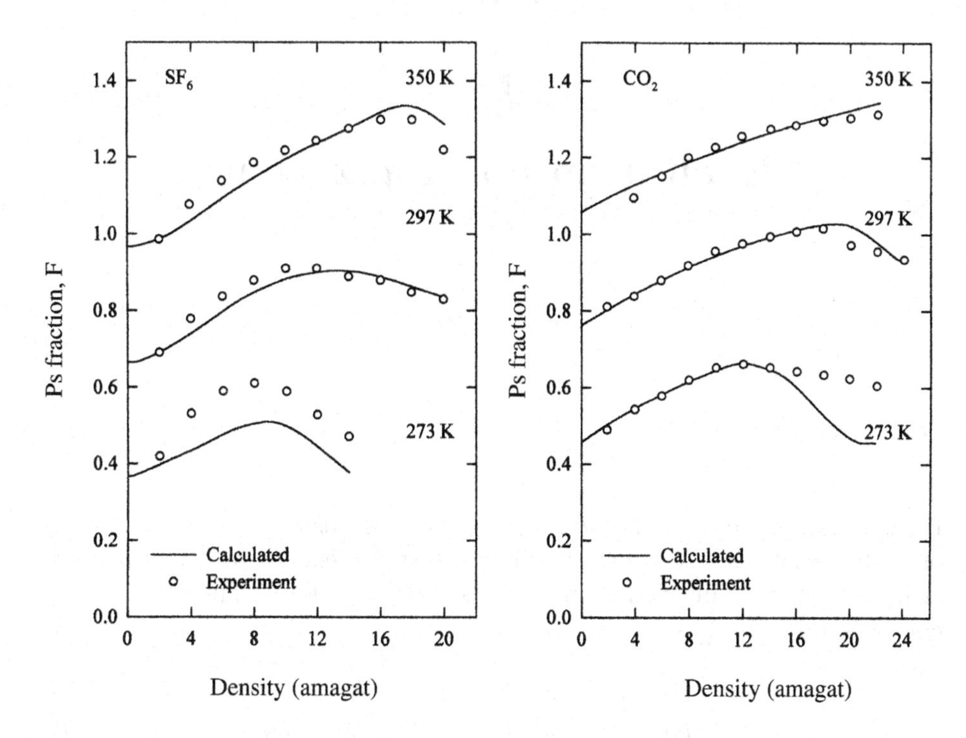

Fig. 4.31. Measured and calculated (see text and Jacobsen, 1986) positronium fractions F for SF_6 (left) and CO_2 (right) gases. The values of the fractions at 297 K and 350 K are displayed offset by +0.3 and +0.6 respectively, for clarity of presentation.

Two other pieces of experimental evidence also add weight to this contention. Curry and Charlton (1985), in a study which shadowed work on liquids performed by Wikander and Mogensen (1982), noted the effect of introducing small quantities of the electron scavengers CCl_4 and CCl_2F_2 into high density CO_2. Briefly, the scavenger can remove free electrons from the spur and prevent positronium formation, the detailed behaviour being contingent upon the electron capture cross section, which governs the capture rate. A definite scavenging effect was observed and such behaviour can occur only if spur processes are an important source of positronium in molecular species. An applied electric field can also deplete the positronium yield in the spur by separating the electron and the positron before they can unite. Early observations of this effect were made by Marder *et al.* (1956) and Obenshain and Page (1962), though at the time its origin was not clear. More recent work to establish the mechanism fully was reported by Charlton and Curry (1985) for CO_2, and by Sharma *et al.* (1985) and Jacobsen, Charlton and Laricchia (1986) for CH_4.

5
Excitation and ionization

In this chapter we consider inelastic collisions of positrons with an atomic or molecular target X which result in electronic excitation or ionization (without positronium formation) of the target system. These processes can be summarized as

$$e^+(E) + X \rightarrow e^+(E - \Delta E) + X^*, \tag{5.1}$$

$$e^+(E) + X \rightarrow e^+(E - \Delta E) + X^{m+} + me^-, \tag{5.2}$$

where X^* is the excited neutral target, E is the kinetic energy of the positron, ΔE is the amount of energy lost by it in the collision and m is the degree of ionization. The value of ΔE is fixed, in the case of reaction (5.1), to be one of the discrete excitation energies of the target levels, termed $E_{\rm ex}$, but is variable in (5.2), the minimum value being the ionization energy $E_{\rm i}$. The 'total' cross sections for these processes will be referred to as $\sigma_{\rm ex}$ and $\sigma_{\rm i}$, with the addition of other notation as necessary to define cross sections for other projectiles, e.g. electrons. The total ionization cross section is the sum of the integrated partial-ionization cross sections, $\sigma_{\rm i}^{m+}$, i.e. $\sigma_{\rm i} = \sum_m \sigma_{\rm i}^{m+}$.

Except for section 5.6, the discussion will centre around these integrated cross sections because for positron impact we cannot as yet distinguish between the various possibilities in reaction (5.2), where X^{m+} may be left in an excited state. Ignorance of the true final state may be particularly serious for collisions with molecules, where fragments may be a combination of charged and neutral species.

In the section on excitation we shall treat only electronic transitions; thus rotational and vibrational processes in molecules are excluded. As will be described in Chapter 6, information on these latter processes has been derived from positron lifetime and other experiments. Our theoretical discussion will mainly concern excitation of the lower levels of hydrogen and helium, which have been the most comprehensively studied target systems. The sparse experimental work is confined to estimates of angle-integrated cross sections, which, owing to poor energy resolution, may contain contributions from more than one channel. Very little is known from either theory or experiment concerning the excitation of molecules by positrons, although some data exist for O_2.

The main body of this chapter deals with ionization, including some of the theoretical approaches and associated results for the simpler targets. There is a growing body of experimental data on various aspects of positron impact ionization, and this will be reviewed in some depth. Comparisons with data for other simple projectiles will be made where appropriate, in an attempt to shed light on the mechanisms involved in ionizing collisions. Work has begun on measurements of energy- and angle-resolved cross sections, and this will be described in section 5.6.

5.1 Excitation

1 Theory

Positron impact excitation of a target system from an initial state (usually the ground state) with energy E_0 to a final excited state with energy E_f is only possible if the energy of the incident positron exceeds the threshold value $E_{\rm ex} = E_f - E_0$. The magnitudes of the initial and final momenta of the positron, k_0 and k_f respectively, are then related through energy conservation by

$$\tfrac{1}{2}k_0^2 + E_0 = \tfrac{1}{2}k_f^2 + E_f. \tag{5.3}$$

In electron scattering, where there is exchange between the projectile and the electrons in the target, all energetically accessible states of the target can be excited, but in positron collisions only those transitions in which the total spin of the target does not change are allowed. As an example, the lowest excited state of helium that can be reached by positron impact is the 2^1S state, with an energy of 20.58 eV, and not the first excited state, 2^3S, with an energy of 19.8 eV.

Consider a transition from an initial state, in which the target system is in its ground state, to a final state, in which the target system has been excited to state f. In terms of the T-matrix element, the differential cross section for scattering through the angle between $\boldsymbol{k}_0$ and $\boldsymbol{k}_f$ is

$$\frac{d\sigma_{f0}}{d\Omega} = \left(\frac{k_0}{k_f}\right)\frac{|T_{f0}|^2}{4\pi^2}, \tag{5.4}$$

where $T_{f0} = \langle\phi_f|V|\varphi_0^+\rangle$. Here, φ_0^+ is the exact total wave function of the positron–target system, ϕ_f is the product wave function for a positron with momentum $\boldsymbol{k}_f$ and the target system in the final state and V is the positron–target interaction potential. The total cross section is then obtained by integration over all directions of $\boldsymbol{k}_f$.

The simplest approximation is to replace φ_0^+ by ϕ_0, the product wavefunction for a positron with momentum $\boldsymbol{k}_i$ and the target system in its ground state; this yields the first Born approximation, $T_{f0} = \langle \phi_f | V | \phi_0 \rangle$. The positron–target interaction potential is equal in magnitude but opposite in sign to the corresponding potential for electrons and therefore, if electron exchange is ignored, the cross sections for electron and positron excitation in the first Born approximation are equal. At sufficiently high energies, several hundred eV for hydrogen, the first Born approximation yields accurate results for the angle-integrated excitation cross sections. However, except for a small angular range about the forward direction, which becomes progressively smaller as the projectile energy is increased, the differential excitation cross section in this approximation falls away much too rapidly with increasing angle of scattering. This deficiency can only be remedied by including some form of second Born amplitude into the calculation of the differential cross section, as was done by Byron, Joachain and Potvliege (1985).

It was shown by Omidvar (1975) that the first Born approximation to the cross section for excitation of atomic hydrogen from its ground state to an excited state with principal quantum number n_{H} is proportional to $1/n_{\mathrm{H}}^3$, where the constant of proportionality depends on the value of the orbital angular momentum quantum l. Strictly, this scaling law is only valid at high incident positron energies and for large values of n_{H}, but it has been used at relatively low energies to estimate the cross sections for excitation to higher excited states when more accurate methods have been used to calculate the cross sections for excitation to a few specific low-lying states (see e.g. the discussion for helium in section 2.7).

Improvements over the first Born approximation at intermediate energies can be made by taking account of the distortion of the positron wave function in both the initial and the final states, as is done in the distorted-wave approximation. The T-matrix element then takes the form $T_{f0} = \langle \varphi_f | V | \varphi_0 \rangle$, where φ_0 and φ_f are wave functions representing elastic scattering of the positron in the potential fields of the target in the initial and final states respectively. In addition to the static interaction, these potentials usually incorporate some allowance for polarization of the target. This approximation has been used by several authors, most notably by McEachran and his collaborators, to determine various excitation cross sections for the noble gases (for helium, Parcell, McEachran and Stauffer, 1983, 1987; for neon, Parcell, McEachran and Stauffer, 1990). Particular attention has been given to calculating the cross section for resonant excitation from the ground state to the lowest optically allowed excited

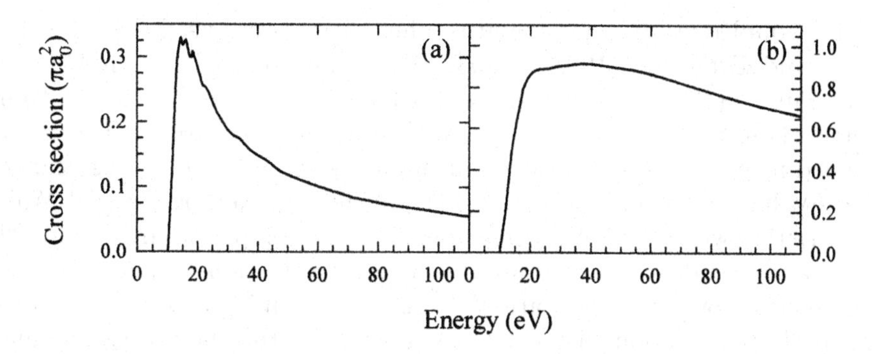

Fig. 5.1. Cross sections for the excitation of atomic hydrogen obtained using the coupled-state approximation with 30 hydrogen states and pseudostates plus three positronium states: (a) 1S–2S; (b) 1S–2P (Kernoghan *et al.*, 1996).

state; this provides the dominant contribution to the total excitation cross section σ_{ex}.

In a complete treatment of positron scattering, all the open channels, including excitation, are coupled together and the cross sections for all the various scattering processes are determined together. Such an approach has been adopted for positron–hydrogen scattering using the coupled-state method, most notably by Kernoghan, McAlinden and Walters (1995), Kernoghan *et al.* (1996) and Mitroy and Ratnavelu (1995). In the intermediate energy region the most accurate results are probably those obtained by Kernoghan *et al.* (1996), using a 33-term expansion of the total wave function which included three positronium states (1s, 2s, 2p) and 30 hydrogen states, some of which were pseudostates. These authors determined the cross sections for excitation to the 2S and 2P states and also for elastic scattering, positronium formation into various states and ionization. The 1S–2S excitation cross section, shown in Figure 5.1(a), exhibits quite a steep rise from the threshold, at 10.2 eV, up to a value of $0.33\pi a_0^2$ at approximately 15 eV. Thereafter it falls steadily, with a roughly exponential shape, to a value of approximately $0.06\pi a_0^2$ at 100 eV. In contrast, the resonant 1S–2P excitation cross section, Figure 5.1(b), rises somewhat less steeply from the threshold but reaches a significantly larger maximum, $0.9\pi a_0^2$, at 20 eV; this is followed by a broad plateau region, after which the cross section falls slowly to $0.7\pi a_0^2$ at 100 eV. At energies beyond 40 eV it is similar in value to the total ionization cross section, and these two cross sections become equally dominant contributions to the total scattering cross section. Calculations of excitation cross sections for hydrogen at higher energies were made by Byron *et al.* (1985). There have been no experiments on the excitation of atomic hydrogen to date.

The coupled-state approximation has also been used by Hewitt, Noble and Bransden (1992b, 1993, 1994) and Kernoghan, McAlinden and Walters (1996) to investigate positron impact excitation of alkali atoms, these being treated as equivalent one-electron atoms. The most detailed investigations of positron–lithium scattering in the energy range 0–60 eV have been probably those of McAlinden, Kernoghan and Walters (1997), who used the three lowest states of positronium and 29 states and pseudostates of lithium: these investigations reveal that the cross section for 2S–2P excitation becomes the dominant contribution to the total cross section just a few eV above the threshold and remains thus throughout the energy range considered. Other applications of the coupled-state approximation to lithium were made by Ward *et al.* (1989), McEachran, Horbatsch and Stauffer (1991) and Khan, Dutta and Ghosh (1987) but these authors excluded all positronium terms from the expansion of the wave function. As stated previously in section 3.2, see equation (3.28), an expansion in terms of the target eigenstates alone is formally complete, but the convergence of the results is poor at energies where positronium formation is significant. At higher energies, however, where positronium formation is less probable, the effect on the elastic scattering and excitation cross sections of neglecting positronium terms in the expansion of the wave function becomes insignificant. These features are clearly illustrated in Figure 5.2, in which various results for the 2S–2P excitation cross section for lithium are displayed. At positron energies below 30 eV an expansion without positronium terms yields results which are significantly larger than those obtained when positronium terms are included, presumably in an attempt to compensate for the lack of flux into the positronium formation channels. This effect is particularly pronounced at energies just a few eV above the excitation threshold.

Similar coupled-state methods, both with and without the inclusion of positronium terms, have been applied to the excitation of other alkali atoms. The results of McAlinden, Kernoghan and Walters (1994, 1997) and Hewitt, Noble and Bransden (1994) for the dominant resonant excitation cross sections for sodium, rubidium and caesium all exhibit a similar energy dependence to that for lithium. Also, the neglect of positronium terms in the expansion, as in the work of McEachran, Horbatsch and Stauffer (1991), again has the effect of increasing the low energy excitation cross sections over those obtained when such terms are included.

Several different approximation methods have been used to investigate excitation in positron–helium scattering, but in all cases rather simple uncorrelated wave functions have been used to represent the ground and excited states of helium. All the reported results, which relate almost exclusively to the excitation of 2^1S and 2^1P states, exhibit a steady rise from the threshold followed by a gentle fall, which continues up to a few

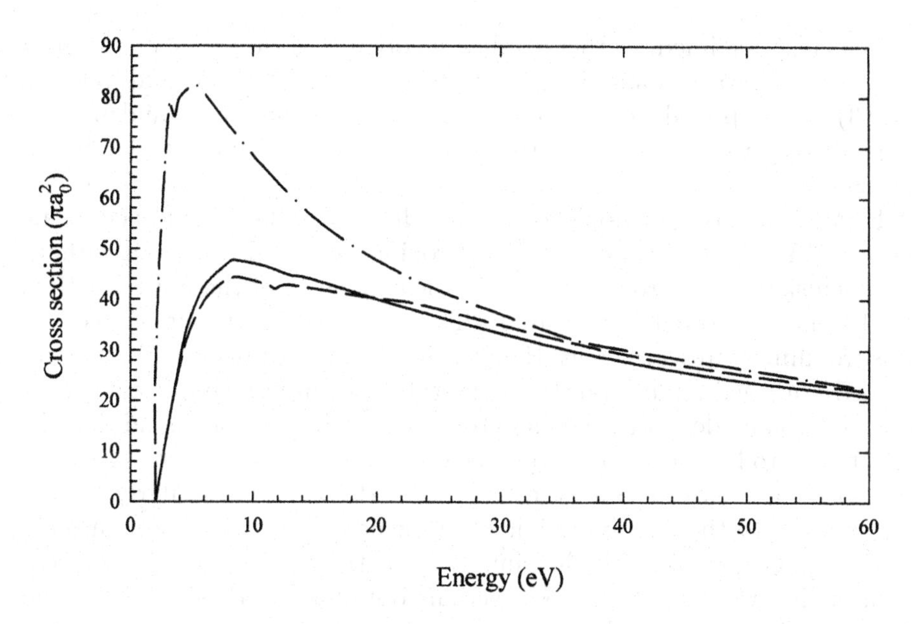

Fig. 5.2. Cross sections for the 2S–2P excitation of lithium obtained using the coupled-state approximation (McAlinden, Kernoghan and Walters, 1997): ——, 29 target states plus three positronium states; – – –, five target states plus three positronium states; — · —, five target states only.

hundred eV, the highest energies considered. All the theoretical results also predict a cross section for 2^1P excitation which is several times larger than that for 2^1S excitation. Buckley and Walters (1975) used various forms of the second Born approximation to calculate the 2^1S excitation cross section in the energy range up to 1000 eV. Beyond approximately 600 eV their results agree quite well with those of the first Born approximation. Saxena, Gupta and Mathur (1984) devised an alternative method of dealing with the failure of the first Born approximation to yield the correct differential excitation cross section at large scattering angles. This method involved using a two-potential modification of the first Born approximation and did indeed give much improved results at large scattering angles.

Various forms of distorted-wave approximation have been used to investigate the excitation cross sections in helium. Parcell, McEachran and Stauffer (1983, 1987) used this approximation to investigate the 2^1S and 2^1P excitation of helium over the energy range from near the threshold up to 150 eV. Their results for the 2^1S excitation were found to be in rather poor agreement with the corresponding experimental results of Sueoka (1982), which prompted Parcell, McEachran and Stauffer (1987) to question whether these measurements were for this transition alone, as had been claimed. Other studies involving the use of some form of the distorted-wave approximation were made by Madison and Winters (1983), who applied both first and second order methods, and by Kumar, Srivastava and Tripathi (1985) and Srivastava, Kumar and Tripathi (1986).

Excitation cross sections for helium have also been obtained using various forms of the coupled-state approximation, both with and without the inclusion of positronium terms in the expansion of the wave function. Willis and McDowell (1982) used a five-term expansion but omitted all positronium states, whereas Hewitt, Noble and Bransden (1992a) included both helium and positronium states but used a simple equivalent one-electron model for the helium atom. The latter authors found that the effect of introducing the positronium states was to lower the cross sections for positron excitation into the 2^1S and 2^1P states, particularly for the latter state in the energy region below which positronium formation is most significant. Despite the simplicity of the helium model used, the sum of the two excitation cross sections obtained by Hewitt, Noble and Bransden is in reasonable accord with the experimental results of Mori and Sueoka (1994) at energies below 36 eV but becomes 50% larger than the experimental results at higher energies, albeit with a similar energy dependence. The most accurate coupled-state results for the excitation of helium to the 2^1P state are probably those obtained by Campbell *et al.* (1998a) in the course of comprehensive investigations of low and intermediate energy positron–helium scattering. Using an expansion of the wave function comprising the first three eigenstates of positronium and 24 states of the helium atom, these authors found that the 2^1P excitation cross section initially rose quite steeply and then retained a fairly constant value of approximately $0.2\pi a_0^2$ up to 150 eV, the highest energy considered. These results are in reasonably good agreement with experiment throughout the energy range. The experimental results are believed to relate to the sum of the 2^1S and the 2^1P excitation cross sections, but the former cross section is expected to be much smaller than the latter. Unlike in positron–alkali atom scattering, where resonant excitation provides the dominant contribution to the total scattering cross section in the intermediate energy range, this cross section for helium only contributes approximately 20% of the total, the dominant contribution now coming from ionization.

Among other methods employed to investigate positron impact excitation of helium, mention should be made of the random-phase approximation used by Ficocelli Varracchio (1990), the results of which are in best overall agreement with the data of Mori and Sueoka (1994). This approximation was also used by Ficocelli Varracchio and Parcell (1992) to determine the 3^1P excitation cross section, which was found to be intermediate in value between the 2^1S and the 2^1P excitation cross sections and to have a similar energy dependence.

The excitation of neon in the energy range up to 40 eV was investigated by Parcell, McEachran and Stauffer (1990) using a similar form of distorted-wave approximation to that used by these same authors for helium. Their results for the resonant excitation cross section are in reasonably good agreement with the measurements of Coleman *et al.* (1982), though they are significantly lower than those of Mori and Sueoka (1994). However, the latter may contain contributions from several excited states of the atom.

2 Experiment

There are few reported experimental studies of reaction (5.1), and most have used a time-of-flight (TOF) energy-loss technique rather than detecting the de-excitation photon which may be emitted after the collision. The only exception is the work of Laricchia, Charlton and Griffith (1988) on simultaneous positronium formation and excitation of the remnant ion in positron scattering by CO_2 and N_2O, as reported in section 4.5. Many important atomic transitions result in the emission of photons in the ultraviolet region of the electromagnetic spectrum; these, though are difficult to detect. Even in the cases where strong optical transitions are expected (e.g. the so-called resonance lines of the alkali metals), there has been no reported work.

The experimental arrangement of Coleman and Hutton (1980) and Coleman *et al.* (1982) was a modified version of that described in section 3.4 and used by Coleman and McNutt (1979) to make the first measurements of differential positron–argon elastic scattering cross sections. A positron beam of the desired energy, produced using a tungsten mesh moderator, was guided by an axial magnetic field. Detection at the end of the flight path was accomplished using a channeltron, and the TOF of the positrons was determined using the technique of Coleman, Griffith and Heyland (1973) (see e.g. section 2.3). A large pressure gradient was maintained between a short gas cell, which was located at the beginning of the flight path, and the remainder of the chamber, so that over 99% of the scattering took place in the cell.

Those positrons which had undergone inelastic collisions, by virtue of either process (5.1) or process (5.2), would have times of arrival at the channeltron that were delayed, the delay being determined by both ΔE and the angle of scattering θ. The ability to resolve individual transitions, and to separate excitation events from those due to ionization and large-angle elastic scattering, would obviously depend upon these parameters and upon experimental considerations such as the timing resolution, the

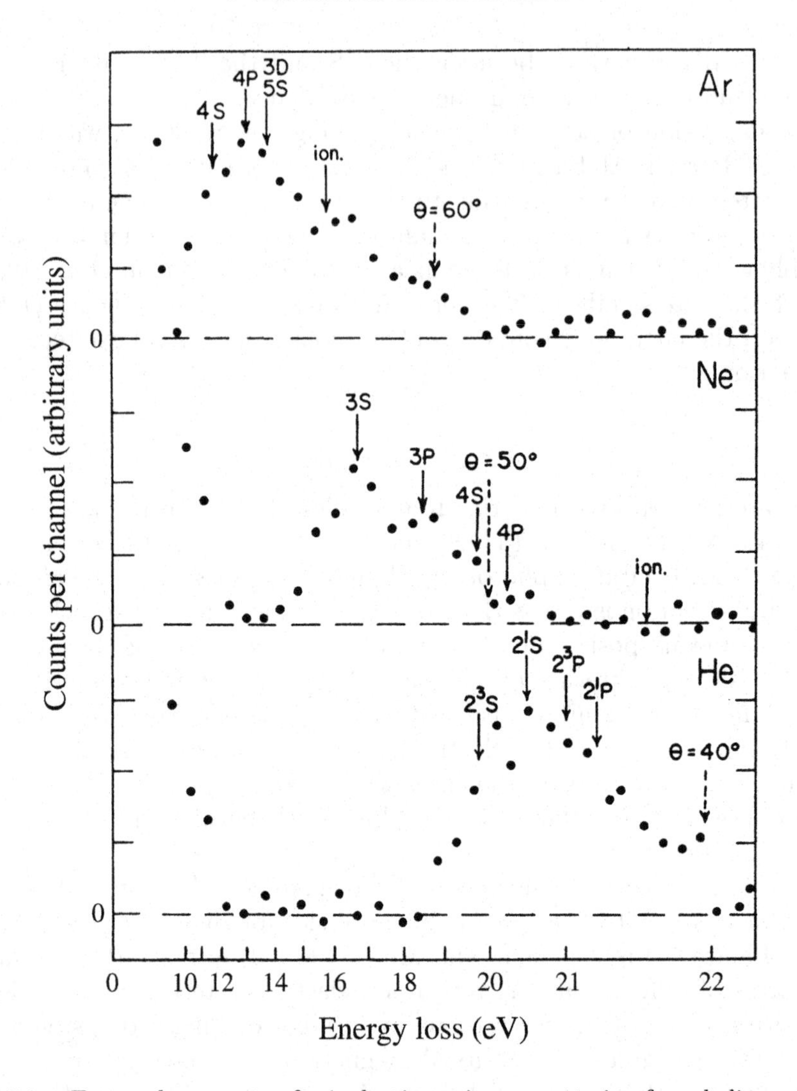

Fig. 5.3. Energy-loss spectra for inelastic positron scattering from helium, neon and argon gases at 24 eV impact energy from Coleman *et al.* (1982). The arrows indicate relevant excitation and ionization thresholds.

signal-to-background ratio and the length of the flight path. The work presented by Coleman and Hutton (1980) and Coleman *et al.* (1982) used this technique, as did later studies by Sueoka and coworkers, which were described in detail by Mori and Sueoka (1994).

Figure 5.3 shows energy-loss spectra derived by Coleman *et al.* (1982) from their measured TOF spectra and the geometry of the apparatus, under the assumption that the scattering took place at $\theta = 0°$ and in the centre of the gas cell. The spectra, each of which contains a prominent secondary peak, are for the same incident energy, 24 eV. The energies of transitions to various excited states are also marked on the spectra along with, for neon and argon, those at which ionization could occur. By assuming that the positrons in the secondary peaks had all lost the same amount of energy, i.e. ΔE is a constant, so that any increased TOF is due to angular deflection, the maximum scattering angles were estimated to be between 30° and 50°, the mean value of θ being much lower. Coleman *et al.* (1982) performed a systematic check which ruled out the possibility that the secondary peaks were caused by positrons which had back-scattered in the gas cell and then been reflected by the moderator potential.

However, the shapes and positions of the peaks suggest that they are caused by positrons which have undergone energy loss in an inelastic collision and have suffered deflection into a narrow range of small forward angles. Coleman *et al.* (1982) noted that even at energies up to ~5 eV above E_i the energy-loss peak appears to be solely due to positrons which have undergone a collision of the type (5.1). The energy-loss values, which correspond to the peak of each spectrum, were found to be independent of impact energy and to be given by 20.6 ± 0.2 eV (He), 16.6 ± 0.3 eV (Ne) and 12.5 ± 0.3 eV (Ar). For helium, this suggests that the 2^1S level is responsible. However, it should be noted that it is difficult to separate higher singlet levels from the 2^1S level since the larger apparent energy loss associated with an enhanced TOF may be in fact the result of a positron being scattered through an angle $\theta > 0°$.

In neon the observed peak at $\Delta E = 16.6$ eV seems to be dominant, though some of the spectra obtained by Coleman *et al.* (1982) contained extra structure; this consisted of a second peak appearing at a TOF corresponding to $\Delta E = 18.5$ eV, which is close to that expected for 3P excitation. For argon, the measured ΔE of 12.5 eV is somewhat above the threshold for excitation at 11.6 eV (4S); however, this could still be consistent with significant contributions from this level, accompanied by some angular deflection, and from the 4P level ($\Delta E = 13.1$ eV). It is clear that more detailed work, with finer energy resolution, is necessary before contributions from individual states can be resolved.

Total cross sections for inelastic scattering, written here as σ_inel since they may contain contributions from several excitation channels and, at certain energies, ionization, were deduced by summing all the events in the secondary peak, N_inel. Assuming no multiple scattering in the target, σ_inel is related to the total scattering cross section σ_T by

$$\sigma_\text{inel} = N_\text{inel}\sigma_\text{T}/N_\text{scatt}, \tag{5.5}$$

where N_scatt, the total number of scattered positrons, was determined, as described by Coleman *et al.* (1982), by comparing beam intensities

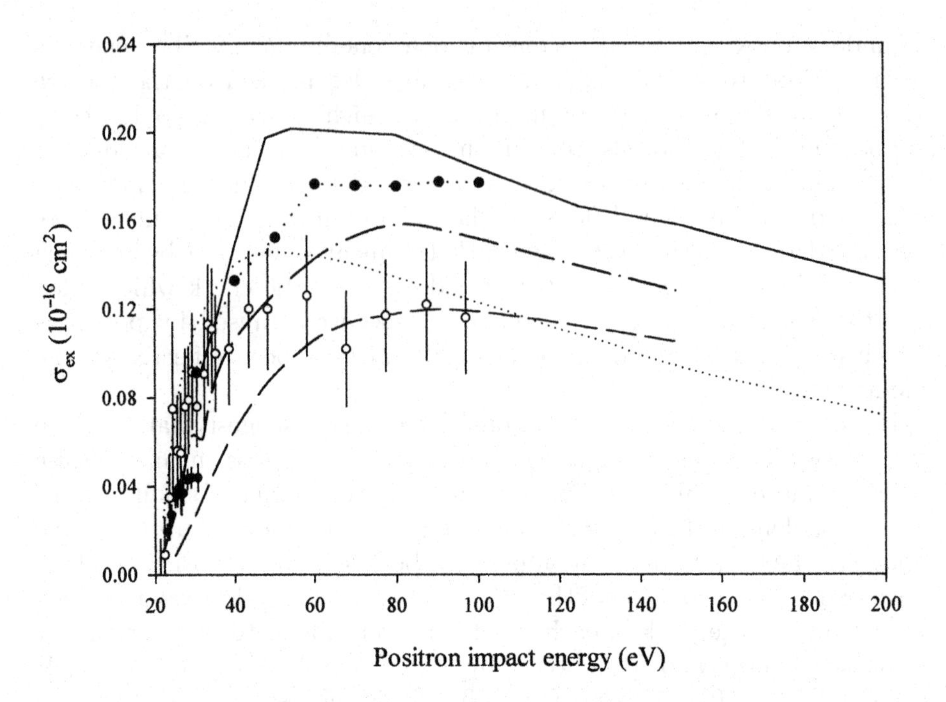

Fig. 5.4. Excitation cross sections for positron–helium scattering. Experiment: •, Coleman *et al.* (1982); ∘, Mori and Suoeka (1994). Theory: ——, Hewitt, Noble and Bransden (1992a); · · · · ·, Ficocelli Varracchio and Parcell (1992); – – –, Parcell *et al.* (1987); — · —, Campbell *et al.* (1998a); ··•··, semi-empirical data for excitation of singlet states only by electron impact, de Heer and Jansen (1977).

recorded with and without gas present in the scattering chamber. An equivalent expression was also given by Mori and Sueoka (1994).

Excitation cross sections obtained by Coleman *et al.* (1982) for energies up to 10 eV above threshold are shown in Figure 5.4 for helium gas. In this range of kinetic energies the excitation signal was found to be distinguishable from that due to ionization, so that $\sigma_{\rm inel} \approx \sigma_{\rm ex}$. Coleman *et al.* (1982) applied small corrections to these data for multiple scattering effects. The errors shown are both statistical and systematic in nature, the latter arising at higher energies from the difficulties of separating the tail of the primary (unscattered) peak from the inelastic events.

Also shown in Figure 5.4 are the data of Mori and Sueoka (1994), which supersede those reported earlier by the same group. Their TOF spectrometer was approximately four times the length of that employed by Coleman *et al.* (1982); this helped extend the impact energy range to around 100 eV. At such energies it is necessary to consider the transport properties of the beam as parameterized by a 'transmission factor' which reflects the ability of the magnetic field to confine scattered positrons so that they are detected and which is a strong function of their kinetic energy and scattering angle. As an example, the positron transmission factor at an energy of 100 eV is about 0.3 for $\theta = 15^\circ$ but falls to practically zero by 30°. Clearly the cross sections measured by this technique depend, to some extent, upon the details of the relevant differential cross section. In addition, the method used by Mori and Sueoka (1994) to separate events due to excitation and ionization does not seem to be unambiguous, and since the cross section for the latter is up to four times that for excitation, at some energies, there are potentially large systematic errors.

Given the experimental limitations regarding the angular range and the lack of timing and energy resolution, it might appear that little meaningful comparison with theory could be achieved. The situation is shown in Figure 5.4, where the results of several calculations are given. The values of Ficocelli Varracchio and Parcell (1992) and Campbell *et al.* (1998a) are in best overall accord with experiment, and these are dominated by excitation to the 2^1P level, contrary to the findings of Coleman *et al.* (1982) described above. Comparison was also made with cross sections for electron impact using data taken from the semi-empirical work of de Heer and Jansen (1977). The positron data are lower than those for electrons, at least for energies above 30–40 eV.

Cross sections for neon and argon have also been presented by Coleman *et al.* (1982) and Mori and Sueoka (1994), though here there are no theoretical data for comparison. The positron and electron cross sections (the latter from the work of de Heer, Jansen and van der Kaay, 1979) are of very similar magnitude, despite the fact that triplet states cannot be excited by positron impact.

In addition to the work on atoms, the study of Katayama, Sueoka and Mori (1987) produced cross sections attributable to excitation of the O_2 molecule by positron impact. The TOF apparatus and the method of analysis were similar to those described above. However, for O_2 a secondary peak was found which, when allowances were made for the energy width of the beam and for positrons which had been scattered through large angles, was concentrated in an energy-loss interval $\Delta E \sim 7$–10 eV. From work on electron and photon impact, this is known to be due to excitation of the Schumann–Runge continuum, an important optically allowed feature in the spectrum of O_2. As such, and following Katayama *et al.* (1987), we denote the derived cross sections as σ_{SR}.

Figure 5.5 shows σ_{SR} along with the electron impact result of Wakiya (1978) obtained by integrating his differential cross sections for forward-scattered electrons only. An interesting feature of the positron data is the presence of a distinct peak just above the threshold, which rises to a maximum at around 12 eV. Katayama *et al.* (1987) speculated that this is

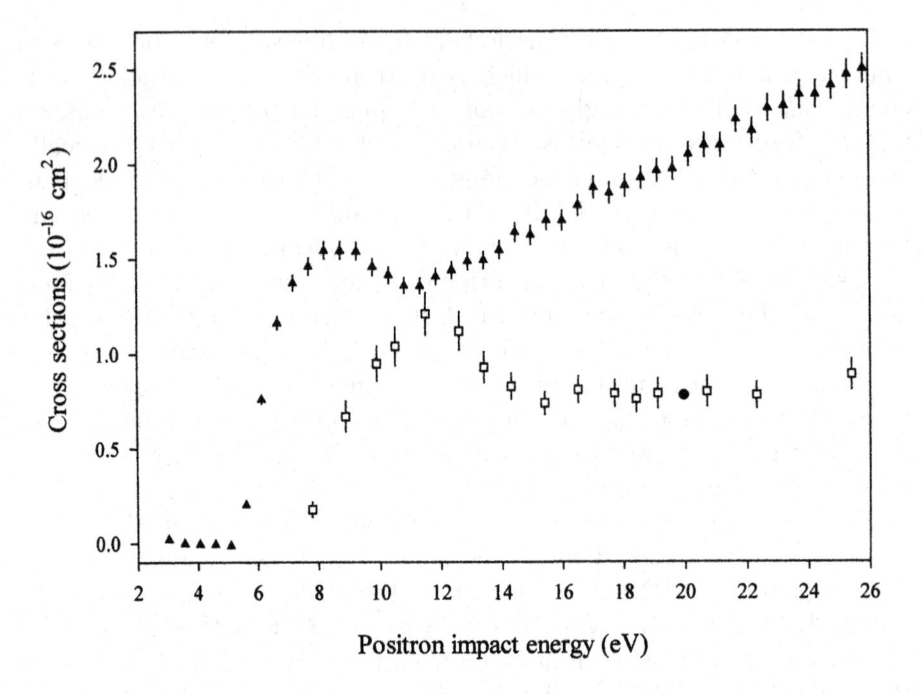

Fig. 5.5. Inelastic scattering cross sections for positron–O_2 collisions. Key: □, σ_{SR} for positron impact, Katayama *et al.* (1987); •, σ_{SR} for electron impact, a single point, Wakiya (1978); ▲, total ion production cross sections (see text), Laricchia, Maxom and Charlton (1993).

a resonance-type contribution to σ_{SR} which is absent in the electron case. Note that the cross sections σ_{SR} for the two projectiles are similar above approximately 16 eV, a fact which prompted Katayama *et al.* (1987) to postulate further that there are 'direct' and 'resonant' (positrons-only) contributions to the cross section.

Katayama *et al.* (1987) offered a possible explanation of the 'resonance' by postulating that the positron can become temporarily attached to the molecule, forming a PsO_2^+ complex. However, another possible explanation was forthcoming from the work of Laricchia, Moxom and Charlton (1993), who studied the total ion production cross section, i.e. the sum of the ionization and positronium formation cross sections. Their data are also included in Figure 5.5, and their technique has been described in detail by Moxom *et al.* (1994). Between the energies E_{Ps} and E_i the total ion yield is due solely to positronium formation and, as shown in Figure 5.5, a distinct peak was found in σ_{Ps} a few eV above threshold. The total ion formation cross section rises again at impact energies above E_i in such a way that the observed trough appears to coincide with the peak in σ_{SR}, which also rises rapidly from threshold but falls once the ionization channel opens. This led Laricchia, Moxom and Charlton (1993) to postulate that the structure in these cross sections is due to a channel-coupling effect; however, it is still possible, as envisaged by Katayama *et al.* (1987), that these structures arise from the formation of a temporary complex involving the positron.

5.2 Ionization – theoretical considerations

In single ionization of a target by positron impact the total energy of the system in excess of the ionization threshold is shared between the two emerging particles, provided that the recoil energy of the residual ion is ignored. If E_1 and E_2 are the energies of the positron and electron respectively, and E is the energy of the incident positron, then conservation of the total energy gives the relationship between these quantities as

$$E = E_1 + E_2 - E_{\rm i}, \tag{5.6}$$

where $E_{\rm i}$ is the ionization energy of the target system. Thus, for a given incident positron energy, the energy of the emerging electron is determined uniquely by that of the scattered positron. The probability that the positron will emerge from the ionizing collision into the solid angle $d\Omega_1$ with an energy in the range E_1 to $E_1 + dE_1$ and that the electron will emerge into the solid angle $d\Omega_2$ is given by the triple differential cross section $d^3\sigma_{\rm i}/d\Omega_1 d\Omega_2 dE_1$. The total ionization cross section is then obtained by integrating the differential cross section over all directions of the two emerging particles and the energy of either one of them.

In electron impact ionization it is not possible to determine which of the two emerging electrons was the incident projectile and which was originally bound in the target. The total wave function of the system should be antisymmetrized with respect to the coordinates of all the electrons and therefore no clear distinction can be made between the 'direct' and 'exchange' amplitudes. In positron impact ionization the distinguishability of the incident positron from the emerging electron avoids any such difficulties. Positron impact ionization, however, is complicated by the fact that the open positronium formation channels also result in the removal of an electron from the target atom or molecule. At incident positron energies several hundred eV above the ionization energy of the target system it is easy to distinguish between positronium formation and ionization: the two emerging particles are likely to have very different energies and momenta, and the positronium formation cross section is much smaller than the ionization cross section. However, at energies close to the ionization threshold of the target no clear distinction can be made between ionization in the usual sense and positronium formation into

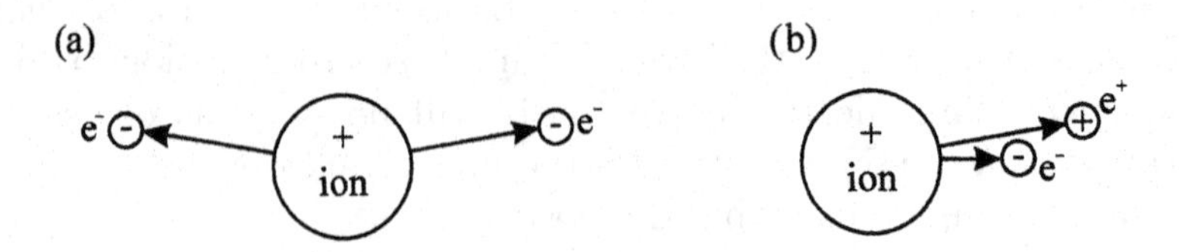

Fig. 5.6. Depiction of near-threshold (a) electron impact ionization and (b) positron impact ionization.

highly excited or continuum states. In both processes the two emerging particles experience an attractive final-state interaction giving rise to highly correlated motion. There must therefore remain some doubt about the validity of treatments of ionization which claim that a clear distinction can be made between 'true' ionization, in the sense of electron ionization, and the formation of positronium into very highly excited, or continuum states.

In the classical treatment of near-threshold electron impact ionization developed by Wannier (1953), the repulsion between the two electrons causes them to emerge with very similar energies but in opposite directions along the so-called Wannier ridge. This effect is depicted in Figure 5.6, where it is contrasted with the case for positron impact described below. According to this theory the energy dependence of the ionization cross section for electron impact is predicted to be

$$\sigma_{\rm i} \propto E_{\rm e}^{1.127}, \tag{5.7}$$

where $E_{\rm e} = E - E_{\rm i}$ is the energy of the projectile in excess of the ionization threshold energy. What cannot be predicted from the theory, however, is the value of the constant of proportionality or the energy range over which this equation is valid.

Klar (1981) derived a threshold law for positron impact ionization using similar classical ideas to those of Wannier. The potential energy function for the system consisting of the positron, electron and residual ion has an unstable saddle structure, and ionization corresponds to trajectories along the Wannier ridge in the potential energy. The positron then emerges in a similar direction to that of the electron, the ratio of the distances of the positron and the electron from the residual ion being approximately 2.15. Trajectories of the system which leave the ridge correspond either to excitation of the target, when the ratio of the positron distance from the ion to the electron distance from the ion tends to infinity, or to positronium formation, when this ratio is approximately 1.0. Klar's theory predicts an energy dependence for the ionization cross section of the form

$$\sigma_{\rm i} \propto E_{\rm e}^{2.651}. \tag{5.8}$$

Again, the value of the constant of proportionality and the range of validity are not given but Klar estimated that this power law was valid over an energy range of a few eV above threshold. As in Wannier's original theory, the energy is predicted to be shared equally between the positron and the electron. Rost and Heller (1994) confirmed this power law using a semi-classical treatment of ionization and they too concluded that the range of validity was probably more than 3 eV above the threshold.

A rather different theory of electron impact ionization was developed by Temkin (1982); it was based on the assumption that one of the electrons remains closer to the core than the other, so that the outer electron moves in the dipole field produced by the inner electron and the core. According to this Coulomb-dipole theory the ionization cross section has a modulated quasi-linear energy dependence of the form

$$\sigma_{\rm i} \propto \frac{E_{\rm e}}{(\ln E_{\rm e})^2}\left[1 + \sum C_l \sin(\alpha_l \ln E_{\rm e} + \mu_l)\right], \tag{5.9}$$

where α_l and μ_l are constants, and this should apply equally to positrons and electrons. In contrast to Wannier's, this theory predicts an unequal division of the energy between the two emerging light particles, the positron energy being the greater. Even if the theory is correct, its range of validity may be too small to be amenable to experimental test.

Ashley, Moxom and Laricchia (1996) measured the positron impact-ionization cross section in helium and found that its energy dependence up to 10 eV beyond the threshold was quite accurately represented by a power law, as in equation (5.8), but with the exponent having the value 2.27 rather than Klar's value of 2.651. This discrepancy prompted Ihra *et al.* (1997) to extend the Wannier theory to energies slightly above the ionization threshold using hidden crossing theory. They derived a modified threshold law of the form

$$\sigma_{\rm i} \propto E_{\rm e}^{2.640} \exp(-0.73 E_{\rm e}^{1/2}), \tag{5.10}$$

which reduces essentially to Klar's power law for sufficiently low energies. As will be shown in subsection 5.4.5, this latter form provides a good fit to the experimental data of Ashley, Moxom and Laricchia (1996) up to 10 eV above the ionization threshold, although the fit is no better than that of the simple power law with a modified exponent.

Although the assumptions made when developing these various theories of threshold ionization may be valid at energies very close to the threshold, there must be some doubt that the range of validity really does extend as far as 10 eV above a threshold.

A variety of approximation methods have been applied to positron impact ionization and, as with other scattering processes, the most attention has been given to atomic hydrogen and helium. The first Born approximation becomes accurate at sufficiently high projectile energies, typically 1 keV for atomic hydrogen, the results for single ionization being the same for both electrons and positrons. In this approximation the scattered positron is assumed to be screened from the residual ion by the slower electron, and its wave function is therefore represented by a plane wave, as also is the wave function of the incident positron. The wave function of the emitted electron, however, has usually been taken to be a Coulomb wave. Improvements over this simple approximation have been made by representing the wave function of the scattered positron in the distorted-wave approximation (Basu, Mazumdar and Ghosh, 1985; Ghosh, Mazumdar and Basu, 1985; Mukherjee, Singh and Mazumdar, 1989; Mukherjee, Basu and Ghosh, 1990). The ionization cross section then has quite a sensitive dependence on the form of the final state, particularly at energies not far above the ionization threshold. Distortion of the wave function of the incident positron from its plane wave form should also be incorporated into the formulation of the ionization process, but Basu, Mazumdar and Ghosh (1985) found that the results are relatively insensitive to this feature.

A more complete formulation of positron and electron impact ionization of hydrogen at intermediate energies was given by Brauner, Briggs and Klar (1989); in this formulation three-body Coulomb functions were used to represent the final state in the asymptotic region. These authors calculated the triple differential cross section for various energies of the incident projectile and the ejected electron. At a given incident energy the magnitude of the cross section decreases rapidly with increasing momentum transfer of the positron; attention was therefore given to the asymmetric kinematics in which the angle of scattering of the positron (or the incident electron) is small and the energy of the ejected electron is much less than that of the scattered projectile. Examples of the results obtained for both positrons and electrons, expressed as functions of the angle of ejection of the electron relative to the incident beam direction, are given in Figure 5.7. Here the incident energy of the projectile is 150 eV, its angle of scattering is 4° and the energy of the ejected electron is 3 eV. There are two maxima in the angular distribution. The binary peak at positive angles (which corresponds to emergence of the initially bound electron on the opposite side of the incident direction to that of the scattered projectile) arises from a direct collision between the projectile and the electron, with the nucleus as a spectator. The recoil peak at negative angles is the result of double scattering, where the bound electron is first struck by the projectile and then scattered by the nucleus.

If the energy of the scattered positron is very similar to that of the ejected electron the two particles may emerge in almost the same direction and in a highly correlated state, which can be considered as a continuum

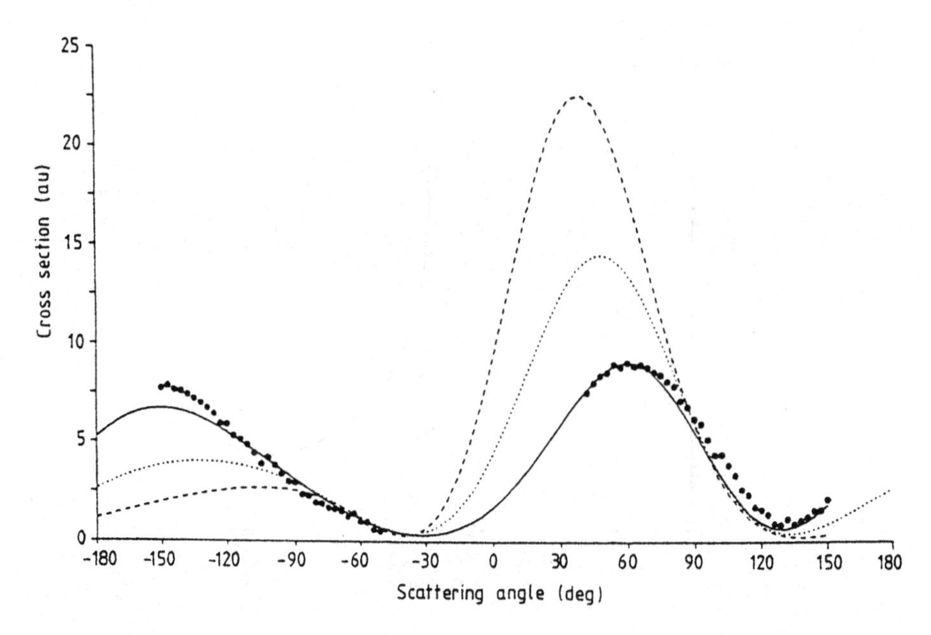

Fig. 5.7. The triple differential cross section for the ejection of electrons from the ionization of atomic hydrogen by positron and electron impact, expressed as a function of the angle made by the ejected electron to the incident beam direction. The energy of the incident projectile is 150 eV, its angle of scattering is 4°, the energy of the ejected electron is 3 eV and coplanar geometry is adopted (Brauner, Briggs and Klar, 1989). Positive angles correspond to the ejected electron emerging on the opposite side of the incident beam direction to that of the scattered projectile, and negative angles correspond to the electron emerging on the same side. ——, theoretical results for electron impact ionization; – – –, theoretical results for positrons; · · · · ·, results of the first Born approximation, which are the same for positrons and electrons; the experimental points are for electrons (Ehrhardt *et al.*, 1985). Reprinted from *Journal of Physics* **B20**, Brauner, Briggs and Klar, Triple differential cross sections for ionization of hydrogen atoms by electrons and positrons, 2265–2287, copyright 1989, with permission from IOP Publishing.

state of positronium. This so-called 'electron capture to the continuum' (ECC) gives rise to a sharply peaked structure in the triple differential ionization cross section when the momenta (and thus the kinetic energies) of the two emerging particles are equal. This feature was first revealed in studies of positron impact ionization by Brauner and Briggs (1986) that used the first Born approximation to the T-matrix element for the transition, $T_{\rm fi} = \langle \Phi_{\rm f}|V|\Phi_{\rm i}\rangle$, where V is the projectile–target interaction potential. For positron impact ionization of atomic hydrogen, the initial state is $\Phi_{\rm i} = (2\pi)^{-3/2}\phi_{\rm H}(\mathbf{r}_2)\exp(i\boldsymbol{k}\cdot\boldsymbol{r}_1)$, i.e. the product of the hydrogen target wave function and a plane wave for the incident positron; the final

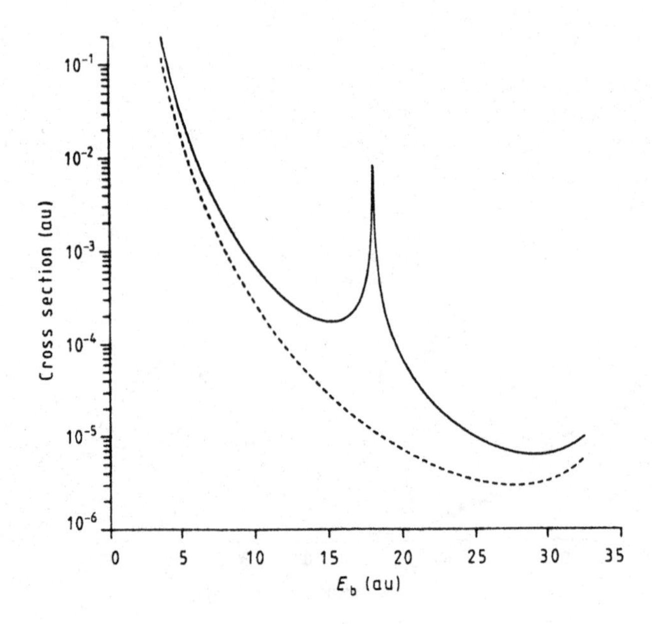

Fig. 5.8. The triple differential cross section for positron impact ionization of atomic hydrogen, expressed as a function of the energy of the ejected electron. The scattered positron and electron both emerge in the direction of the incident positron beam, the energy of which is 1 keV: ——, theoretical result obtained using a Coulomb wave to describe the relative motion of the positron and electron; – – –, result obtained using a plane wave to describe the relative motion (Brauner and Briggs, 1986). Reprinted from *Journal of Physics* **B19**, Brauner and Briggs, Ionization to the projectile continuum by positron and electron collisions with neutral atoms, L325–L330, copyright 1986, with permission from IOP Publishing.

state was taken to be $\Phi_{\rm f} = (2\pi)^{-3/2}\psi^-(\boldsymbol{r}_{12})\exp(i\boldsymbol{K}'\cdot\boldsymbol{\rho})$, i.e. the product of a Coulomb function representing the motion of the positron relative to the electron and a plane wave for the motion of the centre of mass of the unbound positronium relative to the nucleus. An example of the structure found by Brauner and Briggs (1986) is shown in Figure 5.8, which gives the cross section for the ionization of atomic hydrogen by 1 keV positrons when both particles emerge in the direction of the incident beam. The cross section exhibits a singularity when the energy of the ejected electron is equal to that of the scattered positron. Similar singular structure in the triple differential cross section is found at all angles when the momenta of the ejected electron and the scattered positron are equal.

As the energy of the incident positron is increased beyond several keV it becomes increasingly valid to consider the mechanism of electron capture into either bound or continuum states of positronium in terms of double-binary collisions (see Chapter 4), as a result of which the positron and electron are expected to emerge preferentially at a critical angle of 45° to the incident beam direction. There are two such types of double collision, which were illustrated in Figure 4.6, for the case of capture into a bound state; interference occurs between the amplitudes for the two processes, the interference being destructive for electron capture into even-parity states of positronium. Brauner and Briggs (1991, 1993) found that the effect of this interference was to introduce a sharp dip in the triple differential cross section on the low electron energy side of the the ECC singularity structure when the angle of emergence of both particles was 45°. The dip and the singularity tend to coincide as the incident positron energy is increased beyond 10 keV, but at such high energies the ionization cross section is very small.

Some of the most detailed studies of positron impact ionization of hydrogen in the intermediate energy range, 20–120 eV, were made by Kernoghan, McAlinden and Walters (1995) and Kernoghan *et al.* (1996) in the course of their elaborate coupled-state calculations of positron–hydrogen scattering, although no continuum ionization states were explicitly included in their expansion of the wave function. In addition to eigenstates of hydrogen and positronium, their wave functions included several pseudostates, each of which has a non-zero overlap with all the bound and continuum target states. The fraction of a particular pseudostate $|\phi_m\rangle$ that represents the ionization continuum is

$$a_m = 1 - \sum \langle \phi_{n,l} | \phi_m \rangle, \tag{5.11}$$

where the summation is over all bound states. The following ansatz was then used to estimate the ionization cross section:

$$\sigma_\text{i} = \sum_m (1 - a_m) \left[\sigma_\text{H}(m) + \sigma_\text{Ps}(m) \right], \tag{5.12}$$

where $\sigma_\text{H}(m)$ and $\sigma_\text{Ps}(m)$ are the cross sections for excitation of the mth pseudostate of hydrogen and for the formation of positronium into this mth pseudostate respectively. These are calculated in the same way, and at the same time, as are the cross sections for excitation of the target and for positronium formation into the various eigenstates represented in the coupled-state expansion. This procedure, although not completely rigorous, is expected to provide reasonably accurate results if a large number of states and pseudostates are included in the wave function. A similar technique was used by Campbell *et al.* (1998a) to determine the ionization cross section in positron–helium scattering.

The convergent-close-coupling (CCC) method, which was originally developed for electron–hydrogen scattering, has also been applied to positron impact ionization of atomic hydrogen over the energy range from the threshold up to 700 eV (Bray and Stelbovics, 1994). Positronium formation is not explicitly included in this formulation and therefore the results should be interpreted as an approximation to the total break-up cross section, i.e. the sum of the cross sections for ionization and positronium formation. Only at energies beyond 100 eV, where the positronium formation cross section is known to be small, should the CCC results be considered as relating solely to ionization. Beyond 200 eV the CCC results are in good agreement with those of the Born approximation, but this is almost certainly fortuitous and does not imply that the Born approximation is valid at such a low energy. In the energy interval 100–700 eV the CCC results are in good agreement with the experimental measurements; these are summarized in subsection 5.4.1.

Double ionization has been investigated experimentally for positron, electron, proton and antiproton impact but, although the process has received much theoretical attention for electron and proton impact, few theoretical studies have yet been made for positrons.

5.3 Ionization – experimental techniques for integrated cross sections

The first studies of positron impact ionization were based around the TOF systems originally developed for total cross section measurements; see section 2.3 and, for example, Griffith *et al.* (1979b), Coleman *et al.* (1982), Mori and Sueoka (1984, 1994), though only the latter workers, as described in section 5.1, claimed to be able to distinguish between ionization and excitation using the TOF technique.

The first attempts to make direct measurements of ionization cross sections were performed by the Arlington group (Diana *et al.*, 1985) using an apparatus similar to that described in section 4.3 but set to count electrons liberated from helium gas. The results obtained have now been superseded by those obtained using the methods described below.

Detailed studies of σ_i^+ for helium and H_2 were first reported by the Bielefeld group (Fromme *et al.*, 1986, 1988). The arrangement used for this work was described in detail in section 4.3. Here we recall that ionization was distinguished from positronium formation, for which there is also a remnant ion, by detection of the scattered positron and the ion in coincidence. Normalization to electron-impact-ionization cross sections at intermediate energies was used to set the absolute scale since, for helium, convergence to the first Born values had been obtained. This approximation, as described in section 5.2, is independent of the sign of the projectile charge and thus predicts equal cross sections for positrons and electrons.

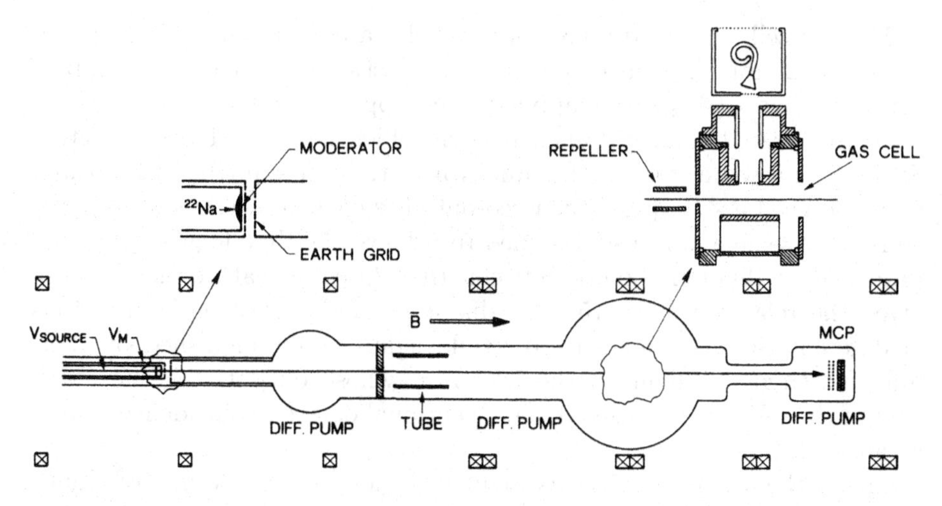

Fig. 5.9. Schematic illustration of the apparatus of Knudsen *et al.* (1990) to measure ionization cross sections. Reprinted from *Journal of Physics* **B23**, Knudsen *et al.*, Single ionization of H_2, He, Ne and Ar by positron impact, 3955–3976, copyright 1990, with permission from IOP Publishing.

In addition to this work, the Bielefeld group have also measured σ_i for atomic hydrogen. The apparatus used by Spicher *et al.* (1990) and later by Hofmann *et al.* (1997) is similar to that employed by Sperber *et al.* (1992) to measure positronium formation cross sections and was described in section 4.3. Coincidences between scattered positrons and ions were used, by virtue of a TOF analysis, to distinguish ions of different charge-to-mass ratios. It is also worth noting here that the scattered positron, which was crucial in identifying the ionization event, had an angular acceptance limited to be less than $\pm 30°$ with respect to the incident beam axis as fixed by the geometry of the detection system.

The apparatus developed for investigations of positron impact ionization by the Aarhus–UCL collaboration (Knudsen *et al.*, 1990), and subsequently used in several adaptations, is illustrated in Figure 5.9. The positron beam was guided by an axial magnetic field of around 5 mT. This was increased to 7.5 mT just in front of the scattering cell and was maintained at this level from the cell to the detector in order to provide more efficient confinement of the scattered positrons. The exit aperture of the scattering cell was chosen to be 25 mm in diameter, and a 40 mm active diameter channel electron multiplier array (CEMA) was used to count the beam, both dimensions having been chosen to ensure that as large a fraction as possible of the scattered positrons struck the detector. A detailed discussion of the transport of scattered particles was given by Knudsen *et al.* (1990).

The gas cell contained a pair of parallel plates (40 mm $\times$ 40 mm) separated by 20 mm, which served as ion extraction electrodes when appropriate voltages, $\pm V_{\rm extr}$ were applied. The upper plate had a grid-covered aperture located centrally and through which ions could be extracted. Such ions passed along a flight tube, where they were further accelerated by a voltage $4.5 \times V_{\rm extr}$ and focussed onto the cone of a ceratron detector. This arrangement was also designed to achieve the time focussing of ions produced at different points between the plates, a feature of some use given the relatively large (5 mm) diameter of the positron beam. Ions of different charge-to-mass ratios could again be distinguished by their different times of flight to the ceratron, whose output, together with that of the CEMA, was used in a conventional delayed-coincidence timing arrangement.

One of the features which distinguished the positron method of Knudsen *et al.* (1990) from earlier work with heavy projectiles (see e.g. Andersen *et al.*, 1987) was the use of pulsed extraction voltages, in order to prevent deleterious effects due to beam deflection. The pulsing system was triggered on detection of a positron at the CEMA. Each pulse had a 10 ns rise time and was held at $V_{\rm extr}$ for a period of between 1 μs and 3 μs, depending upon the target. The voltages were applied as soon as possible after the detection of the positron, the intrinsic delay between firing of the CEMA and application of the pulse to the plate being approximately 220 ns. Added to this was the extra delay caused by the flight time of the scattered positron from the gas cell to the CEMA, which varied with impact energy. Checks were performed by deliberately adding extra delay into the pulsing system, using H_2 gas, the lightest species, which established that the ion extraction efficiency was practically unity. This was also in accord with an estimate of the distance moved by a H_2^+ ion in its random thermal motion.

In order to determine absolute cross sections the ion yield $I_{\rm m+}$, defined as

$$I_{\rm m+} = N_{\rm m+}/(PN_{\rm vac}), \tag{5.13}$$

was measured, where $N_{\rm m+}$ is the number of ions of a particular charge-to-mass ratio created in a gas held at a pressure P by a total number of slow positrons $N_{\rm vac}$. The yield of singly charged ions was found to be pressure dependent, owing to resonant charge-transfer interactions, but extrapolation to $P = 0$ produced a value $I_+(0)$ which was independent of ionic collisional effects. The ionization cross section could then calculated according to

$$\sigma_{\rm i}^+ = I_+(0)\left[P/\left(nl\epsilon_{\rm ex}\right)\right], \tag{5.14}$$

where nl is the target areal density and $\epsilon_{\rm ex}$ is the efficiency of the ion extraction and detection system. Note that the efficiency of the CEMA detector does not enter this expression; because a pulsed extraction system was used, $N_{\mathrm{m}+}$ and N_{vac} are directly proportional to one another. The quantity in square brackets in equation (5.14) was determined by performing a normalization of $I_{+}(0)$ to the quoted value of the electron-impact-ionization cross section for the same target at an energy of 1000 eV, where the first Born approximation was expected to be valid.

A later version of the Aarhus apparatus, used for an extensive study of the ionization of atoms and molecules (Jacobsen *et al.*, 1995a, b; Poulsen, Frandsen and Knudsen, 1994) was similar to the apparatus just described. The important differences were: (i) the inclusion of a pair of $\boldsymbol{E} \times \boldsymbol{B}$ plates in front of the scattering cell to reduce background; (ii) the inclusion of movable apertures in front of, and behind, the scattering cell to achieve precise beam alignment; (iii) the use of a retarding element in front of the scattering cell to narrow the energy width of the beam and also control the beam intensity; and (iv) the use of a 'unit accelerator' behind the scattering region. The purpose of this last device, which consisted of a series of resistively coupled biassed rings with the centre ring held at a high negative potential, was to prevent secondary electrons liberated from the front plate of the CEMA (which was held at −2 kV) from returning down the beamline and causing ionization in the scattering region. This unit replaced a high transmission mesh which performed the same function in the work of Knudsen *et al.* (1990). In this later Aarhus experiment it was also found that by varying the primary beam intensity (thereby altering the mean time between pulses applied to the extraction plates) the measured ion yield, which contained ions extracted randomly by unrelated pulses and mostly originating from positronium formation, could be extrapolated to zero beam intensity in order to give the true yield due to ionization alone. This had the effect of removing the necessity for an empirical correction, as applied by Knudsen *et al.* (1990), which relied upon a detailed knowledge of the behaviour of the positronium formation cross section. Further details were given by Jacobsen *et al.* (1995a).

The final system described here is that of Jones *et al.* (1993), which was developed for positron–hydrogen ionization studies and is illustrated in Figure 5.10. Similar apparatus has been used by Ashley, Moxom and Laricchia (1996) (see subsection 5.4.5 below), Kara *et al.* (1997a, b) and Kara (1999). Several of the basic features, including the pulsed ion extraction and ion transport systems, are similar to those developed by Knudsen *et al.* (1990). $\boldsymbol{E} \times \boldsymbol{B}$ plates were introduced by Jones *et al.* (1993) to remove the slow positrons from the fast β^{+} particles, secondary electrons and gamma-ray flux emanating from the source.

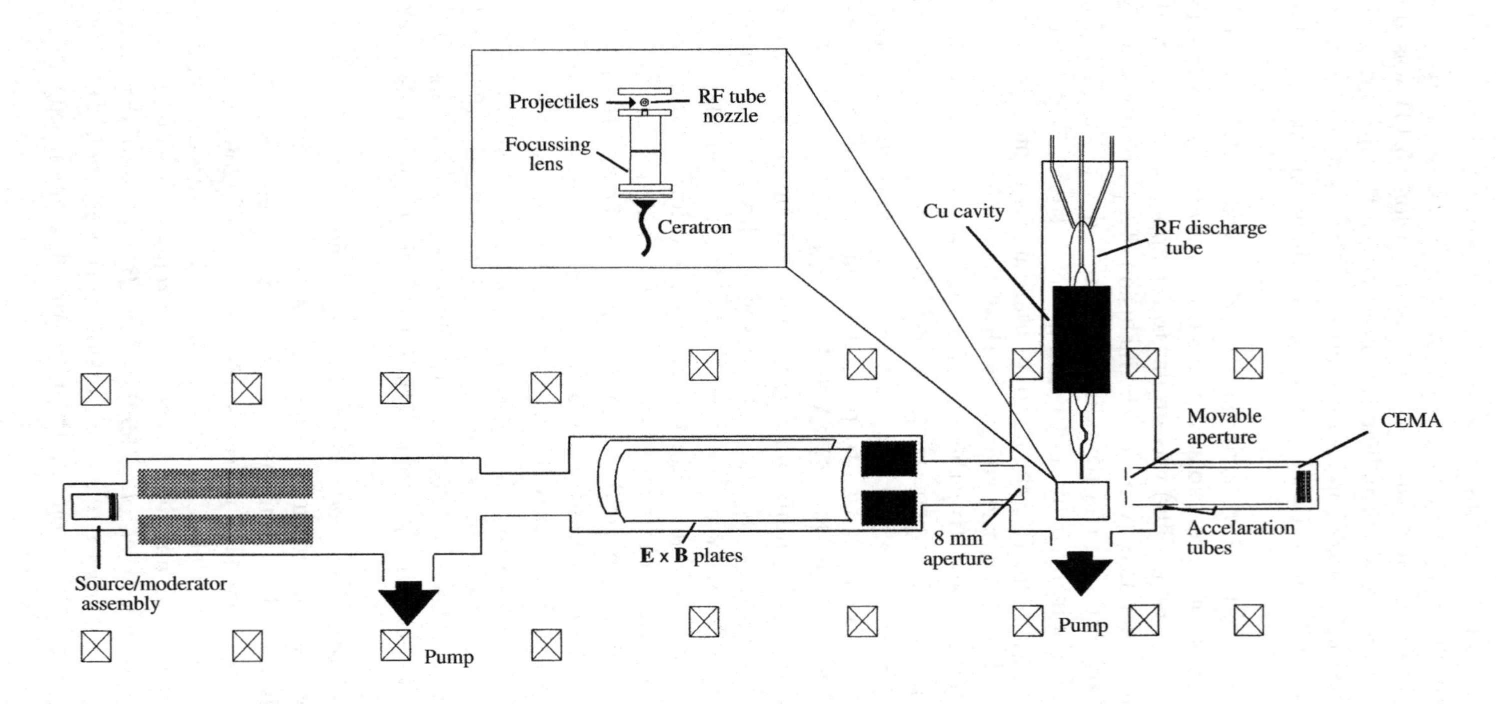

Fig. 5.10. Apparatus, not to scale, of Jones *et al.* (1993) for positron impact ionization of atomic hydrogen. Squares with crosses, Helmholtz coil; shaded rectangles, stainless steel shielding; black rectangles in beam line, lead shielding.

For this experiment the collision region was formed at the output of a radio-frequency discharge tube, which was used as a source of atomic hydrogen (Slevin and Stirling, 1981). This was similar to the arrangement, described in section 4.3, that was used by the Brookhaven and Bielefeld groups (Raith *et al.*, 1996) to measure σ_{Ps} and σ_i. The earthed output nozzle was located as close as possible to the positron beam and, as shown in the inset of Figure 5.10, between the two extraction plates so that the collision geometry was essentially of the beam–beam variety. Due to the directional and dilute nature of the gas beam some ions, once formed, drifted out of the region from which they could be extracted before application of the pulse to the plates occurred. This caused a systematic loss of ion signal at impact energies below 100 eV, owing to the longer time taken by the scattered positrons to reach the CEMA and initiate the ion extraction pulse. This was remedied by inserting a flight tube between the scattering region and the CEMA to accelerate all positrons through more than 100 V on leaving the scattering chamber.

A problem peculiar to this experiment was that some protons (or deuterons if D_2 gas was used) were found to migrate out of the discharge tube; they could then be extracted randomly and appear as an unwanted signal. This signal was subtracted by extrapolating to zero beam intensity (as devised also by Jacobsen *et al.*, 1995a) or by using a signal generator to trigger the extraction pulse at a rate commensurate with the beam intensity.

During the course of their measurements Jones *et al.* (1993) used previously available cross section data for positron and electron impact ionization of molecular hydrogen and electron impact ionization of atomic hydrogen to provide absolute normalization and as internal checks on their data. By invoking the validity of the first Born approximation, the ion yields obtained for each system were independently normalized at high energies to the relevant absolute cross section, though some of these had been previously normalized to existing electron data. The method of normalizing for each target–beam combination was thought to be more reliable, owing to possible differences in gas beam–projectile beam overlap.

5.4 Single ionization – results

1 Atomic hydrogen

The results of Jones *et al.* (1993), Hofmann *et al.* (1997) and Kara (1999) are shown in Figure 5.11, where they may be compared with the electron impact data of Shah, Elliot and Gilbody (1987). The results of calcula-

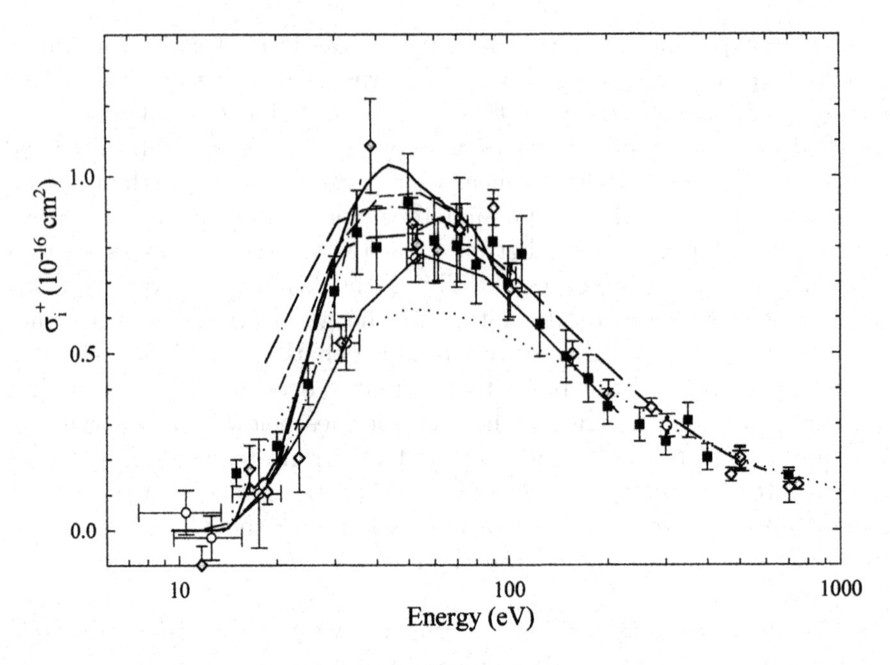

Fig. 5.11. Positron impact ionization of atomic hydrogen. Experiment: ■, Jones *et al.* (1993); ○, Hofmann *et al.* (1997); ◇, Kara (1999). Theory: ——, Kernoghan *et al.* (1996); ——, Ohsaki *et al.* (1985); – – –, Mukherjee *et al.* (1989); — —, Mitroy (1996); — · —, Ratnavelu (1991); — · · —, Janev and Solov'ev (1998); · · · · ·, electron impact (Shah, Elliot and Gilbody, 1987).

tions by Kernoghan *et al.* (1996), Mitroy (1996) and Janev and Solov'ev (1998), as well as those of a number of earlier workers, are also shown. These theoretical results have been selected as a representative subset of what is available, as reviewed in section 5.2. The data of Hofmann *et al.* (1997) supersede those of Spicher *et al.* (1990), which are thought to have a systematic normalization error.

The measured cross sections, which are in good accord with one another and with theory, are found to follow the electron data closely down to an energy of 150 eV whereupon they rise above the latter, the increment being around 30% at energies in the vicinity of the broad maximum in the positron ionization cross section. As described below, this enhancement is typical of that found in other systems and is understood to be a polarization-correlation effect whereby the positron effectively pulls the electron cloud away from the ion, resulting in an enhanced cross section for ionization. The opposite is the case for electron impact, thus depressing the cross section for the negatively charged projectile. These, and related effects in proton and antiproton impact, have been described by Knudsen and Reading (1992) and Paludan *et al.* (1997).

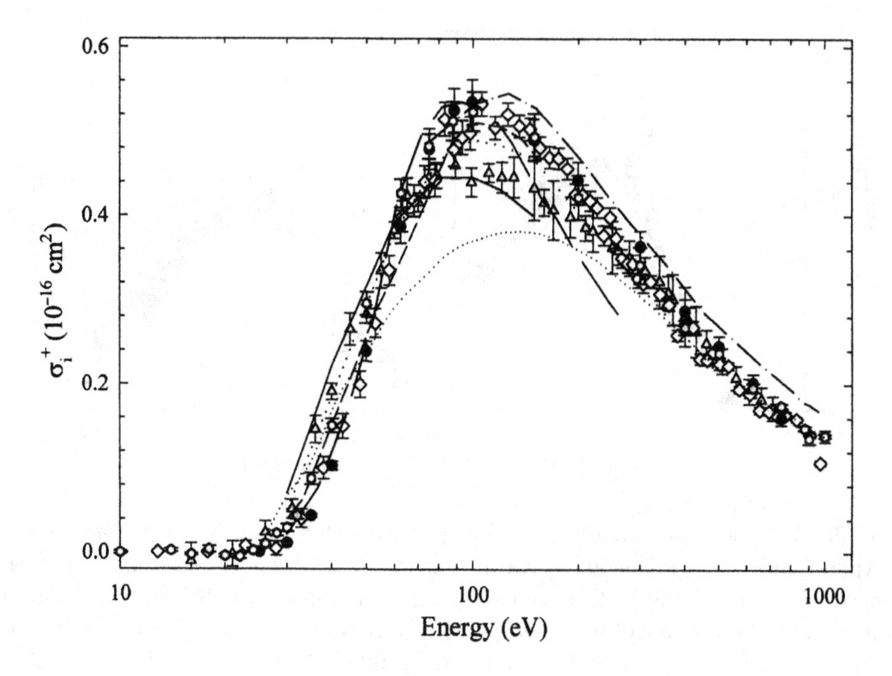

Fig. 5.12. Positron impact ionization of helium gas. Experiment: △, Fromme *et al.* (1986); ◇, Moxom, Ashley and Laricchia, (1996); •, Jacobsen *et al.* (1995b); ○, Knudsen *et al.*, (1990). Theory: — —, Schultz and Olson (1988); ——, Basu, Mazumdar and Ghosh (1985); · · · ·, Campeanu, McEachran and Stauffer (1996); — · —, Chen and Msezane (1994); - - - -, Campbell *et al.* (1998a). · · · · ·, electron impact (Krishnakumar and Srivastava, 1988).

2 The noble gases

All the noble gases up to xenon have been studied; helium, after hydrogen another cornerstone for the meeting of theory and experiment, has received the most attention. The situation for this target is shown in Figure 5.12, which includes the experimental data of Fromme *et al.* (1986), Knudsen *et al.* (1990), Jacobsen *et al.* (1995b) and Moxom, Ashley and Laricchia (1996). The data of Mori and Sueoka (1994), which extend up to 100 eV, are not shown but are in broad agreement, though with large uncertainties. We note that Jacobsen *et al.* (1995b) considered their data to be superior to those of Knudsen *et al.* (1990) at the lower energies, owing to uncertainty introduced by the correction for ion detection from positronium formation (as noted in section 5.3) in the latter. The data of both Fromme *et al.* (1986) and Knudsen *et al.* (1990) are above those of Jacobsen *et al.* (1995b) and Moxom *et al.* (1996) in this energy region. A more detailed look at the near-threshold region is given in subsection 5.4.5. Also shown in Figure 5.12 are the electron

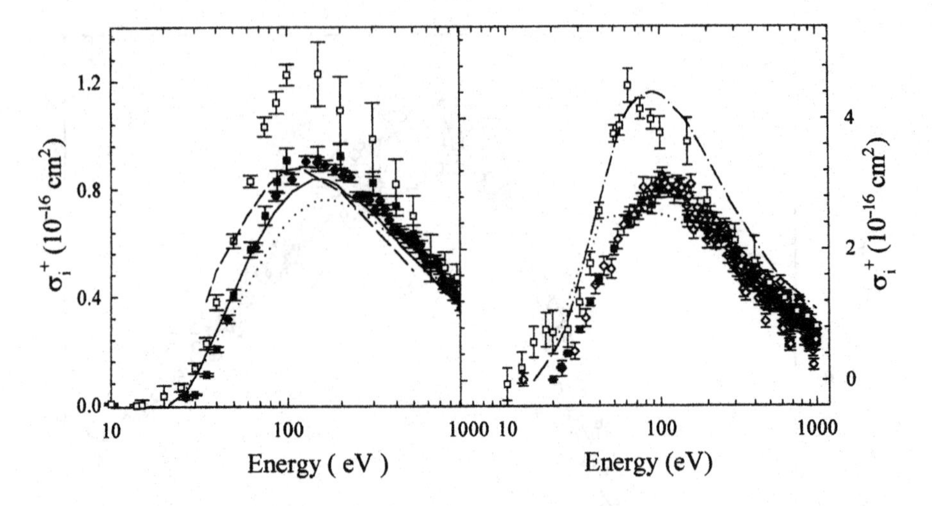

Fig. 5.13. Positron impact ionization for neon (left) and argon (right) gases. Experimental data for neon: ◆, Kara *et al.* (1997a); □, Knudsen *et al.* (1990); ■, Jacobsen *et al.* (1995b). Experimental data for argon: ◇, Moxom, Ashley and Laricchia (1996); ■, Jacobsen *et al.* (1995b); □, Knudsen *et al.* (1990). Theory for neon: – – –, Campeanu, McEachran and Stauffer (1996); ——, Moores (1998). Theory for argon: Moores (1998). In each case the dotted lines are the electron data of Krishnakumar and Srivastava (1988).

data of Krishnakumar and Srivastava (1988), which are in good accord with other modern measurements of this cross section (e.g. Rapp and Englander-Golden, 1965; Montague, Harrison and Smith, 1984) over the entire energy range. In addition, Figure 5.12 shows for comparison the results of various calculations.

The situation for neon and argon targets is shown in Figure 5.13. The electron data are those of Krishnakumar and Srivastava (1988) and they agree with other recent measurements (see e.g. McCallion, Shah and Gilbody, 1992, and references therein for argon data). The positron results are those of Knudsen *et al.* (1990), Jacobsen *et al.* (1995b) and Kara *et al.* (1997a) for neon, and those of Moxom, Ashley and Laricchia (1996) for argon. For neon, the results of Kara *et al.* (1997a) and Jacobsen *et al.* (1995b) are in reasonable accord, but the data of Knudsen *et al.* (1990) are higher over most of the energy range. The same is also true for argon, where the results of Jacobsen *et al.* (1995b) and Moxom, Ashley and Laricchia (1996) are in excellent agreement. Also shown in Figure 5.13 are the results of theoretical work performed by Campeanu *et al.* (1996) and Moores (1998) for neon, who both find reasonable accord with experiment, and by Moores (1998) for argon; these results for argon are closest to the data of Knudsen *et al.* (1990).

Examination of Figures 5.11, 5.12 and 5.13 for all four targets chosen for inclusion here shows that at sufficiently high energies the single-ionization cross sections for electrons and positrons are equal within the accuracy of the data, and are thus in accord with the prediction of the first Born approximation. As the impact energy is lowered, the positron data rises above that for electrons, by a factor of approximately 1.5 at the maximum, for all the targets investigated. The origin of this effect is thought to be similar to that described for atomic hydrogen in subsection 5.4.1. As the impact energy is lowered further towards E_{i}, the positron results tend to fall faster than those for electrons. This is presumably a result of the preponderance of positronium formation at the lower energies. It is notable, though, that the intermediate energy enhancement in σ_{i}^{+} for positron impact is smallest for argon; this trend, observed in helium, neon and argon, was confirmed by the measurements on krypton and xenon of Kara *et al.* (1997a). The effect has been attributed (Laricchia, 1995a; Kara *et al.*, 1997a) to the increasing importance of the static interaction for the more highly charged nuclei. This will tend to repel the positron and thus counterbalance the attractive polarization effect described above.

3 Molecules

A number of molecular targets have been studied in recent years and the effects of direct and dissociative ionization channels have been observed. Positron–methane collisions were also used in the first observation of positronium hydride, as will be described in section 7.5.

The simplest case, shown in Figure 5.14, is the non-dissociative ionization of molecular hydrogen, which displays features similar to those for the noble gases when comparing positron and electron ionization cross sections. The data shown are those of Knudsen *et al.* (1990), Fromme *et al.* (1988), Jacobsen *et al.* (1995a) and Moxom, Ashley and Laricchia (1996) for positrons and those of Rapp and Englander-Golden (1965) for electrons. The agreement between the sets of data for positrons is good at the higher energies, but for energies less than 200 eV the data of Jacobsen *et al.* (1995a) fall markedly below those of the other workers. Between 100 eV and 200 eV the data of Moxom, Ashley and Laricchia (1996) generally lie above those of Fromme *et al.* (1988) and Knudsen *et al.* (1990). The theoretical work is due to Chen, Chen and Kuang (1992).

For electrons, the dissociative ionization of molecular hydrogen was found by Rapp and Englander-Golden (1965) and Rapp, Englander-Golden and Briglia (1965) to contribute only around 6% to σ_{i}. All studies of positron impact have also found the contribution from dissociative processes to be small; however, this may be due in part to the reduced

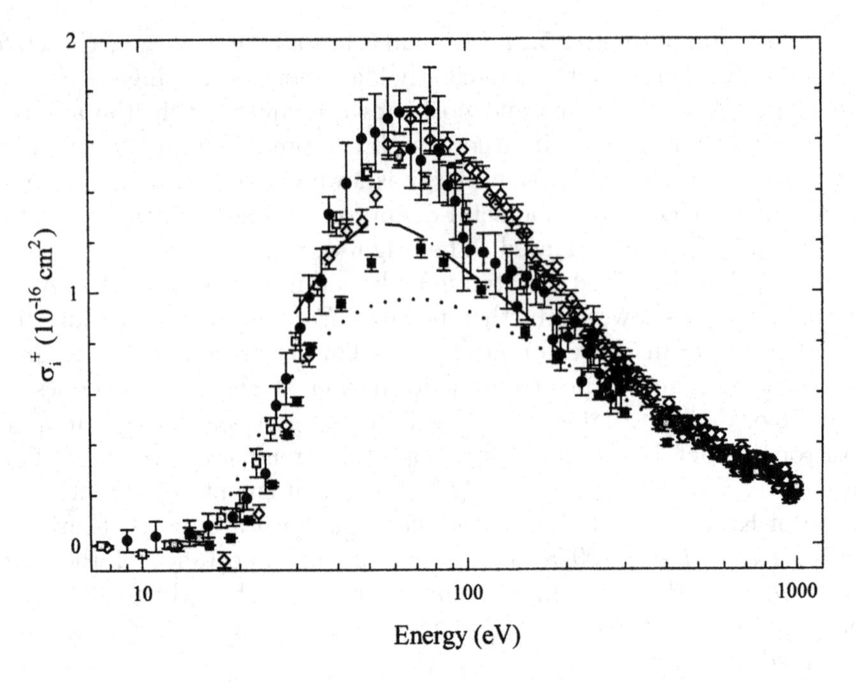

Fig. 5.14. Positron impact ionization of H_2 gas. Experiment: ◇, Moxom *et al.*, (1996); ■, Jacobsen *et al.*, (1995a); □, Knudsen *et al.* (1990); •, Fromme *et al.* (1988). — · —, theory of Chen, Chen and Kuang (1992); · · · · ·, electron data of Rapp and Englander-Golden (1965).

collection efficiency of the ions produced in such collisions since, in contrast to the thermal energies of the ions left behind in the direct process, these species can be liberated with several eV of kinetic energy (see e.g. Massey, 1969).

A comprehensive study of the molecules N_2, CO, CO_2 and CH_4 was undertaken by the Aarhus group (Poulsen *et al.*, 1994), who investigated direct and dissociative ionization. In the former case, trends were found similar to those for H_2 and the noble gases. However, for dissociative ionization of N_2 the cross sections for electron and positron impact do not merge at the highest energies studied, those for electrons being the greater. This is reminiscent of effects found for double ionization of the noble gases (see section 5.5) and is characteristic of processes which involve two active electrons. As the complexity of the target molecule increases, so that dissociative ionization can lead to fragment-ion production in an essentially one-electron transition, the differences between the electron and positron cross sections diminish. There is no theoretical work here for comparison, though related experimental work on antiproton–molecule cross sections was reported by Knudsen *et al.* (1995a).

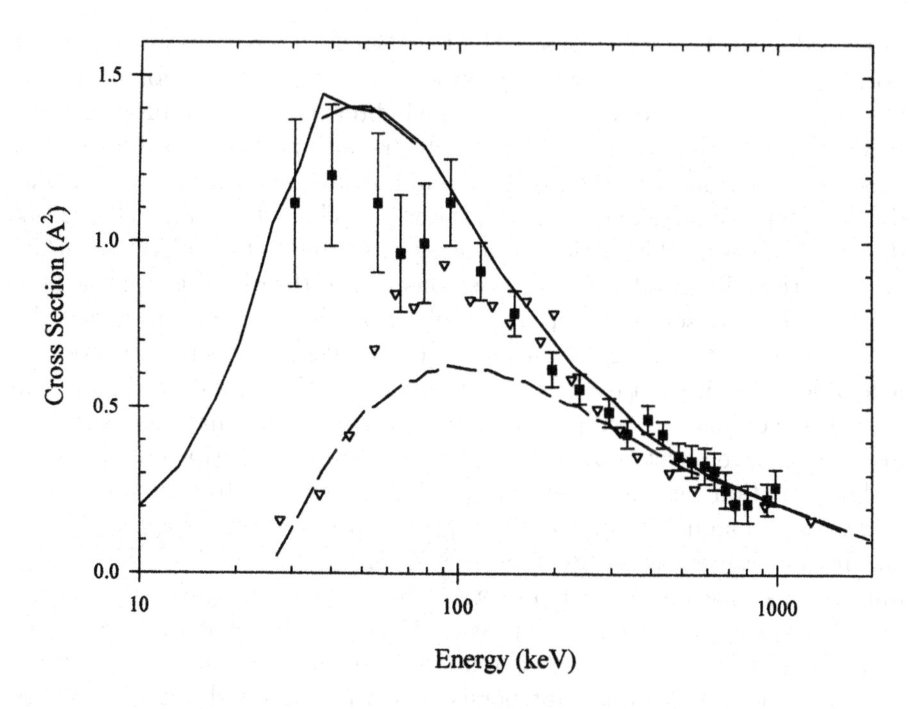

Fig. 5.15. Comparison of the single-ionization cross sections for projectile–hydrogen scattering: ——, proton impact; – – –, electron impact. Key: ■, antiprotons (Knudsen *et al.*, 1995b); ▽, positrons (Jones *et al.*, 1993).

4 Comparisons with heavier projectiles

In the intermediate energy region, where threshold effects are not important, it is instructive to compare positron and electron results for single ionization with those for heavy singly charged projectiles, notably protons and antiprotons. Whilst data for protons have been available for some time, ionization cross sections for antiprotons only appeared recently with work at the low energy antiproton ring (LEAR) at CERN (see e.g. Andersen *et al.*, 1990a, b; Hvelplund *et al.*, 1994; Knudsen *et al.*, 1995a, b). A detailed description of antiproton experiments, together with theories of ionization and comparisons of the atomic scattering data available for the four particles, can be found in the review of Knudsen and Reading (1992). An informative discussion is contained in the short review of Schultz, Olson and Reinhold (1991).

The single-ionization cross sections for the positron–, electron–, proton– and antiproton–hydrogen systems are shown on an equi-velocity scale in Figure 5.15 (Knudsen *et al.*, 1995b). (Note that 1 MeV a.m.u.$^{-1}$ corresponds to a kinetic energy of approximately 544 eV for the lighter projectiles.) At high velocities all cross sections converge and are essentially in accord with the predictions of the first Born approximation. The most notable trend is that below ~200 keV a.m.u.$^{-1}$ (~100 eV for positrons and electrons), the cross sections for the lighter particles fall progressively below those for the heavier particles, which continue to rise. Interestingly, as shown in Figure 5.15 the positron and antiproton data are very similar down to approximately 100 keV a.m.u.$^{-1}$. Thus, the major source of difference between the particles is a mass effect due to the behaviour of the cross sections for positrons and electrons as the threshold is approached. This has been described as a type of 'lack of energy' effect, in which the much lower kinetic energy of the lighter projectiles reduces the phase space available in the final state and thus the cross section. As described in the previous sections, σ_i^+ for positrons generally exceeds that for electrons, and this charge effect is also present for protons and antiprotons.

The situation for helium is similar in most respects to that for atomic hydrogen, though for the former target there is clear evidence that as the heavy projectile energy falls below approximately 50 keV, σ_i^+ for antiprotons exceeds that for protons. Comparisons for the heavier noble gases were compiled by Paludan *et al.* (1997), who noted that the effect of the static interaction on the relative behaviour of the cross sections for the light particles does not persist when protons and antiprotons are compared. The heavier particles are much more immune to this trajectory effect at intermediate speeds, so that the polarization effect dominates even for targets with higher nuclear charge, leading to an enhanced σ_i^+ for protons.

5 Near-threshold studies

As described in section 5.2, there is considerable theoretical interest in the behaviour of σ_i^+ (and of cross sections for multiple ionization) near to their relevant thresholds. This stems, in part, from the different Wannier exponents found for positrons and electrons and also from the general question of the validity of such a classical treatment of ionization. To recap, the Wannier law predicts that σ_i^+ is proportional to E_1^ζ, where $E_1 = E - E_i$, the energy of the positron in excesss of the ionization threshold energy, and $\zeta = 1.127$ for electrons (Wannier, 1953) and 2.651 for positrons (Klar, 1981). We note again that the Wannier theory makes no prediction about the coefficient of this term or the range of excess energy over which the law is expected to be valid. For electron impact there is, as summarized by Read (1985), a great deal of experimental evidence to support the Wannier prediction for a narrow energy range near threshold. The greater Wannier exponent for positrons suggests that positron–electron correlation is more important than that between the two electrons in the electron impact case. Thus, at least near to threshold, positrons seem to offer a more sensitive test of three-body Coulomb systems. Why this is so can be seen qualitatively in Figure 5.6, which gives a schematic illustration of how near-threshold ionization occurs in the Wannier picture.

The first hints that the energy dependence of σ_i^+ near E_i was different for positrons and electrons came from the results of Fromme *et al.* (1986, 1988) for helium and molecular hydrogen, which revealed that $\sigma_i^+(e^+)$ appears to have a steeper energy dependence than $\sigma_i^+(e^-)$ and that the former falls below the latter very close to E_i. This type of behaviour is consistent with the expected Wannier laws for the two projectiles, though the energy width of the positron beam and other instrumental effects (see section 4.3 for a discussion of the operation of the ion extractor in this experiment) meant that the measurements were insufficiently precise for a value of the exponent ζ to be extracted.

Further evidence that the near-threshold energy dependence of the positron and electron ionization cross sections was different came from the work of Knudsen *et al.* (1990) and Jacobsen *et al.* (1995a), although the most detailed study so far has been that reported by Ashley, Moxom and Laricchia (1996). The apparatus they used was a straightforward development from that of Jones *et al.* (1993), shown in Figure 5.10. A retarding field analyser was incorporated before the scattering region and used to reduce the longitudinal energy spread of the beam to around 0.5 eV. The accelerator tubes in the apparatus of Jones *et al.* (1993) were modified to provide a weak electric field penetrating into the interaction region, which aided the extraction of those positrons left with a very low kinetic energy after the collision. In making measurements close to threshold it is important to ensure that all background sources of ionization are accounted for, and correctly subtracted, and that ion and positron detection efficiencies are independent of positron impact energy, which should itself be accurately calibrated. Details of how these difficult objectives were met have been given by Ashley, Moxom and Laricchia (1996).

The latter workers found that their data for both helium and molecular hydrogen could be fitted by power laws of the Wannier type, but in each case the exponent ζ was substantially lower than the value 2.651 predicted by Klar (1981). Fitting over the full energy range of their investigations, up to approximately $E_1 = 10$ eV, they found $\zeta = 2.27 \pm 0.08$ for helium and 1.71 ± 0.03 for molecular hydrogen.

The experimental helium data are shown in Figure 5.16, along with the theoretical results of Ihra *et al.* (1997), and good accord between the two is found. Thus, Ihra *et al.* (1997) deduced that, for most of the energy range investigated, the experiment can be fitted by a threshold law of the form of equation (5.10). This threshold law was derived by

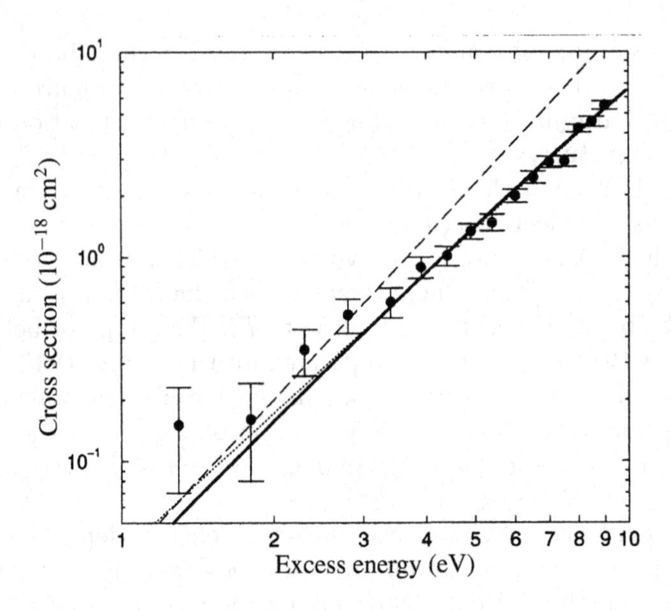

Fig. 5.16. Near-threshold positron impact single ionization of helium gas. The experimental data are from Ashley, Moxom and Laricchia (1996) and the dotted line is their fit to the data with $\zeta = 2.27$: – – –, Wannier prediction of Klar (1981) with $\zeta = 2.65$, equation (5.8); ——, theory of Ihra *et al.* (1997), equation (5.10).

taking account of anharmonicities in the three-particle potential around the normal Wannier configuration. Thus, the true Wannier law appears to be valid over a much narrower energy range for positrons than is the case for electrons.

5.5 Multiple ionization

This section deals predominantly with double ionization, for which, as we will discuss, instructive comparisons of the behaviour of positron cross sections with those for other particles have been made. A comparative study of experimental results for the noble gases was undertaken by Paludan *et al.* (1997). There have been few theoretical studies of multiple ionization dedicated solely to positron impact, though this topic is embraced by the more general theories that have been developed for high-velocity charged-particle impact. Detailed accounts can be found in the reviews of Knudsen and Reading (1992), McGuire (1992) and McGuire *et al.* (1995).

Interest in double ionization stems from the realization of Puckett and Martin (1970) and, particularly, of Haugen *et al.* (1982), that even at very high speeds ($>$ 5 MeV a.m.u.$^{-1}$), the double-ionization cross sections, σ_i^{2+}, for electrons and protons were different by a factor close to two, with that for the former projectile being the greater. This was not in accord with theoretical expectations at the time. McGuire (1982) first suggested that interference between the first and second order amplitudes (or equivalently between mechanisms which involve the projectile in one- and two-electron encounters) would lead to a term in the cross section proportional to the cube of the projectile charge, giving the observed charge dependence of σ_i^{2+}. Most theorists now believe, however, that the effect is caused by electron–electron correlation during the collision, although, as summarized by McGuire (1992), there is still disagreement about the mechanisms by which the correlation operates, with several theories that involve different physical assumptions each finding general agreement with experiment.

To date there have been several experimental measurements of σ_i^{2+} for positron impact. The apparatus and methods used by most of the groups involved are closely related to techniques developed for single-ionization studies, as described in section 5.3, and we will not describe these further. The exception is the work of Hippler and colleagues, whose main focus was multiple ionization. Their technique was centred around the use of a pulsed low energy positron beam, provided by a linear accelerator source (see subsection 1.4.4). The use of a positron pulse enabled a simple ion time-of-flight system to be used for ionic charge-to-mass discrimination. More detailed discussions of the technique were given by Helms *et al.* (1994a, 1995).

Most authors, after appropriate corrections for the relative detection efficiency of the ions (see e.g. Andersen *et al.*, 1987; Helms *et al.*, 1995), have presented their data as cross section ratios, $R_{e+}^{(m)} = \sigma_i^{m+}/\sigma_i^+$; these can then be compared with the corresponding parameter for other, singly charged, projectiles. Values of $R^{(2)}$ for helium are shown in Figure 5.17, where comparisons are made with data for electron, proton and antiproton impact. It is clear from this figure that at high speeds (around 1 MeV a.m.u.$^{-1}$) the positron and proton ratios are in broad agreement but are lower by a factor of two or more than those for their negatively charged counterparts. As the speed of the positrons is reduced, and hence the kinetic energy lowered towards the threshold ($\approx$ 79 eV for helium), the positron and proton data diverge. This can be compared with the effects described in subsection 5.4.4 and can be understood if the similar behaviour for $R_{e-}^{(2)}$ and $R_{\bar{p}}^{(2)}$ is noted; both result from the fact that less kinetic energy is carried by the lighter particles, which causes a decrease in the accessible phase space of the He^{2+} and a consequent drop in the cross section. The fall in $R_{e+}^{(2)}$ and $R_{e-}^{(2)}$ occurs in an energy range in which the value of σ_i^+ does not vary strongly for either projectile, and is rising as E is lowered, so that this fall must be due to a decrease in σ_i^{2+}.

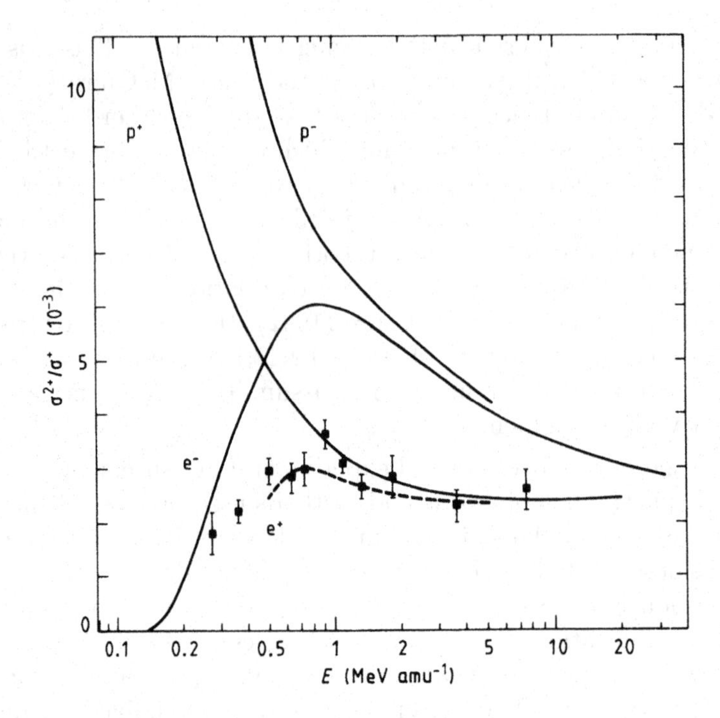

Fig. 5.17. Ratio of the cross sections for double ionization and single ionization, for positrons, electrons, protons and antiprotons on helium gas. The positron data (■) are shown explicitly. The broken line is the 'ratio of ratios' as defined by equation (5.15). Reprinted from *Journal of Physics* **B21**, Charlton *et al.*, Positron and electron impact double ionization of helium, L545–L549, copyright 1988, with permission from IOP Publishing.

Charlton *et al.* (1988, 1989) also constructed the 'ratio of ratios' given by

$$R_{\mathrm{r}}^{(2)} = R_{\mathrm{e}-}^{(2)} R_{\mathrm{p}+}^{(2)} / R_{\bar{\mathrm{p}}}^{(2)}, \tag{5.15}$$

which, it was hoped, would exhibit similar behaviour to $R_{\mathrm{e}+}^{(2)}$. $R_{\mathrm{r}}^{(2)}$ is shown as a broken curve in Figure 5.17, and it can be seen that that this conjecture has some physical basis. Paludan *et al.* (1997) showed how the expression (5.15) gives reasonably good accord also for the heavier noble gases. They noted that this implies that any trajectory effect, noted above with respect to positron and electron single ionization, must cancel out and thus be of equal importance in both single and double ionization. In addition, exchange effects for electron impact must be either negligible or of roughly similar importance for the two processes.

Double-ionization studies of the heavier noble gases have also been reported by Charlton *et al.* (1989), Kruse *et al.* (1991), Helms *et al.* (1994a, 1994b, 1995) and Kara *et al.* (1997a). Although there are small differences between some of the data reported, overall there is reasonable accord. Absolute values for double-ionization cross sections and comparison plots, for both heavy and light projectiles, can be found in the work of Paludan *et al.* (1997).

There are a number of points of note. As the target atomic number increases, $R^{(2)}$ becomes larger, reaching 10% for xenon. In addition, the differences between the electron and positron ratios become smaller and they appear to merge around 1 keV impact energy. Also, the form of $R^{(2)}$ for both positrons and electrons, with a sharp rise from threshold followed by a plateau or minimum and then a further rise, suggests that there is more than one contributory process. Again, considering the behaviour of the cross section for single ionization, the structure must be due to variations in the values of σ_{i}^{2+} for different targets. Helms *et al.* (1994a, 1995) attributed this to the influence of inner shell contributions to this cross section, which become increasingly important as the impact energy is increased. Thus in this picture, double and higher order ionization can be caused by a single inner shell ionizing event followed by shake-off, Auger and Koster–Cronig processes. These authors supported this contention with calculations for argon, based upon a modified Born approximation which takes account of the acceleration (deceleration) of the electron (positron) in the nuclear field of the target. This effect also explains the relative sizes of the electron and positron cross sections. Theoretical results for xenon have also been derived by Helms *et al.* (1995) from the semi-empirical Lotz formula, but with the inclusion of a simple energy shift to allow for the slowing down experienced by the positron during the collision. Further discussion of inner shell ionization can be found in section 5.7.

Triple ionization of the heavier noble gases has been reported by Kruse *et al.* (1991) and Helms *et al.* (1994, 1995a), and again the influence of inner shell effects is notable and quickly becomes the dominant mechanism for this process.

The coming years will see extensions of the work reported in this section to further reactions and cross sections for positron-induced transitions involving more than one electron. In particular, studies of double ionization near to the threshold seem worthwhile, particularly in the light of the results of Helms *et al.* (1994, 1995b) where the electron and positron cross sections appear to intersect. New results for helium gas by Bluhme *et al.* (1998) have been analysed using a modified Rost–Pattard parameterization (Rost and Pattard, 1997), which has shown that a simple unified treatment of double ionization, similar to that developed for single ionization, is applicable. One surprising feature of the study of Bluhme *et al.* (1998) was the strong suppression of the transfer-ionization channel (double ionization involving the formation of positronium).

5.6 Differential cross sections

Understanding the complexities of the few-body physics involving strongly correlated particles which occurs in ionization by charged projectiles remains a fundamental challenge to scattering theory. Over the last three decades or so major advances have been made in this area, particularly for electron impact, through the measurement of differential ionization cross sections and detailed comparisons with ever more sophisticated theories (see e.g. Lahmam-Bennani, 1991, and references therein). Part of the motivation for attempting similar work for positron impact is to gain, by comparisons with both electron data and theory, a greater understanding of collision dynamics, though with the new feature of strong electron–positron correlations and the effects of competition with positronium formation.

Different layers of differential information are available in ionizing collisions. The simplest quantity is the single differential cross section, $d\sigma_{\mathrm{i}}^{+}/d\Omega$, where all the positrons scattered, or the electrons ejected, into a particular solid angle are measured, irrespective of their kinetic energy and the fate of the undetected particle. McDaniel (1989, section 6.9A) stated that no measurements of this cross section have been made for electron impact since it has no physical content, owing to the presence of two electrons in the final state. However, this quantity is of interest in positron collisions.

The collision parameters can be specified further if the double differential cross section is measured. This is usually written as $d^2\sigma_{\mathrm{i}}^{+}/dEd\Omega$, where E and Ω refer to the energy and solid angle of either the scattered positron or the ejected electron. Measurements of this quantity have been made for positron impact and will be described below and compared with data for electrons.

Measurements of $d\sigma_{\mathrm{i}}^{+}/d\Omega$ for positron–argon collisions, in which the scattered positron was detected, have been reported by Finch *et al.* (1996a). This work was undertaken in order to search for structures in this cross section which might reflect those reported previously in $d\sigma_{\mathrm{el}}/d\Omega$ by the Detroit group (see section 3.4). Accordingly, measurements were taken at a fixed angle of 60° over the energy range 40–150 eV, with particular emphasis on the 50–60 eV range where a step-like structure in $d\sigma_{\mathrm{el}}/d\Omega$ was reputed to exist. However, no structure was found in $d\sigma_{\mathrm{i}}^{+}/d\Omega$, which, at 60°, was found to be approximately constant between 40 eV and 60 eV and to fall monotonically at higher energies. A more comprehensive series of measurements of $d\sigma_{\mathrm{i}}^{+}/d\Omega$ for both single and double ionization for argon and krypton targets was reported by Falke *et al.* (1997) at energies between 75 eV and 120 eV.

The first reported study of the behaviour of double differential cross sections for positron impact ionization was that of Moxom *et al.* (1992); these workers conducted a search for electron capture to the continuum (ECC) in positron–argon collisions. In this experiment electrons ejected over a restricted angular range around 0° were energy-analysed to search for evidence of a cusp similar to that found in heavy-particle collisions (e.g. Rødbro and Andersen, 1979; Briggs, 1989, and references therein), which would be the signature of the ECC process.

The scattering cell and ion-extraction system used for this study have been described by Moxom *et al.* (1992) and Moxom, Laricchia and Charlton (1995a). The positron beam was timed by the remoderator-tagging technique, described in section 7.6 (Laricchia *et al.*, 1988; Zafar *et al.*, 1991). The beam was then passed through the scattering cell, and electrons ejected in the forward direction were monitored at the end of the flight path using a ceratron detector. Thus, time-of-flight spectra could be obtained, from which an estimate of the ejected electron energy could be derived. The signal from the ceratron was used to gate on the ion extractor, and the electron signal was only accepted when an ion was found in coincidence. Retarding grids between the scattering cell and the detector were also used to energy-analyse the ejected electrons.

A major limitation of this experiment was that beam confinement was provided by an axial magnetic field. The field at the scattering cell, and beyond to the detector, was deliberately kept low to reduce the angular acceptance, although, as described by Moxom *et al.* (1992), this quantity was dependent upon the energy of the ejected electrons. Experiments at 50 eV, 100 eV and 150 eV positron impact energies revealed that the lower electron energies are most favoured, the number of electrons ejected dropping rapidly to close to zero at around half the impact energy. Close inspection of the energy spectra revealed small bumps in the distributions at 40 eV and 60 eV for 100 eV and 150 eV impact energies respectively, whereas the ECC peak would be expected at around 42 eV and 67 eV.

Whether or not these features can be associated with the ECC process, the experiment found that ECC makes a small contribution to the double differential cross section for positron impact ionization. This was not in accord with quantum mechanical calculations at the time, though in line with results of a classical trajectory Monte Carlo calculation reported by Schultz and Reinhold (1990). The latter workers did not find a sharp cusp in the double differential cross section at the ECC energy but, instead, a broad ridge-like feature, which they attributed to the fact that the light positron can scatter over a large angular range. Although these theoretical data were for the positron–hydrogen system, whereas the experiment was with an argon target, Moxom *et al.* (1992) attempted to compare the shape of their data with the results of Schultz and Reinhold (1990)

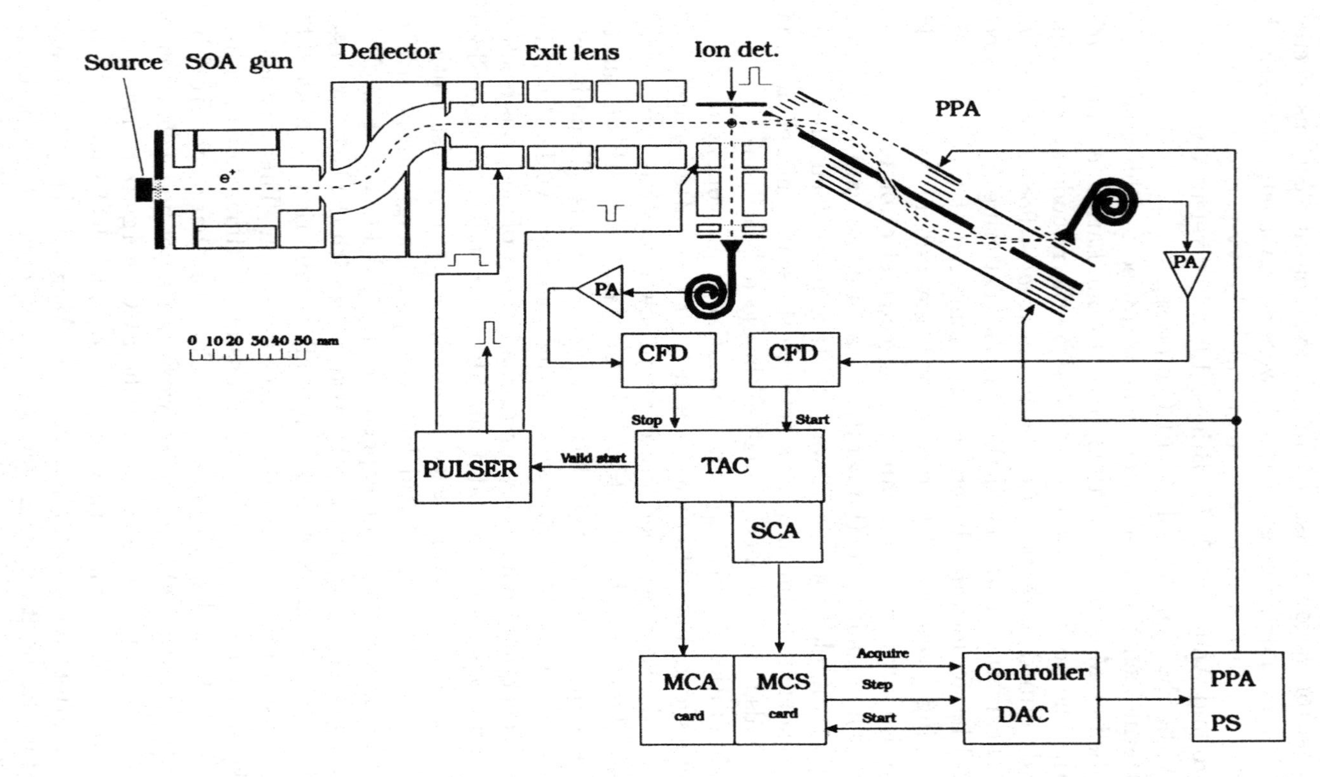

Fig. 5.18. Schematic illustration of the apparatus developed by Kövér and coworkers for studies of positron-impact differential ionization cross sections. Reprinted from *Journal of Physics* **B26**, Kövér, Laricchia and Charlton, Ionization by positrons and electrons at 0°, L575–L580, copyright 1993, with permission from IOP Publishing.

by numerically convoluting the latter with the transmission probability factor of their apparatus. This quantity was obtained by solving the equations of motion for electrons ejected with various energies and angles and over the range of possible starting coordinates in their scattering cell. A small structure was found in this convoluted theoretical energy spectrum located at around 40 eV, though somewhat more pronounced than in the experiment.

In order to probe further the possible role of ECC in positron collisions the UCL group developed the electrostatic system shown in Figure 5.18. The positron beam was formed using a slightly modified version of the Soa system of Canter *et al.* (1986); see also subsection 1.4.3. The positrons were separated from the high-energy β and γ fluxes, using a pair of cylindrical plates, and then focussed using a five-element lens onto a gas target formed by a narrow nozzle. An electron beam could also be derived from this system, though for this a weaker radioactive source was used, which resulted in primary beam intensities similar to those for positrons.

The energy distributions of the scattered positrons (electrons) or ejected electrons were measured using a parallel plate analyser (PPA). The ions were extracted using an electric field which was pulsed on when the PPA detector registered a count. Coincidences of the ions with scattered positrons or ejected electrons were monitored using a standard timing arrangement, whilst a multichannel scaler system was used to ramp the voltage applied to the PPA in order to measure the energy spectra. The positron beam was chopped when the extraction field was present to prevent particles being pulled into the ion detector.

Kövér, Laricchia and Charlton (1993) measured the energy distributions at impact energies of 100 eV, 150 eV and 250 eV for positrons and electrons scattered close to 0°. A surprising feature of these data, shown in Figure 5.19, is that despite the large difference found in the integrated cross sections for positrons and electrons at these impact energies the energy distributions are very similar. In the positron case, no structure was found which could be attributed to the ECC process; the latter, at an impact energy of 100 eV, for instance, should have been present at an energy loss of 58 eV. The level of sensitivity is set by the absolute scale on the 100 eV data to be approximately 10^{-21} m^2 sr^{-1} eV^{-1}. These findings are consistent with those of Moxom *et al.* (1992) described above and with the classical trajectory calculations of Schultz and Reinhold (1990). The quantum mechanical studies of Sil and coworkers (e.g. Bandyopadhyay *et al.*, 1994, and references therein), appear to overestimate the ECC contribution; however, they are for an atomic hydrogen target.

Double differential cross sections for positron–argon scattering at angles other than 0° have been reported by Kövér, Laricchia and Charlton (1994) and Schmitt *et al.* (1994) at 100 eV impact energy and Kövér *et al.* (1997)

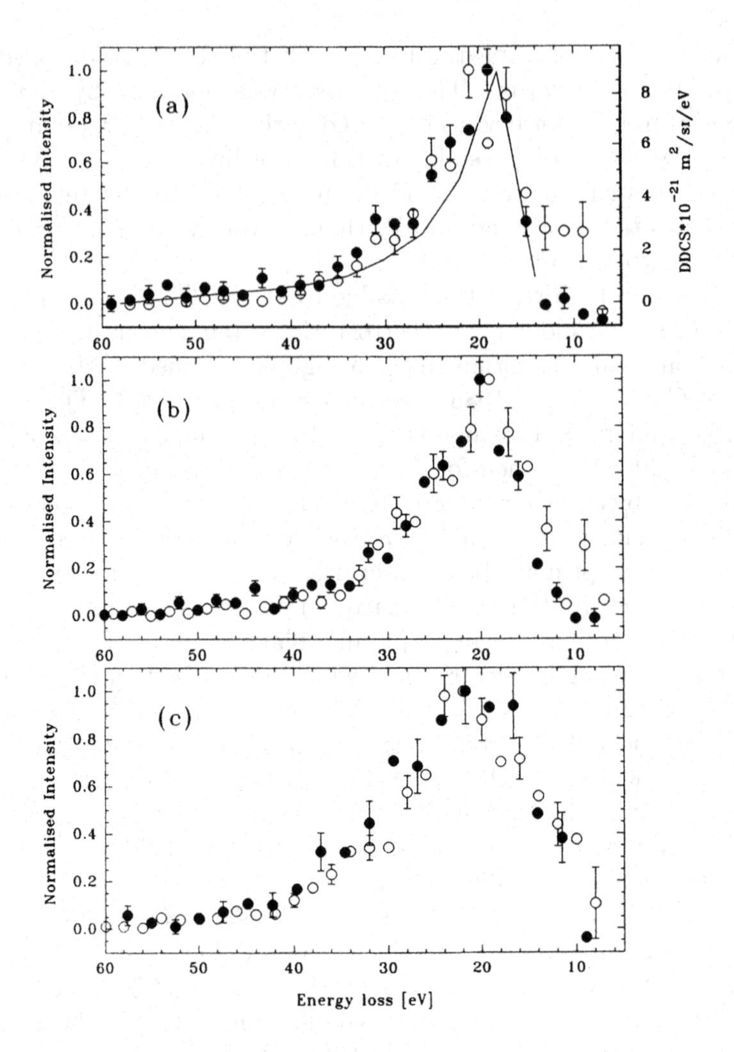

Fig. 5.19. Data for inelastically scattered positrons (•) and electrons (○) around 0° at impact energies of (a) 100 eV, (b) 150 eV and (c) 250 eV. The solid line is from the calculation of Sparrow and Olson (1994) for an incident angle of 3°. Reprinted from *Journal of Physics* **B26**, Kövér, Laricchia and Charlton, Ionization by positrons and electrons at zero degrees, L575–L580, copyright 1993, with permission from IOP Publishing.

at 60 eV. All these studies used positron–ion coincidences to distinguish ionization from other scattering processes, though in the case of Schmitt *et al.* (1994) the ions were extracted using a small d.c. electric field rather than the pulsed field technique. The apparatus employed by Kövér, Laricchia and Charlton (1994) was very similar to this, except that the energy analysis was performed using a simple retarding field analyser

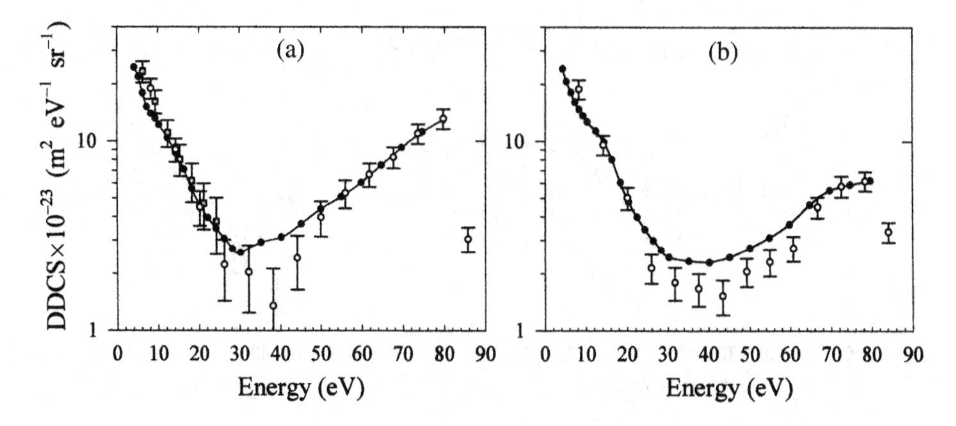

Fig. 5.20. Double differential cross sections (DDCS) for the single ionization of argon gas by impact of 100 eV electrons. The solid circles on a line (the latter serves only to guide the eye) are from the work of DuBois and Rudd (1978) whilst the open circles are from Kövér, Laricchia and Charlton (1994). (a) 30°, (b) 45°.

(RFA), which, though having a poorer energy resolution than the PPA had the virtue of a wider angular acceptance for scattered particles.

The projectile beam was monitored with a separate detector, also having an RFA for diagnostics. The system of Schmitt *et al.* (1994) was similar, except that the scattered positrons or ejected electrons were monitored using a rotatable 90° cylindrical condenser spectrometer.

The electron impact results of Kövér, Laricchia and Charlton (1994) for scattering angles of 30° and 45° are presented in Figure 5.20, where the energy dependences of the data for ejected electrons are in good accord with those found by DuBois and Rudd (1978), to which they were normalized at high energies of ejection. The data at both angles exhibit similar features, namely a preponderance of electrons at both low and high energies. The low energy data are usually attributed to the liberated electron and the high energy data to the scattered projectile, though of course this demarcation cannot be unambiguous, owing to exchange. It is notable that the higher energy electrons are more forward peaked (the cross section for 30° is greater than that for 45° at these energies) and that the probability that the two electrons will be emitted with roughly the same energy is low. There are no theoretical data for comparison.

The positron impact data at 30° and 45° are presented in Figure 5.21. The absolute scale was derived, as described by Kövér, Laricchia and Charlton (1994), by comparison with electron data and positron elastic differential cross sections, and results for both the scattered positrons and the ejected electrons are displayed. Again, the ejected electrons, which

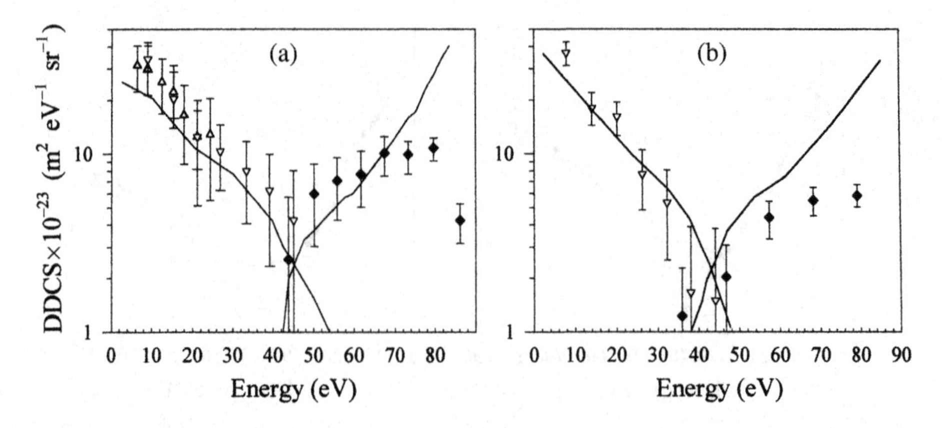

Fig. 5.21. Double differential cross sections for the single ionization of argon gas by impact of 100 eV positrons. Key: ▽, ejected electrons; ▲, scattered positrons; ——, calculation of Sparrow and Olson (1994). (a) 30°, (b) 45°.

can now be unambiguously identified, have mainly low energies, below $(E - E_{\rm i})/2$; the positrons are in the higher energy range and there is no evidence of a substantial contribution to the cross section from ECC. The solid line is the classical trajectory Monte Carlo calculation of Sparrow and Olson (1994), which is in good accord with experiment for the ejected electron spectra but higher than experiment for the scattered positrons, particularly in the low energy-loss region. Similar conclusions have been forthcoming from a further study by Kövér *et al.* (1997), who used the more sensitive PPA to analyse the scattered positron energies. These authors, guided by theoretical work on the positron–hydrogen system by Berakder and Klar (1993), tentatively attributed this effect to the influence of close positron–electron–ion interactions during the collision.

The experiment of Schmitt *et al.* (1994) concentrated on the double differential cross section at an electron ejection energy of 15 eV over the angular range 0°–90° for 100 eV positron–argon collisions. This was chosen because unpublished work for hydrogen by Klar and Berakdar, communicated to Schmitt *et al.* (1994), suggested that for these circumstances the cross section for positron impact would greatly exceed that for electrons. This prediction seems to have been borne out by the results of Schmitt *et al.* (1994), who found e^+/e^- ratios of 16 ± 11 and 5.4 ± 3.4 at 30° and 45° respectively. However, Kövér, Laricchia and Charlton (1994) obtained the values 2.7±1.0 and 1.4±0.6 for this ratio at the same two angles.

Recently, Kövér and Laricchia (1998) reported the first measurement of $d^3\sigma_{\rm i}^+/d\Omega_1 d\Omega_2 dE$, the triple differential cross section for positron collisions. Molecular hydrogen was chosen as the target for positrons at

100 eV impact energy. A small peak was observed close to 42 eV in the ejected electron energy spectrum when both the electron and positron were emitted at small angles to the forward (incident beam) direction. These authors attributed this to the ECC process (see the discussion in section 5.2) and therefore provided the first clear evidence for this mechanism in positron impact ionization.

5.7 Inner shell ionization

In this section we describe experiments whose aim has been to determine explicitly inner shell ionization cross sections (or ratios of electron and positron cross sections). Section 5.5 contained an account of the influence of inner shell processes on multiple ionization of the heavier noble gases.

The first experiments to study inner shell processes, and indeed the first investigations of positron impact ionization, were those of Hansen, Weigmann and Flammersfeld (1964), Hansen and Flammersfeld (1966) and Seif el Nasr, Berényi and Bibok (1974). These workers used β-spectrometers to velocity-select positrons and electrons emitted from radioactive sources, and they measured (by detection of the appropriate X-rays) the ratio of the K-shell ionization cross sections, σ_i^K, for the two projectiles scattering from various targets at high energies, typically hundreds of keV. Within the statistical accuracy of the experiment no deviation from unity was found for the ratio $\sigma_i^K(e^-)/\sigma_i^K(e^+)$. Ito *et al.* (1980), also using β-spectroscopy, measured K- and L-shell cross section ratios for a silver target for a range of positron and electron kinetic energies above 100 keV. Although they found the ratio to be unity for the L-shell processes, $\sigma_i^K(e^-)/\sigma_i^K(e^+)$ was observed to be greater than unity below approximately 150 keV, around six times the K-shell binding energy. A similar trend was observed at relativistic energies by Schneibel *et al.* (1976).

Turning to more recent experiments, a vertical section of the target region used by Schneider, Tobehn and Hippler (1991, 1992) and Schneider *et al.* (1993) is shown in Figure 5.22. The positron beam was provided by the pulsed source at the Giessen electron linear accelerator (see e.g. Ebel *et al.*, 1990, and references therein). An electron beam could also be produced, using a heated filament in close proximity to the target. The kinetic energy of the projectiles was mainly determined by applying an electrical potential to the target, which was tilted at 45° to the beam axis. The targets consisted of various thin silver and gold foils and gold–silver multilayer arrangements (Schneider *et al.*, 1993). The purpose of the latter was to use the silver L-shell X-ray as a standard for comparison of the positron and electron cross sections, since little difference between them is expected in the energy range under investigation. X-rays emitted

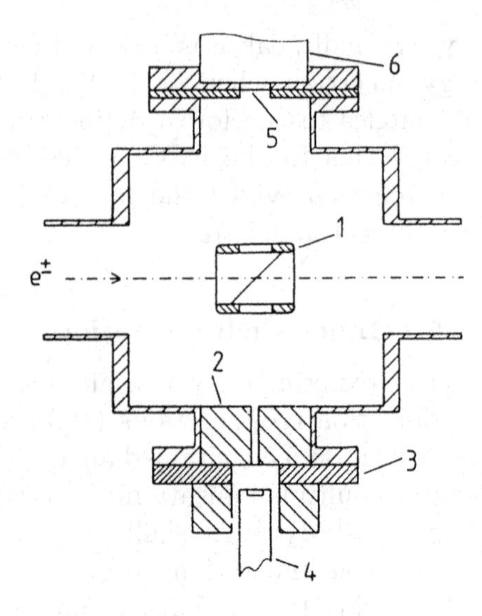

Fig. 5.22. Experimental set-up for studies of positron impact inner shell ionization (Ebel *et al.*, 1989). Key: (1) target support; (2) lead collimator; (3) flange with thin window for the transmission of X-rays to (4) a Si(Li) detector; (5) aluminium window 0.1 mm thick; (6) NaI(Tl) gamma-ray detector.

into a small solid angle passed through a thin mylar window before being registered by a Si(Li) detector. A NaI(Tl) detector was located directly opposite the Si(Li) counter as a veto to suppress the Compton-scattered annihilation γ-rays which may be registered by the latter.

Figure 5.23 shows a summary of what has been obtained for the K-shell ionization of various elements expressed as the ratio $\sigma_i^K(e^-)/\sigma_i^K(e^+)$ versus E/I, the reduced impact energy, where I is the K-shell binding energy. There is good accord between the various experiments in the energy range of overlap, with $\sigma_i^K(e^-)/\sigma_i^K(e^+)$ varying smoothly throughout the range E/I = 2–16. Above $E/I \sim 5$ the ratio is unity, but below this value the electron cross section rises above that for positrons, the ratio reaching a value around eight at the lowest impact energy. This is most clearly seen in the data of Ebel *et al.* (1989) for silver. The steep rise in the ratio at the lower impact energies is reproduced by the results of a plane wave Born approximation with a correction included to take account of modification of the projectile wave function by the Coulomb field of the target nucleus. It is notable that when this correction is omitted the calculation completely fails in the low energy region. As summarized by Knudsen and Reading (1992), the difference between the K-shell ionization probabilities for the two projectiles is mainly governed by the different deflections experienced by the particles in the Coulomb field of the nucleus. The effect of this is expected to lessen with rising

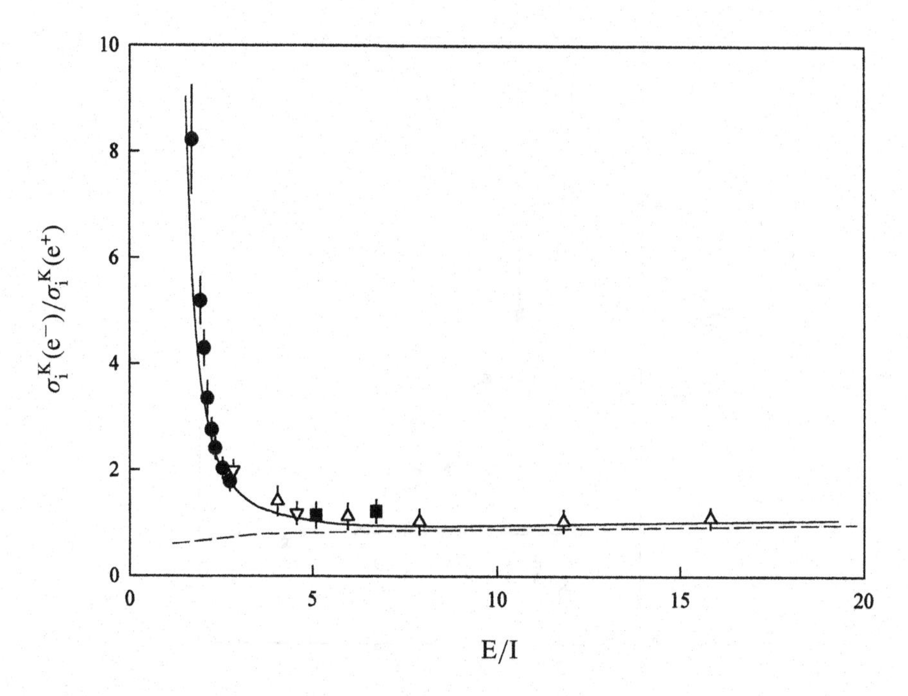

Fig. 5.23. Ratio of K-shell ionization cross sections for electrons and positrons scattering from silver and copper at various impact energies: ——, Born approximation calculation including a Coulomb correction (see text); – – –, Born approximation without the Coulomb correction. The data are: •, Ag (Ebel *et al.*, 1989); ▪, Cu (Ebel *et al.*, 1989); ▽, Cu (Schultz and Campbell, 1985); △, Ag (Ito *et al.*, 1980).

projectile energy (in accord with approach of the cross section ratio to unity) and, as observed, with increasing principal quantum number of the electronic shell.

Calculations by Gryziński and Kowalski (1993) for inner shell ionization by positrons also confirmed the general trend. Theirs was essentially a classical formulation based upon the binary-encounter approximation and a so-called atomic free-fall model, the latter representing the internal structure of the atom. The model allowed for the change in kinetic energy experienced by the positrons and electrons during their interactions with the screened field of the nucleus.

Lennard *et al.* (1988) studied L-shell ionization ratios for gold, where it was hoped that the Coulomb effect would be so reduced as to permit the observation of certain basic differences between positron–electron scattering (Bhabha, 1936) and electron–electron scattering (Møller, 1932), for large energy transfers. Later Schneider, Tobehn and Hippler (1991, 1992), using the apparatus described above, measured the same quantities. As

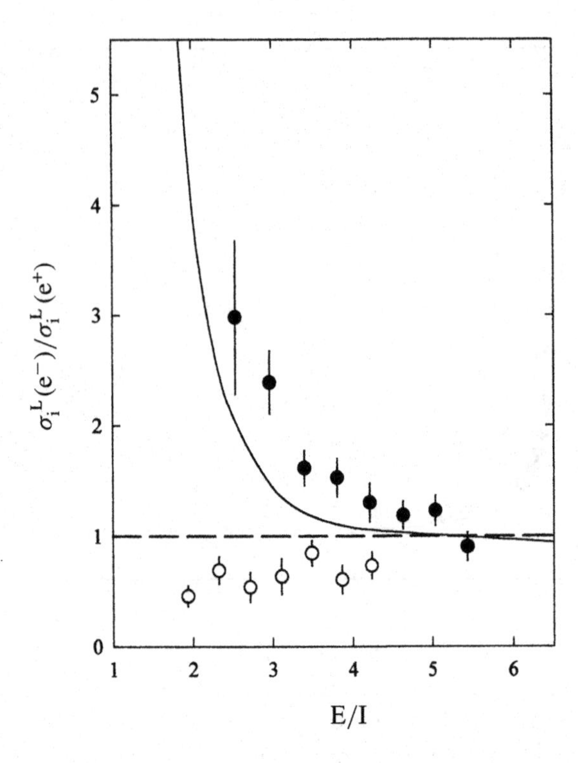

Fig. 5.24. Ratio of L-shell ionization cross sections for electrons and positrons at various impact energies: ——, Born approximation calculation including the Coulomb correction (see text). The data are: •, Schneider, Tobehn and Hippler (1991) for a gold target; ∘, Lennard *et al.* (1988) for a silver target.

shown in Figure 5.24, a major discrepancy exists between the two sets of measurements, which increases as the threshold is approached, with $\sigma_i^L(e^-)/\sigma_i^L(e^+)$ dipping below unity in the data of Lennard *et al.* (1988) whilst exhibiting the opposite trend in the work of Schneider, Tobehn and Hippler (1991). The solid line in this figure, which is in qualitative accord with the Schneider data, was obtained using the plane wave Born approximation, including exchange and allowance for Coulomb effects. The Møller–Bhabha effects mentioned above were not included in this calculation.

Although the general trend of the theories tends to support the work of the Giessen–Bielefeld group, Gryziński and Kowalski (1993) pointed out that the differences between the results from the two experiments may be explained by different target arrangements. They noted that the orientation of the latter with respect to the beam (Ebel *et al.*, 1989, target at 45°; Lennard *et al.*, 1988, target perpendicular) may play a role if there is some anisotropic orientation of the electron orbitals in the foil targets.

This could lead to anisotropies in the emission of the X-rays. Further experimental work is needed to clarify this issue.

Schneider *et al.* (1993) developed their group's technique further in order to assign an absolute scale to the ionization cross sections by electron and positron impact for the K-shell of silver and the L_3-shell of gold. Magnitudes were of the order of 10^{-23} cm^2 and 10^{-22} cm^2 respectively, though strongly energy dependent as the relevant threshold is approached.

6

Positron annihilation

6.1 Introduction and theoretical considerations

Before the advent of low energy beams, the only means of investigating positron interactions with atoms and molecules was to study their annihilation. Information could thereby be obtained directly on the annihilation cross section but only indirectly for other processes such as elastic scattering. In this chapter we consider the annihilation of so-called free positrons in gases. The fate of positrons which have formed positronium prior to annihilation is treated in Chapter 7.

The basic physical principles governing positron annihilation were described in subsection 1.2.1, where the non-relativistic limit of the Dirac cross section for the annihilation of a free positron–electron pair into two gamma-rays was given as $\sigma_{2\gamma} = 4\pi r_0^2 c/v$, equation (1.3). If the electron density in the vicinity of the positron is n_e, then the annihilation rate is $\lambda_{2\gamma} = 4\pi r_0^2 c n_e$. However, as shown in subsection 1.2.2, annihilation into two gamma-rays requires the positron–electron pair to be in a singlet spin state, but only one quarter of the electrons in an unpolarized ensemble would form such a state with the positron. The remaining electrons would form a triplet spin state with the positron, for which annihilation is into three gamma-rays at a much lower rate (less than 1% of the two-gamma rate). Thus, the total free-positron annihilation rate is

$$\lambda_f \simeq \pi r_0^2 c n_e. \tag{6.1}$$

If the electrons are bound in atoms or molecules, each having Z electrons, and the number density of atoms or molecules is n, the electron density is $n_e = nZ$. Therefore, if there were no distortions of the positron–atom system, the free annihilation rate would be given by

$$\lambda_f = \pi r_0^2 c n Z. \tag{6.2}$$

In practice, the positron does influence the charge distribution in the atom or molecule, in such a way as to enhance the electron density in its vicinity. Allowance for this can be made by replacing Z by an effective number of electrons, $Z_{\rm eff}$. Thus, the annihilation rate may be expressed as

$$\lambda_{\rm f} = \pi r_0^2 cnZ_{\rm eff} = 0.201\rho Z_{\rm eff} \qquad (\mu{\rm s}^{-1}), \tag{6.3}$$

where ρ is the gas density in amagat (1 amagat $\equiv 2.69 \times 10^{25}$ m^{-3}), and the annihilation cross section is therefore

$$\sigma_{2\gamma} = \lambda_{\rm f}/(vn) = \pi r_0^2 cZ_{\rm eff}/v. \tag{6.4}$$

Distortion of the target atoms or molecules is particularly pronounced at very low positron speeds, and $Z_{\rm eff}$ may then be considerably larger than Z; however, as the speed increases and the electrons have less time to react to the perturbing field of the positron, $Z_{\rm eff}$ initially decreases.

The value of $Z_{\rm eff}$ is a measure of the probability that the positron is at essentially the same position as any of the electrons in the target, and it can be calculated from the wave function representing elastic positron scattering by the target. If the wave function is $\Psi(\boldsymbol{r}_1, \boldsymbol{r}_2, \ldots, \boldsymbol{r}_{Z+1})$, where $\boldsymbol{r}_1$ is the positron coordinate and $\boldsymbol{r}_2, \ldots, \boldsymbol{r}_{Z+1}$ are the coordinates of the Z electrons in the target system, then

$$\begin{aligned} Z_{\rm eff} &= \sum_{i=2}^{Z+1} \int |\Psi(\boldsymbol{r}_1, \boldsymbol{r}_2, \ldots, \boldsymbol{r}_{Z+1})|^2 \delta(\boldsymbol{r}_1 - \boldsymbol{r}_i)\, d\boldsymbol{r}_1 \cdots d\boldsymbol{r}_{Z+1} \\ &= Z \int |\Psi(\boldsymbol{r}_1, \boldsymbol{r}_2, \ldots, \boldsymbol{r}_{Z+1})|^2 \delta(\boldsymbol{r}_1 - \boldsymbol{r}_2)\, d\boldsymbol{r}_1 \cdots d\boldsymbol{r}_{Z+1}, \end{aligned} \tag{6.5}$$

since the wave function is antisymmetric in all the electron coordinates. In this calculation the positron wave function must be normalized in such a way that its asymptotic form has unit amplitude, i.e.

$$\Psi(\boldsymbol{r}_1, \boldsymbol{r}_2, \ldots, \boldsymbol{r}_{Z+1}) \underset{r_1 \to \infty}{\sim} \exp(i\boldsymbol{k} \cdot \boldsymbol{r}_1)\Phi(\boldsymbol{r}_2, \ldots, \boldsymbol{r}_{Z+1}), \tag{6.6}$$

where $\Phi(\boldsymbol{r}_2, \ldots, \boldsymbol{r}_{Z+1})$ is the wave function of the target atom.

Whereas the error in the calculated value of the elastic scattering phase shift is usually of second order in the error in Ψ, the error in $Z_{\rm eff}$ is of first order; the values of $Z_{\rm eff}$ therefore tend to be rather less accurate than the corresponding phase shifts. Consequently, the value obtained for $Z_{\rm eff}$ provides a sensitive test of the accuracy of a wave function, although admittedly in a very restricted region of configuration space where the positron is close to one of the electrons. Drachman and Sucher (1979) developed an alternative method of calculating $Z_{\rm eff}$ in which the delta function $\delta(\boldsymbol{r}_1 - \boldsymbol{r}_i)$ is replaced by a global operator but, because it is more difficult to implement, the method has had very limited use (Ujc and Stauffer, 1985).

In the first Born approximation the total wave function is taken to be a plane wave for the positron multiplied by the undistorted target wave function, and consequently $Z_{\rm eff} = Z$. This approximation is valid at sufficiently high energies, and one might expect the calculated value of $Z_{\rm eff}$ to tend to Z as the positron energy increases. However, the calculation should only be carried out for positron energies below the positronium formation threshold, $E_{\rm Ps}$. At higher energies, the explicit representation of the open positronium channel in the total wave function, when inserted into equation (6.5), yields an infinite value for $Z_{\rm eff}$, the interpretation of which is as follows. Above $E_{\rm Ps}$, the cross section for positronium formation is several orders of magnitude larger than the annihilation cross section. Once formed, positronium certainly undergoes annihilation, and therefore the positronium formation and positron annihilation cross sections can be considered to be equivalent, implying a very large value of $Z_{\rm eff}$. At energies just below the positronium formation threshold, the positron tends to form virtual positronium with one of the atomic electrons, resulting in an enhanced electron density in its vicinity and, consequently, a very rapid increase in the value of $Z_{\rm eff}$ as the threshold is approached (Van Reeth and Humberston, 1998).

Examples of the energy dependence of $Z_{\rm eff}$ for atomic hydrogen and helium are given in Figure 6.1. These results were obtained using the very accurate elastic scattering wave functions described in detail in subsections 3.2.1 and 3.2.2. The only molecule for which reasonably accurate calculations of $Z_{\rm eff}$ have been made is H_2, where Armour, Baker and Plummer (1990) used the elaborate variational wave functions obtained from their studies of low energy scattering (see subsection 3.2.4). However, such is the sensitivity of the value of $Z_{\rm eff}$ to the quality of the wave function that even this calculation only yielded the value 10.2, compared to the experimental result of 14.8 at room temperature. Nevertheless, this is much closer to the measured value than any other theoretical result for this molecule.

For each of these systems the value of $Z_{\rm eff}$ exceeds the corresponding value of Z by a significant factor, particularly in the case of atomic hydrogen, for which it is almost nine times greater at very low energies, whereas for helium the factor is only two. In a quite highly polarizable atom, such as hydrogen, the outer electrons are readily attracted towards the incident positron, enhancing the probability for annihilation. In an early study of the correlation between the value of $Z_{\rm eff}$ and the dipole polarizability of the target, α, Osmon (1965) found that, for many simple atoms and molecules, a reasonably good fit to the experimental data was

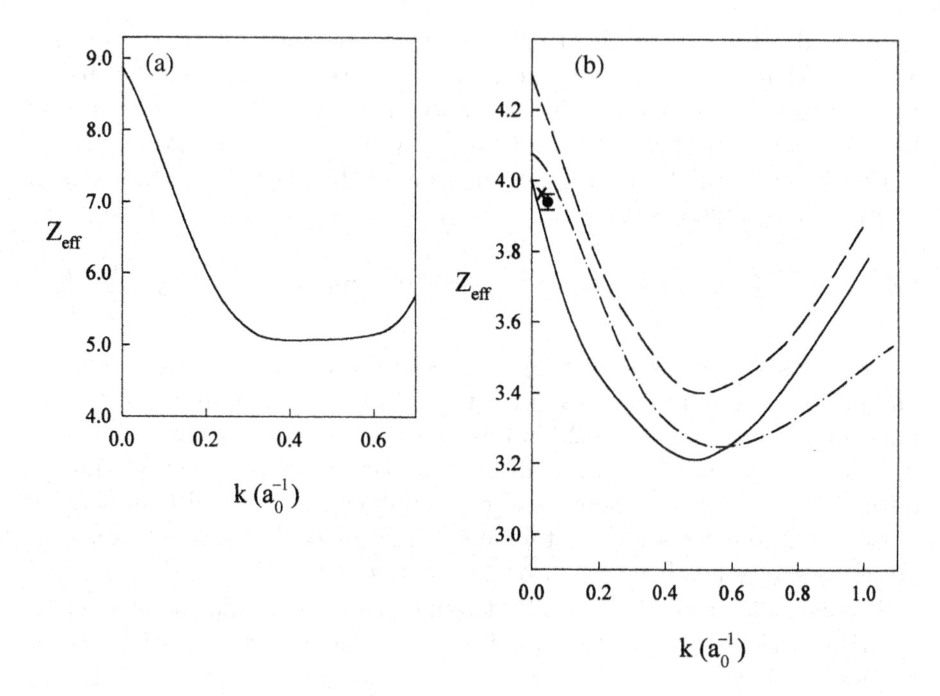

Fig. 6.1. Variation of $Z_{\rm eff}$ with positron momentum. (a) Atomic hydrogen, Humberston and Wallace (1972). (b) Helium: ——, Campeanu and Humberston (1977b) for helium model H5; – – –, Campeanu and Humberston (1977b) for helium model H1; — · —, McEachran *et al.* (1977); ×, Roellig and Kelly (1965) (Fraser, 1968); •, Coleman *et al.* (1975b).

provided by the relationship $\langle Z_{\rm eff}\rangle \propto \alpha^{1.25}$. However, more recent data are consistent with $\langle Z_{\rm eff}\rangle \propto \alpha$ (Wright *et al.*, 1983). Some molecules, particularly large organic molecules (e.g. Iwata *et al.*, 1995) do not fit this pattern and have values of $Z_{\rm eff}$ several orders of magnitude greater than the number of electrons in the molecule; for example, $Z_{\rm eff}$ is around 18 000 for benzene, and greater than 10^6 for anthracene. One theory is that these very large values arise because the positron forms a pseudo-bound-state or resonance with the molecule, in which the positron is trapped in its vicinity for much longer than the usual collision time. Molecules with very large values of $\langle Z_{\rm eff}\rangle$ also have low threshold energies for positronium formation, $E_{\rm Ps}$, the relationship being reasonably well represented by

$$\ln\langle Z_{\rm eff}\rangle \approx A/E_{\rm Ps} + B,$$

where A and B are constants (Murphy and Surko, 1991). This fact prompted Laricchia and Wilkin (1997) to develop an alternative positron-trapping mechanism based on the increasing significance of virtual positronium formation as the threshold energy for real positronium formation is lowered. Exploiting the time–energy uncertainty relationship, they assumed that a positron with an incident energy an amount ΔE below the positronium formation threshold will form virtual positronium and be trapped in the vicinity of the target system for a time $\Delta t \simeq \hbar/\Delta E$, during which it might annihilate with one of the electrons. The value of Z_{eff} is then expressed as

$$Z_{\text{eff}} = \frac{\sigma_{\text{el}} v}{\pi r_0^2 c}\Big(\gamma[1-\exp(-\lambda_{\text{d}} t_{\text{c}})]+(1-\gamma)\{1-\exp[-\Delta t(\lambda_{\text{sa}}+\lambda_{\text{po}})]\}\Big), \quad (6.7)$$

where σ_{el} is the elastic scattering cross section, λ_{d} and λ_{po} are the direct and pick-off annihilation rates, λ_{sa} is the spin-averaged annihilation rate of positronium and $\gamma = \exp(-\Delta t/t_{\text{c}})$, where t_{c} is the collision time. The predictions of this model are in qualitative agreement with the experimental data and also with the theoretical results of Van Reeth and Humberston (1998) mentioned above. Further discussion of positron annihilation on large molecules has been given by Iwata *et al.* (2000).

The wave function of the ion that remains after annihilation is a superposition of eigenstates of the Hamiltonian of the ion, the relative probabilities of which may be determined from the wave function used in the calculation of Z_{eff}. The annihilation process takes place so rapidly, compared with normal atomic processes, that it is reasonable to assume the validity of the sudden approximation. Consequently, the wave function of the residual ion when the positron has annihilated with electron 2 at the position $\mathbf{r}_1 = \mathbf{r}_2$ is

$$F(\mathbf{r}_2;\mathbf{r}_3,\ldots,\mathbf{r}_{Z+1}) = \Psi(\mathbf{r}_1=\mathbf{r}_2,\mathbf{r}_2,\mathbf{r}_3,\ldots,\mathbf{r}_{Z+1}), \quad (6.8)$$

where we have used the same nomenclature as in the positron–atom wave function. The relative probability that the residual ion is in its n_{ion}th eigenstate, with wave function $\Phi^n_{\text{ion}}(\mathbf{r}_3,\ldots,\mathbf{r}_{Z+1})$, is then obtained by projecting the function $F(\mathbf{r}_2;\mathbf{r}_3,\ldots,\mathbf{r}_{Z+1})$ onto this state, squaring the result and integrating over the positron coordinate to give

$$P(n_{\text{ion}}) \propto \int |F(\mathbf{r}_2;\mathbf{r}_3,\ldots,\mathbf{r}_{Z+1})\Phi^n_{\text{ion}}(\mathbf{r}_3,\ldots,\mathbf{r}_{Z+1})\, d\mathbf{r}_3\cdots d\mathbf{r}_{Z+1}|^2\, d\mathbf{r}_2. \quad (6.9)$$

By exploiting closure in the summation over all the states of the ion, it may be shown that the normalized probability of the ion being in the state n_{ion} is

$$P(n_{\text{ion}}) = \frac{n_{\text{ion}}}{Z_{\text{eff}}}\int F(\mathbf{r}_2;\mathbf{r}_3,\ldots,\mathbf{r}_{Z+1}) \times \Phi^n_{\text{ion}}(\mathbf{r}_3,\ldots,\mathbf{r}_{Z+1})\, d\mathbf{r}_3\ldots d\mathbf{r}_{Z+1}|^2\, d\mathbf{r}_2, \quad (6.10)$$

so that

$$\sum_{n_{\text{ion}}} P(n_{\text{ion}}) = 1.$$

Drachman (1966b), using a rather simple representation of positron–helium scattering based on a modified form of the adiabatic approximation, calculated the probabilities of the residual helium ion being in various states after annihilation and found a 95% probability of its being in the ground state.

Traditionally, experimental values of Z_{eff} have been derived from measurements of the lifetime spectra of positrons that are diffusing, and eventually annihilating, in a gas. The lifetime of each positron is measured separately, and these individual pieces of data are accumulated to form the lifetime spectrum. (The positron-trap technique, to be described in subsection 6.2.2, uses a different approach.) An alternative but equivalent procedure, which is adopted in electron diffusion studies and also in the theoretical treatment of positron diffusion, is to consider the injection of a swarm of positrons into the gas at a given time and then to investigate the time dependence of the speed distribution, as the positrons thermalize and annihilate, by solving the appropriate diffusion equation. The experimentally measured Z_{eff}, termed $\langle Z_{\text{eff}} \rangle$, is the average over the speed distribution of the positrons, $y(v,t)$, where $y(v,t)\,dv$ is the number density of positrons with speeds in the interval v to $v + dv$ at time t after the swarm is injected into the gas. The time-dependent speed-averaged Z_{eff} is therefore

$$\langle Z_{\text{eff}}(t) \rangle = \frac{\int_0^\infty Z_{\text{eff}}(v) y(v,t)\,dv}{\int_0^\infty y(v,t)\,dv}. \tag{6.11}$$

Assuming that all positrons in the swarm have energies below the positronium formation threshold and that only elastic collisions and annihilation are possible, the speed distribution may be derived theoretically as the solution of the following diffusion equation (Orth and Jones, 1969):

$$\begin{aligned} \frac{\partial y(v,t)}{\partial t} = {} & \frac{\partial}{\partial v}\left\{ \left[\frac{e^2 \epsilon^2}{3m^2 n v \sigma_{\text{M}}(v)} + \frac{v n \sigma_{\text{M}}(v) k_{\text{B}} T}{M} \right] \frac{\partial y(v,t)}{\partial v} \right. \\ & \left. + \left[\frac{m^2 v n \sigma_{\text{M}}(v)}{M} - \frac{2e^2\epsilon^2}{3m^2 v^2 n \sigma_{\text{M}}(v)} - \frac{2n\sigma_{\text{M}}(v) k_{\text{B}} T}{M} \right] y(v,t) \right\} \\ & - \lambda_{\text{f}}(v) y(v,t), \end{aligned} \tag{6.12}$$

where ϵ is the electric field applied across the gas cell, T is the temperature of the gas, k_{B} is Boltzmann's constant, M is the mass of each gas atom, $\sigma_{\text{M}}(v)$ is the momentum transfer cross section and $\lambda_{\text{f}}(v)$ is the annihilation

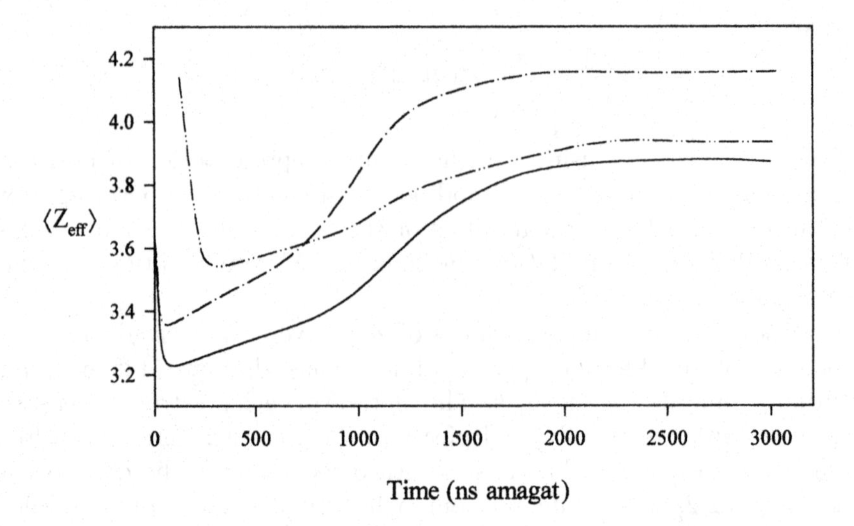

Fig. 6.2. The time dependence of $\langle Z_{\text{eff}} \rangle$ for positrons diffusing in helium gas at room temperature and zero electric field. Experiment: — · · —, Coleman *et al.* (1975b). Theory: ——, helium model H5; — · —, helium model H1, both due to Campeanu and Humberston (1977b).

rate. The momentum transfer cross section is defined as

$$\sigma_{\text{M}}(v) = 2\pi \int_0^{\pi} |f(\theta)|^2 (1 - \cos\theta) \sin\theta \, d\theta, \tag{6.13}$$

which can also be expressed in terms of the partial-wave phase shifts as

$$\sigma_{\text{M}}(v) = \frac{4\pi}{k^2} \sum_{l=0}^{\infty} (l+1) \sin^2(\eta_l - \eta_{l+1}). \tag{6.14}$$

Having calculated $\sigma_{\text{M}}(v)$ and $\lambda_{\text{f}}(v)$ from the elastic scattering wave function and made a suitable choice of the initial speed distribution $y(v, t = 0)$, the diffusion equation can then be solved to obtain the theoretical positron speed distribution for all subsequent times. Provided a reasonable choice of the initial speed distribution is made, the solution to the diffusion equation is not very sensitive to its form, except at very small times after $t = 0$. The time dependence of $\langle Z_{\text{eff}}(t) \rangle$ obtained in this way for positrons diffusing in helium gas in zero electric field is shown in Figure 6.2 (Campeanu and Humberston, 1977b). The input data for equation (6.12), $\sigma_{\text{M}}(v)$ and $\lambda_{\text{f}}(v)$, were derived from the phase shifts and scattering wave functions of Humberston (1973) and Campeanu and Humberston (1975, 1977a).

After a sufficiently long time the positrons reach equilibrium, so that $y(v,t) = f(v)\exp(-\langle\lambda_f\rangle t)$, where $\langle\lambda_f\rangle$ is the equilibrium annihilation rate and $f(v)$ is the associated speed distribution, which is the solution of the time-independent equation

$$\frac{d}{dv}\left\{\left[\frac{e^2\epsilon^2}{3m^2nv\sigma_M(v)} + \frac{vn\sigma_M(v)k_BT}{M}\right]\frac{df(v)}{dv} + \left[\frac{m^2v\sigma_M(v)}{M} - \frac{2e^2\epsilon^2}{3m^2v^2n\sigma_M(\nu)} - \frac{2n\sigma_M(v)k_BT}{M}\right]f(v)\right\} - \lambda_f(v)f(v) = -\langle\lambda_f\rangle f(v). \tag{6.15}$$

The equilibrium value of Z_{eff} is then

$$\langle Z_{eff}\rangle = \frac{\int_0^\infty Z_{eff}(v)f(v)\,dv}{\int_0^\infty f(v)\,dv} = \frac{\langle\lambda_f\rangle}{\pi r_0^2 cn}. \tag{6.16}$$

If the annihilation rate, $\lambda_f(v)$, is approximately constant over the width of the dominant part of the equilibrium speed distribution, so that $\lambda_f(v)f(v) \approx \langle\lambda_f\rangle f(v)$, the solution to equation (6.15) has the form

$$f(v) = Cv^2\exp\left\{-\int_0^v\left[\frac{M}{3}\left(\frac{e\epsilon}{mnv'\sigma_M(v')}\right)^2 + k_BT\right]^{-1} mv'\,dv'\right\}, \tag{6.17}$$

where C is the normalization constant. For zero electric field this reduces to the Maxwell–Boltzmann form

$$f(v) = Cv^2\exp[-mv^2/(2k_BT)].$$

In cases where the annihilation rate, and hence Z_{eff}, is a rapidly varying function of the positron energy, as with xenon (Schrader and Svetic, 1982), the simplification introduced above is not valid and the solution to equation (6.15) must be used. The functional form for $f(v)$ given in equation (6.17) was used by Campeanu and Humberston (1977b) to investigate the variation of the equilibrium value of $\langle Z_{eff}\rangle$ with electric field and temperature, and their results for the former are shown in Figure 6.3.

The experimental techniques involved in measuring the angular correlation and the Doppler broadening of the two annihilation gamma-rays were introduced in section 1.3. These techniques rely on the fact that the motion of the positron–electron pair immediately prior to annihilation causes the two gamma-rays to be emitted in directions differing

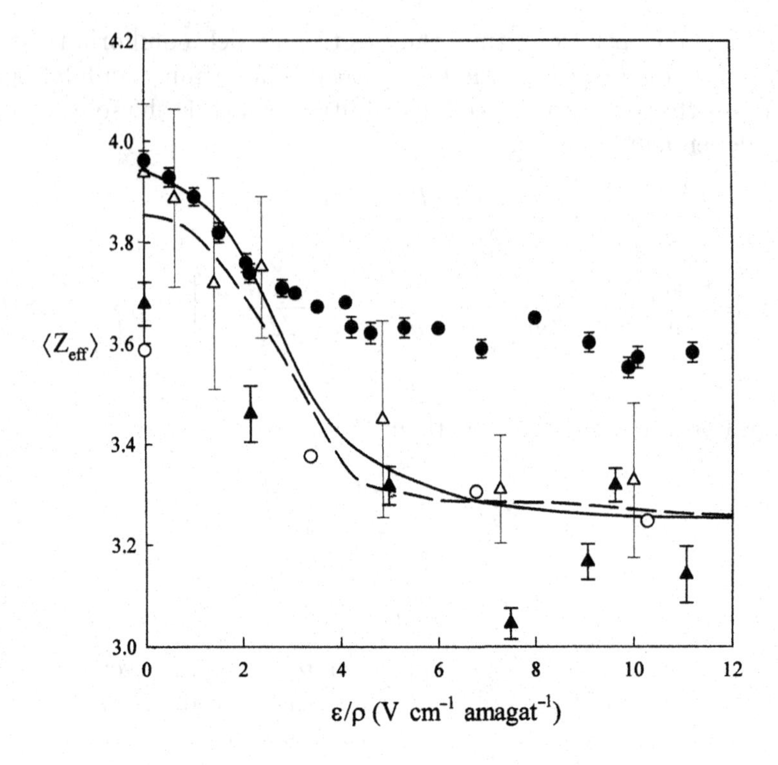

Fig. 6.3. Values of $\langle Z_{\rm eff} \rangle$ for helium gas at various density-normalized electric fields and at (•) 35.7 amagat and (△) 3.5 amagat, from Davies, Charlton and Griffith (1989); ○, Leung and Paul (1969); ▲, Lee, Orth and Jones (1969). The results of the calculations of Campeanu and Humberston (1977b) (——) and Shizgal and Ness (1987) (– – –) are also given.

by an angle θ from exact collinearity and with energies shifted from 511 keV. Thus, measuring the angular correlation between the gamma-rays provides information about the momentum distribution of the annihilating electron–positron pairs. In many cases it can be assumed that the positrons have thermalized before annihilation, so that the momentum of the pair is predominantly that of the electron alone. However, the attractive positron–atom interaction causes the positron to speed up slightly as it approaches an atom, and the electrons also have their velocities modified by the attraction towards the positron. Therefore, the momentum distribution of an annihilating positron–electron pair is not identical to the momentum distribution of the electrons in an undistorted atom.

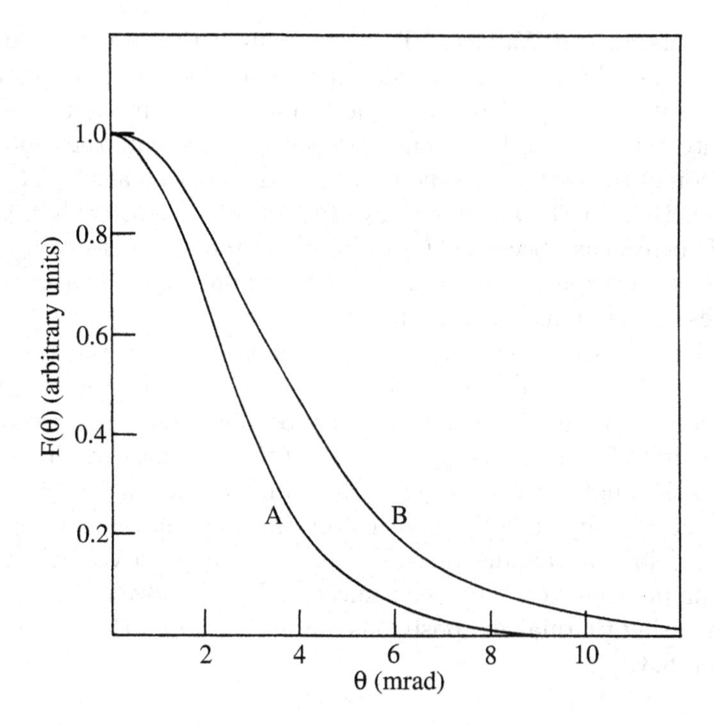

Fig. 6.4. Angular correlation function $F(\theta)$ for the two-gamma annihilation of thermal energy positrons incident on atomic hydrogen: A, $F(\theta)$ obtained using the accurate variationally determined wave function; B, $F(\theta)$ obtained in the Born approximation.

The probability that the annihilating electron–positron pair has a momentum $\boldsymbol{p}$ is proportional to

$$\Gamma(\boldsymbol{p}) = \int |\exp(-i\boldsymbol{p}\cdot\boldsymbol{r}_1)\Psi(\boldsymbol{r}_1, \boldsymbol{r}_2, \ldots, \boldsymbol{r}_{Z+1})\delta(\boldsymbol{r}_1 - \boldsymbol{r}_2)|^2 \, d\boldsymbol{r}_1 \, d\boldsymbol{r}_2 \cdots d\boldsymbol{r}_{Z+1}. \tag{6.18}$$

In many determinations of angular correlation functions for positrons annihilating in gases, only the angle between the two gamma-rays as projected onto a given plane is measured; therefore, in order to make comparison with experiment, the theoretical angular correlation function for the two gamma-rays should be integrated with respect to two components of the momentum to give the distribution function,

$$F(\theta) = \int \Gamma(p_x = mc\theta, p_y, p_z) \, dp_y \, dp_z. \tag{6.19}$$

As an example, the angular distribution function for positrons annihilating in atomic hydrogen, obtained using the accurate variational wave function for zero energy positron–hydrogen scattering described in section

3.2 (Humberston and Wallace, 1972), is shown in Figure 6.4. Also shown there is the distribution function obtained using the Born approximation, in which neither the positron nor the atomic wave function is modified by the interaction. This latter curve therefore represents the momentum distribution of the electron in the undistorted hydrogen atom. The distribution function for the accurate wave function is narrower than that for the undistorted case because the positron attracts the electron towards itself and away from the nucleus, thereby enhancing the probability of low values of the momentum of the pair.

Instead of measuring the angular correlation between the two annihilation gamma-rays, the same information concerning the momentum distribution of the annihilating pair can be obtained by measuring the Doppler shift ΔE_γ in the energy of either of the gamma-rays. This shift is related to the angle, $\pi - \theta$, between the gamma-rays by $\Delta E_\gamma = mc^2\theta/2$, so that the energy in keV of either of the gamma-rays is related to the angle θ in milliradians by $E_\gamma = 511(1 + \theta/2)$ keV. This Doppler shift technique has recently been successfully employed by Surko and coworkers using thermalized positrons annihilating in a Penning trap (see subsection 6.3.5).

6.2 Experimental details

1 The traditional positron lifetime method – analysis and observables

The basic principle behind positron lifetime spectroscopy is simple. A timing signature is obtained when a positron is emitted from a radioactive isotope and also when it annihilates. By measuring the difference between these times for 10^6–10^7 positrons, a lifetime spectrum is obtained from which various annihilation rates and associated parameters can be derived. Schematic illustrations encapsulating the methodology of the most commonly encountered positron lifetime spectrometry were shown in Figure 1.4. The technical level of the discussion in subsection 1.3.1 is sufficient for the present purposes. Further details can be found in the original paper of Coleman *et al.* (1974), who developed a system to facilitate precision measurements of annihilation parameters in low density gases. A concise summary of techniques relevant to positron annihilation in gases was given by Griffith and Heyland (1978). The overall design of lifetime apparatus has not changed significantly since then, though there have been advances in the commercially available timing electronics and in data storage and analysis.

Lifetime spectra obtained for the noble gases argon and xenon are shown in Figures 6.5(a) and (b) respectively. These spectra serve to

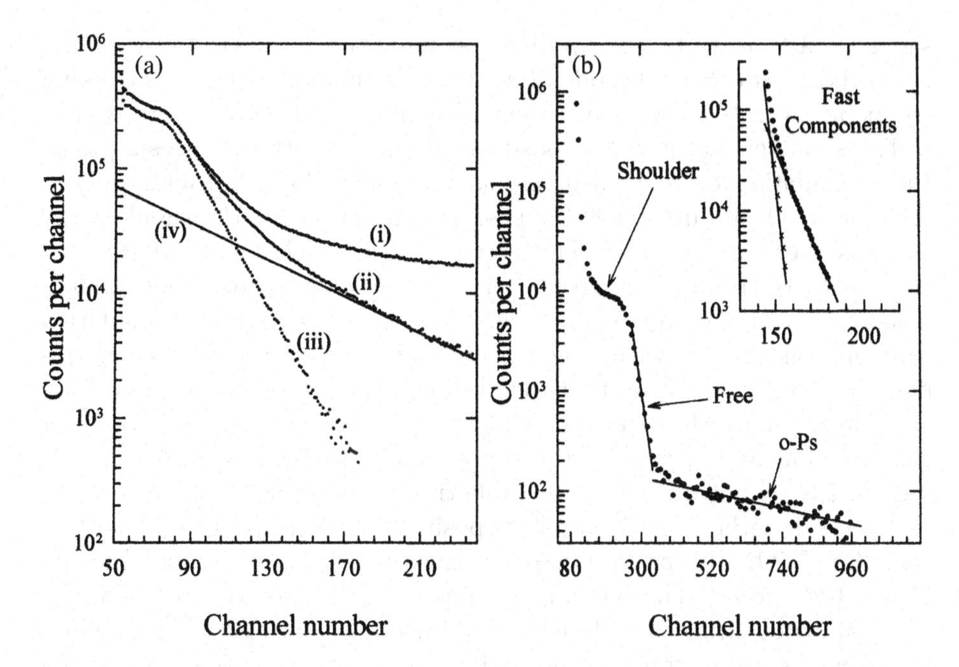

Fig. 6.5. Examples of positron lifetime spectra for (a) argon and (b) xenon gases. The argon data are for a density of 6.3 amagat at 297 K. The channel width is 1.92 ns. In (a), (i) shows the raw data, (ii) shows the signal with background removed, (iii) shows the free-positron component and (iv) shows the fitted ortho-positronium component. In (b), the spectrum for xenon is for room temperature and 9.64 amagat and has a channel width of 0.109 ns. The inset shows the fast components as extracted and discussed by Wright *et al.* (1985).

illustrate both the type of data analysis applied and the parameters which can be derived. The argon spectrum shows the raw data obtained, including the background contribution. Full discussions of the sources of such background events can be found in the works of Coleman, Griffith and Heyland (1974) and Coleman (1979).

Once the background is subtracted, the component of the spectrum due to the annihilation of ortho-positronium is usually visible (see Figure 6.5(a), curve (ii) and the fitted line (iv)). The analysis of the spectrum can now proceed, and a number of different methods have been applied to derive annihilation rates and the amplitudes of the various components. One method, introduced by Orth, Falk and Jones (1968), applies a maximum-likelihood technique to fit a double exponential function to the free-positron and ortho-positronium components (where applicable). Alternatively, the fits to the components can be made individually, if their decay rates are sufficiently well separated, by fitting to the longest component (usually ortho-positronium) first and then subtracting this from the signal data to leave the free-positron signal. The main assumption built into this, as in other methods of analysis, is that the ortho-positronium component can be described by a single exponential back to zero time.

The shoulder width of the positron lifetime spectrum corresponds to the annihilation of free positrons as they slow down through the energy interval from just below the positronium formation threshold energy towards thermal energies. The region of the spectrum in which both the ortho-positronium and free-positron components are exponential is usually termed the equilibrium region, since the measured annihilation rates no longer vary with time. From the discussion given earlier this does not necessarily mean that the positrons (or ortho-positronium) have thermalized, only that the associated annihilation parameters, $\langle Z_{\rm eff} \rangle$ for the positrons and $\langle {}_1Z_{\rm eff} \rangle$ for ortho-positronium, see equation (6.20) and section 7.3, do not have an observable time dependence.

A second method used to analyse positron lifetime spectra is based on the POSITRONFIT programme (see Kirkegaard, Pederson and Eldrup, 1989). Here, too, the individual components of the spectrum are assumed to be of exponential form, but the fit also contains a resolution function which can be a sum of Gaussians and which is normally measured using a well-characterized reference sample. This procedure has not been widely applied in gas lifetime studies because, in most cases, the presence of a large peak in the spectrum at short times, arising from positrons annihilating in the walls of the chamber, precludes a detailed analysis in this time region. This feature, which is commonly referred to as the 'prompt peak', is particularly visible in the xenon spectrum shown in Figure 6.5(b).

The analyses described above can be applied directly to the equilibrium region of a lifetime spectrum. However, in atomic gases, where slowing down below the positronium formation threshold is by elastic collisions only, the positron speed distribution $y(v, t)$ varies relatively slowly with time. Consequently the annihilation rate also varies slowly with time. From Figures 6.5(a) and (b) the existence of a non-exponential, or so-called shoulder, region close to $t = 0$ is evident, and the analysis of this region must be treated separately, as outlined below. Further details of the shape and length of the shoulder can be found in subsection 6.3.1 below.

We now consider the parameters, listed below as (i)–(x) (Heyland *et al.*, 1982), which can be derived from analysis of a gas lifetime spectrum.

(i) The total number of signal events is $N_{\rm S}$, which includes gas events and also, if applicable, events due to annihilation in the source and in the chamber walls (which make up the bulk of the prompt peak).

(ii) Using the method(s) of analysis described earlier, the number of events in each of the resolvable components of the spectrum can be derived by back-extrapolation of the fitted component to $t = 0$. (This is not applicable to the noble gases, which exhibit a non-equilibrium shoulder, but it can be applied to molecular gases where there is usually no such feature.) This can yield the numbers of ortho-positronium and free-positron events, $N_{\text{o-Ps}}$ and N_{f} respectively, plus the number N_{F} due to any other (faster) components, if present. In fact $N_{\text{o-Ps}}$ may have to be corrected to allow for any difference in the detection efficiencies of ortho-positronium and free events that arises from their different annihilation modes. This is important because it allows absolute positronium fractions to be determined; see (vi) below and Coleman *et al.* (1975a).

(iii) The number of para-positronium events which contribute to the gas spectrum but which cannot normally be resolved because of the presence of the prompt peak is assumed to be $N_{\text{o-Ps}}/3$, based upon the spin statistics which govern the relative formation probabilities. Hence the total number of positronium events is $N_{\text{Ps}} = 4N_{\text{o-Ps}}/3$.

(iv) The total number of gas events, N_{G}, is then defined by $N_{\text{G}} = 4N_{\text{o-Ps}}/3 + N_{\text{f}} + N_{\text{F}}$.

(v) The quantity $G = N_{\text{G}}/N_{\text{S}}$, the gas fraction, can be deduced and used to compare the stopping power of different gases for β^+ particles. Furthermore, an unexpectedly low measured value of G could indicate the presence of fast components in the spectrum which cannot be resolved from the prompt peak.

(vi) The positronium fraction F is one of the most widely quoted observables in positron lifetime work and it can be defined as $F = N_{\text{Ps}}/N_{\text{G}}$.

(vii) A mean value, $\langle \lambda_{\text{p}} \rangle$, for the ortho-positronium decay rate can be determined from the equilibrium portion of the spectrum. This represents the sum of the self-annihilation rate, ${}_0\lambda_{\text{o-Ps}}$, and that due to quenching in collisions with the surrounding medium. Thus $\langle \lambda_{\text{p}} \rangle$ is usually written as

$$\begin{aligned} \langle \lambda_{\text{p}} \rangle &= {}_0\lambda_{\text{o-Ps}} + \langle q \rangle \rho \\ &= {}_0\lambda_{\text{o-Ps}} + 0.804\rho \langle {}_1Z_{\text{eff}} \rangle \qquad (\mu\text{s}^{-1}), \end{aligned} \tag{6.20}$$

where ρ was defined in equation (6.3), $\langle q \rangle$ is the so-called quenching coefficient and, following Fraser (1968), $\langle {}_1Z_{\text{eff}} \rangle$ is a measure of the effective number of electrons per atom or molecule responsible for quenching. All averages are over the positronium speed distribution at annihilation. Values of $\langle {}_1Z_{\text{eff}} \rangle$ are discussed in Chapter 7, which deals with positronium and its interactions with other systems, along with the range of validity of equation (6.20) in various gaseous media.

(viii) As mentioned above, the time spectrum for free-positron annihilation, $S_f(t)$, can be obtained from the total gas spectrum $G(t)$ once the ortho-positronium component has been fitted. The subtraction of the assumed exponential ortho-positronium component can be written as

$$S_f(t) = G(t) - \langle\lambda_p\rangle N_{\text{o-Ps}} \exp(-\langle\lambda_p\rangle t). \quad (6.21)$$

The instantaneous free-positron annihilation rate is particularly useful where it is time dependent (i.e. on the shoulder region); it is defined as

$$\langle\lambda_f(t)\rangle = \frac{S_f(t)}{\int_0^\infty S_f(t')\,dt'} = 0.201\rho\langle Z_{\text{eff}}(t)\rangle \qquad (\mu\text{s}^{-1}), \quad (6.22)$$

where $\langle Z_{\text{eff}}(t)\rangle$ is defined by equation (6.11). At the end of the shoulder region $\langle Z_{\text{eff}}(t)\rangle$ tends to $\langle Z_{\text{eff}}\rangle$, the equilibrium value given by equation (6.16). The experimental value of $\langle Z_{\text{eff}}(t)\rangle$ can be compared with the theoretical speed-average of the calculated values of $Z_{\text{eff}}(v)$, equation (6.11), the speed distribution being obtained by solving the diffusion equation (6.12).

(ix) The shoulder width is expressed in density-independent terms as $\tau_s\rho$, usually given in ns amagat, where τ_s was defined by Paul and Leung (1968) as the time for which

$$\langle Z_{\text{eff}}(\tau_s)\rangle = \langle Z_{\text{eff}}\rangle - 0.1\langle\Delta Z\rangle, \quad (6.23)$$

with $\langle\Delta Z\rangle = \langle Z_{\text{eff}}\rangle - \langle Z_{\text{min}}\rangle$, where $\langle Z_{\text{min}}\rangle$ is the minimum value of $\langle Z_{\text{eff}}(t)\rangle$ on the shoulder region; see Griffith and Heyland (1978) for an example.

(x) If there are any fast components in the spectrum which can be resolved from the prompt peak, as, e.g. in the case of xenon, shown in Figure 6.5(b), then their presence can be revealed if $\langle Z_{\text{eff}}(t)\rangle$ rises abruptly before the prompt peak has been reached. The spectrum of such components can only be deduced by subtraction of the free-positron component, which requires assumptions to be made concerning the shape of the shoulder.

2 The positron-trap method for positron–gas studies

Since the mid-1980s a new and radically different method of studying positron annihilation in gases has been developed by Surko and coworkers.

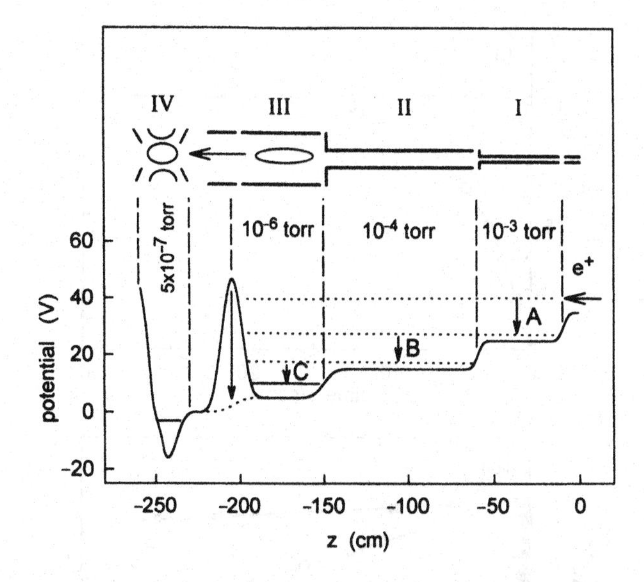

Fig. 6.6. Schematic illustration of the electrode structure of the positron trap of Greaves, Tinkle and Surko (1994). The variation of the electrical potential along the trap, together with the gas pressure in the various regions, is also shown. The letters A, B and C indicate energy-loss collisions of the positrons with the N_2 buffer gas. Reprinted from *Phys. Plasmas* **1**, Greaves *et al.*, Creation and uses of positron plasmas, 1439–1446, copyright 1994, by the American Institute of Physics.

It is based on a trap in which positrons from a low energy beam are confined in a region of low density gas. As will be shown in subsection 6.3.2 below, this approach has been particularly valuable in elucidating positron annihilation on large molecules, where some form of temporary attachment of a positron to the molecule may play an important role.

The operation of the trap was discussed at length by Murphy and Surko (1992), with additional details provided by Greaves, Tinkle and Surko (1994). A schematic illustration of the cylindrically symmetric electrode structure of the trap and the variation of the electrical potential and pressure along the trap are shown together in Figure 6.6. Slow positrons enter the electrode arrangement from the right, pass over an electrostatic barrier (whose height can be adjusted depending upon the initial kinetic energy of the positrons) and into region I, which contains molecular nitrogen gas at a pressure of around 10^{-3} torr. The gas is introduced at the centre of the first electrode, and differential pumping between this region and the remainder of the trap gives rise to the pressure gradient identified in the figure. The positrons are confined radially by an axial magnetic field, which is typically 0.1–0.2 T.

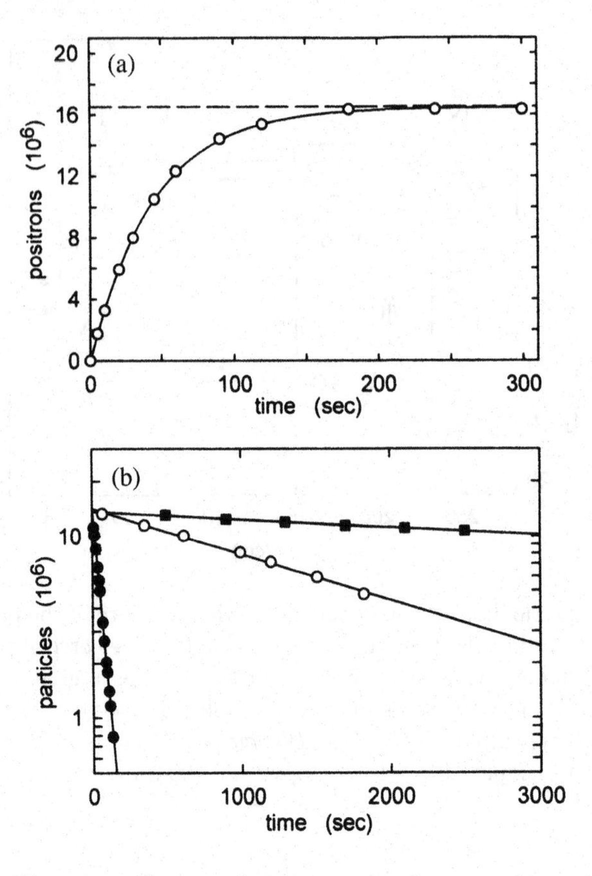

Fig. 6.7. (a) The accumulation of positrons in the trap shown in Figure 6.6. (b) The storage of positrons: •, N_2 gas pressure of 5×10^{-7} torr; ∘, N_2 gas off, base pressure of 7×10^{-10} torr. ▪, electrons stored under the same conditions. Reprinted from *Phys. Plasmas* **1**, Greaves, Tinkle and Surko, Creation and uses of positron plasmas, 1439–1446, copyright 1994, by the American Institute of Physics.

Positrons pass through the regions I, II and III before being reflected by the electrical potential barrier at the end of region III, after which they return towards the entrance of the trap. During the transit there is a reasonable chance, around 30%, that a positron will lose kinetic energy by electronic excitation of the gas. Such positrons are then trapped and eventually lose further energy by exciting vibrational and rotational transitions of the molecule. Thus, they are cooled to room temperature after approximately one second, whereupon they reside in region III at a pressure of 10^{-6} torr.

In this manner positrons can be continuously accumulated, the number entrapped having a time dependence $N(t) = R\tau[1 - \exp(-t/\tau)]$, where

R is the trapping rate and τ is the lifetime in the trap. As shown in Figure 6.7(a), over 10^7 positrons were accumulated in a time of the order of 100 s. If desired, the positrons can then be 'shuttled' into region IV of the trap, which has approximately hyperbolic electrodes, by lowering the potential barrier. In the absence of any added gas, the positron lifetime in region IV is governed largely by annihilation on the N_2 buffer gas. Figure 6.7 shows that under these conditions the lifetime is around 60 s, though this is increased to 30 minutes if the N_2 gas is pumped out. This time is still limited by annihilation on the remaining gas molecules in the trap (at a pressure of 7×10^{-10} torr), since electrons held in the same environment can be confined with a time constant of approximately three hours.

Once positron accumulation has been completed, the gas to be investigated can be added to the system and its effect on the number of trapped particles measured. Thus, positron lifetimes, and the value of the associated $\langle Z_{\rm eff} \rangle$ parameter, can be deduced under conditions which ensure single positron–molecule interactions. Information can also be obtained by measuring the Doppler broadening of the annihilation radiation. In addition, Greaves, Tinkle and Surko (1994) showed how the energy of the trapped positrons can be raised in a controlled manner by the application of an abrupt radio-frequency pulse to one of the electrodes of the trap. This has been used to determine the energy dependence of the annihilation rates for the noble gases (Kurz, Greaves and Surko, 1996), as described in subsection 6.3.2 below.

6.3 Results – positron annihilation

In this section we review the results from positron annihilation experiments, predominantly those performed using the lifetime and positron trap techniques described in section 6.2. Comparisons are made with theory where possible. The discussion includes positron thermalization phenomena and equilibrium annihilation rates, and the associated values of $\langle Z_{\rm eff} \rangle$, over a wide range of gas densities and temperatures. Some studies of positron behaviour in gases under the influence of applied electric fields are also summarized, though the extraction of drift parameters (e.g. mobilities) is treated separately in section 6.4. Positronium formation fractions in dense media were described in section 4.8.

1 Thermalization phenomena

Some positrons undergo annihilation prior to thermalization, provided the slowing down process is not too rapid; this is illustrated by features present in the lifetime data for the noble gases shown in Figures 6.2, 6.5

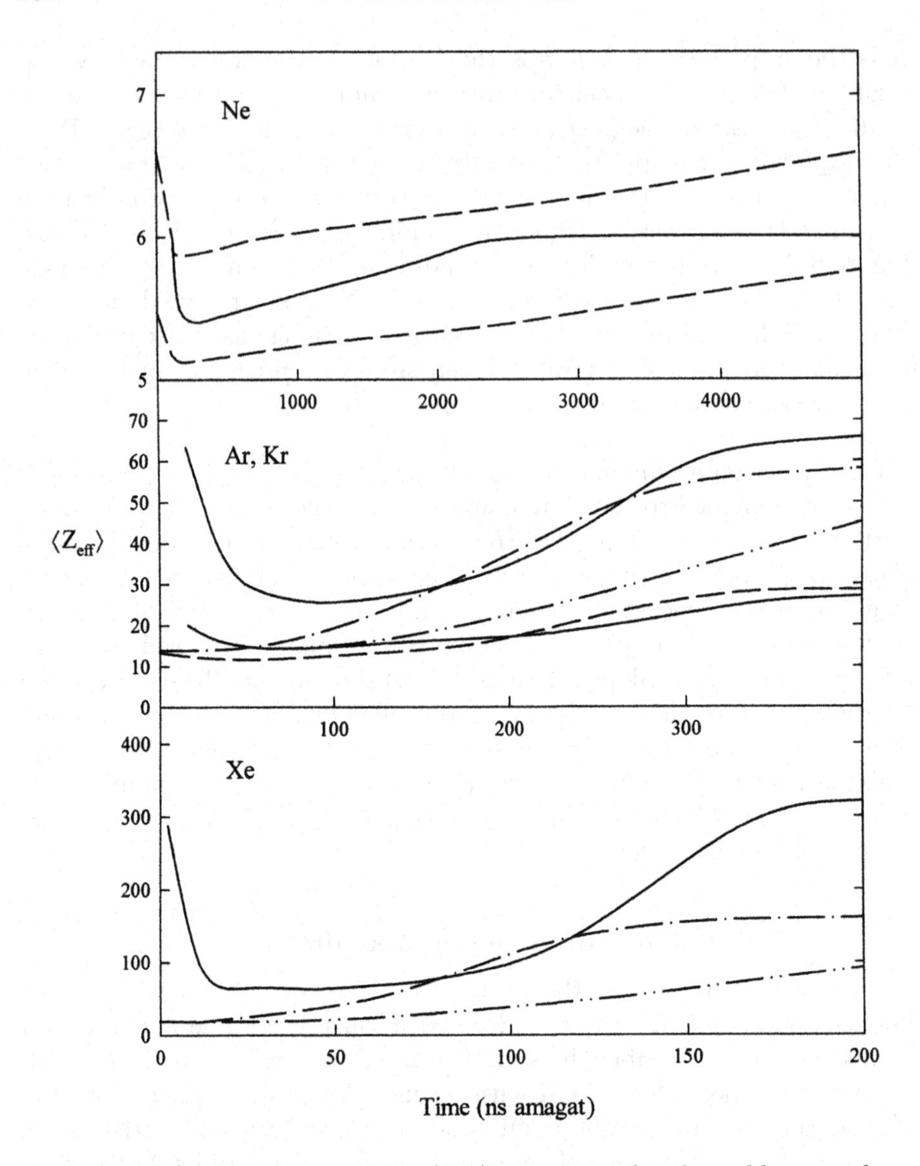

Fig. 6.8. Time dependence of the $\langle Z_{\text{eff}} \rangle$ parameter for the noble gases from neon through to xenon, for experiment (solid lines) and theory (various types of broken-line curve). Neon: ——, Coleman *et al.* (1975b); — —, Campeanu (1981), both curves calculated using the data of McEachran, Ryman and Stauffer (1978) but with different approximations to the positron–neon scattering. Argon (the two lowest curves): ——, Coleman *et al.* (1975b); — —, Campeanu (1981). Krypton (upper three curves): ——, Wright *et al.* (1985); — · —, Campeanu (1982), using data of McEachran, Stauffer and Campbell (1980); — · · —, Campeanu (1982), using momentum transfer cross sections calculated by Schrader (1979). Xenon: ——, Wright *et al.* (1985); — · —, Campeanu (1982), using data of McEachran, Stauffer and Campbell (1980); — · · —, Campeanu (1982), using the momentum transfer cross sections of Schrader (1979).

Table 6.1. Positron thermalization times and shoulder widths for a number of gases. Original references can be found in Charlton (1985a)

Gas	Shoulder width $\tau_s \rho$ (ns amagat)	Gas	Thermalization time (ns amagat)	Gas	Thermalization time (ns amagat)
He	1700 ± 50	H_2	3.3	C_4H_{10}	0.05
Ne	2300 ± 200	D_2	4.4	CCl_4	<0.05
Ar	362 ± 5	N_2	14 ± 2	CCl_2F_2	<0.05
Kr	325 ± 6	CO_2	0.09		
Xe	178 ± 3	CH_4	0.25		

and 6.8. Thermalization shoulders have been observed directly for the noble gases and for N_2, but most molecular gases have thermalization times which are too short for such phenomena to be resolved directly. A list of measured positron thermalization times is given for a variety of gases in table 6.1, along with the shoulder-width parameter, whose definition, for the noble gases, is encompassed in equation (6.23). The molecular thermalization times, except that for N_2, were obtained using the method of Paul and Leung (1968), who noted the reduction in the shoulder width of argon following the addition of small quantities of molecular impurities.

For helium, the momentum transfer and annihilation cross sections calculated using the accurate elastic scattering wave functions described in section 3.2 have enabled detailed comparisons to be made between theory and experiment for the variation of $\langle Z_{\mathrm{eff}}(t)\rangle$ across the shoulder. Campeanu and Humberston (1977b) solved the time-dependent diffusion equation (6.12) for zero electric field and at a temperature of 300 K and obtained the results shown in Figure 6.2. The agreement between theory and the room temperature experiment of Coleman *et al.* (1975b) is good, except at very short times. In the experiment, processes other than direct annihilation of the positrons in the gas contribute to the spectrum close to $t = 0$ where, theoretically, uncertainties relating to the choice of initial speed distribution are most significant. Confirmation that the diffusion and annihilation of positrons in a gas is accurately described by the diffusion equation approach was provided by Farazdel and Epstein (1977, 1978), using a Monte Carlo technique. They used the differential scattering cross sections and annihilation rates calculated by Campeanu and Humberston (1975) and obtained time-dependent annihilation rates in excellent agreement with those of Campeanu and Humberston (1977b).

Similar theoretical investigations of positron diffusion were made by Campeanu (1981, 1982) for the heavier noble gases using momentum transfer and annihilation cross sections obtained in the polarized-orbital approximation. The results are summarized in Figure 6.8, where they may be compared with the experimental data of Coleman *et al.* (1975b), for neon and argon, and Wright *et al.* (1985), for krypton and xenon. In general, the lack of accurate data for $\sigma_M(v)$ and $Z_{eff}(v)$ precludes the attainment of accurate theoretical lifetime spectra. For neon, however, Campeanu (1981) showed that the shoulder length predicted using the theoretical cross sections of McEachran, Ryman and Stauffer (1978) is much longer than that found experimentally, and he concluded that the calculated momentum transfer cross sections were too low at energies below approximately 1 eV. The situation in argon is somewhat better, and Campeanu found that the cross sections of McEachran, Ryman and Stauffer (1979) gave reasonable agreement with experiment.

The theoretical shoulders for krypton and xenon were computed using values of $\sigma_M(v)$ calculated by McEachran, Stauffer and Campbell (1980) and Schrader (1979) and values of $Z_{eff}(v)$ calculated by the former. In the case of krypton, the experimental shoulder length and shape are reproduced well using the data of McEachran, Stauffer and Campbell but the results of Schrader give a much longer shoulder than is observed. For xenon, poorer agreement between theory and experiment has been found for both the shape and the magnitude of $\langle Z_{eff}(t)\rangle$.

In addition to the noble gases, N_2 has also been found to possess an observable shoulder. The most complete study of this system was reported by Coleman, Griffith and Heyland (1981). These workers, using a method to be described below, analysed the $\langle Z_{eff}(t)\rangle$ measurements of Coleman *et al.* (1976a) at a density of 0.84 amagat and at room temperature, and those of Sharma and McNutt (1978) at densities below 2 amagat and at 77 K, to estimate the cross sections for momentum transfer and for rotational excitation and de-excitation. Less detailed analyses for H_2 and D_2 were also performed using the thermalization times given in table 6.1, even though shoulders have not been observed directly for either of these molecules. The long thermalization times in H_2, D_2 and N_2, which were also reported by Paul and Leung (1968) and Tao (1970), are caused by low energy-loss rates below the thresholds for vibrational excitation, E_{vib} = 0.516 eV, 0.360 eV and 0.290 eV respectively. Below these thresholds the energy loss is due to rotational excitation and momentum transfer collisions, with rotational de-excitation providing a heating mechanism that competes when there is a significant population of rotationally excited states. This applies to N_2, which, according to Coleman, Griffith and

Heyland (1981), has 33 rotational levels occupied at room temperature and 16 at 77 K.

Coleman, Griffith and Heyland (1981) used a simple slowing-down equation to analyse the data, rather than the full diffusion equation (6.12). The positrons were assumed to start their slowing down at E_{vib} and then lose energy at a rate

$$-\frac{dE}{dt} = -\left[\left(\frac{dE}{dt}\right)_{\text{r}} + \left(\frac{dE}{dt}\right)_{\text{el}}\right]\left(1 - \frac{3k_{\text{B}}T}{2E}\right), \tag{6.24}$$

where $-(dE/dt)_{\text{r}}$ and $-(dE/dt)_{\text{el}}$ are the energy-loss rates for rotational and elastic collisions respectively and the factor $1 - 3k_{\text{B}}T/2E$ allows, to first order, for the energy gained from the gas molecules at each collision. The individual energy-loss rates were deduced from

$$-\left(\frac{dE}{dt}\right)_{\text{el}} = \frac{2m}{M}nE\left(\frac{2E}{m}\right)^{1/2}\sigma_{\text{M}}(E) \tag{6.25}$$

and

$$-\left(\frac{dE}{dt}\right)_{\text{r}} = \left(\frac{2E}{m}\right)^{1/2}\sum_J n_J[\sigma_{J,J+2}(E)(E_{J+2} - E_J) - \sigma_{J,J-2}(E)(E_J - E_{J-2})]; \tag{6.26}$$

here the number density of gas molecules in the Jth rotational state is n_J and $\sigma_{J,J+2}(E)$ and $\sigma_{J,J-2}(E)$ are the cross sections for transitions between the energy levels E_J and $E_{J\pm 2}$.

Further details of these calculations are given by Coleman, Griffith and Heyland (1981), although it is not difficult to see that equations (6.25) and (6.26) can be used to derive the time variation of the positron energy. This was then employed, with an expression for $Z_{\text{eff}}(E)$ taken from the work of Darewych and Baille (1974) but fitted to the results of Sharma and McNutt (1978) and Coleman *et al.* (1976a), at 77 K and 300 K respectively, to deduce values of $\langle Z_{\text{eff}}(t)\rangle$. The results compare favourably with experiment. The values of $\sigma_{\text{M}}(E)$ obtained from different calculations did not agree, and therefore Coleman, Griffith and Heyland (1981) varied this parameter to produce the best fit to the data of Coleman *et al.* (1976a), which were, statistically, the more accurate set. The reader is referred to the original work for further details and for information, derived from the study, regarding positron behaviour in H_2 and D_2.

Thermalization in N_2 gas has also been studied using the positron-trap apparatus developed by Surko and coworkers and described in subsection 6.2.2. By storing positrons in the trap at a known pressure for various lengths of time before ejecting them and measuring their mean

Table 6.2. Experimental and theoretical values for $\langle Z_{\text{eff}} \rangle$ at low densities and room temperature for the noble gases. See the work of Griffith and Heyland (1978) for a review of earlier results

Gas	$\langle Z_{\text{eff}} \rangle$, experiment	$\langle Z_{\text{eff}} \rangle$, theory
He	3.94 ± 0.02	3.88
Ne	5.99 ± 0.08	7.0 ± 0.3
Ar	26.77 ± 0.09	27.6
Kr	65.7 ± 0.3	57.6 ± 2.9
Xe	320 ± 5	217 ± 11

energy, the rate of slowing down could be obtained (Murphy and Surko, 1992). Analysis reveals the thermalization time to be in fair accord with the results of Coleman and coworkers (1976a, 1981) and Paul and Leung (1968), given the very different nature of the experiments.

2 Equilibrium phenomena at low gas densities

Measurements of the equilibrium annihilation-rate parameter, $\langle Z_{\text{eff}} \rangle$, have been made for a plethora of gases over a wide range of densities. In this section we confine ourselves to the low density and high temperature region, in which many-body effects can be ignored and where the results may be compared with those obtained from scattering theory, as outlined in section 6.1.

Table 6.2 presents theoretical and experimental values of $\langle Z_{\text{eff}} \rangle$ for the noble gases helium to xenon. Traditional lifetime experiments have found $\langle Z_{\text{eff}} \rangle$ to be density dependent, particularly for argon, krypton and xenon; the values quoted in table 6.2 correspond to low gas densities. The results obtained from theory and experiment are in reasonable agreement, the theoretical results for the heavier targets having been derived from the results of McEachran and coworkers (1978, 1979, 1980), who performed a systematic study of all the noble gases, using the polarized-orbital approximation. For helium, the most accurate theoretical value, 3.88, obtained by Van Reeth *et al.* (1996), is in good accord with the accurate experimental result, 3.94 ± 0.02, of Coleman *et al.* (1975b). A more detailed survey of the experimental work on helium gas was given by Griffith and Heyland (1978).

The theoretical values for $\langle Z_{\text{eff}} \rangle$ quoted in table 6.2 have been speed-averaged to correspond to the room-temperature positron speed distribution. This is particularly important, as pointed out in section 6.1, for xenon, where the theoretical results vary rapidly at low energies. In their work on xenon, Wright *et al.* (1985) reported that their measured 'equilibrium' value, $\langle Z_{\rm eff}\rangle = 320 \pm 5$, may be an underestimate, owing to incomplete thermalization of the positrons in this gas and the high probability of epithermal annihilation. This tentative conclusion was reached on the basis of measurements with small quantities of H_2, He or N_2 added to the xenon, which effected more rapid thermalization; $\langle Z_{\rm eff}\rangle$ was found to increase to around 400. This value was corroborated by Murphy and Surko (1990), who used the positron-trap method to determine the annihilation rate in xenon at very low pressures (around 10^{-4} Pa). The accord between the results obtained using the traditional lifetime and positron-trap methods is also reasonable for argon, although, in the case of krypton, Iwata *et al.* (1995) found $\langle Z_{\rm eff}\rangle \approx 90$, previous values being in the range 65–66, as given in table 6.2. Evidence presented by Wright *et al.* (1985) from studies of krypton–gas mixtures suggests that this cannot be due to the effects of incomplete thermalization in the traditional lifetime experiments.

Many positron annihilation experiments have also been performed on molecular gases using the traditional lifetime technique. Table 6.3 gives a selection of values of $\langle Z_{\rm eff}\rangle$ for a range of molecular species (Heyland *et al.*, 1982; Wright *et al.*, 1983). Heyland *et al.* (1982) grouped these gases into two broad categories: those for which $\langle Z_{\rm eff}\rangle \approx Z$ and those for which $\langle Z_{\rm eff}\rangle \gg Z$. It is now thought that the formation of temporary bound states is responsible for the very large values of $\langle Z_{\rm eff}\rangle$ in the latter category (see below).

In the case of H_2, the low density measurements of $\langle Z_{\rm eff}\rangle$ by McNutt, Sharma and Brisbon (1979) are in accord with those of Wright *et al.* (1983), though until fairly recently all the theoretical results were much closer to $Z = 2$ (see Wright *et al.*, 1983, for a detailed discussion). However, as mentioned in section 6.1, Armour, Baker and Plummer (1990) investigated annihilation as part of their general study of positron–H_2 scattering using the Kohn variational method, see subsection 3.2.1, and by using increasingly elaborate trial wave functions, which allow explicitly for positron–electron correlations, they obtained the zero energy value $Z_{\rm eff} = 10.2$. This is in much better agreement with the room temperature experimental value, 14.8; it is likely that further theoretical advances will lead to even better accord.

Theory and experiment can also be compared in the case of N_2. Darewych and Baille (1974) calculated $Z_{\rm eff} = 22$ and 20 at 9 MeV and 38 MeV respectively, in reasonable accord with the experiments of Coleman *et al.* (1976a) and Tao (1970) at 297 K, and of Sharma and McNutt (1978) at 77 K. Methane is also of interest, with a measured value for $\langle Z_{\rm eff}\rangle$

Table 6.3. Measured values of $\langle Z_{eff} \rangle$ at low gas densities and 297 K, for various molecular gases. Values of Z are included for comparison

Gas	Z	$\langle Z_{eff} \rangle$
H_2	2	14.7
D_2	2	14.7
N_2	14	30.5
CO	14	38.5
NO	15	34
O_2	16	26
CO_2	22	53
N_2O	22	78
SF_6	70	97
CH_4	10	140
C_2H_6	18	660
C_3H_8	26	3 500
C_4H_{10}	34	15 000
CH_3Cl	26	15 000
CCl_2F_2	58	700
NO_2	23	1 090
NH_3	10	1 300

of 140, which has been considered to be too large to arise from direct scattering alone (see e.g. McNutt *et al.*, 1975, and references therein). However, the only calculation of this quantity, by Jain and Thompson (1983), found $Z_{eff} \approx 100$ at 297 K, in fair accord with the experimental value considering the complexity of the target.

The positron-trap technique has been used to measure the annihilation rate of positrons interacting with a wide variety of molecules. The species investigated by Iwata *et al.* (1995) include many hydrocarbons, substituted (e.g. fluorinated and chlorinated) hydrocarbons and aromatics; as mentioned in section 6.1, large values of $\langle Z_{eff} \rangle$ (in excess of 10^6) were found for some molecules. Several distinct trends are exhibited in the data of Iwata *et al.* (1995). Though much of the detailed physics involved in the annihilation process on these large molecules is still unclear, the model of Laricchia and Wilkin (1997), described in section 6.1, may offer a qualitative explanation of the observations.

The temperature, or energy, dependence of the annihilation rate, or $\langle Z_{eff} \rangle$, has also been investigated using a positron trap. In this technique positrons are first accumulated at room temperature, and then their

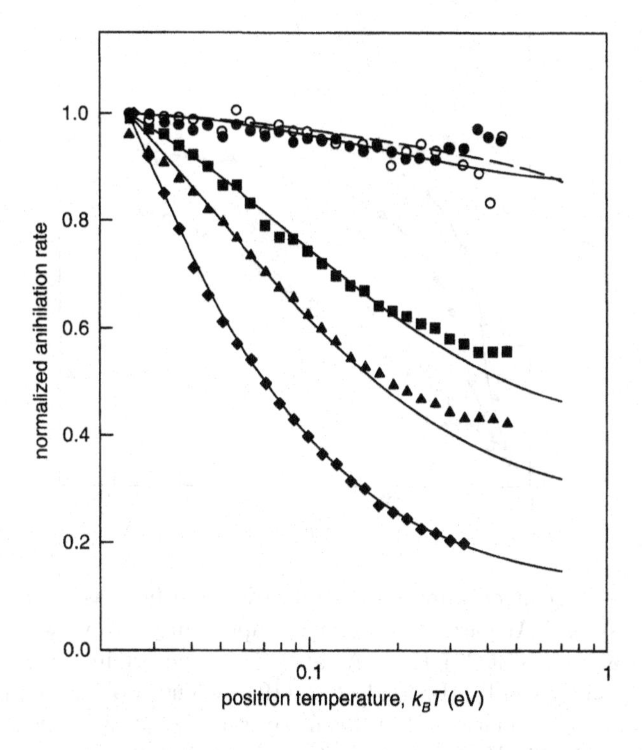

Fig. 6.9. Positron annihilation rates at various temperatures (normalized to unity at room temperature) for the noble gases, from Kurz *et al.* (1996). Key: ∘, He; •, Ne; ▪, Ar; ▲, Kr; ◆, Xe. The curves are from the theoretical work of McEachran and coworkers (1977, 1978, 1979, 1980) for helium (— —) and neon, argon, krypton and xenon (——). Reprinted from *Physical Review Letters* **77**, Kurtz, Greaves and Surko, Temperature dependence of positron annihilation rates in noble gases, 2929–2932, copyright 1996, by the American Physical Society.

energy is increased by applying radio-frequency noise to one of the trap electrodes. The annihilation rate on the relevant gas, and the positron temperature, are then measured with time as the positrons cool in the gas. Figure 6.9 shows the plots of normalized annihilation rate versus temperature obtained by Kurz, Greaves and Surko (1996) for the noble gases. In general, excellent agreement is obtained with the theoretical work of McEachran and coworkers. Also, the results of Van Reeth *et al.* (1996) for helium cannot be distinguished from experiment at the level of accuracy displayed in Figure 6.9. Good agreement with the older data of Lee and Jones (1974), which were taken using the traditional lifetime method, was found by Kurz *et al.* (1996) in the temperature region of overlap.

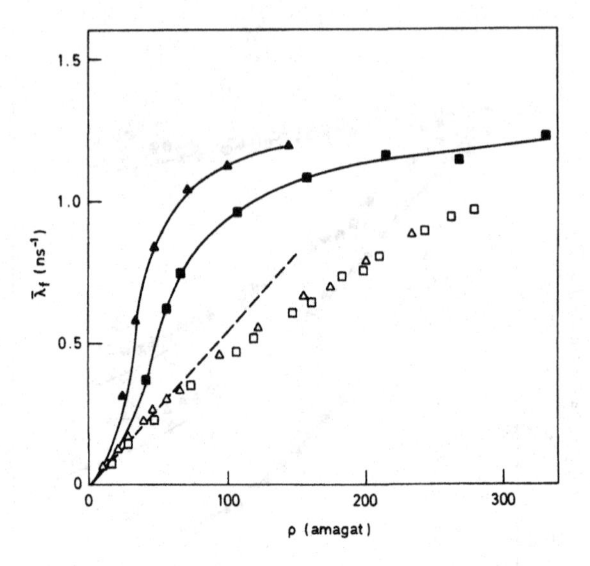

Fig. 6.10. Mean positron annihilation rate (denoted here as $\bar{\lambda}_f$) at various gas densities for N_2 and Ar gases at different temperatures. Key: ▲, N_2 at 130 K; ■, Ar at 160 K; △, N_2 at 297 K; □, Ar at 297 K. The original sources for these measurements are given by Heyland *et al.* (1986). The broken line indicates the linear rise expected, equation (6.3), for a constant $\langle Z_{eff} \rangle$ of 27. Reprinted from *Physics Letters* **A119**, Heyland *et al.*, On the annihilation rate of thermalized free positrons in gases, 289–292, copyright 1986, with permission from Elsevier Science.

3 Equilibrium phenomena – many-body effects

As the density of a gas is increased and/or its temperature is lowered towards or below the critical temperature, T_c, new phenomena associated with the trapping and localization of positrons are sometimes encountered, indicating that many-body processes affect positron annihilation. We briefly describe these phenomena here, but a much more detailed treatment can be found in the review of Iakubov and Khrapak (1982).

Two seemingly distinct types of behaviour occur. Figure 6.10 (Heyland *et al.*, 1986), shows how both types are manifest for N_2 and argon gases. The behaviour at room temperature, when $\langle \lambda_f \rangle$ rises more slowly with density than the linear increase predicted by equations (6.16) and (6.3), is contrasted with that in the approximate temperature range $T < 2T_c$, when $\langle \lambda_f \rangle$ rises more rapidly than expected.

The density effects evident in Figure 6.10 for room temperature argon and N_2 were discussed by Iakubov and Khrapak (1982), and Nieminen (1980) carried out a similar exercise for helium. These authors attributed this behaviour to the effects of multiple scattering at high densities, when the positron is considered to be interacting continually with the medium. Heyland *et al.* (1986), however, made some simple observations concerning this behaviour, which has also been observed in room temperature data for O_2 and CO (Griffith and Heyland, 1978), in H_2 (Wright *et al.*, 1983; McNutt, Sharma and Brisbon, 1979), and in helium gas at 77 K and selected lower temperatures (Fox, Canter and Fishbien, 1977; Hautojärvi *et al.*, 1977). Heyland *et al.* (1982) had previously suggested that the density dependence of $\langle Z_{\rm eff}\rangle$ in the high temperature regime could be represented by the form $\langle Z_{\rm eff}(\rho)\rangle = \langle Z_{\rm eff}(0)\rangle/(1+\beta\rho)$, where $\langle Z_{\rm eff}(0)\rangle$ is the zero density limit of $\langle Z_{\rm eff}\rangle$ and β is a constant. This expression can be rearranged using equation (6.3) to yield

$$\frac{1}{\langle\lambda_{\rm f}\rangle} = \frac{1}{0.201\rho Z_{\rm eff}(0)} + \frac{1}{\langle\lambda_l\rangle}, \tag{6.27}$$

where $\langle\lambda_l\rangle = \beta/[0.201\langle Z_{\rm eff}(0)\rangle]$, implying that a plot of $1/\langle\lambda_{\rm f}\rangle$ versus $1/\rho$ should be linear with a slope of $1/[0.201\langle Z_{\rm eff}(0)\rangle]$ and an intercept of $1/\langle\lambda_l\rangle$. Examples of such plots for a variety of gases, as given by Heyland *et al.* (1986), are shown in Figure 6.11. The derived values of $\langle Z_{\rm eff}(0)\rangle$ are found to be in close accord with the extrapolation of $\langle Z_{\rm eff}\rangle$ to zero density, with a non-zero intercept close to $\langle\lambda_l\rangle = 2\ {\rm ns}^{-1}$ at the high density limit in all cases. This implies a positron lifetime of the same order as that characteristic of spin-averaged positronium and roughly that expected for a positron bound to an atom or molecule with configuration approximating to that of a positronium bound to the corresponding ion (see section 7.5). The lines drawn for helium and neon in Figure 6.11 are essentially predictions from the approach of Heyland and coworkers, and the reader is referred to the original paper for further discussion.

At lower temperatures other phenomena are found. The basic behaviour is shown in Figure 6.10, with a further selection given in Figure 6.12 for the important case of helium gas (the data shown are those of Hautojärvi *et al.*, 1977) where this phenomenon was first observed (Roellig and Kelly, 1965; Canter and Roellig, 1970). Similar features have been reported in a variety of other species, e.g. H_2 (Laricchia *et al.*, 1987a; McNutt, Sharma and Brisbon, 1979), CO_2 and SF_6 (Heyland *et al.*, 1985), argon (Canter and Roellig, 1975; Tuomisaari, Rytsölä and Hautojärvi, 1985) and CH_4 (McNutt *et al.*, 1975), and it is now considered that in almost all gases over certain density and temperature ranges positrons annihilate after self-trapping in clusters of atoms or molecules.

In helium, at low densities $\langle\lambda_{\rm f}\rangle$ increases linearly with density before rising rapidly to an approximately constant value, the magnitude of which is dependent upon the temperature of the gas and is characteristic of the particular clustered state. The transition is abrupt in this case, but is shown to be 'softer' for N_2 since the rise in $\langle\lambda_{\rm f}\rangle$ occupies a much broader

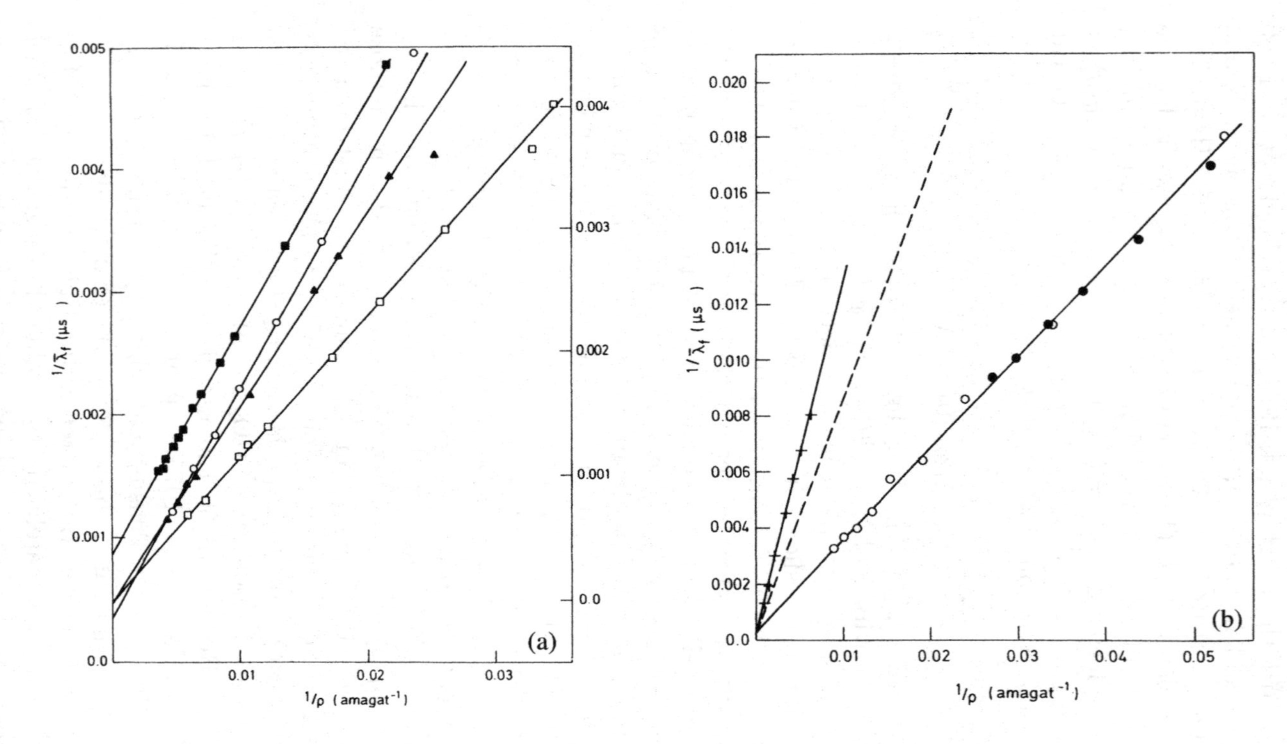

Fig. 6.11. (a) Plots of the inverse of the mean positron annihilation rate versus inverse density for (□) CO, (▲) N_2, (○) O_2 and (■) Ar at 297 K. The offset scale on the right is for argon only. (b) Similar plots for H_2 (○ and •) at room temperature. The solid line with crosses is for helium with $Z_{\rm eff}(0) = 4$ and $\langle\lambda_l\rangle = 2$ ns^{-1} and the broken line is for neon with $Z_{\rm eff}(0) = 6$ and $\langle\lambda_l\rangle = 2$ ns^{-1}. Reprinted from *Physics Letters* **A119**, Heyland *et al.*, On the annihilation rate of thermalized free positrons in gases, 289–292, copyright 1986, with permission from Elsevier Science.

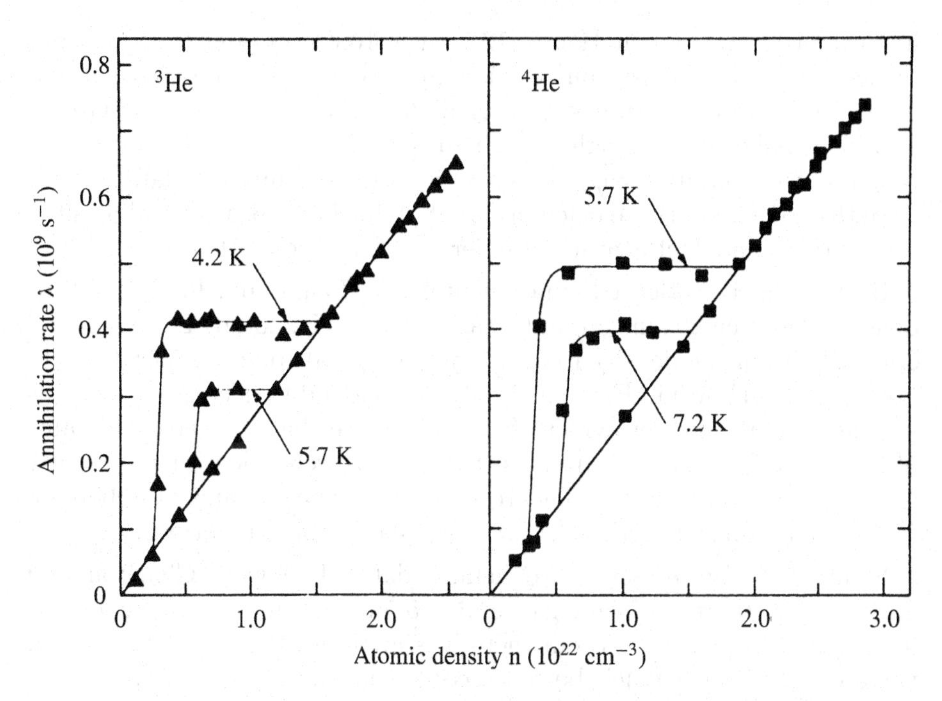

Fig. 6.12. An example of the effect of self-trapping in clusters on the free-positron annihilation rate. The right-hand side is for ^{4}He at 5.7 K and 7.2 K, whilst the left-hand data are for ^{3}He at 4.2 K and 5.7 K (Hautojärvi *et al.*, 1977).

range of densities. Rytsölä, Rantapuska and Hautojärvi (1984) accounted for this difference by showing that the potential energy gained by the positron in making the transition to the (positron + cluster) state is much larger for helium than for N_2, the value for N_2 being comparable to the thermal energy and so resulting in an ill-defined cluster. The behaviour of $\langle\lambda_f\rangle$ as a result of cluster formation was reproduced by Hautojärvi and Rytsölä (1979) and Rytsölä, Rantapuska and Hautojärvi (1984), using a density functional formalism which incorporates a calculated static density profile of the cluster and known values of $Z_{\rm eff}$ to compute the average annihilation rate at each density and temperature.

4 Electric fields

Several studies have been made of the behaviour of low energy positrons in gases under the influence of a static electric field ϵ. The broad aim of this work has been to study the diffusion and drift of positrons in order to understand better the behaviour of the momentum transfer and annihilation cross sections at very low energies. The theoretical background has been given in section 6.1, and the diffusion equation with an applied electric field, equation (6.12), was solved accurately for positrons in helium gas by Campeanu and Humberston (1977b) and Shizgal and Ness (1987). Similar studies were made for other noble gases by Grover and his collaborators (Singh and Grover, 1987; Sinha and Grover, 1987; Singh, 1989), mainly using the scattering and annihilation data derived from the polarized-orbital calculations of McEachran, Ryman and Stauffer (1978, 1979) and McEachran, Stauffer and Campbell (1980).

Helium was investigated experimentally by Leung and Paul (1969) at densities between 20 amagat and 30 amagat at 77 K, and by Lee, Orth and Jones (1969) in the density range 30–40 amagat at room temperature. In both cases, as shown in Figure 6.3, $\langle Z_{\rm eff} \rangle$ was found to decrease as ϵ/ρ was increased. It is however evident from this figure that the zero field value of 3.63 for $\langle Z_{\rm eff} \rangle$ obtained from both studies is well below that currently accepted (see table 6.2). The results of Campeanu and Humberston (1977b) and Shizgal and Ness (1987) are shown for comparison.

Figure 6.3 also presents the helium data of Davies, Charlton and Griffith (1989) at 35.7 amagat and 3.5 amagat in the range ϵ/ρ = 0–12 V cm^{-1} amagat^{-1}. From there it is apparent that (i) the zero-field value of $\langle Z_{\rm eff} \rangle$ is in much better accord with theory, and with the experiment of Coleman *et al.* (1975b), than were the earlier measurements and (ii) there is a density effect, with the measured value of $\langle Z_{\rm eff} \rangle$ at the higher density being substantially greater than theory. The lower density data are in much better accord but are subject to much larger errors, primarily as a result of the difficulty in collecting statistically accurate helium lifetime spectra. Davies, Charlton and Griffith (1989) gave some discussion of the density effect (a similar phenomenon was also found for argon gas), although no firm conclusion as to its origin was reached.

These workers were also able to extract some information on the non-equilibrium behaviour of positrons in helium and argon by determining the shoulder lengths at various electric fields, and qualitative agreement with theory was found. In addition to their work on the noble gases, Davies, Charlton and Griffith (1989) also presented data for four molecular species in which $\langle Z_{\rm eff} \rangle$ was found to decrease linearly by around 10%–15% over the ϵ/ρ range investigated.

The effect of an electric field on positrons annihilating in dense helium gas at low temperatures was investigated by Canter and coworkers (Ruttenberg, Tawel and Canter, 1985; Tawel and Canter, 1986). This work centred on the localization phenomena responsible for the clustering behaviour of helium atoms around the positron in dense helium gas, as illustrated in Figure 6.12. The aim of the studies was to test the hypothesis of Canter *et al.* (1980), supported by Azbel and Platzman

(1981), that the clustering phenomenon was preceded by localization of the positron when its kinetic energy fell below a certain value, E_c^+. This energy is termed the mobility edge and marks the boundary between the extended (i.e. freely diffusing) states and the Anderson-localized states (see e.g. Mott and Davis, 1979) of the positron, which can occur at densities when its de Broglie wavelength and mean free path are comparable.

Ruttenberg, Tawel and Canter (1985) studied the behaviour of the lifetime spectrum under the influence of a near-uniform static electric field. They found best agreement with the Monte Carlo calculations of Farazdel and Epstein (1978) and a later extension by Farazdel (1986) in which 'sink-like' behaviour below a certain energy was incorporated into the simulation. Furthermore, Ruttenberg, Tawel and Canter (1985) established that neither the energy threshold E_R for positron self-trapping (i.e. formation of a positron–helium cluster state) nor the positron annihilation rate in that state were affected by an electric field of 52 V cm^{-1} at a density of 129 amagat. Both of these observations agree with expectations since, once trapped in a cluster, the positron is immobile. Note that there is a subtle but very important difference between the mobility edge E_c^+, at which the positron is first immobilized, and E_R, with $E_c^+ > E_R$.

Tawel and Canter (1986) extended these investigations using a pulsed electric field of varying amplitude, which was applied at a minimum time of 26 ns after the emission of a positron from the ^{22}Na radioactive source. The field was triggered by detecting the 1.274 MeV gamma-ray. Their work established that the two thresholds, E_R and E_c^+, were distinct, which led to observable differences between the lifetime spectra when the electric field was pulsed on at different times after positron emission.

A lifetime spectrum obtained with a pulsed electric field is shown in Figure 6.13, superimposed upon a spectrum taken at zero field. The field was turned on after $\tau_d = 26 \pm 0.5$ ns and removed after 56 ± 5 ns. The two spectra agree up to $\tau_c = 28.7 \pm 0.5$ ns. The part of the peak which is missing when the field is present appears in the heated component and is attributed to positrons with energies greater than E_c^+ at τ_d. Some of these are slowed to E_R once the field has been removed, resulting in the small peak observed at 50–60 ns. The unheated, or localized, group of positrons continues to lose energy towards E_R and is not affected by the electric field.

The two-threshold model, i.e. a model with distinct values for E_c^+ and E_R, can account for the data, notably that the pulsed field spectrum is identical to that at zero field until a time $\Delta\tau = \tau_c - \tau_d$ when the distribution begins to reach E_R. Figure 6.14 shows pulsed field spectra (and also an example with a constant field) taken at different times τ_d

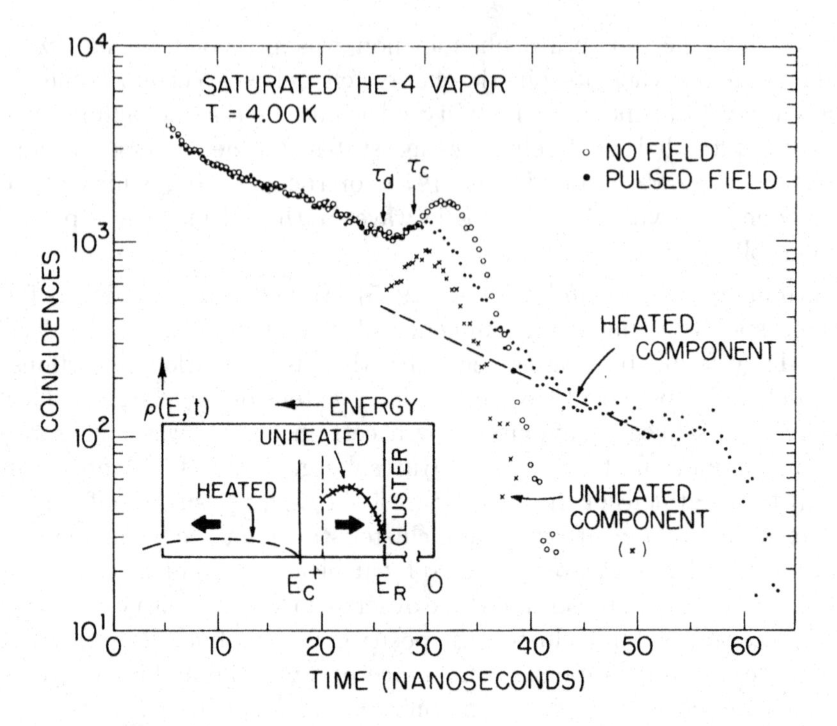

Fig. 6.13. Superimposed zero field and pulsed field (81 V cm^{-1} peak amplitude) positron lifetime spectra. The pulsed field spectrum has been decomposed into heated components (broken line) and unheated components (crosses) to illustrate how the electric field splits up the positron ensemble. This is also illustrated by the inset, which shows, schematically, the energy distribution $\rho(E,t)$ of the positron ensemble in the two-threshold model (see text). Reprinted from *Physical Review Letters* **56**, Tawel and Canter, Observation of a positron mobility threshold in gaseous helium, 2322–2325, copyright 1986 by the American Physical Society.

covering the range from below 26.0 ns, where the spectra are similar to the constant field case in which all the positrons are heated by the field, through to around 35 ns. There is then no difference between the spectra with the field on and off since all the positrons have had time to slow down below E_c^+ and are thus not affected by the field. Tawel and Canter (1986) estimated that $E_R = 5.5 \pm 0.5$ meV from the position of the peak in the lifetime spectrum shown in Figure 6.14 and deduced that $E_c^+ = 15 \pm 3$ meV by relating these two energies, the observed $\Delta\tau$ and the energy-loss rate in the gas. They also pointed out the utility of studying the relatively simple positron–helium system and the 'striking manner in which the mobility and the cluster formation thresholds manifest themselves'.

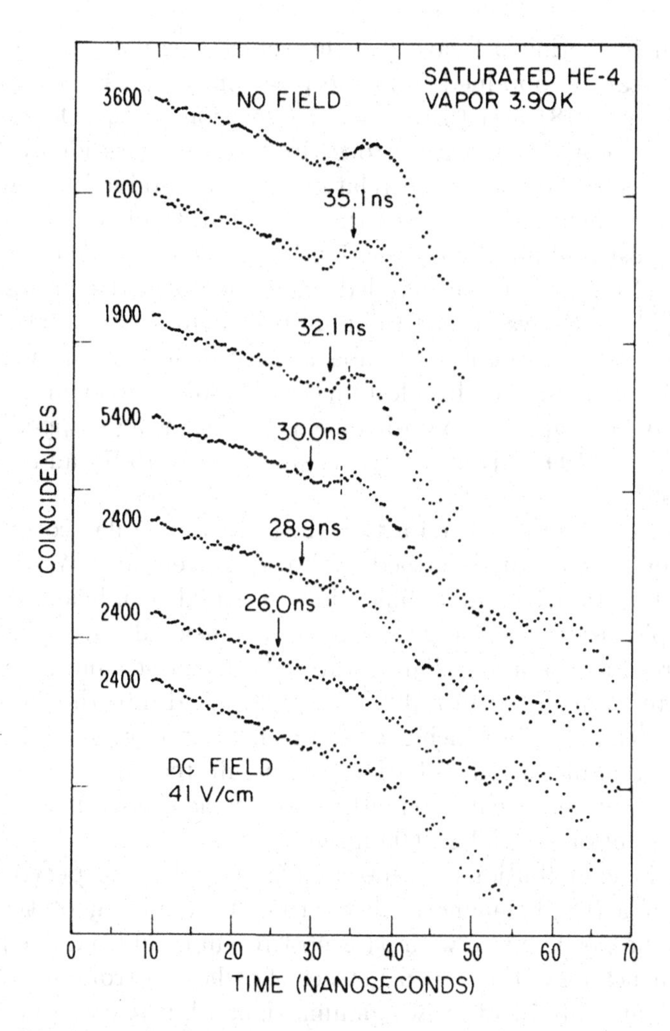

Fig. 6.14. Semilogarithmic pulsed field positron lifetime spectra corresponding to different values of τ_d (indicated by the arrows) for a mean field of 41 V cm^{-1}. (A spectrum with a d.c. field is also shown.) The numbers on the left are the numbers of coincidences for the first plotted point and the broken lines indicate the cut-off times τ_c where the spectra depart from the zero-field spectrum. Reprinted from *Physical Review Letters* **56**, Tawel and Canter, Observation of a positron mobility threshold in gaseous helium, 2322–2325, copyright 1986 by the American Physical Society.

5 Angular correlation and Doppler broadening studies

The ability of angular correlation and Doppler broadening techniques to provide information concerning the momentum of an annihilating electron–positron pair was briefly discussed in section 1.3. Also, it was shown in section 6.1 how the theoretical angular correlation function, $F(\theta)$, can be obtained from the positron-scattering wave function, see equations (6.18) and (6.19), and its relation to the Doppler-shifted gamma-ray energy spectrum was described. Early work identifying some basic features of the angular correlation spectra obtained in the presence of a magnetic field, which mixes the $m = 0$ state of ortho-positronium with para-positronium (see e.g. section 7.4), was undertaken by Heinberg and Page (1957). A more detailed experimental investigation of $F(\theta)$ for the noble gases was carried out by Coleman *et al.* (1994), with the data of Stewart, Briscoe and Steinbacher (1990) for the condensed gases available for comparison. In addition, the Doppler broadening technique has recently been applied to positron–gas studies using trapped positrons by Tang *et al.* (1992), Iwata, Greaves and Surko (1997) and Van Reeth *et al.* (1996).

Coleman *et al.* (1994) used a two-dimensional angular correlation apparatus similar to that described by West, Mayers and Walters (1981). Although the traditional one-dimensional method is sufficient to measure the isotropic distribution in gases, a two-dimensional apparatus provides a much greater data accumulation rate. Positrons produced from a ^{22}Na source were transported in a 0.8 T magnetic field into the small central volume of a chamber in which the gas was held at a pressure of one atmosphere. The annihilation radiation was monitored by position-sensitive gamma-ray counters located on either side of the chamber and having an angular resolution of 3.50 ± 0.03 mrad.

The main contributions to the angular correlation spectrum in the presence of a 0.8 T magnetic field arose from mixing of the $m = 0$ state of ortho-positronium with para-positronium and from free positrons. Pick-off quenching of the $m = \pm 1$ states of ortho-positronium (see section 7.3) also contributed to the two-gamma signal, but it could be neglected at the low gas pressure used. Unfortunately, the events arising from para-positronium, which are due to its annihilation before appreciable slowing down occurred, resulted in a broad component in the spectrum which could not be distinguished from that for the free positrons. This is the major drawback of this technique. However, the mixed-state ortho-positronium, with a vacuum lifetime of 9.7 ns in a magnetic field of 0.8 T, has a longer time in which to slow down in the gas (thus reducing the centre-of-mass momentum of the annihilating pair) and this was in some cases discernible as a narrow component in the spectra.

Figure 6.15 shows the cylindrically averaged angular distributions for the five noble gases, helium through to xenon. The mixed-state ortho-positronium was only clearly separable in helium, and a free fit to the data using Gaussian components was performed in this case. For neon and argon, the mixed-state ortho-positronium was fitted by constraining

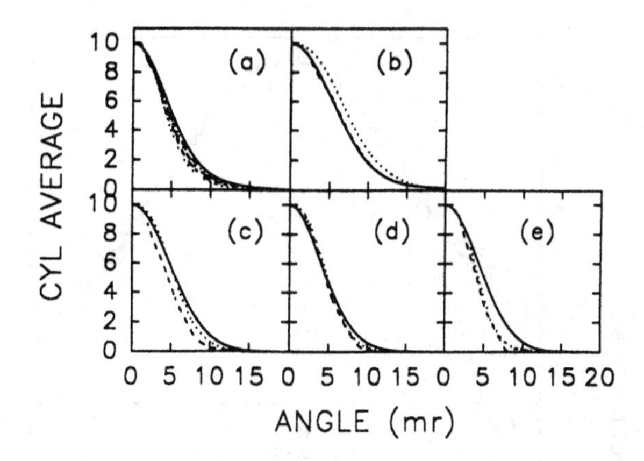

Fig. 6.15. Cylindrically averaged angular correlation of annihilation radiation (ACAR) distributions for positron annihilation in the noble gases, (a) helium, (b) neon, (c) argon, (d) krypton and (e) xenon, from the work of Coleman *et al.* (1994). Reprinted from *Journal of Physics* **B27**, Coleman *et al.*, Angular correlation studies of positron annihilation in the noble gases, 981–991, copyright 1994, with permission from IOP Publishing.

the amplitude of this component, using the known positronium fraction as deduced from lifetime studies (see section 4.8) and the known extent of the mixing of the ortho-positronium component. The krypton and xenon data, however, could be adequately described by a single Gaussian and thus the results of the fits cannot be solely due to free-positron annihilation.

There is no rigorous justification for assuming a Gaussian fit to the data, and accurate theoretical and experimental results for helium clearly show the deviations from the Gaussian form out on the wings of the distribution; see Figure 6.16 and the discussion at the end of this section. Nevertheless, this form does provide a reasonably good approximation to most of the data.

The positron-trap technique has been used by Surko and coworkers to measure the Doppler broadening of the 511 keV line for positrons in helium gas. This method does not have the drawback of the experiment described above, in which both positronium and free-positron events overlap on the angular distribution curves; here the positrons are thermalized prior to the introduction of the gas and therefore cannot form positronium. A comparison of the theoretically predicted and experimentally measured Doppler spectra (Van Reeth *et al.*, 1996) is shown in Figure 6.16. The theoretical results were obtained from the variational wave functions for low energy positron–helium scattering calculated by Van Reeth and Humberston (1995b); see equations (3.75) and (3.77).

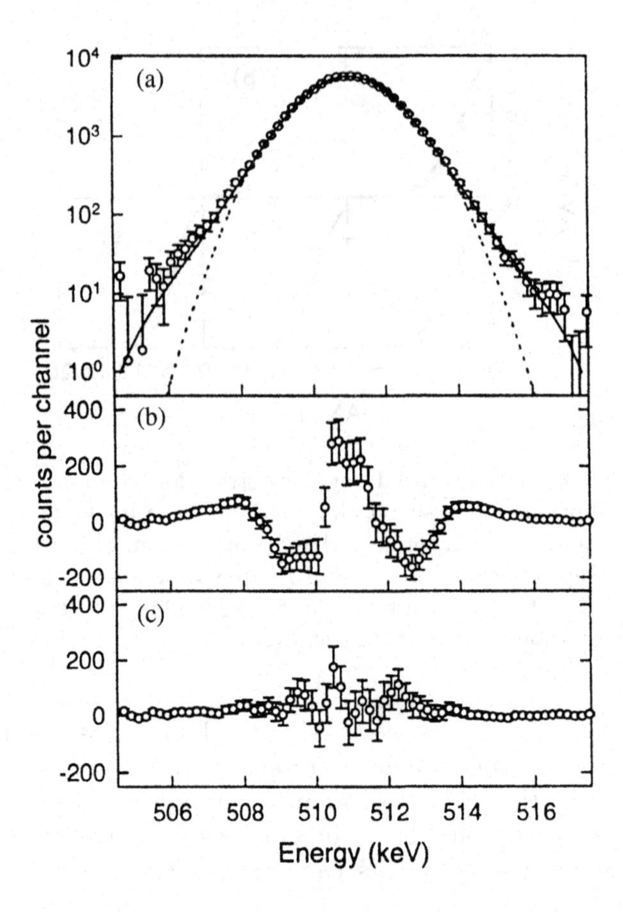

Fig. 6.16. (a) Annihilation gamma-ray spectrum for positrons interacting with helium atoms. The full curve is the theoretical prediction (see text) convoluted with the detector response function, whilst the dotted curve is a Gaussian fit to the experimental data (o). (b) Residuals from the Gaussian fit and (c) residuals from the theory. Reprinted from *Journal of Physics* **B29**, Van Reeth *et al.*, Annihilation in low energy positron–helium scattering, L465–L471, copyright 1996, with permission from IOP Publishing.

Before comparing with experiment, however, the theoretical results at an energy of 40 meV (equivalent to room temperature) were convoluted with the energy resolution function of the detector used for the measurement. This procedure was adopted because deconvolution of the experimental data was found to be numerically unstable. The convoluted theoretical data were then normalized to the experimental data at zero Doppler shift to yield the results shown. The agreement between the convoluted theoretical results and experiment is extraordinarily good, extending as it does over more than three orders of magnitude. These results also reveal that the shape of the gamma-ray spectrum is not Gaussian, in contrast to what has frequently been assumed in the past (see above). Nevertheless, Van Reeth *et al.* (1996) also fitted a Gaussian to the data and derived a Doppler FWHM of 2.53 ± 0.03 keV, corresponding to an angular width of 9.90 ± 0.12 mrad. This is to be compared with the values 10.30 ± 0.05 mrad obtained by Coleman *et al.* (1994) and 9.4 ± 0.5 mrad found by Stewart *et al.* (1990), for liquid helium. Iwata, Greaves and Surko (1997) also studied Doppler broadened spectra for argon, krypton and xenon and were able to observe contributions from annihilation with inner shell electrons.

6.4 Positron drift

The study of positron drift in gases has been very limited, in contrast to the vast literature available for electrons (e.g. Huxley and Crompton, 1974). The reason for this lies in the experimental difficulties in obtaining a drift time and distance such that the drift speed for positrons, v_+, can be computed. An ideal positron-drift experiment would involve the injection of a burst of low energy positrons into a scattering cell at a precisely known time; transport of the positrons through the cell and on to a detector would then be accurately evaluated to yield the mobility, $\mu_+ = v_+/\epsilon$ (Charlton and Jacobsen, 1987). Unfortunately an experiment of this kind has not yet been carried out. In this section we give brief descriptions of two experimental arrangements which have been employed to measure positron drift, though so far with a limited degree of applicability.

The first observations of positron drift were made by Rodionov, Sannikov and Solodov (1969, 1971) for helium gas. Their method was later adopted and improved by Paul and coworkers (Paul and Tsai, 1979; Paul and Böse, 1982; Böse, Paul and Tsai, 1981) and we present here a description of the latter apparatus only; this is shown schematically in Figure 6.17 (Paul and Böse, 1982). Positrons from two ^{58}Co sources were confined to the axis of a 4 cm long chamber by a uniform magnetic field of approximately 0.12 T. A small fraction of the positrons thermalized in the gas, uniformly throughout the length of the chamber owing to the low pressures used (typically well below 10^5 Pa). These positrons could then be drifted along the chamber by application of an electric field, using a ring and grid arrangement and the equipotential target foil which served to terminate the drift. This foil, which was chosen to be a 600 Å carbon film in order to reduce annihilation of the fast positrons, was viewed by a pair of detectors located behind the apertures in the lead shielding.

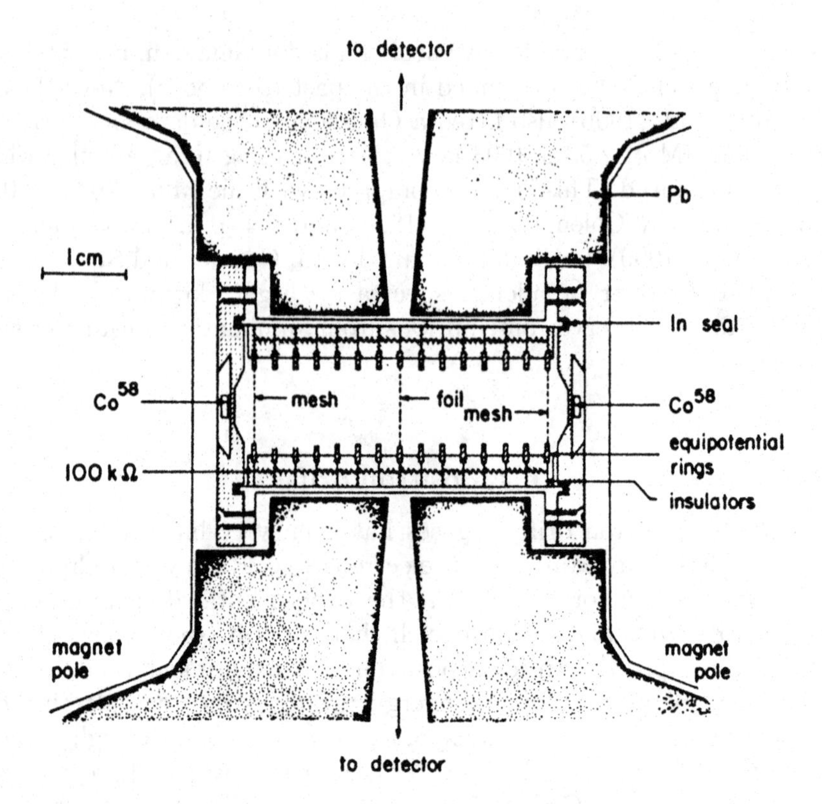

Fig. 6.17. Schematic illustration of the positron-drift apparatus used by Paul and coworkers.

Paul and Tsai (1979) showed that the fraction $F_{\rm d}$ of positrons which drift into the foil after stopping in the gas is given by

$$\begin{aligned} F_{\rm d} &= \frac{v_+}{\langle\lambda_{\rm f}\rangle l}\left[1 - \exp\left(-\frac{\langle\lambda_{\rm f}\rangle l}{v_+}\right)\right] \\ &= X\left[1 - \exp\left(-\frac{1}{X}\right)\right], \end{aligned} \tag{6.28}$$

where l is the maximum drift distance and $\langle\lambda_{\rm f}\rangle$ is the annihilation rate in the gas. At higher fields, when positronium formation is possible, $\langle\lambda_{\rm f}\rangle$ will be augmented by the rate at which this latter process occurs. An additional assumption made in arriving at equation (6.28) was that the diffusion length is much smaller than the drift length. This, in general, may not be true, and a more sophisticated analysis of the drift and diffusion problem was given by Paul and Tsai (1979).

Coincidences due to positron annihilation at the foil were recorded for various values of ϵ/ρ and at several pressures of molecular hydrogen in the range 10–200 torr. From these the data at 100 torr were selected for

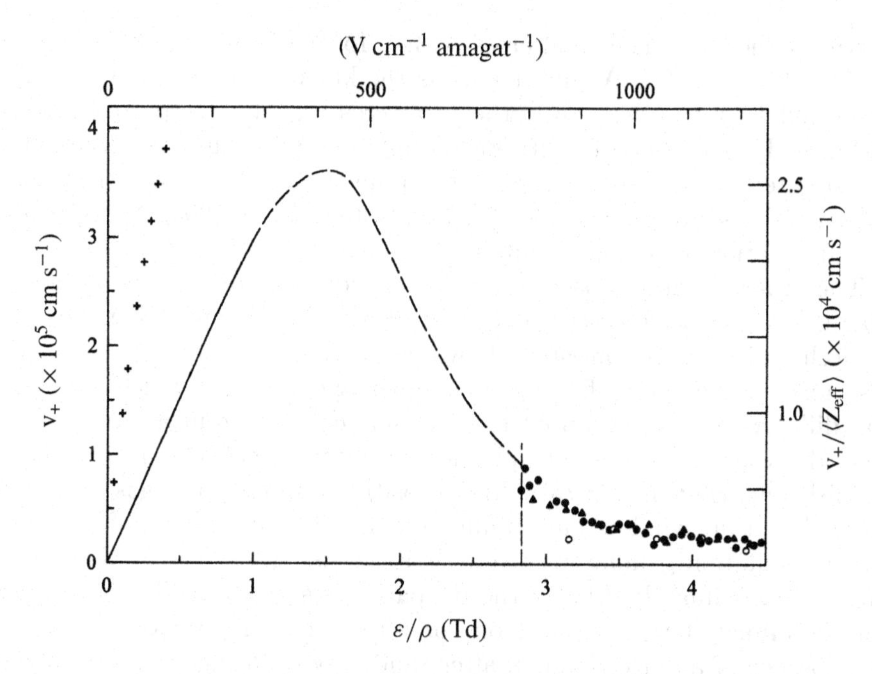

Fig. 6.18. Results from the positron-drift experiments of Paul and coworkers. The full curve is $v_+/\langle Z_{\text{eff}}\rangle$, obtained from a fit to all the data; see text for details. Electron drift velocities (+) are shown for comparison on the left-hand scale. The points to the right of the vertical broken line were taken from runs at molecular hydrogen pressures of 50 torr (•), 25 torr (▲) and 10 torr (○).

an analytic fit to F_{d}, equation (6.28), in terms of the parameter X. This parameter was density scaled, setting $X_{\text{c}} = X\rho/\rho_{\text{c}}$, to give the shape of the other curves from $F_{\text{d}}^{\text{c}} = X_{\text{c}}[1 - \exp(-1/X_{\text{c}})]$; X was then varied to optimize the fits at all the other pressures. Using values for X and $\langle Z_{\text{eff}}\rangle$ (the latter assumed to be independent of electric field), a curve of $v_+/\langle Z_{\text{eff}}\rangle$ was obtained, and this is shown in Figure 6.18 along with the derived values of v_+ and the corresponding values for electrons (Huxley and Crompton, 1974).

The broken-line portion of the $v_+/\langle Z_{\text{eff}}\rangle$ curve, which attains a maximum and then falls, was explained by Böse, Paul and Tsai (1981) in terms of the formation of positronium due to positron heating in the electric field, so that the apparent value of $\langle Z_{\text{eff}}\rangle$ rises as the amount of positronium formation increases. At high electric fields nearly all the positrons form positronium and do not annihilate at the foil.

As shown in Figure 6.18, electron drift velocities below $\epsilon/\rho = 1$ Td ($\equiv 10^{17}$ V cm^2) are at least four times larger than those for positrons. Böse, Paul and Tsai (1981) attributed this difference to higher momentum transfer cross sections for positrons than for electrons at very low (i.e.

thermal) energies. The total cross section for electrons is approximately 9×10^{-16} cm^2 at 0.2 eV and it falls as the kinetic energy is lowered. As described in subsection 2.6.1, the total cross section for positron scattering in H_2 rises rapidly with decreasing energy at the lowest energies investigated, lending some plausibility to the proposed explanation. Other measurements have been made with this system (Paul, 1993, 1995, private communications), though results for v_+ have not been extracted.

The other apparatus used to study positron drift in a range of molecular gases has been described by Charlton (1985b) and Charlton and Laricchia (1986). It consisted of two electrodes 10.35 mm apart, which also formed the walls of the gas chamber and between which a potential difference could be applied. Positrons emitted from a ^{22}Na source were detected using the thin plastic scintillator method (e.g. Coleman, Griffith and Heyland, 1973). Most positrons entered the gas chamber through a thin window and annihilated in the metal walls of the cell. The annihilation gamma-rays were detected using a large plastic scintillator. Approximately 0.1% of the β^+ particles stopped in the gas, where they annihilated as free positrons or after forming positronium or, in the absence of an electric field, after randomly diffusing to a wall of the chamber.

When an electric field was applied across the chamber some positrons annihilated prematurely, following field-induced drift to one of the electrodes. In this case the free-positron component of the lifetime spectrum was field dependent; the maximum drift time, $\tau_{\rm md}$, was given by the end-point of the lifetime spectrum and was due to thermalized positrons which had traversed the entire drift length l. The drift speed was then $v_+ = l/\tau_{\rm md}$ and the mobility could be found from

$$\mu_+ = \epsilon l/\tau_{\rm md}. \tag{6.29}$$

Charlton (1985b) selected O_2 and CO_2 gases for investigation. Experiments had shown (e.g. Wright, 1982) that the ortho-positronium component is rapidly quenched in the former, so that at 400 torr the lifetime is only around 30 ns. Thus, the free-positron spectrum is well separated, and despite the low signal-to-background ratio, $\tau_{\rm md}$ is relatively easy to discern. The latter gas was selected because, even at low densities, there is sufficient stopping power to ensure adequate statistical accuracy of the data. Examples of truncated lifetime spectra are given in Figure 6.19(a). Nevertheless, the major source of error in these measurements still lay in the determination of $\tau_{\rm md}$, which could only be evaluated with approximately 10%–15% accuracy.

The drift velocities for O_2 and CO_2 at various fields and at two different pressures are shown in Figures 6.19(b) and (c). The velocities were measured at values of ϵ/ρ in the ranges 1200–4500 V cm^{-1} amagat^{-1} for CO_2

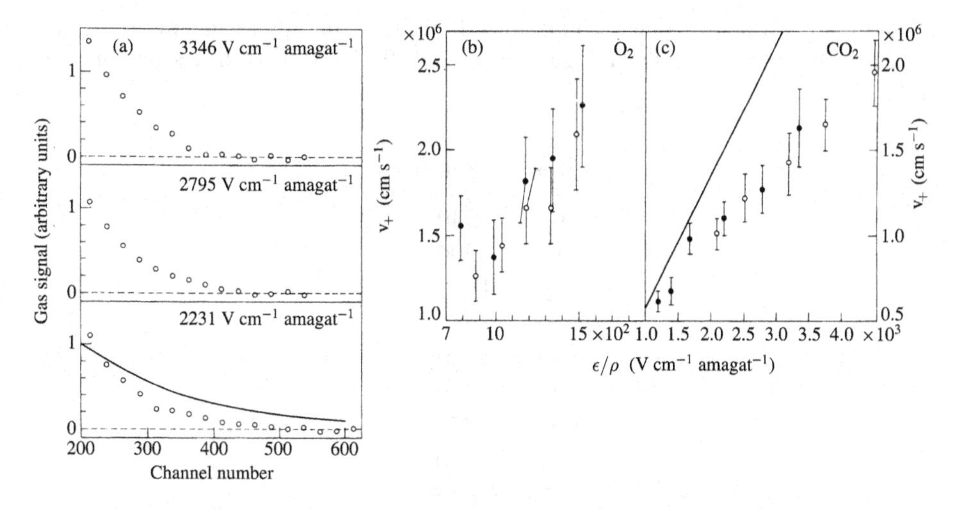

Fig. 6.19. (a) The free-positron component with electric fields applied to 0.26 amagat of CO_2, illustrating the truncation of the spectra. The solid curve is a schematic illustration of this component with zero applied field. (b) Positron drift velocities for O_2 at various density-normalized electric fields: •, 600 torr; ∘, 400 torr. (c) As (b), but for CO_2: •, 200 torr; ∘, 100 torr. The solid line represents the corresponding electron drift velocities.

and 800–1700 V cm^{-1} amagat^{-1} for O_2. It was assumed, with reference to the corresponding work over these ranges of ϵ/ρ for electrons, that the positrons remain in thermal equilibrium with the gas. The density-normalized values of the mobilities were found to be $(1.44 \pm 0.19) \times 10^3$ cm^2 V^{-1} s^{-1} amagat for O_2 and $(4.78 \pm 0.42) \times 10^2$ cm^2 V^{-1} s^{-1} amagat for CO_2. These results correspond to $(3.9 \pm 0.5) \times 10^{22}$ (V cm s)$^{-1}$ and $(1.29 \pm 0.11) \times 10^{22}$ (V cm s)$^{-1}$ respectively.

Also shown in Figure 6.19(c) is the line which corresponds to the average electron mobility for CO_2 in the same ϵ/ρ range. These data were extracted from the results of Peisert and Sauli (1984) and Christophorou (1971), and correspond to a density-normalized mobility of 600 cm^2 V^{-1} s^{-1} amagat. In the case of O_2 this quantity for electrons is approximately 4.5×10^3 cm^2 V^{-1} s^{-1} amagat (Crompton and Elford, 1973) at low ϵ/ρ, though this falls sharply to 2.8×10^3 cm^2 V^{-1} s^{-1} amagat as ϵ/ρ is increased to 10^3 V cm^{-1} amagat^{-1}. If one considers the ϵ/ρ-independent values for electrons, then in each case the mobility of electrons is greater than that of positrons. As was the case for H_2, this can probably be attributed to the relative behaviour of the momentum transfer cross sections at very low energies.

Several other molecular gases have been investigated using this technique, and the results presented by Charlton and Laricchia (1986) are

Table 6.4. Values of density-normalized positron mobility (in $cm^2\ V^{-1}\ s^{-1}$ amagat) for various molecular gases at $T = 297$ K. Uncertainties are around 10%–15%

Gas	Mobility
H_2	1940
D_2	2200
O_2	1440
N_2	1560
CO	1430
CO_2	480
CH_4	370
SF_6	300

given in table 6.4. In the case of H_2, we can compare the value 1940 $cm^2\ V^{-1}\ s^{-1}$ amagat for the mobility with the value 1100 $cm^2\ V^{-1}\ s^{-1}$ amagat derived from the work of Böse, Paul and Tsai (1981). The discrepancy between the two measurements is greater than the combined errors, though given the very different nature of the experiments and the difficulty in extracting drift velocities for positrons it is perhaps gratifying to find agreement within a factor of two. Further work on positron drift is desirable since it can complement beam measurements, particularly at energies below approximately 1 eV.

7
Positronium and its interactions

In this chapter we consider the physics of the positronium atom and what is known, both theoretically and experimentally, of its interactions with other atomic and molecular species. The basic properties of positronium have been briefly mentioned in subsection 1.2.2 and will not be repeated here. Similarly, positronium production in the collisions of positrons with gases, and within and at the surface of solids, has been reviewed in section 1.5 and in Chapter 4. Some of the experimental methods, e.g. lifetime spectroscopy and angular correlation studies of the annihilation radiation, which are used to derive information on positronium interactions, have also been described previously. These will be of most relevance to the discussion in sections 7.3–7.5 on annihilation, slowing down and bound states. Techniques for the production of beams of positronium atoms were introduced in section 1.5. We describe here in more detail the method which has allowed measurements of positronium scattering cross sections to be made over a range of kinetic energies, typically from a few eV up to 100–200 eV, and the first such studies are summarized in section 7.6.

Important advances continue to be made in measurements of the intrinsic properties of the positronium atom, e.g. its ground state lifetimes (Rich, 1981; Al-Ramadhan and Gidley, 1994; Asai, Orito and Shinohara, 1995) and various spectroscopic quantities (Berko and Pendleton, 1980, Mills, 1993; Hagena *et al.*, 1993). These are reviewed in section 7.1.

7.1 Fundamental studies with the positronium atom

1 Lifetimes against annihilation

Since the earliest work with positronium by Deutsch and coworkers (e.g. Deutsch, 1951; Deutsch and Brown, 1952) its annihilation lifetimes, or decay rates, have been studied both theoretically and experimentally.

The decay rate of ground state ortho-positronium in vacuum, ${}_0\lambda_{\text{o-Ps}}$, has been investigated more often than any other property of positronium, because it is relatively straightforward to produce the atom in abundance and the accurate measurement of its 142 ns lifetime, corresponding to an annihilation rate of 7.04 μs^{-1}, is well within technological capabilities. In contrast, the much shorter 125 ps lifetime of ground state para-positronium, corresponding to an annihilation rate of 8.0 ns^{-1}, has only been measured twice with a precision of better than 1%; this has been achieved by utilizing the mixing of the $m = 0$ states of ortho-positronium and para-positronium in a uniform magnetic field and assuming a value for the ground state hyperfine splitting. To our knowledge there have been no measurements of the decay rates for excited state positronium.

In Chapter 1, the first order contributions to the annihilation rates from the dominant modes of decay of the S-states of both ortho- and para-positronium (for arbitrary principal quantum number n_{Ps}) were given as equations (1.5) and (1.6). These contributions are included in the following equations for the rates for the two ground states, which also contain terms of higher order in the fine structure constant, α:

$$\begin{aligned} {}_0\lambda_{\text{o-Ps}} &= 2\alpha^6 mc^2 \frac{\pi^2 - 9}{9\pi\hbar}\left[1 - \frac{A\alpha}{\pi} - \frac{1}{3}\alpha^2 \ln\alpha^{-1} + B\left(\frac{\alpha}{\pi}\right)^2 \right. \\ &\qquad \left. - \frac{3\alpha^3}{2\pi}(\ln\alpha)^2 + \cdots\right], \qquad (7.1) \\ {}_0\lambda_{\text{p-Ps}} &= \frac{\alpha^5 mc^2}{2\hbar}\left[1 - \frac{\alpha}{\pi}\left(5 - \frac{\pi^2}{4}\right) + \frac{2}{3}\alpha^2\ln\alpha^{-1} + B\left(\frac{\alpha}{\pi}\right)^2 + \cdots\right]. \qquad (7.2) \end{aligned}$$

Note that, as can be seen from the discussion in subsection 1.2.1, the contributions from the higher order annihilation modes are negligible at the present levels of precision. Thus, the rate for the annihilation of ortho-positronium into five gamma-rays is only 10^{-6} of that for three gamma-rays, with a similar value for the ratio of the rates for para-positronium annihilation into four and two gamma-rays.

The most accurate determination of the coefficient A in the ortho-positronium decay rate has yielded the value 10.2866 (Adkins, Salahuddin and Schlam, 1992), giving a 2.3% change in the lowest order annihilation rate. For para-positronium, the corresponding first order correction is only 0.6%. The coefficient B multiplying the term $(\alpha/\pi)^2$ has been determined by Mil'stein and Khriplovich (1994) to have the value 46 for ortho-positronium (and 40 for para-positronium), producing a further change of approximately 250 ppm in the annihilation rate. Taking all these corrections into account, the most accurate theoretical value of the ortho-positronium annihilation rate is 7.0420 μs^{-1}, in very good agreement with the experimental value of Asai, Orito and Shinohara (1995); see below.

In the early 1970s theory and experiment seemed to be converging on a value for the ortho-positronium decay rate, ${}_0\lambda_{\text{o-Ps}}$, of approximately 7.24 μs^{-1}. However, this was called into doubt later in that decade by the work of Gidley and colleagues (Gidley, Marko and Rich, 1976; Gidley and Zitzewitz, 1978) who obtained values below 7.1 μs^{-1}. This discrepancy was partly resolved by the calculations of Caswell, Lepage and Sapirstein (1977) and Caswell and Lepage (1979), and the experimental measurement of Griffith *et al.* (1978b). However, as more accurate measurements were performed by the Michigan group during the next ten years or so, including some in vacuum using a positron-beam technique, a small but significant discrepancy with theory arose again. This would now appear to be resolved by the aforementioned measurements of Asai, Orito and Shinohara (1995). A similar conclusion had previously been obtained by Hasbach *et al.* (1987). These authors created positronium in vacuum in a manner similar to that developed by Gidley and Zitzewitz (1978) and latterly applied by Nico *et al.* (1990), but they used a very different counting technique. Asai, Orito and Shinohara (1995) formed ortho-positronium in low density silica powder, and derived a value for ${}_0\lambda_{\text{o-Ps}}$ using a novel method to account for pick-off annihilation. We describe some of these measurements below.

The apparatus used by Westbrook *et al.* (1987, 1989) is shown in Figure 7.1 and consists of a cylindrical gas chamber placed between the pole pieces of an electromagnet. The latter was used to provide a field of 0.68 T across the interaction region to increase the signal rate by causing all positrons with a forward momentum component to follow helical paths through the region viewed by the gamma-ray detectors. Although this field mixed the $m = 0$ substates, it did not alter the decay rate of the $m = \pm 1$ ortho-positronium states.

The β^+ particles were derived from a ^{22}Na source deposited onto a thin plastic scintillator coupled by a light pipe to a phototube. This provided start signals for the timing sequence with high efficiency. The stop was furnished by the annihilation gamma-rays detected using two semi-annular scintillators surrounding the gas chamber. The combined detection efficiency for the three-gamma ortho-positronium decay was found to be in the range 25%–50%. Tungsten annuli inside the chamber shielded the stop detector from the ^{22}Na source and from annihilations on the opposite walls of the chamber.

Gas could be admitted to the chamber through the tubing shown, which also served for evacuation, and the static gas sample was pumped out, flushed and recharged on a daily basis. Absolute pressures and temperatures were recorded every hour. Various other tests were made

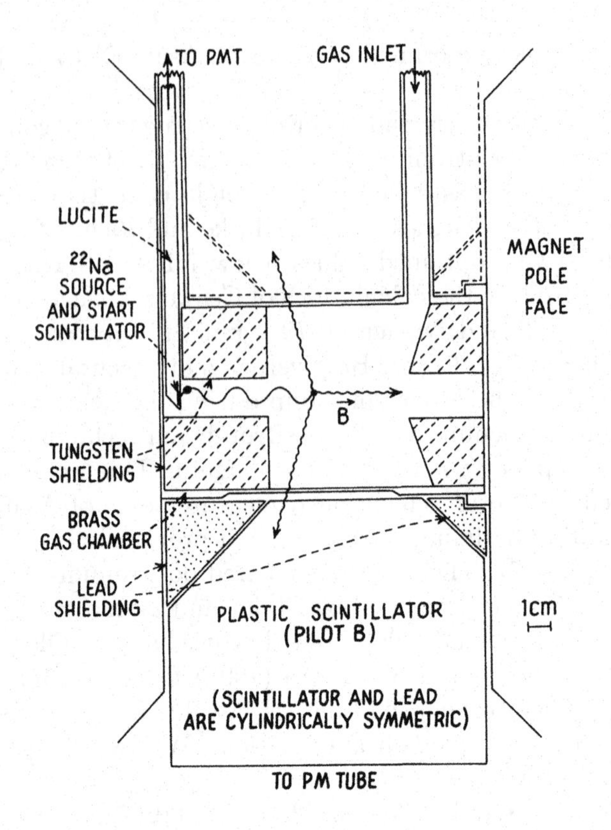

Fig. 7.1. Schematic illustration of the positronium formation chamber and detector arrangement used by Westbrook *et al.* (1987, 1989). Reprinted from *Physical Review* **A40**, Westbrook *et al.*, Precision measurement of the ortho-positronium vacuum decay rate using the gas technique, 5489–5499, copyright 1989 by the American Physical Society.

for gas leaks, contaminants from outgassing and for the thoroughness of preparation of gas mixtures (used in the cases of N_2 and neon buffer gases); the gas densities were computed from the relevant measurements with appropriate virial coefficient corrections. The electronics used for this experiment incorporated fast and slow timing systems, each with its own deliberately imposed dead time. Details are given by Westbrook *et al.* (1989).

In common with all recent measurements of ${}_0\lambda_{\text{o-Ps}}$, the raw data were corrected for background and then carefully analysed (see also e.g. Griffith *et al.*, 1978b; Hasbach *et al.*, 1987). Westbrook *et al.* (1989) used data from the spectrum which corresponded to times between, typically, 180 ns and 930 ns, the exponential fit being repeated at regular intervals out from the 100 ns point in steps of 8–10 ns. Two examples of the variation of the fitted decay rate with starting channel are shown in Figure 7.2 for

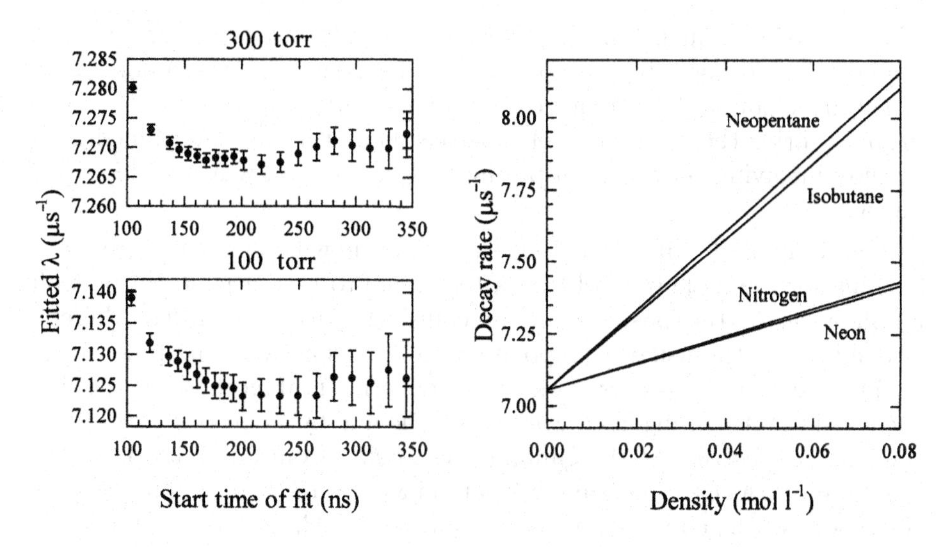

Fig. 7.2. The left-hand boxes show the fitted ortho-positronium decay rate at two values of isobutane pressure, for various start times of the fit to the component. The right-hand plot shows the observed decay rates, and their extrapolation to zero density, of Westbrook *et al.* (1989). The error bars on the individual points are approximately equal to the thickness of the line.

isobutane gas. Note that the fitted decay rate appears to settle down to a value independent of the starting channel by approximately 180 ns. The final fitted decay rates are also shown for four gases, together with the extrapolations to zero gas density. The final result of 7.0514 ± 0.0014 μs^{-1} is more than six standard deviations above the best theoretical value.

Support for these extrapolations also came from 'joining' the higher pressure points taken by Westbrook *et al.* (1989) for neon and N_2 to data taken much earlier for the same gases by Coleman *et al.* (1976a), using a completely different apparatus and data reduction technique, and over a range of much higher densities. Very good agreement was found with the extrapolations to zero density deduced from the 'low density' results.

Westbrook *et al.* (1989) considered a number of possible systematic effects which may have affected their data and caused the extrapolated value of ${}_0\lambda_{\text{o-Ps}}$ to be higher than the theoretical prediction. The two most difficult possibilities to account for were those related to the production of excited state positronium, Ps*, in the gas (see section 4.5 for a discussion of investigations of this phenomenon by Laricchia *et al.*, 1985) and those related to positronium thermalization (see section 7.4). At the time both were ruled out, but subsequent work on the latter effect (Skalsey *et al.*, 1998) has shown that the positronium was not completely thermalized, so that the energy dependence, or equivalently the temperature dependence, of the collisional quenching rate becomes an important parameter. The implications for possible corrections to the result of Westbrook *et al.* (1989) have not yet been finalized, although indications are that they will probably bring the value into better agreement with that determined from a study involving positronium production in vacuum, which is described below.

Asai, Orito and Shinohara (1995) used silica powder as the positronium-forming medium; they developed a technique to take account of pick-off annihilation of the positronium in collisions with the grains, thereby circumventing the usual extrapolation to zero density of the powder (or gas) inherent in most other work. Their experimental method consisted of the usual timing arrangement, but supplemented by a detector to measure the energy spectrum of the gamma-rays arising from annihilation of the positronium. As described in more detail in section 7.2 below, pick-off processes lead to the emission of two gamma-rays, each with 511 keV energy, in contrast to the three-gamma-ray vacuum decay of ortho-positronium, which has the continuous energy distribution illustrated in Figure 1.3. By using this energy distribution and a Monte Carlo simulation to allow for all gamma-ray interactions, both in the detector and the materials making up the apparatus, Asai, Orito and Shinohara (1995) were able to show that the expected and measured three-gamma-ray distributions were nearly indistinguishable. The ratio of the three-gamma-ray and two-gamma-ray events could thus be derived from the measured energy spectra and then time-correlated with the signal from the trigger detector to yield the ratio $\lambda_{po}(t)/{}_0\lambda_{o\text{-}Ps}$, where $\lambda_{po}(t)$ is the time-dependent pick-off annihilation rate. Once this parameter was derived, it could be used to fit the time spectrum, which had the form

$$N(t) = N_0 \exp\left[-{}_0\lambda_{o\text{-}Ps} \int_0^t \left(1 + \lambda_{po}(t')/{}_0\lambda_{o\text{-}Ps}\right) dt'\right]. \tag{7.3}$$

Background contributions and relative detection efficiencies had to be taken into account also; see Asai, Orito and Shinohara (1995). The final result, 7.0398 ± 0.0025 (stat.) ± 0.0015 (sys.) μs^{-1}, obtained for ${}_0\lambda_{o\text{-}Ps}$ is in agreement with the latest theoretical value but differs from the gas extrapolation result from Michigan. Asai, Orito and Shinohara (1995) offer some speculations as to potential inaccuracies in the gas extrapolation method but, given the new results on positronium thermalization and related phenomena (see section 7.4), it is not appropriate to add further comment here. Furthermore, the results for ${}_0\lambda_{o\text{-}Ps}$ from the Michigan gas experiments have, to date, been validated by experiments performed by the same group in vacuum using a low energy positron beam.

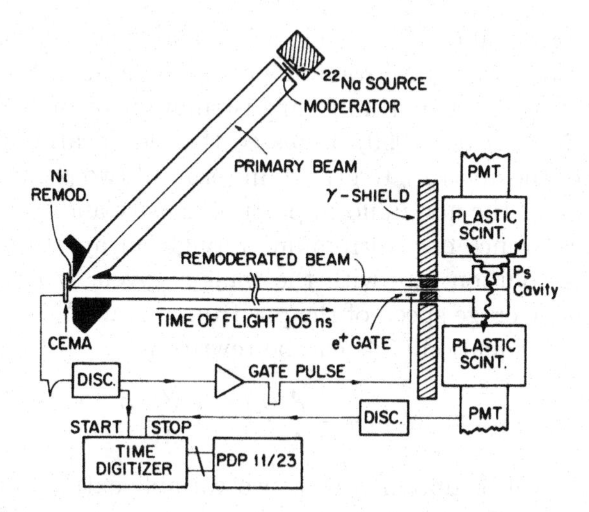

Fig. 7.3. Schematic illustration of the timed and gated slow positron beam used by Nico *et al.* (1990) to measure the vacuum decay rate of ortho-positronium. Reprinted from *Physical Review Letters* **65**, Nico *et al.*, Precision measurement of the ortho-positronium decay rate using the vacuum technique, 1344–1347, copyright 1990 by the American Physical Society.

The Michigan apparatus (Nico *et al.*, 1990), shown in Figure 7.3, consisted of a time 'tagged' positron beam, in which positronium was formed in a cavity lined with magnesium oxide. The primary low energy positrons were focussed onto a nickel foil remoderator, and secondary electrons liberated in this process were detected by the CEMA and used to start the timing sequence. This signal also opened an electrostatic gate to allow only the timed positrons to enter the cavity, thereby eliminating background from the untagged remoderated beam (85% of the total). A quarter of the positrons which entered the cavity at 700 eV formed positronium on the magnesium oxide surface, which also served to minimize quenching in collisions with the wall. The annihilation gamma-rays were detected by two semi-annular fast plastic scintillation counters arranged around the cavity. Details of the electronics used, and analysis of the results, were given by Nico *et al.* (1990).

Systematic effects arising from the disappearance of ortho-positronium through the cavity entrance aperture, and the rate of annihilation by collisions with the cavity walls, were taken into account by expressing the measured annihilation rate as

$$\lambda = {}_0\lambda_{\text{o-Ps}} + c_{\text{e}}(A'/S)\nu + P_{\text{a}}\nu, \tag{7.4}$$

where S is the cavity surface area, A' is the effective area of the cavity entrance, ν is the collision rate of the ortho-positronium with the wall, on which it has a probability P_a of annihilation, and c_e is the probability that a gamma-ray from ortho-positronium which has escaped from the cavity will not be detected. For a uniform distribution of ortho-positronium in the cavity, Nico *et al.* (1990) discussed the modification of the actual physical area of the aperture to give A' in terms of two parameters, one of which accounts for the non-uniform particle density and the other for the random disappearance of positronium through an aperture of non-zero thickness. The collision rate with the walls is given by $\nu = \langle v \rangle S/(4V)$, where $\langle v \rangle$ is the average speed of the positronium and V is the volume of the cavity. Thus, equation (7.4) can be rewritten as

$$\lambda = {}_0\lambda_{\text{o-Ps}} + \frac{c_e A' \langle v \rangle}{4V} + \frac{P_a \langle v \rangle S}{4V}, \tag{7.5}$$

and ${}_0\lambda_{\text{o-Ps}}$ can then be obtained by a two-variable extrapolation in A'/V and S/V. Figure 7.4 shows the results of these extrapolations. The two intercepts, 7.0497 ± 0.0013 μs^{-1} and 7.0482 ± 0.0015 μs^{-1}, are in good accord but the theoretical value, shown on each plot for comparison, is around five standard deviations lower. The possibility that various systematic effects influence the results, most notably the formation at the surface of excited state positronium, was considered, though all were discounted as having a negligible effect on the final values.

It should be clear from the preceding discussion of the experimental situation that there are still unresolved issues that warrant further work on the measurement of ${}_0\lambda_{\text{o-Ps}}$ and on consideration of the related experimental systematic errors.

In addition to the decay rate for the triplet state of positronium, it is important to consider the rate for the singlet state, although, as mentioned above, its 125 ps lifetime precludes a direct determination of ${}_0\lambda_{\text{p-Ps}}$ at present. This is mainly due to the difficulty in isolating the para-positronium component from the signal obtained from other rapid positron annihilation mechanisms. Other techniques, however, have been successfully applied, the first determination being by Theriot *et al.* (1970); they derived a value 7.99 ± 0.11 ns^{-1} from the width of the radio-frequency resonance in a measurement of the positronium ground state hyperfine splitting (see subsection 7.1.2 below). The latest theoretical results are 7989.5 μs^{-1} (Khriplovich and Yelkhovsky, 1990) and 7986.7 μs^{-1} (Caswell and Lepage, 1979), the difference being due to a discrepancy in the calculated coefficient of the $\alpha^2 \ln \alpha^{-1}$ term.

The most recent experimental determination of ${}_0\lambda_{\text{p-Ps}}$ is that of Al-Ramadhan and Gidley (1994). The apparatus and analysis techniques are similar to those of Westbrook *et al.* (1989) and will therefore not be described here. Their method used the effect of singlet–triplet mixing in a static magnetic field (Gidley *et al.*, 1982); this allowed ${}_0\lambda_{\text{p-Ps}}$ to be

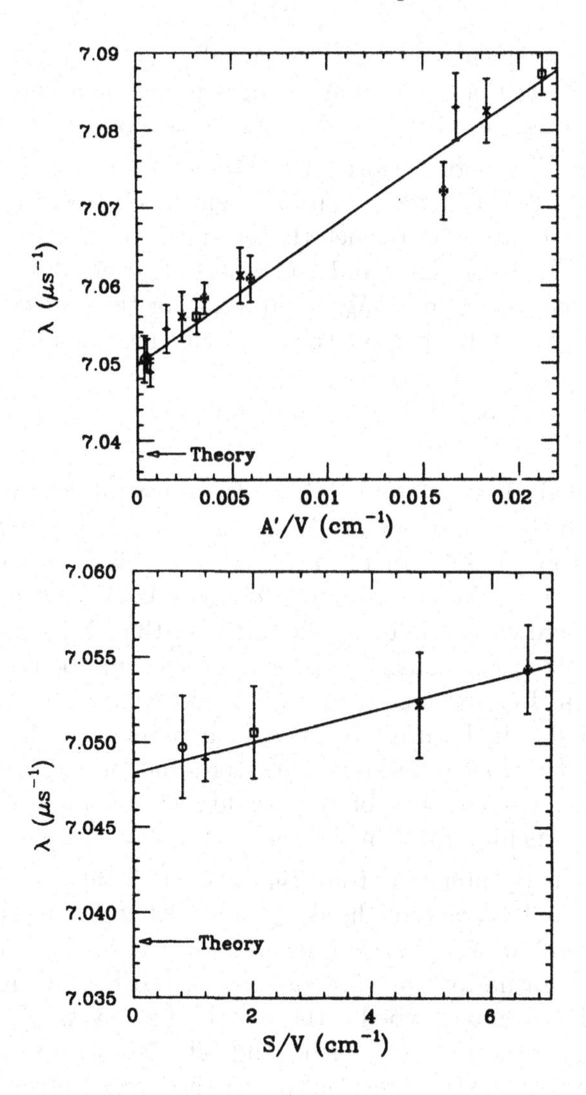

Fig. 7.4. Extrapolations of the measured ortho-positronium decay rates in A'/V and S/V (see text). Reprinted from *Physical Review Letters* **65**, Nico *et al.*, Precision measurement of the ortho-positronium decay rate using the vacuum technique, 1344–1347, copyright 1990 by the American Physical Society.

extracted from ${}_0\lambda'_{\text{o-Ps}}$, the decay rate of the mixed $m = 0$ states. Rich (1981, and references therein) showed that the perturbed ortho-positronium vacuum decay rate can be written as

$$ {}_0\lambda'_{\text{o-Ps}} = (1 - b^2)\,{}_0\lambda_{\text{o-Ps}} + b^2\,{}_0\lambda_{\text{p-Ps}}, \tag{7.6}$$

where $b = y^2/(1 + y^2)$, $y = x/[1 + (1 + x^2)^{1/2}]$, $x = 2g'\mu_0 B/h\Delta\nu_{\text{hfs}} \approx B/3.65$ tesla, $g' = g(1 - 5\alpha^2/24)$ and $\Delta\nu_{\text{hfs}}$ is the ground state, zero field, hyperfine frequency interval. A typical value for ${}_0\lambda'_{\text{o-Ps}}$ is 30 μs^{-1} in a field of 0.4 T, so that ${}_0\lambda_{\text{p-Ps}}$ can be determined at a known fixed field by measuring this quantity and assuming values for ${}_0\lambda_{\text{o-Ps}}$ and $\Delta\nu_{\text{hfs}}$.

The experiments were performed at two values of the magnetic field, 0.375 T and 0.425 T, and at various densities of N_2 gas with small admixtures of isobutane to quench the free-positron component (see subsection 6.3.2). Al-Ramadhan and Gidley (1994) derived a quantity $\Lambda(\rho)$ from their measured values of $\lambda'_{\text{o-Ps}}$ and $\lambda_{\text{o-Ps}}$, for the mixed and unmixed ortho-positronium states respectively, at a gas density ρ given by

$$\Lambda(\rho) = [\lambda'_{\text{o-Ps}}(\rho) - \lambda_{\text{o-Ps}}(\rho)]/b^2 + {}_0\lambda_{\text{o-Ps}}, \tag{7.7}$$

where b, defined above, parameterizes the degree of mixing induced by the magnetic field. The value of ${}_0\lambda_{\text{o-Ps}}$ was taken to be 7.0482 ± 0.0016 μs^{-1} (Nico *et al.*, 1990). In the absence of a collisional spin-exchange mechanism the quenching coefficients for the mixed and unmixed ortho-positronium were expected to be identical, so that $\Lambda(\rho)$ could be taken as a measure of ${}_0\lambda_{\text{o-Ps}}$. Thus, the measurements could be biassed towards higher gas densities for increased statistical accuracy. A plot of $\Lambda(\rho)$ versus ρ is shown in Figure 7.5. The straight line which was fitted to these data had a slope consistent with zero and provided experimental confirmation of the equality of the perturbed and unperturbed ortho-positronium quenching rates in the gas.

The zero-density intercept from Figure 7.5 is 7990.3 ± 3.1 μs^{-1}, but a simple weighted average of the data gave 7990.9 ± 1.0 μs^{-1}, with the error due to statistics only. This procedure was justified by obtaining independent information on the equality of the quenching coefficients for perturbed and unperturbed ortho-positronium by searching for a difference in $\lambda_{\text{o-Ps}}(\rho)$ with the field on and off. No significant effect was found, although an extra contribution to the overall error was thereby incorporated. Other errors included in the final analysis were due to time calibration and linearity of the lifetime spectra, the magnetic field measurement, averaging over the magnetic fields sampled by the ortho-positronium (necessary because of slight positional variations) and a small correction for the possibility that there was more than one positronium atom in the chamber following the start of the timing system. The final result was given by Al-Ramadhan and Gidley (1994) as ${}_0\lambda_{\text{p-Ps}} = 7990.9$ ± 1.7 μs^{-1}, which, with an error of 215 ppm, is of similar accuracy to the precision measurements of ${}_0\lambda_{\text{o-Ps}}$ and capable of distinguishing between the two calculated values of the coefficient of the $\alpha^2 \ln \alpha^{-1}$ term in equation (7.2): the result finds in favour of the later work, that of Khriplovich and Yelkhovsky (1990).

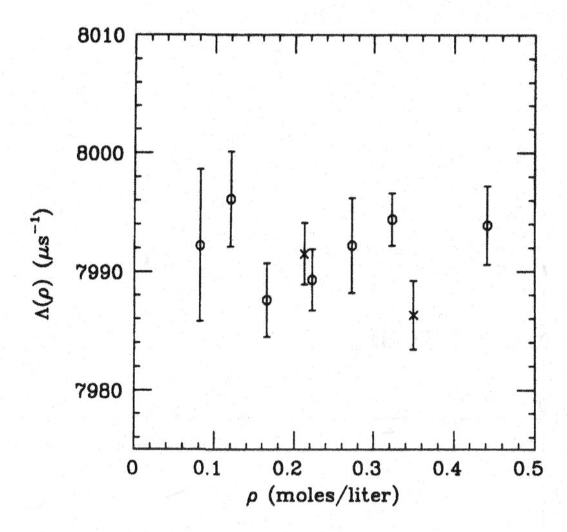

Fig. 7.5. Plot of the measured $\Lambda(\rho)$, as defined by equation (7.7), in N_2–isobutane mixtures. Data were taken at magnetic fields of 0.375 T ($\times$) and 0.425 T ($\circ$). Reprinted from *Physical Review Letters* **72**, Al-Ramadhan and Gidley, New precision measurement of the decay rate of singlet positronium, 1632–1635, copyright 1994 by the American Physical Society.

2 Spectroscopic properties

Experimentally, the first property of positronium to be investigated was the ground state hyperfine structure (hfs), as described in the excellent review of Rich (1981). The historic measurements of Deutsch and Dulit (1951) first verified the necessity of incorporating the virtual annihilation term into the calculation of the energy separation, and soon afterwards Deutsch and Brown (1952) made a relatively precise measurement of $\Delta\nu_{\mathrm{hfs}}$ using a technique employed later in more accurate experiments by groups at Brandeis and Yale Universities. The most recent values for this quantity are 203.3875 ± 0.0016 GHz (Mills and Bearman, 1975) and 203.389 10 ± 0.000 57 ± 0.000 43 GHz (Egan, Hughes and Yam, 1977). See also Hughes (1998), for a recent review. The above values can be compared with the theoretical result (Karplus and Klein, 1952; Bodwin and Yennie, 1978; Caswell and Lepage, 1979)

$$\Delta\nu_{\mathrm{hfs}} = \frac{\alpha^4 mc^2}{4\pi\hbar}\left[\frac{7}{3} - \frac{\alpha}{\pi}\left(\frac{32}{9} + 2\ln 2\right) + \frac{5}{6}\alpha^2 \ln(\alpha^{-1}) + O(\alpha^2)\right], \quad (7.8)$$

which yields 203.400 GHz. Rich (1981) stated that uncalculated terms of order α^2 would contribute around 7 MHz to this value if their coefficients were unity. This has been borne out by recent work; see e.g. Czarnecki, Melnikov and Yelkhovsky (1999) and Pachucki and Karshenboim (1998).

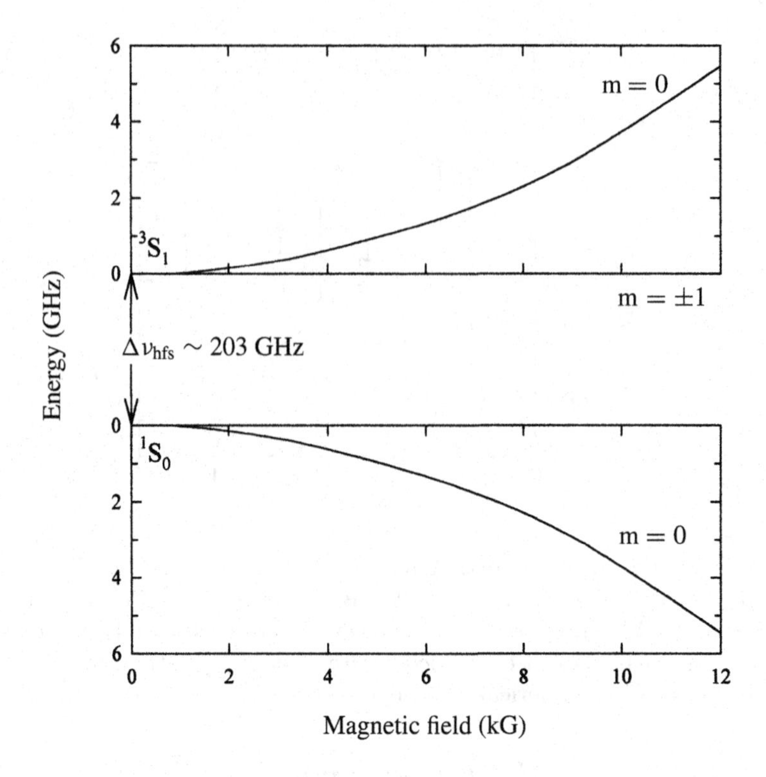

Fig. 7.6. The energy levels of ground state positronium in a magnetic field.

The value found is 203.392 01(46) GHz, around three standard deviations from experiment. Here we will describe only the Yale experiment in detail, since that of the Brandeis group is similar in many respects. A summary of the latter can be found in the review of Berko and Pendleton (1980). Note that the values quoted above were both corrected upwards by Mills (1983c), who, acting on a suggestion of Rich (1981), properly took into account the deviation, caused by the effects of annihilation, of the line shape from a true Lorentzian. Shifts of order $({}_0\lambda_{\text{p-Ps}}/4\pi\Delta\nu_{\text{hfs}})^2 \approx 10^{-5}$ were found, resulting in corrections of 2.5 ppm and 21 ppm to the values of Mills and Bearman (1975) and of Egan, Hughes and Yam (1977) respectively.

An illustration of the principle behind the technique is given in Figure 7.6, which shows the ground state positronium energy levels in a static magnetic field. In the experiment a field of around 0.7–1.0 T was chosen, and transitions between the unperturbed $m = \pm 1$ states and the $m = 0$ state were induced by applying a radio frequency magnetic field perpendicular to the static field. For convenience, the magnitude of the static field was varied, rather than the radio frequency, in the search for

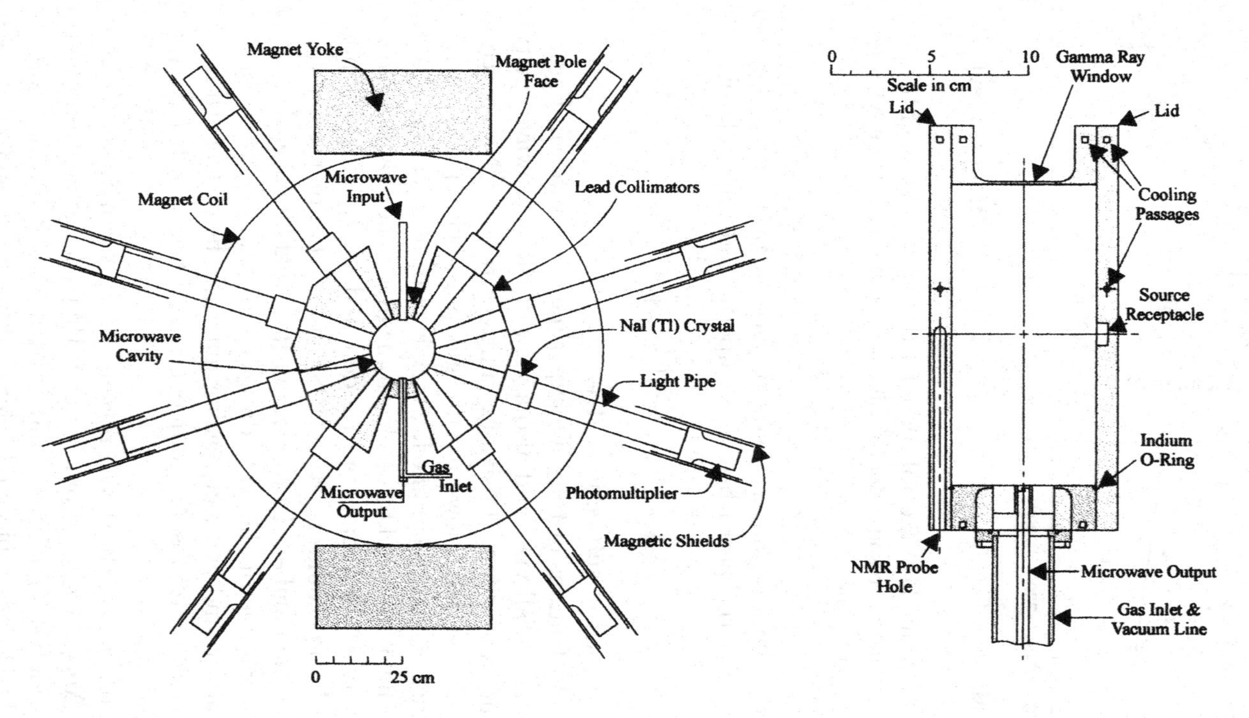

Fig. 7.7. Schematic of the Yale experiment to measure the hyperfine interval of ground state positronium. On the left there is an overall view of the apparatus, whilst the right-hand figure shows the microwave cavity and gas cell in which the positronium is formed.

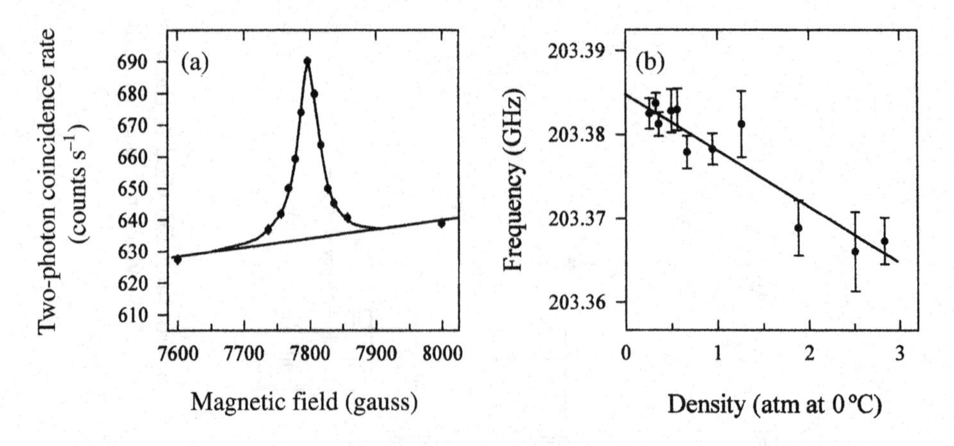

Fig. 7.8. (a) An example of the Yale data for the determination of the positronium hyperfine structure resonance. (b) Linear extrapolation to zero gas density of the measured hyperfine interval.

the resonance. The location of this resonance is governed by the energy difference between the perturbed and unperturbed triplet levels and is given by

$$\frac{1}{2}\Delta\nu_{\rm hfs}[(1+x^2)^{1/2}-1], \tag{7.9}$$

where x was defined in subsection 7.1.1.

Schematic illustrations of the Yale experiment are given in Figure 7.7, which shows an overall view and also a view of the microwave cavity and the source and gas region. Positrons emitted from a ^{22}Na source form positronium in a suitable low-pressure gas (e.g. helium or neon). The volume in which the gas is confined, which is also the microwave cavity, is located in the field produced by an electromagnet. The quantity measured is the number of 511 keV gamma-rays detected using pairs of oppositely situated scintillation counters. Rich (1981) noted that on passing through the resonance, the number of two-gamma-ray events increased by around 10%, and an example of the Yale data is shown in Figure 7.8(a). The rise in the 'background' with increasing magnetic field strength is due to the small increase in off-resonance quenching of the $m = 0$ $1^3\mathrm{S}_1$ state of ortho-positronium.

Inspection of Figure 7.8(a) also reveals one of the major difficulties in determining $\Delta\nu_{\rm hfs}$ to a few ppm, namely that the natural line width is large, approximately 6200 ppm, and extremely fine line splitting must be accomplished. This topic, and related systematics, were discussed in detail by Rich (1981), who noted that the accord between the two groups, who used different data analysis techniques, is important confirmation of the reported accuracy.

The other major uncertainty in the measurements arises from collisions of the ortho-positronium with the buffer gas, the main effect of which is to lead to a shift in the value of $\Delta\nu_{\mathrm{hfs}}$. In fact, the dominant mechanism involved is the attractive long-range van der Waals force, which tends to increase the positron–electron separation in positronium, thus lowering the splitting. This is shown explicitly in Figure 7.8(b) (Egan, Hughes and Yam, 1977), where a linear extrapolation to zero gas density was made (see also Mills and Bearman, 1975).

Laser spectroscopy of the 1S–2S transition has been performed by Mills and coworkers at Bell Laboratories (Chu, Mills and Hall, 1984; Fee *et al.*, 1993a, b) following the first excitation of this transition by Chu and Mills (1982). Apart from various technicalities, the main difference between the 1984 and 1993 measurements was that in the latter a pulse created from a tuned 486 nm continuous-wave laser with a Fabry–Pérot power build-up cavity, was used to excite the transition by two-photon Doppler-free absorption, followed by photoionization from the 2S level using an intense pulsed YAG laser doubled to 532 nm. Chu, Mills and Hall (1984), however, employed an intense pulsed 486 nm laser to photoionize the positronium directly by three-photon absorption from the ground state in tuning through the resonance. For reasons outlined by Fee *et al.* (1993b), it was hoped that the use of a continuous-wave laser to excite the transition would lead to a more accurate determination of the frequency interval than the value $1\,233\,607\,218.9 \pm 10.7$ MHz obtained in the pulsed 486 nm laser experiment (after correction by Danzmann, Fee and Chu, 1989, and adjustment consequent on a recalibration of the Te_2 reference line by McIntyre and Hänsch, 1986).

The experimental arrangement of Fee *et al.* (1993a, b) is shown schematically in Figure 7.9. A 15 ns burst of approximately 10^4 low energy positrons, produced by the bunched output of a microtron-based beam (Mills *et al.*, 1989b), was incident upon a heated single crystal of aluminium. Here it produced positronium in vacuum, mainly by the thermal desorption of surface-trapped positrons, as described in subsection 1.5.3. The background at the CEMA detector, which registered the laser-ionized positrons, was reduced by pre- and post-skimmers. The pre-skimmer was positioned so that ionized positrons would be extracted only from the region below the target to where the neutral 2S positronium could migrate and still interact with the YAG laser. The ionized positrons were drifted in the magnetic field to the CEMA, where they could be distinguished from background by their time of flight relative to the pulsed YAG laser. Figure 7.10 shows the positronium 1S–2S resonance curve, along with the Te_2 resonance line used for calibration. A fit to these data yielded a value $1\,233\,607\,216.4 \pm 3.2$ MHz for the frequency interval (Fee *et al.*, 1993a, b). This is in good accord with the value $1\,233\,607\,221.7$ MHz

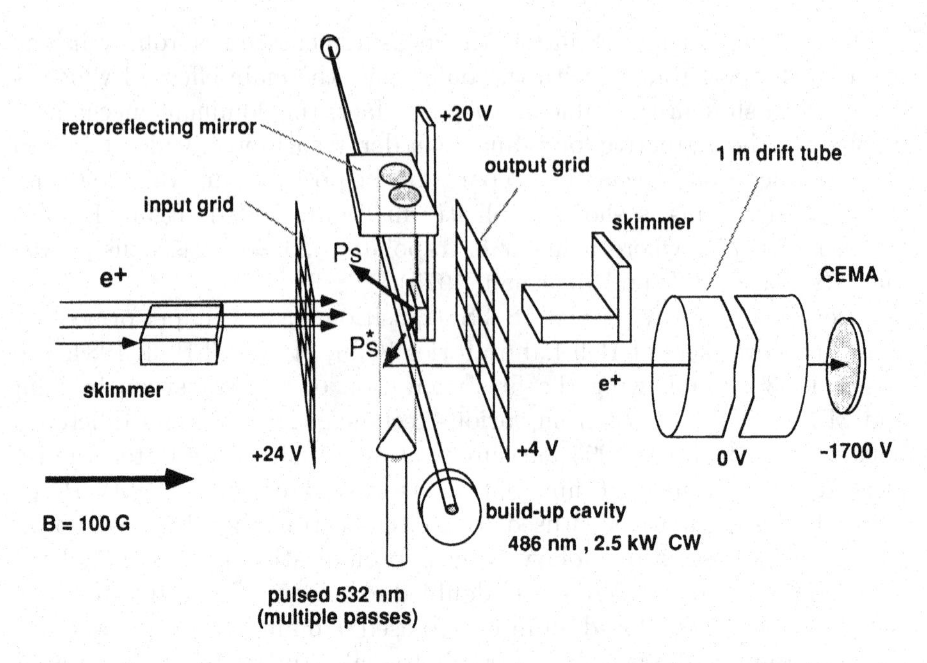

Fig. 7.9. The positronium source and laser interaction region used by Fee and coworkers. A magnetic field of 100 G (0.01 T) guides the incident beam onto the target and also transports the positrons liberated from the photoionized 2S positronium to the detector. Reprinted from *Physical Review Letters* **70**, Fee *et al.*, Measurement of the positronium 1^3S_1–2^3S_1 interval by continuous wave two-photon excitation, 1397–1400, copyright 1993 by the American Physical Society.

(Fell, 1992; Khriplovich, Milstein and Yelkhovsky, 1992) obtained from the theoretical expression for the frequency:

$$\begin{aligned}\nu_{21} &= [E(2^3\mathrm{S}_1) - E(1^3\mathrm{S}_1)]h^{-1} \\ &= cR_\infty \left[\frac{3}{8} - \frac{719}{1536}\alpha^2 + \frac{\alpha^3}{\pi}\left\{ \frac{161}{960} - \frac{21}{16}\ln 2 - \frac{21}{16}\ln \alpha^{-1} \right.\right. \\ &\qquad \left.\left. - \frac{1}{6}\ln 2.811\,77 + \frac{4}{3}\ln 2.984\,129 \right\} - \frac{7}{48}\alpha^4 \ln \alpha^{-1} \right]. \end{aligned} \tag{7.10}$$

The term of order α^4 has been calculated by Pachuki and Karshenboim (1998) and has shifted ν_{21} downwards by 0.7 MHz (with an estimated error of 1.0 MHz) from the value quoted above.

The first crude spectroscopic measurement performed on excited state positronium was the identification of the 243 nm Lyman-α radiation emitted in the 2P–1S transition. The first observation of this line was

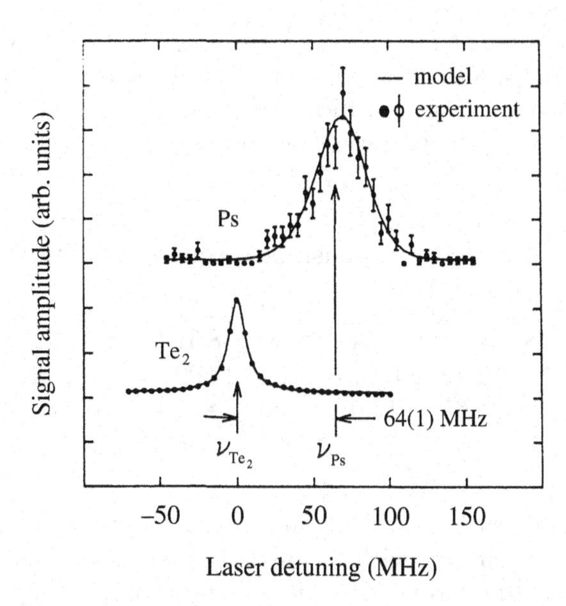

Fig. 7.10. The positronium 1^3S_1–2^3S_1 resonance curve, together with the Te_2 calibration line. Details of the model fit to the experimental data can be found in Fee *et al.* (1993b).

obtained in the slow-positron-beam studies of Canter, Mills and Berko (1975), which followed two decades of unsuccessful attempts, by a variety of workers, based upon traditional methods, e.g. positrons stopping in solids and high density gases. These measurements were catalogued by Day (1993) and need not concern us further here.

The basic principle of the experiment of Canter, Mills and Berko (1975) was to collide low energy positrons with a surface and to look for coincidence between a Lyman-α photon and a delayed gamma-ray arising from the subsequent annihilation of a 1^3S positronium. The presence of the Lyman-α signal was verified by the use of three interference filters with pass bands centred on, just above, and just below, 243 nm. An enhanced coincidence rate was found with the 243 nm filter in place. A similar Lyman-α gamma-ray technique has been adopted by all subsequent workers in this field (e.g. Laricchia *et al.*, 1985; Hatamian, Conti and Rich, 1987; Ley *et al.*, 1990; Schoepf *et al.*, 1992; Steiger and Conti, 1992; Hagena *et al.*, 1993; Day, Charlton and Laricchia, 2000).

To date, it has been found that the positronium formation potential, see equation (1.14), is positive for $n_{\rm Ps} > 1$, so that excited state positronium emission by a work function process is forbidden. Only positrons which reach a surface epithermally (which occurs predominantly at low impact energies where they do not penetrate far into the solid) may

Table 7.1. Summary of the available theoretical and experimental data for the 2^3S_1–2^3P_J transitions of positronium. Key: experiment, (a) Hagena *et al.* (1993), (b) Hatamian, Conti and Rich (1987) (the first-quoted errors are statistical whilst the second-quoted are systematic); theory, (c) Pachuki and Karshenboim (1998)

Transition	Transition frequencies (MHz)		
	(a)	(b)	(c)
$2^3S_1 \rightarrow 2^3P_2$	8 624.38 ± 0.54 ± 1.40	8 619.6 ± 2.7 ± 0.9	8 626.87
$2^3S_1 \rightarrow 2^3P_1$	13 012.42 ± 0.67 ± 1.54	13 001.3 ± 3.9 ± 0.9	13 012.58
$2^3S_1 \rightarrow 2^3P_0$	18 499.65 ± 1.20 ± 4.00	18 504 ± 10.0 ± 1.7	18 498.42

form positronium with $n_{Ps} > 1$. Such positronium atoms are probably emitted from the surface, with a range of kinetic energies which reflects the energy of the positrons as they return to the surface, and also the energy dependence of the capture process; it is expected to be several eV wide. This energy spread influences spectroscopic investigations, owing to the Doppler effect. Nevertheless, frequencies for the 2^3S_1–2^3P_J ($J =$ 0, 1, 2) transitions have been measured, and table 7.1 summarizes the available experimental and theoretical data.

Since the principle of the experiments is the same in all cases, we will restrict the discussion to the work of Hagena *et al.* (1993); a schematic illustration of the apparatus they used to determine the 2^3S_1–2^3P_J transitions is shown in Figure 7.11(a). Positrons produced by pair production at the Giessen Electron Linac (Faust *et al.*, 1991) were, after moderation, incident at 100 eV upon a molybdenum surface inside a microwave guide. To reduce motional Stark effects, this part of the apparatus was removed from the axial magnetic guiding field used at the linac beam, so that the residual field was around 0.3 ± 0.1 mT. Lyman-α photons were observed through the grids using a solar blind phototube, located at the end of a 25 cm light guide, to reduce background from the detection of annihilation gamma-rays.

The signal, S, for the fine structure transition was obtained as a ratio involving the numbers of counts with the microwave power on and off: $S = (R_{on} - R_{off})/R_{off}$. The value of R_{off} included contributions from annihilation photons and from Lyman-α photons produced from direct 2P de-excitation, with a maximum increase of around 7% due to the 2^3S_1–2^3P_J microwave-induced transition (followed by Lyman-α emission). A representative example, shown in Figure 7.11(b), is that for the 2^3S_1–2^3P_0 transition at 18 500 MHz. Discussion of the observed width of the lines was given by Hagena *et al.* (1993).

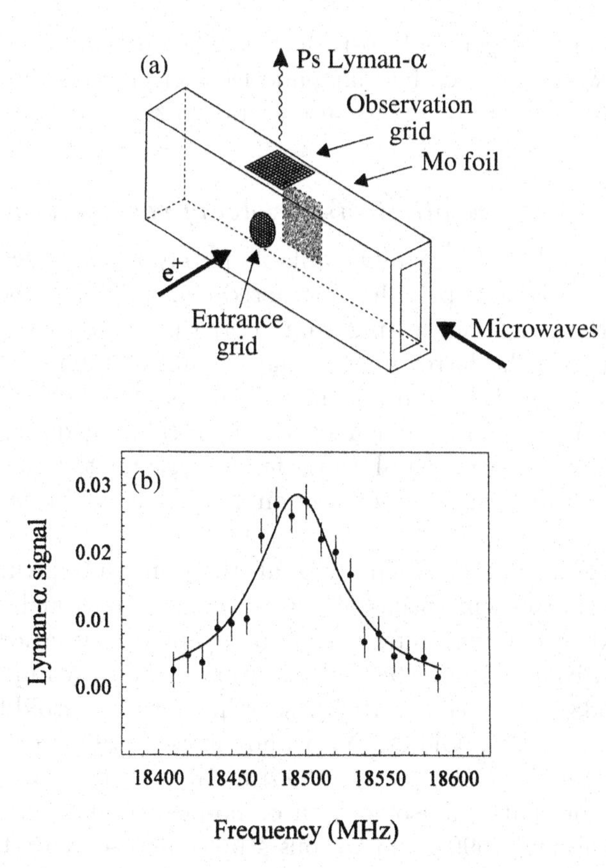

Fig. 7.11. (a) Schematic illustation of the apparatus of Hagena *et al.* (1993) used to observe the 2^3S_1–2^3P_J transitions in positronium. (b) Resonance curve of the 2^3S_1–2^3P_0 transition.

3 Non-spectroscopic laser studies of positronium

Ziock *et al.* (1990a), using the pulsed positron source at the Lawrence Livermore Laboratory, USA, optically saturated the one-photon (1^3S–2^3P) transition. One notable feature of this work was the use of a dye laser modified to obtain light with a bandwidth sufficient to cover a significant fraction of the Doppler profile of the positronium, which also allowed all 2^3P_J states to be accessed. This accomplishment paved the way for Ziock *et al.* (1990b) to make the first observation of the resonant excitation of a high-n_{Ps} Rydberg state of positronium. This was achieved by shining a second laser, again with a large bandwidth, onto the pumped atoms. The laser could be tuned into the red region of the spectrum, and evidence for excitation of some of the levels with n_{Ps} = 13–19 was obtained. Finally, it is noted that optical saturation of the 1S–2P transition is also the basis for a scheme to laser-cool positronium. The desirability and feasibility of so doing were discussed by Liang and Dermer (1988), who also considered appropriate cooling rates; see also section 8.2.

4 Exotic tests involving positronium

Positronium, being a readily available purely leptonic system and also a particle–antiparticle pair, has attracted considerable experimental interest over the years as a testing ground for the existence of exotic particles or couplings. The latter may perhaps manifest themselves in the decay properties of positronium, so that attempts have been made to observe forbidden modes. In particular, the longstanding discrepancy between the Michigan experimental value for ${}_0\lambda_{\text{o-Ps}}$ and the results from QED calculations, described in subsection 7.1.1, has acted as a spur to such investigations.

Although an in-depth survey of all these measurements is beyond the scope of the present discussion, we present a partial list to which the interested reader may refer. A summary of the situation up to around 1980 was given by Rich (1981). Later work includes symmetry tests (Arbic *et al.*, 1988; Conti *et al.*, 1993), searches for the forbidden two-gamma (Asai *et al.*, 1991; Gidley, Nico and Skalsey, 1991; Nico *et al.*, 1992) and four-gamma (Yang *et al.*, 1996) annihilations of ortho-positronium, a search for spatial anisotropy in ortho-positronium annihilation (Mills and Zuckerman, 1990), and various studies and searches for the emission of particles other than gamma-rays in the decay of ortho-positronium (Gninenko *et al.*, 1990; Gninenko, 1994; Orito *et al.*, 1989; Mitsui *et al.*, 1993; Adachi *et al.*, 1994; Asai *et al.*, 1994, and references therein). It is sufficient for the present purpose to say that no evidence for any forbidden decay modes or new particles has been forthcoming within the limits set by the relevant experiments.

7.2 Theoretical aspects of annihilation and scattering in gases

1 Annihilation

During the collision of positronium with an atom or molecule, the positron finds itself in close proximity to the target electrons as well as to its companion electron. This results in an enhanced probability of annihilation with an electron in a relative spin singlet state, leading to the emission of two gamma-rays. This is termed quenching, and a number of different processes may occur, depending upon the chemical nature of the atom or molecule. For ground state para-positronium, with its lifetime of only 125 ps, the enhancement is very small, but this is not the case for ortho-positronium. As outlined in section 6.2, where the observable parameters in a lifetime experiment were discussed, the annihilation rate of ortho-positronium, $\langle\lambda_{\rm p}\rangle$, is often written as

$$\langle\lambda_{\rm p}\rangle = {}_0\lambda_{\text{o-Ps}} + \langle q\rangle\rho = {}_0\lambda_{\text{o-Ps}} + 4\omega\rho\langle{}_1Z_{\rm eff}\rangle, \tag{7.11}$$

where $\langle q\rangle$ is the quenching coefficient, which may be a function of the density and temperature of the gas, and $\langle{}_1Z_{\rm eff}\rangle$ is conventionally interpreted as the effective number of electrons per atom or molecule available for annihilation with the positron by pick-off quenching (see below). The angle brackets denote that in an experiment the measured value is an average over the energy distribution of the positronium at the time of annihilation, and the factor of four arises (Fraser, 1968) from the fact that only one quarter of the target electrons are in a singlet spin state relative to the positron.

The three mechanisms which dominate quenching in collisions with an atom or molecule, denoted by X, can be summarized as follows.

(i) *Pick-off quenching* via the reaction

$$\text{ortho-Ps} + X \to X^+ + \mathrm{e}^- + \text{two } \gamma\text{-rays}$$

for atoms or molecules which have filled outer shells of electrons. The cross section for this annihilation process, which is typically less than 10^{-25} m^2, can be expressed as

$$\sigma_{\rm a} = 4\pi r_0^2(c/v)\,{}_1Z_{\rm eff}. \tag{7.12}$$

The parameter ${}_1Z_{\rm eff}$ is a measure of the probability that the positron is at the same position as one of the target electrons and in a singlet spin state with it. Its value can be determined by projecting the wave function representing ortho-positronium scattering by the target system onto the singlet spin state representing the positron and a target electron. Thus, if the total wave function is $\Psi(\boldsymbol{r}_1, s_1; \boldsymbol{r}_2, s_2; \cdots; \boldsymbol{r}_i, s_i; \cdots; \boldsymbol{r}_{Z+2}, s_{Z+2})$, where $\boldsymbol{r}_1, s_1$ are the position and spin coordinates of the positron and $\boldsymbol{r}_i, s_i$ are the corresponding coordinates of the ith electron, the projection onto the singlet spin state of the positron and the ith electron, $\chi_0(s_1, s_i)$, is

$$\Phi_i(\boldsymbol{r}_1 = \boldsymbol{r}_i, \boldsymbol{r}_2, s_2; \cdots; \boldsymbol{r}_i; \cdots; \boldsymbol{r}_{Z+2}, s_{Z+2}) = \langle\chi_0(s_1, s_i)|\Psi(\boldsymbol{r}_1 = \boldsymbol{r}_i)\rangle, \tag{7.13}$$

and so

$${}_1Z_{\rm eff} = \sum_{i=2}^{Z+2}\sum_{\rm spins}\int|\Phi_i|^2\,d\boldsymbol{r}_2\cdots d\boldsymbol{r}_{Z+2}. \tag{7.14}$$

The summation over spins in equation (7.14) does not, of course, include the spins of the positron and the ith electron. The form of Ψ for positronium–hydrogen scattering in the static-exchange approximation is given in equation (7.18) below.

Comparisons of calculated and measured quenching rates provide a useful measure of the accuracy of the wave function used for the system. As an example, the value of ${}_1Z_{\rm eff}$ for helium calculated from the zero energy static-exchange wave function of Barker and Bransden (1968) is 0.0347, or 0.0445 when the van der Waals potential is added to the static-exchange equation; however, the experimental value obtained by Coleman *et al.* (1975b) at room temperature is 0.125 ± 0.002 (see section 7.3). This rather large discrepancy, a factor of three, shows that the static-exchange wave function provides a poor representation of the electron–positron correlations in this system.

(ii) *Exchange, or conversion, quenching* occurs according to

$$\text{ortho-Ps} + X(\uparrow) \rightarrow \text{para-Ps} + X(\downarrow),$$

where the electron in the ortho-positronium exchanges with an atomic electron to produce para-positronium, which then usually undergoes two-gamma annihilation before the reverse process can occur. The final state of the target, represented above by $X(\downarrow)$, may be an excited state but if the target atom has an unpaired electron, as in hydrogen, the conversion of ortho-positronium to para-positronium can occur at all energies and no excitation of the atom need take place. If, however, the atom has no unpaired electrons, as in helium, exchange quenching can only occur at positronium energies above the first triplet excitation threshold of the atom, which for helium is at 20.58 eV.

Calculations of the ortho-para conversion cross section have been made for positronium scattering by electrons (Ward, Humberston and McDowell, 1987) and by hydrogen atoms (Hara and Fraser, 1975; Drachman and Houston, 1976), as described in the next subsection. In both cases the conversion cross section is only a few per cent of the elastic positronium scattering cross section, but it is several orders of magnitude larger than the direct pick-off annihilation cross section. As an example, the elastic and conversion cross sections for positronium–hydrogen scattering in the static-exchange approximation (Hara and Fraser, 1975) are shown in Figure 7.12.

(iii) *Chemical quenching.* Here the ortho-positronium is quenched after forming a chemical complex, which may occur by means of the addition or substitution reactions

$$\text{ortho-Ps} + X \rightarrow \text{ortho-Ps}X + \text{energy},$$

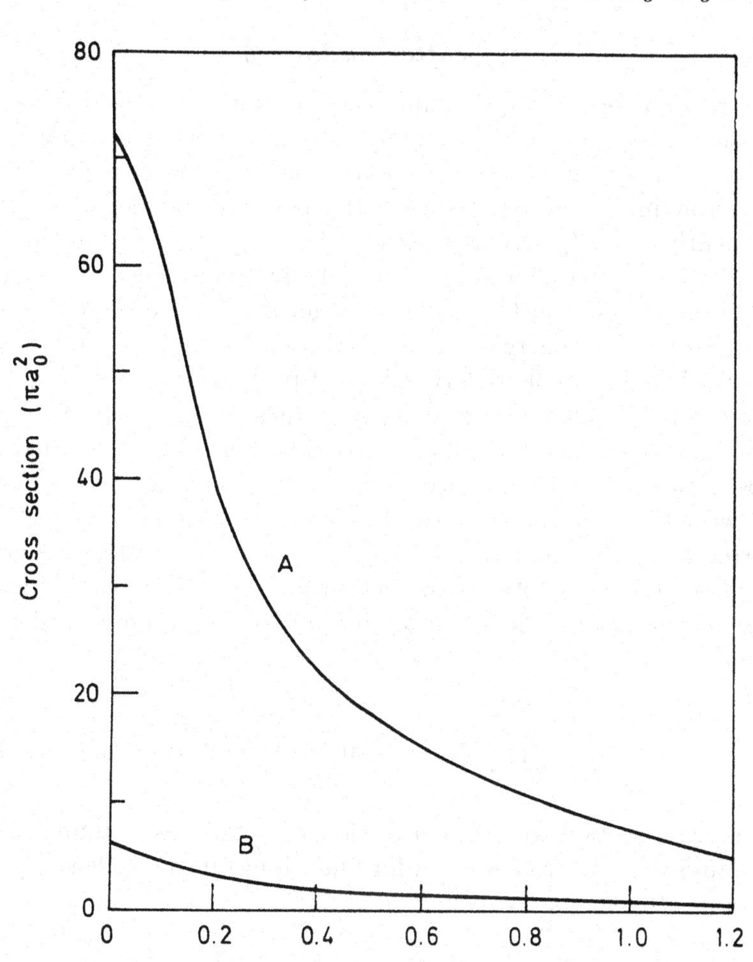

Fig. 7.12. Cross sections for positronium–hydrogen scattering in the static-exchange approximation (Hara and Fraser, 1975). A, elastic scattering; B, ortho- to para-positronium conversion.

$$\text{ortho-Ps} + XY \rightarrow \text{ortho-Ps}X(Y) + Y(X),$$

both of which lead to rapid annihilation of the positron. Cross sections for these reactions vary depending upon the process and the species involved, but they are typically much higher than those for pick-off quenching. Such reactions properly belong to positronium chemistry, a field which is covered for the liquid phase by the monograph of Mogensen (1995) (see particularly Chapter 8 of that work). Examples of gas-phase studies, which imply the existence of stable positronium–atom (or molecule or radical) bound states can be found in section 7.5.

2 Scattering theory

The interaction between positronium and a target system, whether charged or neutral, is, as outlined in subsection 1.6.2, somewhat unusual because the static component is zero; the reason is that the centre of mass of the positronium is midway between the positive and negative charges. Consequently, the direct elastic scattering amplitude in the first Born approximation is zero. Exchange effects between the electron in the positronium and the electrons in the target system are therefore important, at least at low energies, as also are polarization effects because of the relatively large dipole polarizability of positronium ($\alpha = 72a_0^3$).

Elastic ortho-positronium scattering from an atom with an unpaired electron may occur for either a singlet or a triplet state of the electron in the positronium and the unpaired electron in the target atom. For each partial wave there are therefore two phase shifts, η_l^1 and η_l^3, corresponding to scattering in the singlet and triplet states respectively. In terms of these phase shifts the total cross section for the elastic scattering of an unpolarized beam of positronium by an unpolarized target is, in units of πa_0^2,

$$\sigma_{\rm el} = \frac{1}{k^2}\sum_{l=0}^{\infty}(2l+1)(\sin^2\eta_l^1 + 3\sin^2\eta_l^3), \tag{7.15}$$

and the cross section for the conversion of ortho-positronium to para-positronium (i.e. the cross section for quenching) in the collision is (Hara and Fraser, 1975)

$$\sigma_{\rm c} = \frac{1}{4k^2}\sum_{l=0}^{\infty}(2l+1)\sin^2(\eta_l^1 - \eta_l^3). \tag{7.16}$$

Positronium scattering by a single charged particle, being a process involving a three-body system, has been studied using similar techniques to those employed for positron scattering by atomic hydrogen (see section 3.2). Indeed, positronium–proton scattering has been investigated within the wider context of positronium formation in positron–hydrogen scattering. A complete description of this two-channel process requires the incorporation of elastic positronium–proton scattering and hydrogen formation into the formulation, and all four cross sections (for elastic positron–hydrogen scattering, positronium formation, elastic positronium–proton scattering and hydrogen formation) can then be obtained simultaneously. Little interest has been shown in elastic positronium–proton scattering as such, but the formation of hydrogen in such a collision has received considerable attention because the charge-conjugate process, antihydrogen

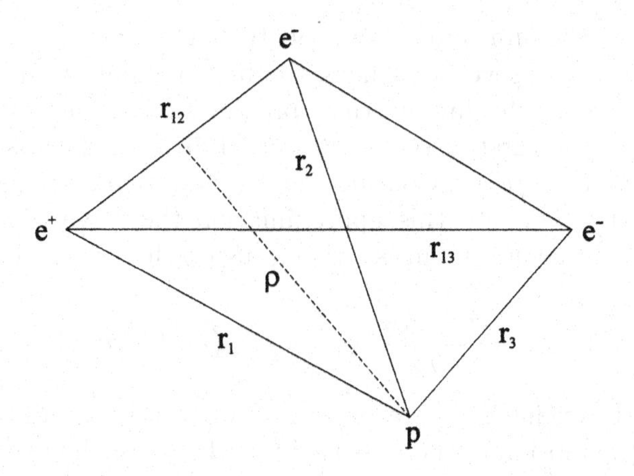

Fig. 7.13. The coordinates of the positronium–hydrogen system.

formation in positronium–antiproton scattering, has been proposed as a means of producing antihydrogen (see section 8.3).

Investigations have also been made of low energy positronium scattering by electrons and positrons, which, according to invariance under charge conjugation, should be identical. The motivation for much of this work has been to obtain accurate p-wave elastic scattering wave functions for use in determining the photo-detachment cross section of the positronium negative ion, Ps^-; this will be discussed in section 8.1. One of the most detailed studies was made by Ward, Humberston and McDowell (1985, 1987), who used similar variational methods to those employed by Humberston (1982) and Brown and Humberston (1984, 1985) in their studies of positronium formation in positron–hydrogen scattering (see section 4.2). Ward *et al.* (1985, 1987) also calculated the scattering lengths by fitting the low energy s-wave phase shifts to the effective range formula

$$\tan \eta_0 = -ak + Bk^2 + Ck^3 \ln k, \tag{7.17}$$

where a is the scattering length. The results obtained in this way for the singlet and triplet scattering lengths are ${}^1a = (12.0 \pm 0.3)a_0$ and ${}^3a = (4.6 \pm 0.4)a_0$ respectively. Similar values, ${}^1a = 11.98a_0$ and ${}^3a = 4.78a_0$, were obtained by Kvitsinsky, Carbonell and Gignoux (1992) using the Fadeev equations in configuration space.

Positronium scattering by other atomic ions should also in principle be considered as part of a complete description of positron scattering by the relevant atom when the positronium formation channel is open, but few such calculations have been made.

Detailed investigations of positronium scattering by atoms have been confined mainly to hydrogen, although some studies have also been made of scattering by helium and argon. Scattering by atomic hydrogen has been investigated by several authors, the first of whom were Massey and Mohr (1954), using the first Born approximation. The static-exchange approximation was first used by Fraser (1961a) and subsequently by Hara and Fraser (1975) to obtain singlet and triplet phase shifts for several partial waves. In this approximation the wave function of the positronium–hydrogen system is written, using the nomenclature of Figure 7.13, as

$$\Psi^{\pm} = \frac{1 \pm \mathrm{P}_{12}}{\sqrt{2}} \left[\varphi_{\mathrm{Ps}}(r_{12})\varphi_{\mathrm{H}}(r_3) f^{\pm}(\boldsymbol{\rho})\right] \tag{7.18}$$

where P_{12} is the exchange operator for the two electrons and the functions $f^{\pm}$ describe the motion of the positronium relative to the hydrogen atom, the spatially symmetric combination corresponding to a singlet spin state of the two electrons and the antisymmetric combination corresponding to a triplet spin state. Hara and Fraser (1975) also calculated the scattering lengths, obtaining the values ${}^1a = 7.275a_0$ and ${}^3a = 2.476a_0$. The quenching cross section calculated by Hara and Fraser was found to be little more than 10% of the elastic scattering cross section (see Figure 7.12).

Some allowance for the effects of distortion of the positronium and the target atom could be made by introducing the van der Waals interaction potential into the static-exchange equation for the scattering function, and Martin and Fraser (1980) and Au and Drachman (1986) calculated this potential with such an aim in mind. Its form, as determined by these latter authors, and also by Manson and Ritchie (1985), is

$$U(\rho) = \frac{-69.6702}{\rho^6} + \frac{503.626k^2 - 237.384}{\rho^8}. \tag{7.19}$$

Ray and Ghosh (1996) used a momentum-space formulation of the static-exchange approximation, and Sinha, Chaudhury and Ghosh (1997) used a somewhat similar approach but with more terms in the coupled-state expansion. They included the terms H(1s) and Ps(1s, 2s, 2p), both with and without exchange, thereby allowing for some distortion and possible excitation of the positronium but not of the hydrogen, the justification being that the dipole polarizability of positronium is 16 times that of hydrogen. In addition to the singlet and triplet elastic scattering cross sections, the latter authors calculated cross sections for quenching and for positronium excitation to the $n_{\mathrm{Ps}} = 2$ states. As expected, the neglect of exchange was found to have a large effect on the elastic and total scattering cross sections at low energies, the results without exchange being much smaller than those with exchange included.

The most accurate values of the singlet and the triplet positronium–hydrogen scattering lengths are probably those calculated by Page (1976) using the Kohn variational method with trial functions of the form

$$\Psi^{\pm} = \frac{1 \pm \mathrm{P}_{23}}{\sqrt{2}} \left\{ \varphi_{\mathrm{Ps}}(r_{12})\varphi_{\mathrm{H}}(r_3) \left[1 - \frac{a_{\mathrm{t}}}{\rho}(1 - e^{-\delta\rho}) + \sum_{j=0}^{2} b_j \rho^j e^{-\delta\rho} \right] \right.$$
$$\left. + \sum_i c_i e^{-(\alpha r_1 + \beta r_2 + \gamma r_3)} r_1^{k_i} r_2^{l_i} r_{12}^{m_i} \right\}, \qquad (7.20)$$

where the second summation includes all terms with $k_i + l_i + m_i \leq 4$, a total of 35 terms. The results obtained were ${}^1a = 5.844a_0$ and ${}^3a = 2.319a_0$, both of which are rigorous upper bounds on the exact values. The triplet result is an upper bound because there is no triplet bound state of the total system, but so also is the singlet result because the short-range terms in the trial function are sufficiently flexible to represent the positronium–hydride bound state, PsH (see section 7.5). Page's values are less positive, and therefore more accurate, than those of Hara and Fraser, which, being derived from the results of the static-exchange approximation, are also rigorous upper bounds. Page also determined the value of the singlet effective range as $r_0^+ = 2.90a_0$, using the relationship

$$({}^1a^+) = (2E_{\mathrm{B}})^{1/2} - {}^1r_0 E_{\mathrm{B}}, \qquad (7.21)$$

where E_{B} is the binding energy of PsH with respect to break up into positronium and hydrogen.

A somewhat less conventional technique was used by Drachman and Houston (1975) to extract the singlet s-wave phase shift from wave functions obtained in the course of investigating the positronium–hydride bound state, PsH. The eigenvectors of the total Hamiltonian matrix of the positronium–hydrogen system, obtained in a normalizable basis, were used to generate approximate wave functions for the system. The eigenvector corresponding to the lowest eigenvalue provides an approximation to the wave function of PsH, but the eigenvectors corresponding to higher energy eigenvalues represent states in the positronium–hydrogen scattering continuum. These wave functions do not, of course, have the correct asymptotic form for true scattering states but, nevertheless, for intermediate values of ρ, the coordinate between the centre of mass of the positronium and the hydrogen atom (see Figure 7.13), each wave function should approximate to the form of the true scattering function at the energy of the eigenvalue. The phase shift was obtained from the total wave function, Ψ, by projecting out the wave function $f(\rho)$ describing the motion of the positronium relative to the hydrogen atom. Thus

$$f(\rho) = \iint \Psi \varphi_{\mathrm{Ps}}(\boldsymbol{r}_{12})\varphi_{\mathrm{H}}(\boldsymbol{r}_3)\, d\boldsymbol{r}_{12}\, d\boldsymbol{r}_3, \qquad (7.22)$$

the 'asymptotic' form of which was then fitted to

$$f(\rho) \sim \frac{A \sin(k\rho + \eta_0)}{\rho}. \tag{7.23}$$

Here k, the wave number of the positronium, is related to the total energy of the system, E_{T}, by $k^2/4 - 0.75 = E_{\mathrm{T}}$. Using these phase shifts in the effective range formula

$$k \operatorname{ctn} \eta_0 = -a^{-1} + \tfrac{1}{2} r_0 k^2 + O(k^4), \tag{7.24}$$

Houston and Drachman also calculated the singlet scattering length, ${}^1a = 5.3a_0$, and the effective range, ${}^1r_0 = 2.5a_0$, obtaining reasonably good agreement with the more accurate results of Page. The singlet scattering length obtained by Houston and Drachman is less positive than that of Page, but their method of calculation does not yield an upper bound and their result is almost certainly not as accurate. A similar technique was used by Drachman and Houston (1976) to determine the triplet s-wave phase shift and the ortho–para conversion cross section.

McAlinden, MacDonald and Walters (1996) conducted an extensive investigation of positronium–hydrogen scattering over a wide energy range, 0–150 eV, taking excitation and ionization of the target hydrogen atom and excitation of the positronium into account as well as elastic scattering. However, they only used the first Born approximation and ignored exchange between the two electrons. Their cross sections for elastic scattering and excitation of the positronium to any state of even parity were therefore identically zero at all energies, but the cross sections for other inelastic processes were expected to be reasonably accurate at energies beyond 20 eV. Campbell *et al.* (1998b) greatly improved on these calculations by including several states and pseudostates of positronium, but only the ground state of hydrogen, in a coupled-state formulation. They confirmed the existence of the S-state resonance first predicted by Drachman and Houston (1975) and they also found resonances in several other partial waves. Their results for the total and various partial cross sections are shown in Figure 7.14. As can be seen, the dominant process at higher energies is the ionization of positronium.

Positronium–helium scattering has attracted theoretical interest for some time because, although measurements of the total scattering cross section have only recently been achieved, the quenching of ortho–positronium diffusing in helium gas has been investigated experimentally for many years. The first theoretical investigation of positronium-helium scattering was made by Fraser (1961b) using the coupled-static approximation with a simple uncorrelated wave function for the helium atom. Fraser only considered s-wave scattering, but higher-partial-wave contributions were

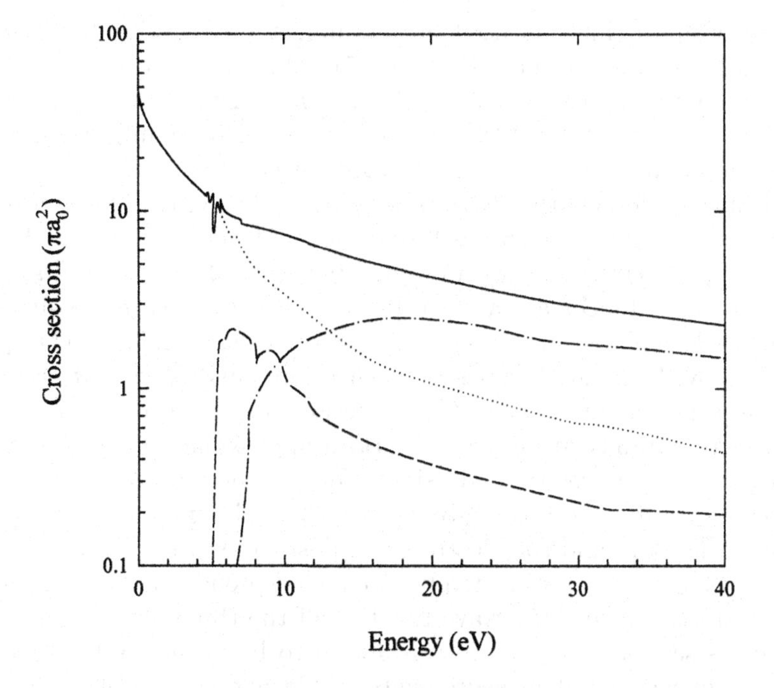

Fig. 7.14. The spin-averaged total cross section, and its various components, for positronium–hydrogen scattering, as calculated using a 22-term (21 Ps, 1 H) coupled-pseudostate approximation (Campbell *et al.*, 1998b): ——, total cross sections; · · · · ·, elastic scattering; – – –, positronium excitation to the $n_{\mathrm{Ps}} = 2$ states; — · —, ionization of positronium.

calculated later by Fraser and Kraidy (1966) using the same approximation. Barker and Bransden (1968) also used the static-exchange approximation, but they attempted to represent distortion by introducing a van der Waals interaction term into the equations for the scattering function.

The static-exchange approximation yields a scattering cross section of $13\pi a_0^2$ at zero energy, dropping to $7.7\pi a_0^2$ at a positronium energy of 13.6 eV. Adding the long-range van der Waals interaction changed the former values to $16.9\pi a_0^2$ and $7.6\pi a_0^2$ respectively. The theoretical results for the scattering cross sections are likely to be much closer to the exact results than might be inferred from the poor agreement between the experimental and theoretical values of ${}_1Z_{\mathrm{eff}}$ (mentioned briefly in subsection 7.2.1) because, as stated in section 6.1 when discussing the direct positron annihilation parameter Z_{eff}, the error in this parameter is only of first order in the error in the wave function whereas the error in the cross section is of second order.

The sensitivity of the low energy elastic positronium–helium scattering cross sections to the quality of the target wave function and to the method of approximation used in obtaining the scattering function was investigated by Sarkar and Ghosh (1997). They used the static-exchange and Born approximations with Hylleraas and Hartree–Fock helium wave functions and found that the two sets of static-exchange results were quite similar to each other. The two sets of Born results, however, were significantly different from each other and also from the static-exchange results, particularly at energies below 5 eV. At energies beyond 150 eV, however, good agreement was obtained between all sets of results.

McAlinden, MacDonald and Walters (1996) investigated positronium scattering by helium and argon, using a somewhat similar technique to that employed in their studies of positronium–hydrogen scattering and over the same energy range. Cross sections involving excitation of the target were obtained using the first Born approximation, but excitation of the positronium was treated using a frozen atom approximation which reduced the system to a three-body problem. The interaction potentials between the atom and the electron and positron were taken to be of the forms $U(r)$ and $-U(r)$ respectively. A coupled-pseudostate expansion was then used to represent the wave function of the three-body system. The total cross sections for helium were found to be in moderate agreement with the experimental measurements of Garner *et al.* (1996) at intermediate energies beyond 25 eV, and the corresponding results for argon agree reasonably well with the measurements of Garner *et al.* (1998) at energies greater than 70 eV. At lower energies, however, where the use of the Born approximation and the neglect of exchange cannot be justified, the theoretical results agree significantly less well with the experimental measurements.

7.3 Experimental studies of positronium annihilation in gases

Numerous measurements of $\langle {}_1Z_{\text{eff}} \rangle$ for a variety of gases have been performed throughout the last 30 years, and a selection of values obtained at low gas densities and room temperature is provided in table 7.2. An extensive review of early measurements in this field was given by Goldanskii (1968). For most gases there is no theoretical work for comparison, the exception being helium, as described above. The possibility that quenching mechanisms other than direct pick-off have been observed in krypton and xenon gases was described in subsection 4.8.2, in relation to the effect the associated lifetime components have upon the determination of positronium fractions.

Recently Vallery *et al.* (2000) have investigated the dependence of $\langle \lambda_{\text{p}} \rangle$ on temperature, in the range between room temperature and 300 °C and for a number of gases, including He, Ne, Ar and N_2. The authors found that the pick-off rate, see equations (7.11) and (7.12), normalized to that

Table 7.2. A selection of $\langle {}_1Z_{\rm eff} \rangle$ values for various gases at low densities and room temperature. The original sources for these values are given in Charlton (1985a). For butane, C_4H_{10}, the upper value refers to n-butane and the lower to isobutane

Gas	$\langle {}_1Z_{\rm eff} \rangle$
He	0.125 ± 0.002
Ne	0.235 ± 0.008
Ar	0.314 ± 0.003
Kr	0.478 ± 0.003
Xe	1.26 ± 0.01
H_2	0.186 ± 0.001
N_2	0.260 ± 0.005
CH_4	0.446 ± 0.01
CO_2	0.500 ± 0.001
CO	0.285 ± 0.010
SF_6	0.52
C_2H_6	0.625 ± 0.020
C_4H_{10}	0.772 ± 0.009 0.729 ± 0.002

at 22 °C, rose linearly across the temperature range investigated for all gases, the rise being most pronounced for the heavier gases. These data can be interpreted in terms of a velocity dependence of $\langle {}_1Z_{\rm eff} \rangle$ or equivalently as a departure from the $1/v$ dependence of $\sigma_{\rm a}$, equation (7.12), predicted by s-wave scattering. Complementary theoretical work has been performed recently by Miller, Reese and Worrell (1996) and Worrell, Miller and Reese (1996).

One of the most interesting gases for which low energy positronium interactions have been studied is O_2. This has been the subject of many experimental studies, and has recently been shown to exhibit both elastic and inelastic exchange quenching. Furthermore, a discussion of the methodology used in these studies (Kakimoto *et al.*, 1987; Kakimoto, Hyodo and Chang, 1990) will serve to introduce the topic of the slowing down of positronium in the following section.

The long-slit, i.e. one-dimensional, angular correlation (ACAR) apparatus used for these studies is similar to that shown in Figure 1.6. Silica aerogel (see subsection 1.5.2) was employed as a convenient source of positronium, emitted into the space between the pores. The chamber could be evacuated and the high purity gases under investigation then

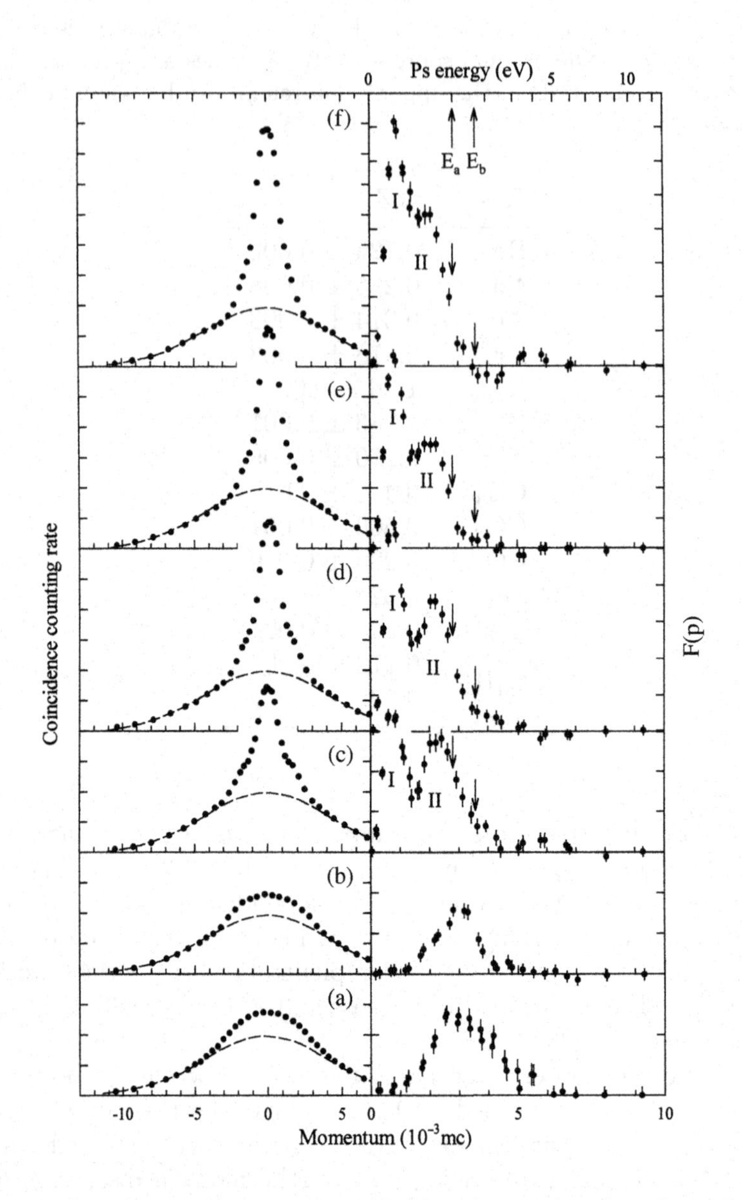

Fig. 7.15. Angular correlation spectra (left) and corresponding derived positronium momentum distributions $F(p)$ (see the text for details) for silica aerogel under the following conditions: (a) vacuum; (b) 1 atmosphere of N_2 gas; (c) 0.1 atmospheres of O_2 gas; (d) 0.2 atmospheres of O_2 gas; (e) 0.4 atmospheres of O_2 gas; (f) 0.8 atmospheres of O_2 gas. The arrows on the right-hand diagrams indicate the momenta corresponding to the excitation energies of the $a^1\Delta_g$ and the $b^1\Sigma_g$ states of O_2. A discussion of the components marked I and II can be found in the text.

introduced. The momentum resolution of the apparatus was given by Kakimoto *et al.* (1987) as $0.27 \times 10^{-3}mc$.

The ACAR data of Kakimoto *et al.* (1987) are shown in Figure 7.15, which illustrates the progression from vacuum up to an O_2 pressure of 0.8 atmospheres and also includes, for comparison, a spectrum taken at 1 atmosphere of N_2, a gas for which only pick-off quenching can occur. The curves for N_2 and the vacuum could be decomposed into two components, and those for O_2 into three. The broader or broadest component, $B(p)$, shown in each case by a broken line, was independent of gas pressure and was attributed to positron annihilation within the silica grains, whilst the lower momentum components were due to the annihilation of positronium between the grains. The momentum distribution of these latter components, $F(p)$, was obtained from the entire angular correlation curve $N(p)$, according to $F(p) = -2pd[N(p) - B(p)]/dp$, and these data are shown in Figure 7.15 in the right-hand diagrams. The positronium momentum corresponding to the first and second electronic excitation energies of O_2, E_a ($a^1\Delta_g$ at 0.98 eV) and E_b ($b^1\Sigma_g^+$ at 1.62 eV) are indicated. The ground state is a triplet with configuration $X^3\Sigma_g^-$.

As described by Kakimoto *et al.* (1987), the vacuum spectrum is the result, to a good approximation, of the distribution in the para-positronium momentum on emission from the grains. The N_2 result, which is only slightly lower in momentum, shows the moderating effect of a gas in which the electronic excitation energy is much higher than the initial kinetic energy of the positronium, so that slowing down (see section 7.4) can only occur by elastic scattering and rotational and vibrational excitation of the molecule. This has hardly any effect on the short-lived para-positronium. The situation for O_2 is markedly different: two components, denoted by I and II, are clearly visible, the narrower component, I, being responsible for the lower momentum peak in Figure 7.15.

Kakimoto *et al.* (1987) attributed component I to the conversion of ortho-positronium, with initial kinetic energy below E_a, to para-positronium by elastic exchange collisions. The momentum is smaller than that of component II since the lower energy ortho-positronium has a sufficiently long lifetime to undergo appreciable slowing down. Analysis of the spectra yielded an elastic conversion cross section of 1.5×10^{-19} cm^2, in accord with other measurements (Klobuchar and Karol, 1980; Kiefl, 1982).

Turning to component II, if the initial kinetic energy of the ortho-positronium is above E_a then it can undergo inelastic conversion by excitation of the $^1\Delta_g$ level of O_2. However, Figure 7.15 shows that as the O_2 pressure increases so the peak of component II increases at the expense of the tail (at energies $>$ 1 eV). This tail is due to the initial momentum of the short-lived para-positronium, implying that this process has a much higher cross section than that responsible for component I. Kakimoto *et al.* (1987) therefore ascribed component II to the slowing down of para-positronium by excitation of the O_2, though they noted that in so doing it converts to ortho-positronium owing to the conservation of spin configuration. Thus, to treat the data properly, ortho-positronium to para-positronium conversion must be accounted for. The shape of component II also indicates a threshold energy close to 1 eV, in accord with that for the $^1\Delta_g$ level. By noting that the peak of component II is already present at 0.1 atmospheres of O_2, Kakimoto *et al.* (1987) estimated the inelastic ortho-positronium conversion cross section, applicable to a positronium kinetic energy of 1.5 eV, to be $\geq 2 \times 10^{-17}$ cm^2, or more than a hundred times that for the elastic process.

A follow-up study at O_2 pressures below 0.05 atmospheres (Kakimoto, Hyodo and Chang, 1990), where the para-positronium to ortho-positronium conversion is suppressed because it occurs at a rate lower than that for para-positronium annihilation, yielded cross sections in good accord with the estimates given above.

Returning to the discussion of $\langle {}_1Z_{\rm eff} \rangle$ from pick-off quenching, we now turn to gas density and temperature regimes where significant departures from the linear rise predicted by equation (7.11) have been observed. As reviewed by Iakubov and Khrapak (1982), the first evidence for anomalous effects in the pick-off quenching rate in gases came from the low temperature helium work of Daniel and Stump (1959). The most detailed study of this system was performed by Hautojärvi and Rytsölä (1979), who found that the ortho-positronium lifetime in low temperature, high density helium gas is substantially longer than that predicted using the value of $\langle {}_1Z_{\rm eff} \rangle$ given in table 7.2. This effect is illustrated in Figure 7.16, which shows their data for the pick-off quenching rate $\langle q \rangle \rho$ plotted against gas density ρ for a variety of temperatures.

It is now considered that this type of behaviour is caused by the self-trapping of ortho-positronium in bubbles in the low temperature gas. The bubbles are thought to form because the ortho-positronium–atom interaction at low energies is dominated by repulsive exchange forces, and this effect results in a lowering of the annihilation rate; the bubble is so rarified in some cases that $\langle q \rangle \rho$ approaches zero.

The behaviour shown in Figure 7.16 is, in some respects, typical of other gases investigated. In particular, the behaviour found for helium at 77 K by Fox *et al.* (1977), in which $\langle \lambda_{\rm p} \rangle$ falls gradually from the extrapolated low density and high temperature line, has been observed for a number of other gases at moderate densities. Examples include the work of McNutt

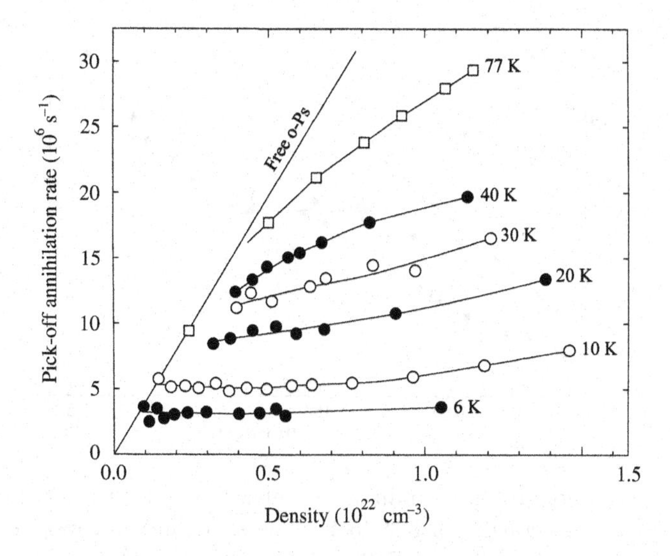

Fig. 7.16. The pick-off annihilation rate $\langle q \rangle \rho$, see equation (7.11), for ortho-positronium in ^{4}He gas at various temperatures, observed by Hautojärvi and Rytsölä (1979). At the lowest temperature $\langle q \rangle \rho$ is almost independent of density, indicating stable bubble formation. The behaviour gradually changes to that of free ortho-positronium, indicated by the straight line whose slope corresponds to $\langle {}_1Z_{\rm eff} \rangle = 0.125$ (see table 7.2). The data at 77 K are due to Fox *et al.* (1977).

and Sharma (1978, CH_4), McNutt *et al.* (1979, H_2) and Wright *et al.* (1983, CO_2) and the data presented for argon and krypton at 297 K by Griffith and Heyland (1978).

Sharma, Eftekhari and McNutt (1982) and Sharma, Kafle and Hart (1984) found particularly striking effects for C_2H_6, which has a critical temperature just above room temperature, and some of their data are presented in Figure 7.17. McNutt and Sharma (1978) and Sharma, Eftekhari and McNutt (1982) attempted to account for this behaviour using a model in which the positronium preferentially annihilates in regions of lower than average density caused by naturally occurring rarefactions in the gas. This approach is analogous to the free-volume model of positronium annihilation in condensed media and is qualitatively different from the bubble model in that it ignores the positronium–atom(molecule) interactions and attributes the anomalous behaviour of $\langle \lambda_{\rm p} \rangle$ solely to the thermodynamic properties of the gas. Some support for the free-volume approach at moderate densities of N_2 gas was provided by Kawaratani, Nakayama and Mizogawa (1985). Further aspects of many-body phenomena involving positronium can be found in the reviews of Iakubov and Khrapak (1982) and Sharma (1988, 1992).

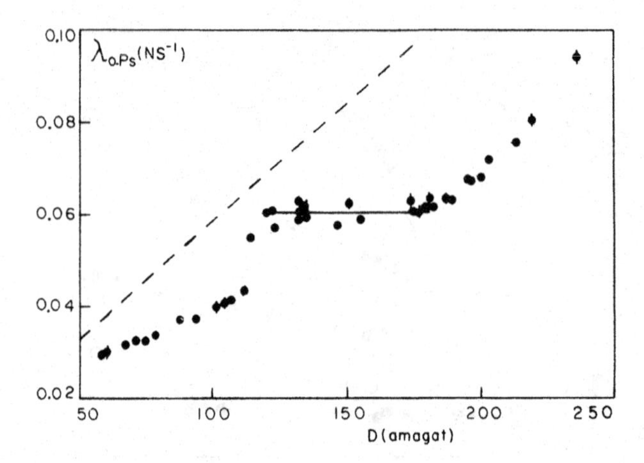

Fig. 7.17. Ortho-positronium annihilation rates at various values of ethane gas density D, at a temperature of 305.45 K. The solid line is a weighted average of the annihilation rates between 120 and 180 amagat. The broken line is the prediction for free ortho-positronium. The data are due to Sharma, Kafle and Hart (1984). Reprinted from *Physical Review Letters* **52**, Sharma, Kafle and Hart, New features in the behaviour of ortho-positronium annihilation rates near the vapour–liquid critical point of ethane, 2233–2236, copyright 1984 by the American Physical Society.

7.4 Slowing down

Recent experimental advances involving the use of the ACAR technique and positronium formation in silica aerogel or powder, and time-resolved Doppler broadening studies (TRDBS) of positronium annihilation in gases, have meant that it has been possible to observe the effects of the slowing down of positronium in collisions with gas atoms and molecules, allowing average momentum transfer cross sections to be derived. One advantage of the aerogel method is that it allows the positronium to interact with the low density gas in the large spaces between the grains, whilst the β^+ particles from the source are stopped and form positronium efficiently. A potential disadvantage is that interactions of the positronium with the surface of the gel have to be accounted for. This feature can, however, be circumvented by the TRDBS technique.

The first discussion of the thermalization of positronium appears to have been that of Sauder (1968), who derived a general (classical) expression for moderation by elastic collisions of a particle in a medium, allowing for the thermal motion of the atoms or molecules of the medium. By assuming that the momentum transfer cross section, σ_M, is a constant he found that the time dependence of the mean positronium kinetic energy,

E, was given by

$$E/E_{\rm th} = \coth^2(\beta + \rho\Gamma t), \tag{7.25}$$

where $E_{\rm th}$ is the thermal energy and $\Gamma = \sigma_{\rm M}(2E_{\rm th}m)^{1/2}/M$, with m the positronium mass and M the mass of the gas atoms. The parameter β is defined at $t = 0$, when $E = E_0$, as $\coth^{-1}(E_0/E_{\rm th})^{1/2}$. It is relatively straightforward to show that this expression is equivalent to that derived by Hyodo *et al.* (1989) and applied by Kakimoto *et al.* (1989) and Nagashima *et al.* (1995), though in these cases an extra term due to positronium interactions with the aerogel had to be introduced.

To our knowledge, the first study of gases in powders was that of Fox and Canter (1978), who investigated the effects of adding high pressure helium to silica and magnesium oxide powders. Behaviour inconsistent with a simple single-exponential ortho-positronium component was observed in the lifetime spectrum with the gas evacuated, and this was substantially modified by the introduction of the gas. Indeed the approach to the exponential (and assumed equilibrium) behaviour of the ortho-positronium component was hastened by the helium, and this was attributed to moderation of the kinetic energy of the positronium by the gas.

We now proceed to describe in more detail the ACAR-aerogel work of Hyodo and coworkers, and the TRDBS measurements of Skalsey *et al.* (1998). It is worth noting, following Hyodo (1992), the physical processes which contribute to the two-gamma events registered by the ACAR technique in this type of experiment:

(i) positron and positronium annihilation in the grains of the aerogel;

(ii) self-annihilation of para-positronium between the grains;

(iii) pick-off annihilation of ortho-positronium in collisions with the grain surface or the added gas;

(iv) two-gamma self-annihilation of the $m = 0$ mixed ortho-positronium state in a magnetic field;

(v) the self-annihilation of para-positronium resulting from exchange quenching of ortho-positronium in a paramagnetic gas.

The key to the ACAR technique is process (iv), since by application of a magnetic field the lifetime of the mixed state is changed and the time-dependence of the positronium momentum distribution can be accessed.

ACAR studies have been made of all the noble gases and some molecules (Hyodo, 1992; Nagashima *et al.*, 1995; Coleman *et al.*, 1994) at a fixed magnetic field; by Hyodo and coworkers at 0.29 T and by Coleman *et al.*

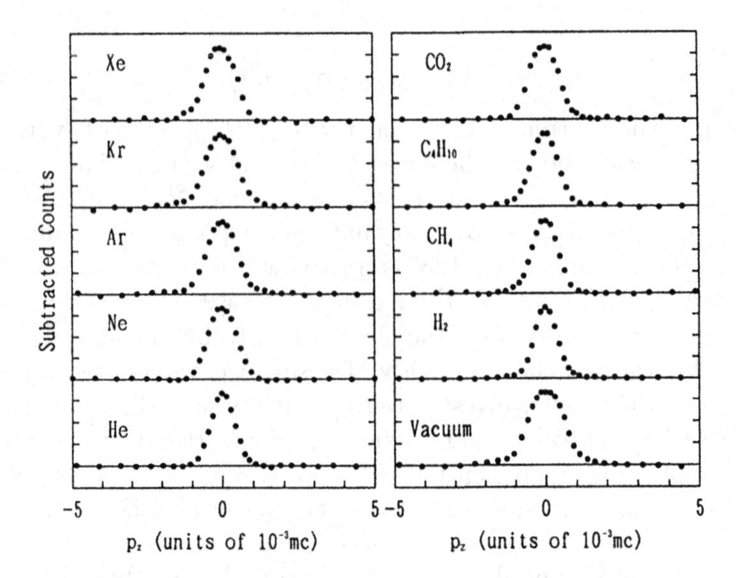

Fig. 7.18. Derived momentum distributions for the perturbed $m = 0$ state of ortho-positronium, for various gases (Nagashima et al., 1995) in an applied static magnetic field of 0.29 T. Reprinted from *Physical Review* **A52**, Nagashima *et al.*, Thermalization of free positronium atoms by collisions with silica-powder grains, aerogel grains and gas molecules, 258–265, copyright 1995 by the American Physical Society.

(1994) at 0.8 T. The results of Nagashima *et al.* (1995) are shown in Figure 7.18 and, interestingly, exhibit a progressive broadening of the narrow component of the distribution as the atomic or molecular mass increases. Similar effects to these were observed by Coleman *et al.* (1994) in their study of the noble gases, although their momentum resolution was inferior to that of Nagashima *et al.* (1995), so that the narrow positronium component could only be separated in helium, neon and argon.

The derived momentum transfer cross sections, typically in the range 50–150 $\times$ 10^{-16} cm^2 (dependent upon the gas) at positronium energies below 100 meV, have been given by Nagashima *et al.* (1995), who also compared their data with geometric cross sections taken from viscosity measurements. They found that the ratio of these two cross sections does not increase when going from atoms to molecules and they argued that this is evidence for the near absence of low energy inelastic (rotational and vibrational) effects in the interactions of positronium with molecules.

Hyodo (1992) and Nagashima *et al.* (1995) justified this by noting that there is a mismatch between the period, ν^{-1}, of internal excitation of the molecule and the positronium collision time. The latter is determined by

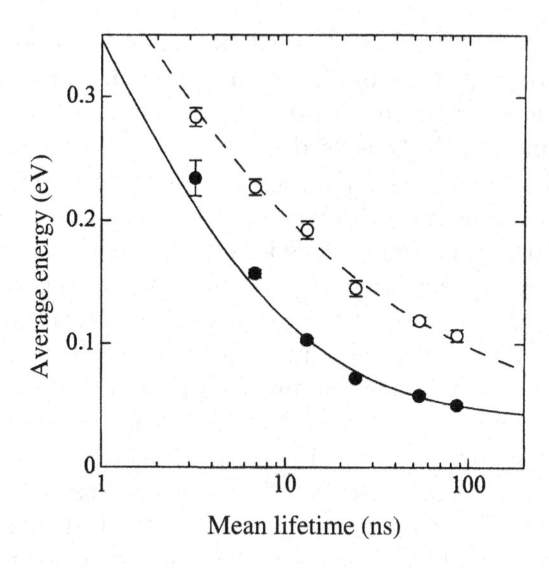

Fig. 7.19. The time dependence of the mean ortho-positronium kinetic energy measured in silica aerogel under vacuum (∘) and with 0.92 amagat of helium gas added (•). The two curves were simultaneously fitted to the data by Nagashima *et al.* (1998), to obtain estimates of the Ps–He momentum transfer cross section. Reprinted from *Journal of Physics* **B31**, Nagashima *et al.*, Momentum transfer cross section for slow positronium–He scattering, 329–339, copyright 1998, with permission from IOP Publishing.

short-range van der Waals forces and is approximately a_0/v_{Ps}, where v_{Ps} is the speed of the positronium and a_0 is an estimate of the molecular size. That is, the excitation is only likely when $\nu^{-1} \approx a_0/v_{Ps}$. Inserting values for the spacings between the rotational and vibrational energy levels, approximately 10^{-2} eV and 0.1 eV respectively, it is found that ν^{-1} is of order 4×10^{-12} s and 4×10^{-14} s respectively. However, the positronium interaction time is typically one or two orders of magnitude smaller, effectively suppressing these processes. This information can also be gleaned qualitatively from Figure 7.18, where it can be seen that H_2 is a much more efficient positronium moderator than the heavier and more complex molecules CO_2 and C_4H_{10}.

Nagashima *et al.* (1998) re-investigated helium using a variable field, from 0.16 to 1.5 T, resulting in a lifetime of the $m = 0$ 1^3S_0 state in the range 86–3.2 ns. The ACAR spectra of the mixed component were isolated by subtracting the suitably normalized $B = 0$ spectrum from those for $B \neq 0$. Broadening of the the mixed component was evident as the magnetic field was increased, a clear manifestation that the ortho-positronium had had less time to moderate before conversion (and rapid annihilation) occurred.

Nagashima *et al.* (1998) performed an analysis improved over that used in their previous work by using a Boltzmann-equation approach in which an average momentum transfer cross section was considered as an adjustable parameter; this was used to fit the average positronium kinetic energy derived from the ACAR spectra. However, they had to build into this approach an energy-dependent, and therefore time-dependent, term to allow for the effects of collisions with the grains. Values for the time dependence of the average positronium kinetic energy are shown in Figure 7.19, which gives separate results for the aerogel when evacuated and with the helium added. The lines show simultaneous fits to the data for the two cases, which yielded a value of σ_M equal to $(11\pm3)\times10^{-16}$ cm^2 below 0.3 eV. This is much larger than the value derived from a lifetime experiment by Spektor and Paul (1975) (see Charlton and Laricchia, 1991, for a discussion of that work) but is in much closer accord with the available theoretical estimates (see section 7.2). Coleman *et al.* (1994) reported a lower-bound value of 7.9×10^{-16} cm^2 from a simple analysis of their ACAR data. This is similar in magnitude to the values obtained at somewhat higher positronium kinetic energies using the beam technique (see section 7.6).

The TRDBS technique (Skalsey *et al.*, 1998) involves measuring the Doppler broadening of the 511 keV photons emitted as a result of Ps annihilation using a high-resolution Ge detector (see subsection 1.3.2 and Figure 1.5) in a particular time interval. The γ-ray energy information, which is a measure of the longitudinal positronium momentum upon annihilation, was recorded in the time range 30–50 ns for the perturbed $m = 0$ ortho-positronium component, which had a vacuum lifetime of 52 ns in the magnetic field of 0.285 T applied to the gas. Skalsey *et al.* (1998) discussed how this component of the γ-ray spectrum was isolated from other contributions, notably that due to free-positron annihilation. The essence of the technique is that the ortho-positronium energy is measured in the fixed time interval at different gas densities, so that a measure of the thermalization, and hence the momentum transfer cross sections, can be obtained. It is noteworthy that only collisions with the gas need to be considered; thus the thermalization model is independent of effects due to the aerogel, which was used in most of the ACAR work.

Accordingly, from equation (7.25) of Sauder (1968), Skalsey *et al.* (1998) plotted arccoth $[(E/E_{\rm th})^{1/2}]$ against $\rho\langle t\rangle$, where $\langle t\rangle = 38$ ns is the weighted average of the delayed time window. Data for several gases are shown in Figure 7.20, where qualitative accord with the elastic scattering model in the energy range from approximately 0.3 eV to 2 eV is evident. All gases approach an intercept as $\rho\langle t\rangle \to 0$ of around 2–4 eV, which is

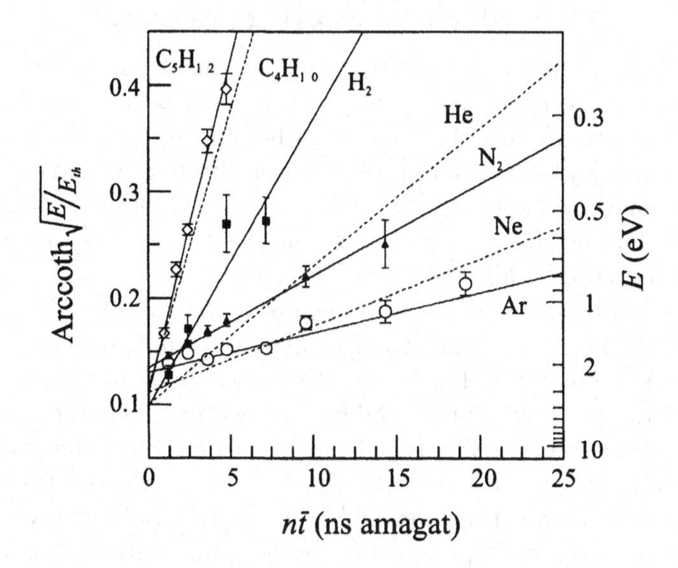

Fig. 7.20. Fits performed by Skalsey *et al.* (1998) to Sauder's positronium thermalization model using their TRDBS data. The slopes yield the positronium thermalization rate, whilst the intercepts give the average initial energy. (For clarity only the fitted lines are shown for He, Ne and iso-C_4H_{10}.) Reprinted from *Physical Review Letters* **80**, Skalsey *et al.*, Thermalization of positronium in gases, 3727–3730, copyright 1998 by the American Physical Society.

consistent with the Ore model of positronium formation in dense gases described in section 4.8. The value of Γ and hence σ_M can be derived from Figure 7.20, and Skalsey *et al.* (1998) found that σ_M/M varies by a factor of around 20 for the gases investigated. Their values of σ_M range from 2.3 ± 0.4 Å^2 for He to 208 ± 17 Å^2 for C_5H_{12}. Thus, the value for He, though at a somewhat higher average energy than that for the study of Nagashima *et al.* (1998), is not in accord with the latter. It is also evident that the conclusions of Skalsey *et al.* (1998) are not in accord with those of Nagashima *et al.* (1995), noted above; these workers found that the thermalization rate, as determined in their ACAR study, was the same for He, Ne and iso-C_4H_{10}.

Although simplifying assumptions have had to be made in analysing the data obtained from both types of experiment described in this section, it is clear that valuable information on positronium scattering and interactions at low energies can be obtained, particularly from the TRDBS technique. It is hoped that this work will complement the experimental data which are becoming available at higher kinetic energies from positronium-beam studies and which offer a stimulus to renewed theoretical efforts.

7.5 Bound states involving positronium

A bound system containing a positron may be considered either as positronium bound to a positive ion or an atom or as a positron bound to the corresponding atom or negative ion. It is therefore appropriate to consider here all bound states containing positrons. Bound systems containing more than one positron, however, such as $e^+e^+e^-$ (the anti-system of the positronium negative ion Ps^-) and the positronium molecule Ps_2, are considered in more detail in sections 8.1 and 8.2.

Until recently there was no firm theoretical evidence that a positron could bind to any atom other than positronium; it had been rigorously proved by Armour (1978, 1982) that it cannot bind to atomic hydrogen, and the evidence that it cannot bind to helium is overwhelming. The most likely candidates were the highly polarizable alkali atoms, and states of the positron–atom system below the positron–atom scattering threshold do indeed exist. However, they were all believed to lie above the threshold for positronium scattering by the corresponding positive ion, and were therefore not true bound states.

Evidence of true bound states then began to emerge, possible candidates being magnesium, zinc, cadmium and mercury (Dzuba *et al.*, 1995) and the HF molecule (Danby and Tennyson, 1988); nevertheless doubts remained, in view of the fact that the models of the systems to which positrons were supposed to be bound were not exact. Recently, however, it has been rigorously established by Ryzhikh and Mitroy (1997) that a true bound state of the positron–lithium atom exists below the threshold for positronium–Li^+ scattering. This state is therefore more appropriately described as positronium bound to the Li^+ ion, and the binding energy with respect to break-up into positronium and Li^+ was calculated to be 0.065 eV. Ryzhikh and Mitroy used the Rayleigh–Ritz variational method with the full non-relativistic Hamiltonian and a suitably antisymmetrized trial wave function; this consisted of the sum of a large number of Gaussoid basis functions, each containing several non-linear variational parameters and of the form

$$G_i = \exp\left(-\tfrac{1}{2}\sum_{\mu,\nu=1}^{N-1} A^i_{\mu\nu} x_\mu s_\nu\right), \tag{7.26}$$

where the x_μ are Jacobi coordinates of the system and the $A^i_{\mu\nu}$ are variational parameters. Minimization of the lowest energy eigenvalue with respect to these parameters was achieved using the stochastic variational method. Because the variational method yields an upper bound on the energy of the system, any value obtained for this quantity that is below the lowest scattering threshold provides rigorous proof of binding.

Similar methods were used by Ryzhikh, Mitroy and Varga (1998a, b) to establish that a positron can bind to beryllium and also, probably, to sodium and magnesium, although the evidence is not so conclusive because a frozen-core model was used to represent the atom. The binding of a positron to magnesium had previously been predicted by Gribakin and King (1996) using many-body theory.

A positron might also be expected to bind to a negative atomic ion, the Coulomb interaction giving rise to a Rydberg series of levels of the composite system. However, in all cases the electron affinity of the ion is less than the binding energy of positronium (6.8 eV) and therefore the ground state of the composite system should more properly be considered as positronium interacting with the neutral atom and possibly binding to it. Higher energy states of the composite system have energies greater than the minimum for positronium scattering by the atom, and therefore manifest themselves as Feshbach resonances in positronium–atom scattering.

Positronium can bind to a positively or negatively charged particle if the mass of that particle is not too large. Thus, positronium cannot bind to a proton or a positive muon (Armour, 1983) but it can bind to an electron or positron provided the two identical particles in the combination are in a singlet spin state. Positronium can bind to itself to form the positronium molecule, Ps_2, again provided each pair of identical particles is in a singlet spin state, and it can also bind to a hydrogen atom to form the positronium–hydride molecule, PsH, provided the two electrons are in a singlet spin state. There are, however, no bound excited states. Several calculations of the binding energy of this system with respect to dissociation into positronium and hydrogen have been made (Page and Fraser, 1974; Ho, 1986b; Frolov and Smith, 1997; Ryzhikh, Mitroy and Varga, 1998b). The most accurate value, 1.066 eV, was obtained by Ryzhikh, Mitroy and Varga (1998b) using a 500-term Gaussoid basis. These authors also calculated the electron–positron annihilation rate to be 2.452 ns^{-1}, which is quite close to the weighted mean of the singlet and triplet rates for free positronium, implying that the structure of PsH is essentially that of positronium rather weakly bound to a hydrogen atom. In addition to the bound state, there is a rich structure of Feshbach resonances associated with the Coulomb interaction of the positron with the residual negative ion.

Ryzhikh, Mitroy and Varga (1998b), using similar techniques to those employed by these same authors in the investigations mentioned above of positron binding to atoms, showed that positronium can almost certainly form bound states with lithium and sodium atoms, the binding energies being 0.33 eV and 0.15 eV respectively. Karl, Nakanishi and Schrader (1984) found evidence that positronium could bind to approximately half the atoms they investigated, and also to a few light negative ions, but

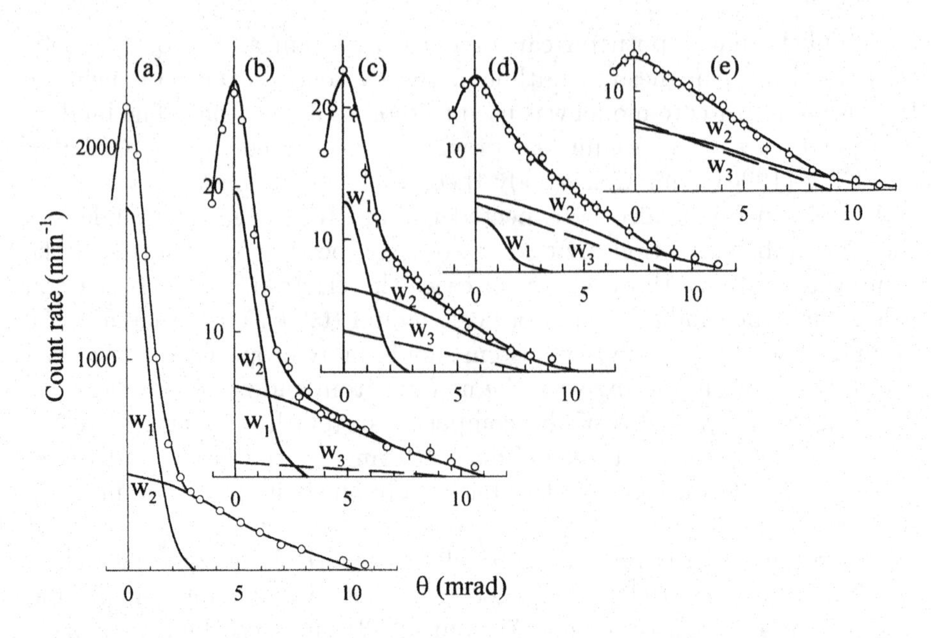

Fig. 7.21. Angular correlation curves for mixtures of O_2 and Cl_2 gases with an overall pressure of 120 atmospheres. (a) Pure O_2, (b) O_2 with 0.02 atmospheres of Cl_2, (c) O_2 with 0.05 atmospheres of Cl_2, (d) O_2 with 0.2 atmospheres of Cl_2 and (e) O_2 with 1 atmosphere of Cl_2. Goldanskii and Mokrushin (1968) attributed the components labelled W_1, W_2 and W_3 to the annihilation of thermalized para-positronium atoms (W_1, the narrow component), the annihilation of free positrons in O_2 (W_2) and the annihilation of positrons in the PsCl compound (W_3). The intensity of the last, i.e. W_3, grows progressively with the addition of Cl_2 to the O_2 buffer.

to no positive ions. A review of bound molecular systems containing positrons has been given by Schrader (1998).

We now briefly review experimental evidence for the existence of some simple positronium compounds; more detailed accounts have been given for early lifetime experiments by Goldanskii (1968) and for the liquid phase by Mogensen (1995). In the case of PsCl we shall see how traditional positron experiments using lifetime and ACAR techniques have provided strong evidence for the stability of this compound, in accord with theory. The first direct experimental evidence of the existence of PsH came from a positron-beam experiment (Schrader *et al.*, 1992).

One of the first studies of PsCl was that of Tao (1965), who used a lifetime experiment (subsection 1.3.1); in this experiment positronium was formed in argon or N_2 gases to which small quantities of Cl_2 vapour had been added. The intensity of the long-lived ortho-positronium component was found to decrease as the Cl_2 concentration was increased, and an extra fast component (similar to those found later in krypton and xenon, see subsection 4.8.2) arose whose lifetime decreased and intensity increased with the amount of added chlorine. Tao (1965) interpreted these observations in terms of the reaction

$$Ps + Cl_2 \rightarrow PsCl + Cl, \tag{7.27}$$

for which a threshold energy of approximately 0.5 eV was extracted. From this work Goldanskii and Mokrushin (1968) deduced a rate constant (equivalent to the product of the reaction cross section and the positronium speed) of 4×10^{-9} cm^3 s^{-1}. In the same study Goldanskii and Mokrushin observed a Ps–Cl_2 reaction using the ACAR technique. They measured spectra, similar to those shown in section 7.4, for high pressure O_2 with Cl_2 added. Their curves are shown in Figure 7.21 for an overall pressure of 120 atmospheres, with the Cl_2 pressure in the range 0–1 atmosphere. It is clear that the narrow component, attributed to exchange quenching with the unpaired electrons in the O_2, falls in intensity to practically zero as the Cl_2 pressure is raised. The component labelled W_3 in Figure 7.21(b)–(e) is broad, owing to the nature of the chemical quenching whereby the positron annihilates with a molecular electron; consequently the angular deviation reflects mainly the momentum of the electron. The fact that the narrow component appears to fall to zero is somewhat at odds with the existence of the threshold energy as derived by Tao (1965) and the energy range over which the positronium is assumed to form. Nevertheless, both of these early gas-phase experiments provided strong evidence for the existence of the PsCl bound state.

Later work by Mogensen and Shantarovich (1974) and Mogensen (1979), although again not direct observations, removed any remaining doubts about the formation of Ps–halide bound states. These experiments were carried out in the liquid phase and again involved the ACAR technique. Here, halide ions in the form of aqueous solutions were studied in such a way that reactions between a positron and Cl^-, Br^- or I^- were observed. The results of the analysis performed by Mogensen (1979) show excellent agreement with the theoretical values of Farazdel and Cade (1977) except in the case of F^-, where no experimental evidence for a bound state could be found. Nevertheless, the overall accord between theory and experiment is all the more remarkable when it is remembered that the former is for an e^+X^-, or PsX, bound state in vacuum. Mogensen (1979) suggested that this may be caused by the formation of a bubble around the positron complex, similar to the many-body phenomena outlined in section 7.3, so that it may be viewed as being in isolation.

Having described investigations performed with traditional positron techniques, we now pass on to a very different type of study based upon low energy beams. This type of experiment is similar to the charge

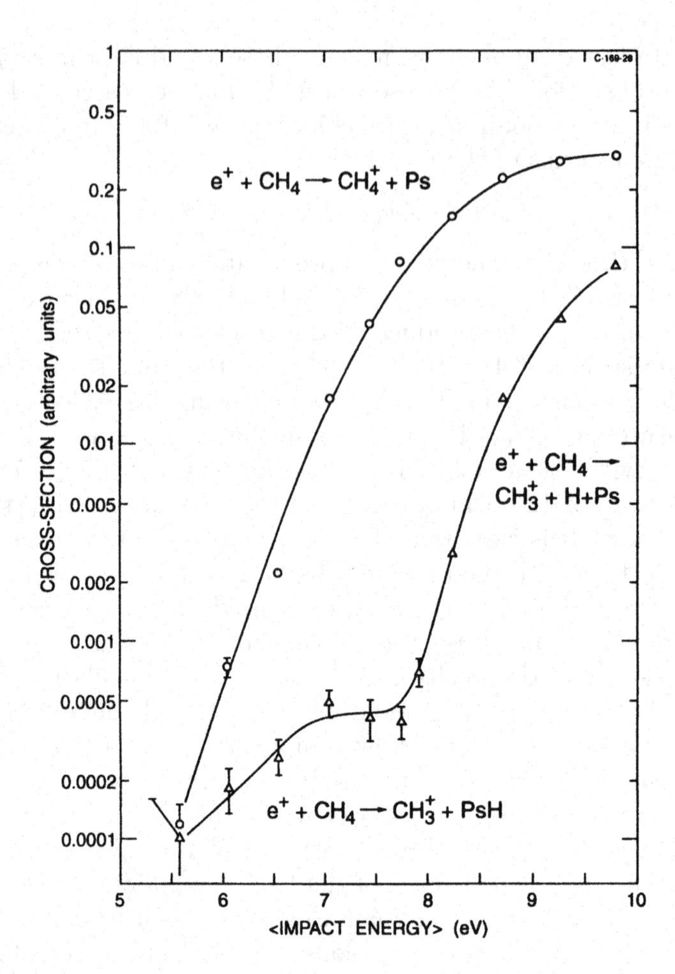

Fig. 7.22. Cross sections (given in arbitrary units) for the production of CH_4^+ and CH_3^+ ions in collisions of positrons with CH_4 gas. The small step in the CH_3^+ yield between 6 eV and 8 eV is direct evidence for the formation of PsH. Reprinted from *Physical Review Letters* **69**, Schrader *et al.*, Formation of positronium hydride, 57–60, copyright 1992 by the American Physical Society.

transfer and ionization studies reported in Chapters 4 and 5. The beam is passed through a scattering cell and the impact energy is carefully swept over a well-defined energy range where the onset of various channels leading to ion production may occur. By correlating signals from the annihilating positron and the ion using a time-of-flight technique, it is possible to identify the ion and deduce a value for the onset threshold. This method was used by Schrader *et al.* (1992) in their detection of PsH.

The measured variations of the yields of the ions in coincidence with a gamma-ray are shown in Figure 7.22. The two major contributing channels are given below, with their respective threshold energies in brackets:

$$e^+ + CH_4 \to CH_4^+ + Ps \qquad (6.18\ \text{eV})$$

and

$$e^+ + CH_4 \to CH_3^+ + H + Ps \qquad (7.55\ \text{eV}).$$

A third process, that for PsH formation, is given by

$$e^+ + CH_4 \to CH_3^+ + PsH \qquad (7.55\ \text{eV} - E_{\text{PSH}}).$$

Using the theoretical value, 1.07 eV, for the PsH binding energy given earlier in this section, the threshold for the last process should be 6.48 eV. Schrader *et al.* (1992) argued that the small increase in CH_3^+ production around 6 eV impact energy is due to this reaction. They expected the cross section for this process to rise rapidly from the threshold and then fall off quickly, since the positron must come almost to rest before it can form PsH. The more gradual increase observed was attributed to the 1 eV energy spread of the beam as it passed through the scattering region. Based upon the onsets for CH_3^+ and CH_4^+ production, Schrader *et al.* (1992) deduced a value for E_{PSH} of 1.1 ± 0.2 eV. Whilst this is not sufficiently accurate to challenge theory, it does constitute an enormous step forward and indicates the direction in which future advances in our understanding of bound states involving positronium might be made; see also Schrader, Laricchia and Horsky (1993) and an update on related recent experimentation by Moxom *et al.* (1998).

7.6 Studies with positronium beams

Most studies of positronium interactions have depended upon monitoring the annihilation process after positronium has been formed by β^+ particles stopping in relatively dense media (e.g. sections 7.3 and 7.4). Fortunately, as introduced in subsection 1.5.3 and described in more detail below, the availability of positron beams has made it possible to create variable energy positronium atoms under controlled conditions in vacuum. In this section we discuss the development of such beams, in which the positronium atom is considered as a swift atomic projectile.

Positronium beams may have a variety of applications. In surface physics the interest in positronium diffraction from crystals, as pointed out by Canter (1984), arises mainly from the fact that the relatively long de Broglie wavelength of positronium at intermediate energies enables the surface layers to be probed more deeply than is possible with traditional-atom diffraction. Weber *et al.* (1988) carried out a study of positronium reflection from a single crystal. Surko *et al.* (1986) proposed the injection of positronium atoms into a tokamak plasma to act as a diagnostic tool for investigating charged-particle transport in the plasma. If the positronium is injected at sufficiently high energies ($\geq$ 20 eV) it is likely to break up in collision, thus liberating a positron. The transport of the positron to the walls of the tokamak could then be detected by monitoring the emission of annihilation radiation.

One further example, and the topic of most interest in the present context, is positronium scattering from atoms and molecules as a probe of many-body atomic physics. For such studies the positronium beam must be well collimated, have a known, and tunable, energy distribution (or be quasi-monoenergetic) and be in a known quantum state. In the following section we describe efforts to achieve these aims, concentrating on the production of positronium using charge exchange in positron–gas collisions.

1 The production of positronium beams by positron–gas collisions

In the positron–gas method the positronium is formed when a beam of positrons of well-defined energy collides with a dilute gas target of atoms or molecules according to $e^+ + X \rightarrow Ps + X^+$. This reaction has been the subject of detailed discussion in Chapter 4, where, in section 4.7, it was noted that the differential cross section is peaked around the incident beam direction, so that the emerging positronium can be considered as being naturally collimated. Since the recoil energy of the target is small, the kinetic energy distribution of the positronium in a state of given principal quantum number n_{Ps} is determined by the energy of the incident positron and is therefore tunable.

In principle, the positronium may be formed in any of the energetically allowed states, though the state with $n_{Ps} = 1$ is found to dominate in the cases studied so far for beam production. The major contributions from excited species are expected to come from the 2^3P and 2^3S states. The former state decays to the ground state with a radiative lifetime of 3.2 ns, thus effectively producing a second ground state beam with a kinetic energy approximately 5 eV below that of the ground state. The 2^3S positronium is metastable against radiative decay and, with a 1.1 μs lifetime against three-gamma annihilation, can travel appreciable distances in its original state. Again, the energy of this beam is approximately 5 eV below the energy of the ground state beam. These considerations suggest that, if useful scattering cross sections are to be measured with positronium beams, some means must be developed of distinguishing between the quantum states, assessing the amount of each state present and perhaps eliminating unwanted states from the beam.

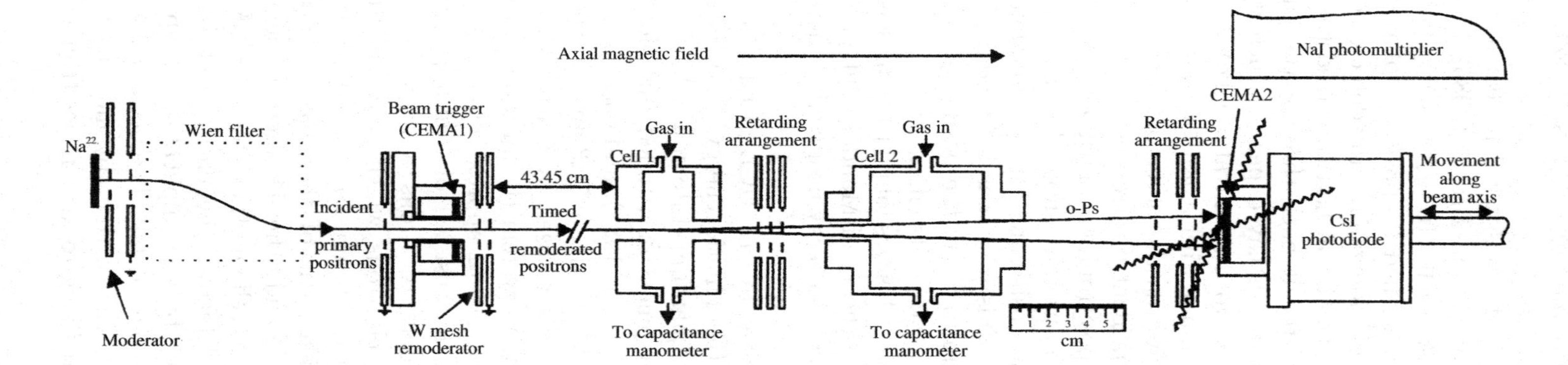

Fig. 7.23. Schematic illustration of the positronium-beam apparatus used for studies of positronium–atom(molecule) collisions (Garner, Özen and Laricchia, 1998).

The first evidence that useful positronium beams could be formed by positron–gas scattering came from Brown (1985, 1986) and, independently, from Laricchia *et al.* (1986). In the former experiment the annihilation of para-positronium formed by positrons of various energies was monitored using a high resolution gamma-ray detector located on the axis of a positron beam. Once the beam had traversed the scattering region a retarding potential reflected it back through the gas. The resulting gamma-ray energy spectrum displayed peaks which were red and blue Doppler-shifted with respect to the central 511 keV line and which were attributed to para-positronium with a narrow range of kinetic energies moving away from and towards the detector respectively. The shift of the two side peaks with incident positron energy was a clear suggestion of the tunability of the positronium. Further information came from Laricchia *et al.* (1986), who used a simple channel electron multiplier to detect forward-going ortho-positronium in a preliminary study involving positron–argon collisions. This work was followed up by Laricchia *et al.* (1987b) who found that up to 4% of the positrons colliding with helium gas could be detected as ortho-positronium emitted into a 6° cone about the incident beam direction. This was found to be in reasonable agreement with expectations from the theory of Mandal, Guha and Sil (1979), even though the efficiency of the low energy positronium detector could not be quantified.

The next step undertaken by the UCL group was to incorporate a timing system whereby the kinetic energy distribution of the ortho-positronium and the gross quantum state in which it was formed could be discerned. The first development of a timed tunable positronium beam was reported by Laricchia *et al.* (1988), and ensuing advances have been described by Laricchia (1995b). Figure 7.23 gives a schematic illustration of the positronium-beam system used by Garner *et al.* (1998). In one mode of operation the timing system (beam tagger) could be employed. Positrons with kinetic energies of approximately 400 eV were incident upon a remoderator, which consisted of overlapping tungsten meshes. Around 12% of the incident positrons were remoderated, and about half produced secondary electrons which were counted by the tagger (CEMA1). The remoderated beam was guided by a magnetic field to the first gas cell, in which some of the beam formed forward-peaked positronium. All positrons, and other charged particles, were prevented from reaching the second gas cell and the detector (CEMA2 in coincidence with a CsI photodiode detector, the entire unit being mobile along the axis of the beamline) by the application of appropriate voltages to the retarding grid arrangements. The second gas cell was used as the positronium scattering cell so that, by measuring the pressure and temperature and by normalizing to known positron–gas total scattering

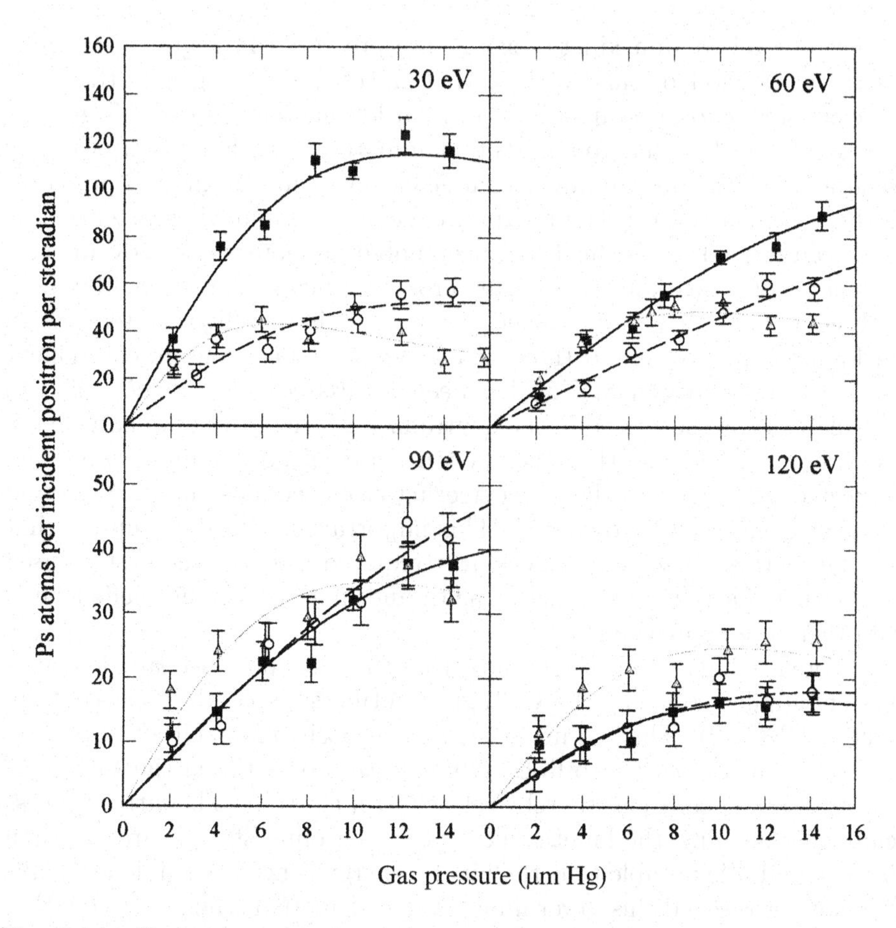

Fig. 7.24. Positronium-beam production efficiencies versus gas pressure at positronium kinetic energies of 30 eV, 60 eV, 90 eV and 120 eV for (○) helium, (△) argon and (■) molecular hydrogen gases. The curves are polynomial fits to the data, which were performed by Garner, Laricchia and Özen (1998).

cross sections to obtain the absolute value of the product of the cell length and gas density (see e.g. section 2.3), total positronium scattering cross sections could be obtained using the attenuation technique. The distance between the positronium scattering cell and the CEMA2 CsI detector could be varied to allow tests of the possible influence of forward scattering on the measured cross sections (see section 2.4 for a discussion relevant to positron cross sections) and also tests for the possible presence of an excited state component in the beam.

Garner, Laricchia and Özen (1996) have published the most detailed study of the production of positronium beams by the gas collision method, and their results for molecular hydrogen, helium and argon gases at four values of positronium kinetic energy are presented in Figure 7.24. The broad conclusion of this work is that molecular hydrogen is the most efficient source for a positronium beam at low energies (see also Tang and Surko, 1993), though argon tends to dominate at the higher energies. The saturation of the positronium beam efficiency at higher gas-cell pressures is due to the increasing likelihood that the positronium will scatter once it has been formed. This effect had been noted previously and used in early estimates of positronium scattering cross sections (Zafar *et al.*, 1991).

It is also notable that most studies using the time-of-flight system shown in Figure 7.23 have found that the excited state positronium component in the beam is negligible. This has been attributed (Zafar *et al.*, 1996) to the high gas pressure used in the neutralizer cell (in order to maximize the positronium yield), so that the excited states of positronium, which are expected to have much larger scattering cross sections than the ground state, are effectively attenuated. This important advance has meant that the beam tagging section is no longer required (see e.g. the system used by Garner, Laricchia and Özen, 1996), and inherent timing inefficiencies can therefore be avoided.

Before leaving this section we note that the positronium-beam detection system applied so far by the UCL group utilizes a secondary electron detector (CEMA2). This requires the incident positronium either to possess sufficient kinetic energy to liberate an electron on striking the surface or to break up on collision, thereby releasing an electron. The latter process cannot occur until the kinetic energy of the ground state is greater than 6.8 eV, and it is notable that the kinetic energy range accessible to beams has not yet reached this lower limit. A new detection scheme based either entirely upon the detection of positronium annihilation or on the use of surfaces with low work functions would seem to be appropriate in seeking to extend the measurement range to lower energies.

2 Scattering experiments with positronium beams

To date, only total positronium scattering cross sections have been measured using positronium beams, and here we concentrate on the most recent results of the UCL group. Figure 7.25 shows a compendium of experimental data for helium, argon, H_2 and O_2 from Garner, Özen and Laricchia (1998). The cross sections are quoted at a fixed solid angular resolution of 2.15 msr, which was set by the gas-cell geometry and the distance of the detector from the cell. The reason is that Garner and his coworkers noted in some instances variations in the measured cross sections when this solid angle was changed; they have described how corrections may need to be applied to the measured data in order to extract the true cross sections. It is even possible to extract information

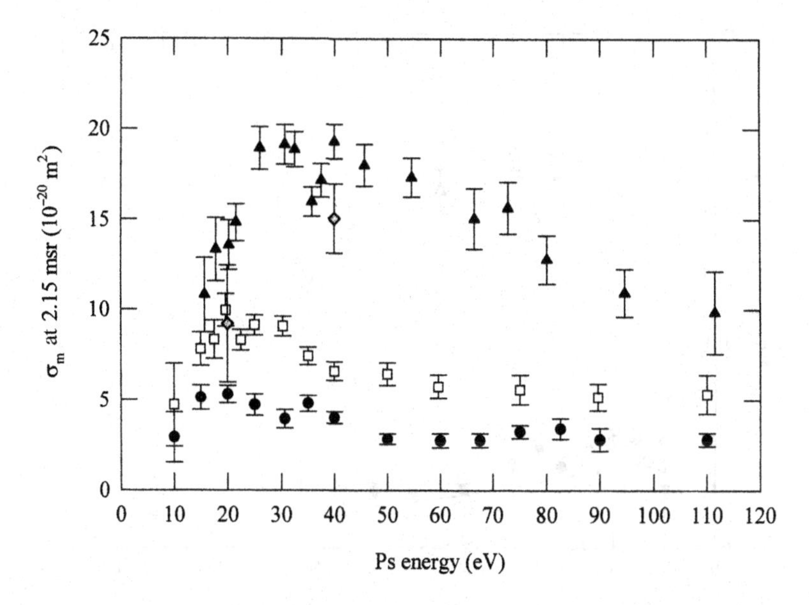

Fig. 7.25. Compendium of positronium total scattering cross sections from Garner, Özen and Laricchia (1998): •, helium; ▲, argon; □, molecular hydrogen; ◇, molecular oxygen.

concerning the angular dependence of the positronium elastic scattering cross section.

It is notable that the behaviour for all the gases is broadly similar, though only two energies have been investigated for molecular oxygen. The total cross section increases as the positronium energy is raised from its lowest value and then passes through a broad maximum before decreasing again at higher energies. The maximum is most pronounced for argon. The rise in the cross section may be due to positronium break-up, which is believed from theory to be an important channel; see e.g. the summary by Charlton and Laricchia (1991).

The results for helium and argon targets are shown in more detail in Figure 7.26, along with the relevant theory (see subsection 7.2.2). For helium, the data of Garner, Laricchia and Özen (1996) supersede those of Zafar *et al.* (1991), which were obtained using an indirect method. The results of Coleman *et al.* (1994) and Nagashima *et al.* (1998) for the average momentum transfer cross section at low energies, derived using the ACAR technique as described in section 7.4, are of comparable magnitude. The theoretical results of McAlinden, MacDonald and Walters (1996) are in reasonably good agreement with experiment above approximately 60 eV, but they markedly disagree at low energies, probably owing to neglect of exchange. The semi-classical calculations of Peach

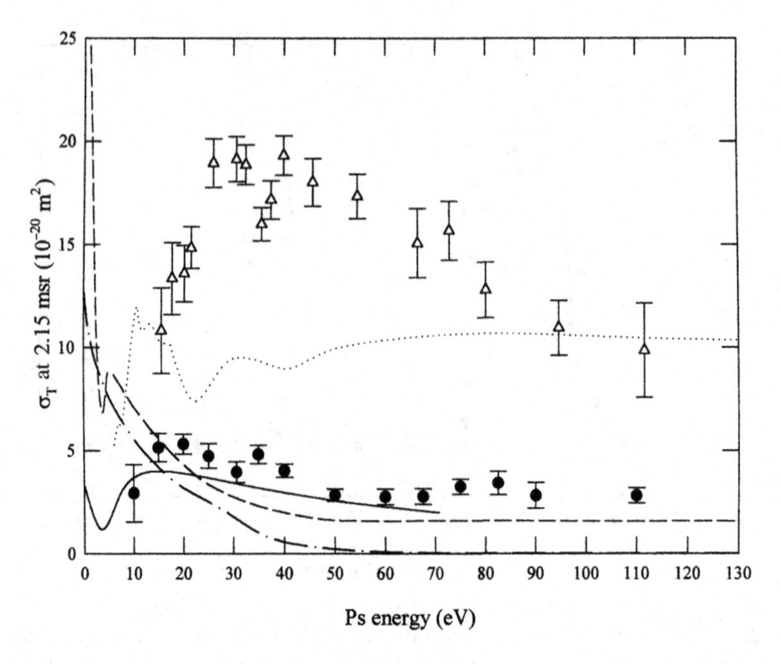

Fig. 7.26. Positronium total scattering cross sections for helium and argon gases. Experiment: Garner, Laricchia and Özen (1996); •, He; △, Ar. Theory for helium: ——, Peach (1993, private communication to G. Laricchia); – – –, McAlinden, McDonald and Walters (1996); — · —, Sarkar and Ghosh (1997), elastic component only. Theory for argon: · · · · ·, McAlinden, MacDonald and Walters (1996).

(1993, private communication to G. Laricchia) are in better accord with experiment, both in shape and magnitude, at least up to 70 eV. For argon the results of Garner, Özen and Laricchia (1998) are somewhat higher that those of Zafar *et al.* (1996) (not shown in Figure 7.26), owing to lack of angular resolution in the latter. The theory of McAlinden, MacDonald and Walters (1996) again yields results in poor accord with experiment at lower energies, though tending towards the experimental values as the energy is increased.

The results described in this section mark only the beginnings of positronium–atom(molecule) collision studies. Investigations in the intermediate energy range and the measurement of total cross sections of comparable, or better, accuracy should soon be available for a variety of targets. Extensions to both higher and lower positronium energies await developments in beam production and detection techniques. With an eye to the future, Charlton and Laricchia (1991) identified a number of positronium reactions which would be of interest to study, and we reproduce their list here:

$$
\begin{aligned}
\mathrm{Ps} + X &\rightarrow \mathrm{Ps} + X && \text{elastic scattering;} \\
&\rightarrow \mathrm{Ps}^* + X && \text{projectile excitation;} \\
&\rightarrow \mathrm{Ps} + X^* && \text{target excitation;} \\
&\rightarrow \mathrm{e}^+ + \mathrm{e}^- + X && \text{projectile ionization or break-up;} \\
&\rightarrow \mathrm{Ps} + X^+ + \mathrm{e}^- && \text{target ionization.}
\end{aligned}
$$

8
Exotic species involving positrons

We now consider several of the more exotic systems in which one or more positrons may be involved, some of which were introduced in subsection 1.2.3. The positronium negative ion ($e^-e^+e^-$), Ps^-, has been observed in the laboratory (Mills, 1981) and its lifetime against annihilation determined experimentally (Mills, 1983b). We discuss these experiments and the relevant theory in section 8.1. Observation of the positronium molecule, Ps_2, and other systems containing more than one positron or positronium atom (as yet unrealized) depends upon the generation of large instantaneous densities of positrons. The situation here is more encouraging than might be expected, owing to progress in developing very intense brightness-enhanced and time-focussed beams, as summarized in subsection 1.4.4. Many-positron systems and how they may be observed are described in section 8.2.

Antihydrogen, as discussed in subsection 1.2.3, has recently been observed in the laboratory, although only at relativistic speeds. However, progress with the trapping of cold antiprotons and positrons, and the production of positronium in a cryogenic environment, leads us to anticipate the synthesis of antihydrogen atoms with very low kinetic energies (or temperatures); thus it may be possible to trap them, and perform precision spectroscopy upon them. The motivation for the production of low temperature antihydrogen is described in section 8.3, along with the mechanisms and methodologies involved in some of the proposed formation processes.

8.1 The positronium negative ion

An electron can bind to positronium to form Ps^-, provided that the two electrons are in a singlet spin state. This system, and its charge conjugate counterpart consisting of two positrons and one electron, was predicted to be bound by Wheeler (1946), and was first detected experimentally by Mills (1981). One of the most accurate values of the energy of the bound state with respect to break-up into positronium plus an electron (or positronium plus a positron for the charge conjugate system) is −0.326 677 eV, which was obtained by Bhatia and Drachman (1983) using the Rayleigh–Ritz variational method with a 220-term Hylleraas trial function of the form

$$\Psi_{\mathrm{Ps}^-} = \sum_i c_i(1 + \mathrm{P}_{12}) \exp[-(\alpha_1 r_1 + \alpha_2 r_2)]\, r_1^{k_i} r_2^{l_i} r_{12}^{m_i}, \qquad (8.1)$$

where $\boldsymbol{r}_1$ and $\boldsymbol{r}_2$ are the position vectors of the two identical particles relative to the third one, and $r_{12} = |\boldsymbol{r}_1 - \boldsymbol{r}_2|$. Very similar results have been obtained by Ho (1983, 1993), also using a Hylleraas trial function, and by Petelenz and Smith (1987) and Frolov and Yeremin (1989), using exponential variation expansions. No other bound states of Ps^- exist, but the system does possess several autoionizing states, or resonances, in the electron–positronium continuum, the energies and widths of which were studied extensively by Ho (1984), using the complex coordinate rotation method.

The rate of annihilation of the positron with one of the electrons into two gamma-rays is given in terms of the bound state wave function, Ψ_{Ps^-}, as

$$\begin{aligned} \Gamma &= 2\pi\alpha^4 \left(\frac{c}{a_0}\right) \left[1 - \alpha\left(\frac{17}{\pi} - \frac{19\pi}{12}\right)\right] \frac{\langle \Psi_{\mathrm{Ps}^-}|\delta(\boldsymbol{r}_1)|\Psi_{\mathrm{Ps}^-}\rangle}{\langle \Psi_{\mathrm{Ps}^-}|\Psi_{\mathrm{Ps}^-}\rangle} \\ &= 100.617 \frac{\langle \Psi_{\mathrm{Ps}^-}|\delta(\boldsymbol{r}_1)|\Psi_{\mathrm{Ps}^-}\rangle}{\langle \Psi_{\mathrm{Ps}^-}|\Psi_{\mathrm{Ps}^-}\rangle} \quad (\mathrm{ns}^{-1}). \end{aligned} \qquad (8.2)$$

Using their most accurate wave function, Bhatia and Drachman (1983) obtained $\Gamma = 2.0861\ \mathrm{ns}^{-1}$, in excellent agreement with the experimental value of $2.09 \pm 0.09\ \mathrm{ns}^{-1}$; see below. This value is also close to the weighted average of the annihilation rates for the singlet and triplet states of free ground state positronium (see subsection 7.1.1), as would be expected for a system having the configuration of positronium weakly bound to an electron. Ho (1983) obtained similar theoretical values and also calculated the angular correlation function for the two gamma-rays created in the annihilation process (see subsection 1.3.3), obtaining a full width at half-maximum of 1.4 mrad.

The photodetachment of Ps^- was suggested by Mills (1981) as a potential source of a tunable-energy positronium beam. Once produced and accelerated electrostatically to the required energy, the Ps^- would undergo photodetachment to form the desired beam according to

$$\mathrm{Ps}^- + \text{photon} \rightarrow \mathrm{Ps} + \mathrm{e}^-. \qquad (8.3)$$

The photodetachment cross section for a photon of angular frequency ω is, in the length formulation,

$$\sigma_\omega(\mathrm{L}) = \tfrac{2}{9}k\omega\alpha a_0^2 \langle \Psi_k | Q_\mathrm{L} | \Psi_{\mathrm{Ps}^-} \rangle, \qquad (8.4)$$

where Ψ_k is the wave function for p-wave electron–positronium elastic scattering at electron wavenumber k and $Q_\mathrm{L} = \frac{2}{3}\boldsymbol{k}\cdot(\boldsymbol{r}_1 + \boldsymbol{r}_2)$ is the dipole transition operator. Energy conservation gives the relationship between ω and k as

$$\hbar\omega + E_{\mathrm{Ps}^-} = E_\mathrm{Ps} + \tfrac{3}{2}k^2, \qquad (8.5)$$

allowing for the recoil of the positronium. An alternative expression for the photodetachment cross section, in the so-called velocity formulation, is

$$\sigma_\lambda(v) = \frac{2k\alpha a_0^2}{9\omega} \langle \Psi_k | Q_\mathrm{V} | \Psi_{\mathrm{Ps}^-} \rangle, \qquad (8.6)$$

where $Q_\mathrm{V} = 4\boldsymbol{k}\cdot(\nabla_{\boldsymbol{r}_1} + \nabla_{\boldsymbol{r}_2})$. The difference between the results obtained using these two expressions, which would be zero if the exact initial and final wave functions could be used, provides a measure of their actual quality.

The photodetachment process was first investigated theoretically by Bhatia and Drachman (1985), using a simple asymptotic form for Ψ_{Ps^-} and the p-wave component of a plane wave for the electron in the final state. Much more detailed studies have since been made by Ward, Humberston and McDowell (1987). These authors calculated accurate singlet p-wave electron–positronium scattering wave functions using the Kohn variational method in a similar manner to that described previously in section 3.2 for positron–hydrogen scattering; they used a very accurate Hylleraas wave function for the ground state wave function of Ps^-. Their results for both the length and velocity formulations are given in Figure 8.1, together with the results of Bhatia and Drachman (1985). The sharp rise in the photodetachment cross section found by Ward, Humberston and McDowell (1987) for wavelengths less than 4000 Å is caused by a series of Feshbach resonances just below the $n_\mathrm{Ps} = 2$ excitation threshold of positronium.

The apparatus used by Mills (1981) for the first observation of Ps^- was similar to that shown in Figure 8.2. Slow positrons were guided by an axial magnetic field onto a 40 Å thick carbon film G_2. The kinetic energy of the positrons was adjusted so that some could penetrate the foil and emerge bound to two electrons as Ps^-. The geometry and method are analogous to the production of H^- by proton bombardment of thin foils (see e.g. Allison, 1958).

The grid G_3 located behind the carbon film was biassed positively so as to accelerate the Ps^- but return any transmitted positrons to the foil.

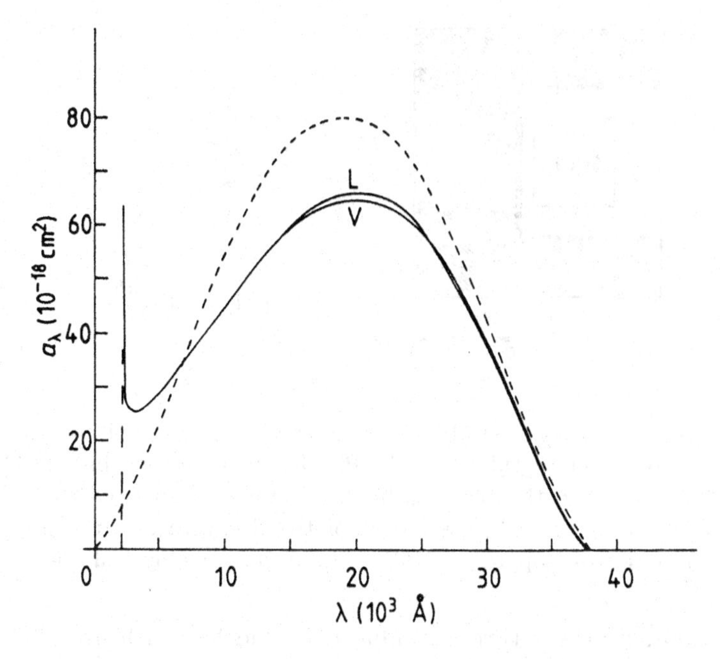

Fig. 8.1. The photodetachment cross sections for Ps^-: ——, results obtained with very accurate variational wave functions; L, length formulation; V, velocity formulation (Ward, Humberston and McDowell, 1987); – – –, plane wave approximation (Bhatia and Drachman, 1985). The vertical broken line marks the position of the $n_{Ps} = 2$ excitation threshold of positronium.

The gamma-ray energy spectra recorded by the Ge(Li) detector, shown in Figure 8.3, illustrate that as the voltage difference, $W = V_G - V_C$, between the grid and the carbon foil was increased, a small peak was discerned which was blue-shifted with respect to the 511 keV annihilation line, the shift being dependent on W. This is the signature of the Doppler-shifted two-gamma annihilation of Ps^- moving towards the detector. The arrows in Figure 8.3 indicate the expected positions of the peak, whose shift from an energy of 511 keV is given by

$$\Delta E_\gamma = [\lambda + (2\lambda + \lambda^2)^{1/2} \cos\phi] mc^2, \tag{8.7}$$

where $\lambda = eW/(3mc^2)$ and ϕ is the angle between the direction of gamma-ray emission and the Ps^- velocity.

An averaged value of $\cos\phi$ (ϕ is dependent on the detection geometry) was used in evaluating ΔE_γ. A detailed analysis performed by Mills (1981) allowed for the fact that the Ps^- underwent acceleration for a significant proportion of its lifetime. From this he was able to deduce that the shifted annihilation line arose from a system with a mass-to-charge

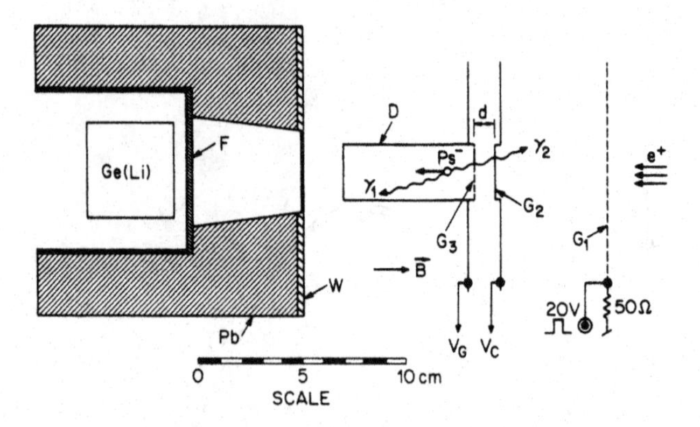

Fig. 8.2. Apparatus used by Mills for studies of the positronium negative ion; G_1 is a pile-up-reducing grid, G_2 is the Ps^--forming carbon film and G_3 is the acceleration grid. See the text for further details. Reprinted from *Physical Review Letters* **50**, Mills, Measurement of the decay rate of the positronium negative ion, 671–674, copyright 1983 by the American Physical Society.

ratio of approximately three, produced in the foil with an efficiency of 3×10^{-4} of the incident positrons.

Following this observation, Mills (1983b) measured the decay rate of Ps^- using the apparatus of Figure 8.2. Positrons were guided through the grid G_1, where they were accelerated by the voltage applied to the carbon film G_2. As before, Ps^- was formed on the transmission side of the foil and accelerated by the potential applied to G_3, which was separated from the carbon film by a distance d. Again, the blue-shifted gamma-rays from the in-flight annihilation of Ps^- moving towards the detector were distinguished from the 511 keV gamma-rays due to positron annihilation in the film. The region beyond G_3 was field-free.

Mills argued that the intensity of the Doppler-shifted gamma-ray peak is proportional to the probability that the Ps^- reaches the grid G_3, and that the time interval, t, between emission and arrival at G_3 is proportional to d for a given potential difference W between G_3 and G_2. It follows that the intensity of the Ps^- peak will be exponentially dependent on d so that, by measuring the intensity at various values of d, the lifetime of Ps^- can be determined.

Figure 8.4 shows a plot of the logarithm of f^-, which is the ratio of the Ps^- and 511 keV photopeaks versus the parameter $t/g(\lambda, \epsilon)$. This is essentially the time, evaluated from

$$t = \left(\frac{d}{\lambda c}\right) \ln[1 + \lambda + (2\lambda + \lambda^2)^{1/2}] g(\lambda, \epsilon), \tag{8.8}$$

where λ is as defined after equation (8.7), $\epsilon = T/(3mc^2)$ and the function

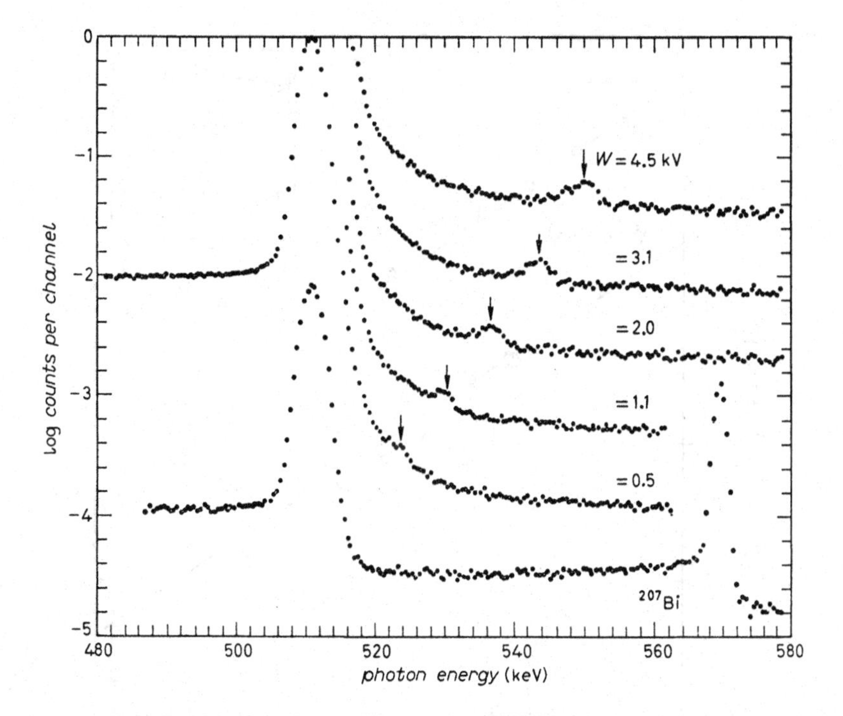

Fig. 8.3. Annihilation gamma-ray spectra showing the blue-shifted photo-peak due to Ps^- at various values of W, the acceleration potential. The arrows indicate the expected positions of these peaks. The ^{207}Bi line (at approximately 569.6 keV) used for calibration is also shown. Reprinted from *Positron Solid-State Physics*, Proceedings of the International School of Physics 'Enrico Fermi', Course 83, Mills, Experimentation with low energy positron beams, 432–509, copyright 1983, with permission from Elsevier Science.

$g(\lambda, \epsilon)$ is a correction for the non-zero initial kinetic energy, T, of the Ps^-; this was written by Mills as

$$g(\lambda, \epsilon) = 1 - \left(\frac{T}{W}\right)^{1/2} + \frac{1}{2}\left(\frac{T}{W}\right) + \cdots. \tag{8.9}$$

The data are fitted by straight lines which yield values of $g\Gamma$, where Γ is the desired Ps^- decay rate: $g\Gamma = 1.851 \pm 0.058$ ns^{-1} for $W = 1000$ eV and 1.971 ± 0.034 ns^{-1} for $W = 3936$ eV. These values are plotted against $W^{-1/2}$ in the inset of Figure 8.4; this, from equation (8.9), should give an approximately linear dependence, according to $g\Gamma = \Gamma(1 - (T/W)^{1/2})$. Extrapolation to infinite W gives $\Gamma = 2.09 \pm 0.09$ ns^{-1}, which is in good agreement with, but much less precise than, the calculated values of Bhatia and Drachman (1983) and Ho (1983, 1993). The energy of emission of the Ps^-, T, is found to be in the range 3–32 eV, with a mean value

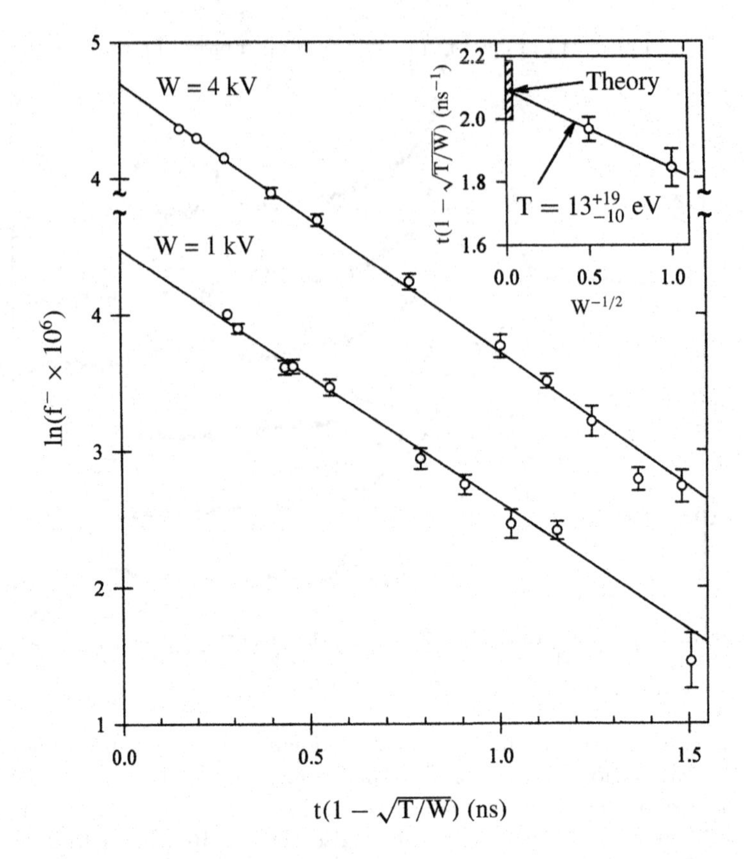

Fig. 8.4. Plot of the logarithm of f^- versus $t/g(\lambda, \epsilon)$; see text. The inset shows the extrapolation to infinite acceleration potential to yield the decay rate of Ps^-. Here the small hatched rectangle covers the experimental error, and the value calculated by Ho (1983, 1993) is indicated.

of 13 eV. The level of precision in this experiment was dominated by the counting statistics, and a preliminary account of an improved experiment (Mills, Friedman and Zuckerman, 1990) has been given, though no new results for Γ have been reported.

8.2 Systems containing more than one positron

Although positron plasmas can be considered to be systems containing many positrons, and as such technically fall within the scope of this section, we will not consider them here. Rather, we will concentrate on the theory of, and the possibilities of observing, assemblages of particles containing both positrons and electrons. These include the positronium molecule and a Bose–Einstein (BE) condensate of positronium atoms.

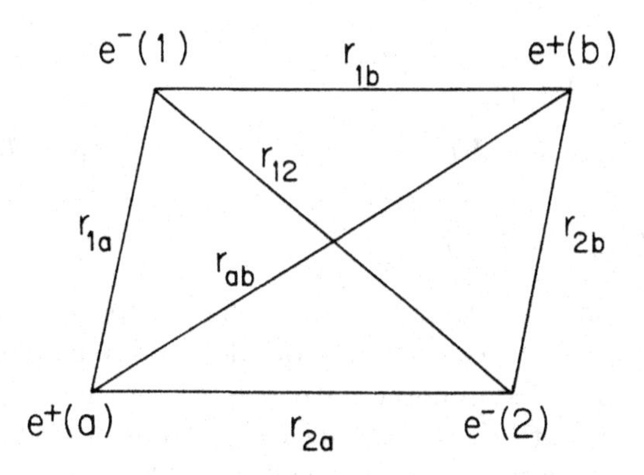

Fig. 8.5. The coordinate system for the positronium molecule. Particles a and b are the two positrons and particles 1 and 2 are the two electrons.

A discussion of some applications of positron plasmas in atomic-physics investigations is given in Chapter 6, and more general information can be found in the review of Greaves, Tinkle and Surko (1994).

Mills (1984) pointed out that it might soon be experimentally feasible to realize systems containing positrons which overlapped one with another, both spatially and temporally. This was due to the production of time-focussed beams and the prospects (since demonstrated) for brightness-enhanced, highly focussed beams.

Apart from the system consisting of two positrons and one electron, which is merely the charge conjugate of Ps^-, the simplest bound system containing two positrons is the positronium molecule, Ps_2. In order that binding can take place, the two electrons must be in a singlet spin state, and the two positrons likewise. The wave function is therefore symmetric under the interchange of the spatial coordinates of the electrons and the positrons separately.

Wheeler (1946) had speculated that a bound state of two positronium atoms might exist, but he was unable to prove it. The first successful attempt to establish theoretically that Ps_2 is bound was made by Hylleraas and Ore (1947), who calculated the binding energy with respect to break-up into two positronium atoms as 0.116 eV. Later calculations, by Lee, Vashista and Kalia (1983) and Ho (1986a), gave a binding energy of approximately 0.41 eV, but significantly larger values, 0.978 eV and 0.846 eV, were obtained by Sharma (1968) and Huang (1973) respectively. There is, however, some doubt about the reliability of these larger values, and the result obtained by Ho (1986a) is probably more accurate, although his wave function was of a somewhat restricted form. A trial function of the form

$$\begin{aligned}\Psi_{\mathrm{Ps}_2} = &\ (1+P_{12})(1+P_{ab}) \\ &\times \sum_i c_i \exp[-(d_1 r_{1a} + d_2 r_{2a} + d_3 r_{ab} + d_4 r^r_{1b} + d_5 r_{2b} + d_6 r_{12})] \\ &\times r_{1a}^{k_i} r_{2a}^{l_i} r_{ab}^{m_i} r_{1b}^{n_i} r_{2b}^{p_i} r_{12}^{q_i}, \qquad (8.10)\end{aligned}$$

where the nomenclature is that of Figure 8.5, incorporates all the required spatial symmetry for the two pairs of identical particles, but in fact Ho ignored the spatial symmetry with respect to the two positrons and also set the coefficients d_4, d_5 and d_6 equal to zero. There is no valid justification for treating the pair of positrons differently from the pair of electrons, but his Ps_2 wave function was a simple modification of the trial wave function he had used for positronium hydride, PsH, where the two positively charged particles, the positron and the proton, are distinguishable; see section 7.5. Other calculations have been made, by Kinghorn and Poshusta (1993) and Kozlowski and Adamowicz (1993), using trial wave functions which incorporated spatial symmetry with respect to both pairs of identical particles. Up to 300 correlated Gaussian functions were included, and a binding energy 0.435 eV was obtained, slightly larger than the value of Ho.

The most recent calculation, and probably the most accurate, is that of El-Gogary *et al.* (1995), who obtained the value 0.573 eV using a fully symmetrized trial wave function containing only 22 terms, each being of the Hylleraas type but expressed in terms of prolate spheroidal coordinates. These authors attributed the fact that they obtained a result significantly improved over most of the previous values with a trial wave function containing far fewer terms to the better physical structure of their wave function. In addition to providing a detailed account of their own calculations, El-Gogary and coworkers also gave a comprehensive review of other calculations of the binding energy of Ps_2.

Electron–positron annihilation in Ps_2 was investigated by Tisenko (1981), who, using the relatively simple wave function of Hylleraas and Ore (1947), obtained the annihilation rates into two and three gamma-rays as 16 ns^{-1} and 0.043 ns^{-1} respectively. No such calculations have been performed using the more elaborate wave function of Ho.

As with Ps^-, there is only one bound state of Ps_2 but there exist Rydberg series of autodissociating states arising from the attractive interaction between one of the positrons and the residual Ps^- (or between one of the electrons and the charge conjugate of Ps^-). The positions and widths of several of these states were determined by Ho (1989) using the complex coordinate rotation method. To date Ps_2 has not been observed in the laboratory.

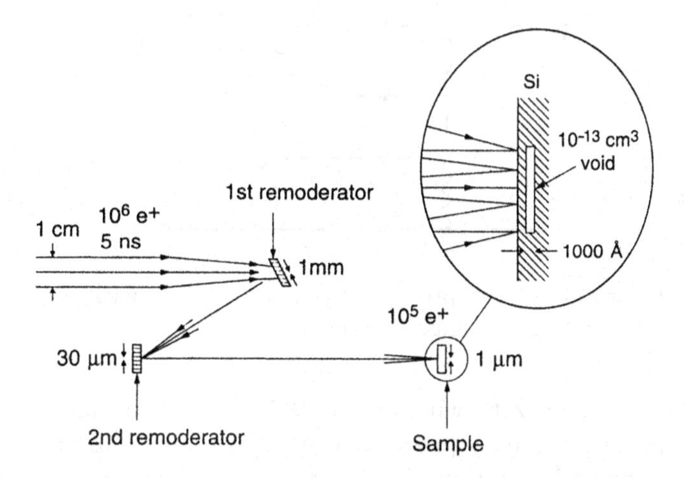

Fig. 8.6. Schematic of a proposed configuration for the production of conditions for Bose–Einstein condensation of positronium using a pulsed, brightness-enhanced positron beam (see text for details). Reprinted from *Physical Review B* **49**, Platzman and Mills, Possibilities for Bose condensation of positronium, 454–458, copyright 1994 by the American Physical Society.

The possibility of producing a system of positronium atoms at a sufficiently high density and low temperature to produce Bose–Einstein (BE) condensation has been raised by Liang and Dermer (1988) and Platzman and Mills (1994). The former authors outlined a scheme in which positronium atoms are laser-cooled in vacuum, which seems feasible despite their short lifetimes because of their low mass. The required temperature of the positronium is around 0.1 K and the density is 10^{15} cm^{-3}. Liang and Dermer (1988) argued that the overall scheme appears possible, but hitherto neither the temperature nor the density condition has been approached and laser cooling of positronium has not yet been attempted.

Platzman and Mills (1994) proposed a completely different approach to BE condensation. A schematic illustration encapsulating their suggestion is reproduced in Figure 8.6. It comprises a beam of 5 ns pulses, each containing 10^6 positrons, which was brightness-enhanced twice at remoderators (Mills, 1980, and subsection 1.4.4) before being focussed in a 1 µm spot on to a target. The latter was a cold silicon sample with a deliberately introduced void, 1 µm in diameter and 1000 Å deep, located at a similar depth below the surface. Platzman and Mills estimated that around 25% of the implanted positrons would reach the cavity and be re-emitted as positronium with kinetic energies in the eV range. The para-positronium will decay rapidly, leaving the 'hot' ortho-positronium in the cavity at a density of around 10^{18} cm^{-3}. Platzman and Mills (1994) also identified the need to use a polarized low energy positron beam, such

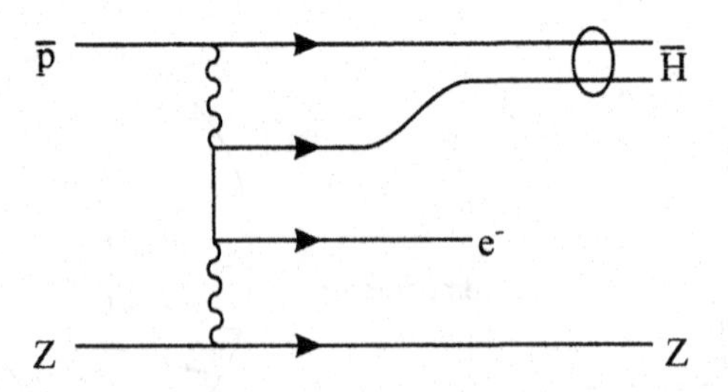

Fig. 8.7. Illustration of the process of pair production with capture for the formation of antihydrogen at high kinetic energies.

as that described by Zitzewitz *et al.* (1979), which utilizes the natural polarization of the β-decay process, resulting in polarization of the beam in the direction of its motion. This is to avoid rapid annihilation of the positronium atoms arising from spin exchange during collisions with one another. These authors then showed, using relatively simple arguments about the nature of the interactions of positronium atoms with the silicon surface, that the slowing-down rate corresponds to a temperature drop of around 20 K every 10^{-10} s, so that the positronium temperature should fall below the BE condensation temperature (around 20 K at the expected positronium densities) on a nanosecond time scale. The presence of the condensate could be monitored by inducing triplet-to-singlet transitions, e.g. by turning on a magnetic field, and using the angular correlation technique, subsection 1.3.3, to monitor the positronium momentum distribution. The condensate signature should be a strong peak at almost zero momentum. Further interesting observations which may be made using a system of positronium atoms trapped in such a small volume, such as of Ps–Ps collisions and the temperature and density dependences of the condensate, were also noted by Platzman and Mills (1994).

Other speculations on uses for dense positronium systems exist in the literature, such as in the creation of a gamma-ray laser (e.g. Liang and Dermer, 1988, and references therein; Vlasov, Gadomskii and Shageev, 1990), but further discussion is beyond the scope of the present treatment.

8.3 Antihydrogen

1 Introduction

Antihydrogen was recently observed at CERN by Baur *et al.* (1996) and at Fermilab by Blanford *et al.* (1998). The production mechanism relied upon pair production with capture during the interaction of an energetic antiproton ($\bar{\rm p}$) with an atomic nucleus, Z: $\bar{\rm p} + Z \rightarrow \bar{\rm H} + {\rm e}^-$. The Feynman diagram illustrating this process is shown in Figure 8.7.

Unfortunately, the high speed ($\simeq 0.9c$) of the few antihydrogen atoms which were created meant that they were destroyed as they struck the first solid object in their path. Although, with increased production rates, it may be feasible to attempt some experiments on relativistic antihydrogen produced in this way (Munger, Brodsky and Schmidt, 1994), detailed and challenging comparisons between hydrogen and antihydrogen are probably not possible. Thus, the rest of this section is devoted to the motivation for, and the progress towards, the production of antihydrogen at low energies.

The motives for producing and studying antihydrogen have been summarized by Hughes (1993a), Charlton *et al.* (1994) and Holzscheiter and Charlton (1999) and centre mainly around tests of CPT invariance and the weak equivalence principle (WEP). As pointed out by Hughes (1993a), CPT is the minimal invariance condition for the existence of antiparticles within quantum field theory, since there is no proof of invariance under the individual C, P and T operations. Some of the implications of CPT invariance are that particle and antiparticle should have equal but opposite charges and equal masses, lifetimes and gyromagnetic ratios. In addition, the CPT symmetry of QED means that the spectra of hydrogen and antihydrogen are expected to be identical. In order to test this prediction it is necessary to make high-precision comparisons of the various transition frequencies. In particular, the metastable 2S level, with a lifetime of 0.125 s, offers the eventual possibility of fractional precision in the range 10^{-15}–10^{-18} of the frequency of the 1S–2S two-photon transition. Progress with the spectroscopy of hydrogen (summarized by Hänsch and Zimmermann, 1993) has allowed a precision of better than 1 part in 10^{11} to be achieved, and further improvements are forseen. A more detailed description of the potential for spectroscopic investigations of antihydrogen, including both trapped and untrapped atoms, was given by Charlton *et al.* (1994); they pointed out that due to the low excitation rates the Doppler-free two-photon 1S–2S transition seems only to be feasible if trapping is implemented. Techniques which may make this possible have been described by Walraven (1993) and Hänsch and Zimmermann (1993), and Cesar *et al.* (1996) have recently performed spectroscopy on trapped hydrogen atoms.

The other main potential testing ground for antihydrogen is, as mentioned above, the WEP. Several authors (e.g. Gabrielse, 1988; Beverini *et al.*, 1988; Phillips, 1997) have suggested schemes for making 'direct' measurements of the gravitational acceleration of antihydrogen, thereby providing a WEP test for the antiproton, although the level of precision which might be obtained is difficult to assess. Hughes and Holzscheiter (1992) pointed out that a WEP test for the positron could be obtained from high-precision antihydrogen spectroscopy, owing to the variation in the red shift of the transition frequencies at different locations in a gravitational potential. For instance, if the 1S–2S transition frequencies were measured for both hydrogen and antihydrogen as the gravitational potential at the surface of the Earth changed due to its eccentric orbit around the Sun and if they were found not to be equal but to vary according to the strength of the potential, then the WEP would be invalid for antihydrogen. (Tests of the WEP for matter were summarized by Hughes, 1993b.) An estimate of the precision necessary for such a test suggests that it is rather daunting, e.g. if no hydrogen–antihydrogen frequency difference emerged after three months at a level of 1 part in 10^{18}, the WEP would have been tested to around one part in 10^9 for the positron. The level of precision would be reduced by around three orders of magnitude if the daily variation in the gravitational potential were used instead.

As well as the fundamental-physics reasons for producing antihydrogen, there are also others based upon understanding its interaction with hydrogen (see Campeanu and Beu, 1983, and references therein), which has been a topic of debate for some time due to a possible cosmological significance. Finally, by creating small amounts of antihydrogen and learning how to manipulate these anti-atoms, the way will be paved for the creation of heavier antimatter systems.

2 Antiproton deceleration, trapping and cooling

The production and manipulation of antiprotons is routine at some high energy accelerator laboratories. In the following discussion we concentrate on the machines at CERN, where the low energy antiproton ring (LEAR) (which closed in December 1996) was a unique facility for the storage of antiprotons at low energies (≥ 6 MeV). It was by extracting bursts of these particles from LEAR that advances in their deceleration and subsequent trapping and cooling have been possible. A brief review of the operation at LEAR and associated machines can be found in Charlton *et al.* (1994).

The first successful capture of antiprotons in a trap was achieved by Gabrielse *et al.* (1986), with some later improvements and modifications described by Gabrielse *et al.* (1990) and Holzscheiter *et al.* (1996). In a typical trapping sequence, antiprotons in a 200 ns burst were moderated through their interactions with matter as they entered a vacuum chamber containing a Penning trap. Fine tuning of the amount of moderating material is necessary to optimize the yield of captured particles, and this was achieved using gas cells. Radial confinement of the antiprotons was provided by the 6 T magnetic field of a superconducting solenoid.

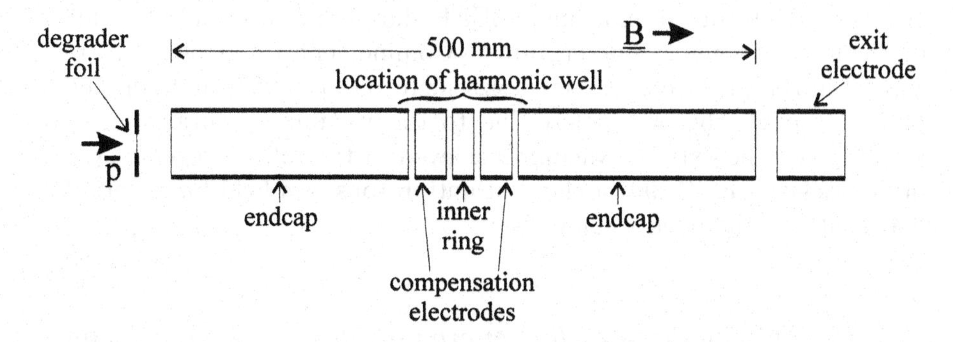

Fig. 8.8. Schematic illustration of the electrode structure of an antiproton catching and cooling trap. The dimension shown is approximate and is illustrative of the trap size used by Holzscheiter *et al.* (1996).

The electrostatic elements of a typical trap are, as schematically illustrated in Figure 8.8, an arrangement of cylindrical tubes located along the axis of the magnetic field to form a so-called open-endcaps Penning trap (Gabrielse, Haarsma and Rolston, 1989). Appropriate potentials applied to the inner elements of this array form a harmonic trap for long-term confinement of the antiprotons. Superimposed on this is a catching trap comprising an exit electrode, to which a large negative d.c. potential is applied, and an aluminium degrader foil at the trap entrance. Some of the antiprotons which leave the degrader foil are repelled by the endcap voltage and returned towards the entrance, where the sudden application of a potential to the foil completes the trapping procedure. Overall capture efficiencies (the fraction of the positrons ejected by LEAR that was trapped) were as high as 0.5% for a 15 kV trapping potential.

In the absence of a rapid cooling process, the antiprotons travel back and forth in the trap maintaining their initial kinetic energy to a high degree. (Synchrotron cooling is a slow process for heavy particles, with a time constant for antiprotons of around 5×10^8 s in a 6 T field). Fortunately, electrons can also be introduced into the trap in large numbers, where they rapidly cool to the ambient temperature (close to 4.2 K) and settle in the small axial harmonic well shown in Figure 8.8. As the antiprotons pass to and fro, they cool by dissipating their kinetic energy to the trapped electrons, to which they are strongly coupled by the Coulomb interaction. Within a time period of around tens or hundreds of seconds they reach thermal equilibrium with the electrons and thus are also trapped in the well. The electrons can be removed without disturbing the antiprotons by, for instance, selectively exciting the axial component of their energy or simply by rapidly raising and relowering the harmonic potential well. By 1996 it had become routine to be able to capture

10^5–10^6 antiprotons from a single LEAR shot and to cool more than 90% of them to cryogenic temperatures. Confinement times of hours were typical in the work reported by Holzscheiter *et al.* (1996) and Feng *et al.* (1997), though this was solely due to the vacuum conditions in their trap. Under the extreme vacuum achieved in the fully cryogenic system of Gabrielse *et al.* (1990) a cloud of antiprotons was held for a period of 3.4 months without noticeable loss.

3 Possible methods of low energy antihydrogen production

Here we briefly describe five reactions, summarized by the following equations, which have been proposed for the production of low energy antihydrogen:

$$\bar{\mathrm{p}} + \mathrm{e}^+ \rightarrow \bar{\mathrm{H}} + h\nu \tag{8.11}$$

$$\bar{\mathrm{p}} + \mathrm{e}^+ + mh\nu \rightarrow \bar{\mathrm{H}} + (m+1)h\nu \tag{8.12}$$

$$\bar{\mathrm{p}} + \mathrm{Ps} \rightarrow \bar{\mathrm{H}} + \mathrm{e}^- \tag{8.13}$$

$$\bar{\mathrm{p}} + \mathrm{e}^+ + \mathrm{e}^+ \rightarrow \bar{\mathrm{H}} + \mathrm{e}^+ \tag{8.14}$$

$$\begin{aligned} \bar{\mathrm{p}} + \mathrm{He} \rightarrow\ & \alpha\mathrm{e}^-\bar{\mathrm{p}} + \mathrm{e}^- \\ & \searrow + \mathrm{e}^+(\text{or Ps}) \rightarrow \bar{\mathrm{H}} + \mathrm{He}^+(\mathrm{He}). \end{aligned} \tag{8.15}$$

In the last method the antiproton forms an exotic metastable antiprotonic helium atom, which then reacts with deliberately introduced positrons (or positronium atoms).

The basic spontaneous radiative capture reaction (8.11) was first suggested as a source of antihydrogen by Budker and Skrinsky (1978); in it the photon carries away the excess energy. The matter equivalent of (8.11) has been studied for many years (see e.g. Bethe and Salpeter, 1977) and has been of interest recently since it is a loss mechanism which can occur when electrons are used to cool protons and positive ions in storage rings.

An illustration of the capture process is given in Figure 8.9, which shows the positron (or electron) occupying a state in a narrow temperature or energy band in the ionic continuum. In effect the antiprotons(protons) are virtually at rest in the positron(electron) gas. This is the case, for instance, for near equi-velocity particle beams, in which the kinetic energy, E_e, of the positrons(electrons) in the rest frame of the antiprotons(protons) is much less than the binding energy, E_0, of the lowest atomic bound state. Under these conditions the cross section for reaction (8.11) has been given in analytic form (Bethe and Salpeter, 1977) for

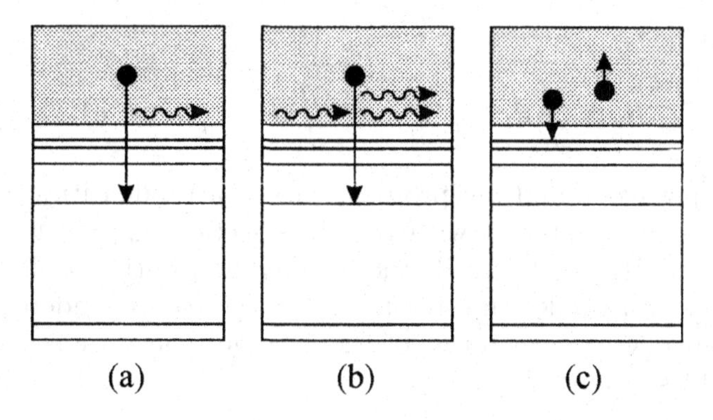

Fig. 8.9. Schematic energy diagrams illustrating recombination mechanisms. The ionization continuum is shown shaded. (a) Spontaneous radiative capture, reaction (8.11); (b) stimulated radiative capture by irradiation with laser light, reaction (8.12); (c) three-body recombination in which the excess energy is removed by an extra positron, reaction (8.14).

capture to a state of principal quantum number $n_{\bar{\mathrm{H}}}$, as

$$\sigma_{n_{\bar{\mathrm{H}}}} = 2^5 \frac{\pi\alpha^3}{3\sqrt{3}} a_0^2 E_0 \frac{1}{n_{\bar{\mathrm{H}}} E_{\mathrm{e}}(1 + n_{\bar{\mathrm{H}}}^2 E_{\mathrm{e}}/E_0)}; \tag{8.16}$$

thus

$$\sigma_{n_{\bar{\mathrm{H}}}} = 2 \times 10^{-22} E_0 \frac{1}{n_{\bar{\mathrm{H}}} E_{\mathrm{e}}(1 + n_{\bar{\mathrm{H}}}^2 E_{\mathrm{e}}/E_0)} \qquad (\mathrm{cm}^2), \tag{8.17}$$

where α is the fine structure constant and a_0 is the Bohr radius. We note that this cross section favours capture into low $n_{\bar{\mathrm{H}}}$ states. In practice, however, capture to a range of $n_{\bar{\mathrm{H}}}$ states is possible up to some cut-off value, $n_{\bar{\mathrm{H}}}^{\mathrm{cut}}$, which will be defined by experimental constraints. These are typically set by field ionization of nascent Ryberg atoms or even re-ionization by collisions with positrons which remain in the continuum. Here it is appropriate to define a total antihydrogen formation cross section, $\sigma_{\bar{\mathrm{H}}}$, as

$$\sigma_{\bar{\mathrm{H}}}(E_{\mathrm{e}}) = \sum_{n_{\bar{\mathrm{H}}}=1}^{n_{\bar{\mathrm{H}}}^{\mathrm{cut}}} \sigma_{n_{\bar{\mathrm{H}}}}(n_{\bar{\mathrm{H}}}, E_{\mathrm{e}}), \tag{8.18}$$

and Müller and Wolf (1997) have evaluated this cross section with $n_{\bar{\mathrm{H}}}^{\mathrm{cut}} =$ 2000, which is an essentially complete determination.

The most general expression for the recombination rate, $R_{\bar{\mathrm{H}}}$, involves a double integration over the phase space overlap of the clouds, thus incorporating terms arising from their spatial overlap and the velocity dependence:

$$R_{\bar{H}} = \int d\boldsymbol{r}\, n_{e^+}(\boldsymbol{r}) n_{\bar{p}}(\boldsymbol{r}) \int d\boldsymbol{v}\, \sigma_{n_{\bar{H}}}(v) v f(\boldsymbol{v}), \tag{8.19}$$

where the positrons and antiprotons, with spatial densities $n_{e^+}(\boldsymbol{r})$ and $n_{\bar{p}}(\boldsymbol{r})$ respectively, interact with the cross section $\sigma_{n_{\bar{H}}}(v)$, v being their relative speed; $f(\boldsymbol{v})$ is effectively the velocity distribution of the positrons. This expression includes the simplification that $f(\boldsymbol{v})$ is independent of $\boldsymbol{r}$ and is often rewritten in terms of the average relative speed, v_r, of the two ensembles of particles as

$$R_{\bar{H}}(v_r) = \alpha(v_r) \int n_{e^+}(\boldsymbol{r}) n_{\bar{p}}(\boldsymbol{r})\, d\boldsymbol{r}, \tag{8.20}$$

where $\alpha_r = \int \sigma(v) v f(\boldsymbol{v}) d\boldsymbol{v}$. Müller and Wolf (1997) described various approximations to $f(\boldsymbol{v})$ which allow $\alpha(v_r)$ to be determined. However, the simplest approximation, which does not allow for the overlap efficiency of the ensembles, is given by

$$R_{\bar{H}} = N_{\bar{p}} n_{e^+} \langle v_r \sigma(v_r) \rangle, \tag{8.21}$$

where v_r is now an average relative speed and $N_{\bar{p}}$ is the number of antiprotons.

Also illustrated in Figure 8.9 is the process whereby capture is stimulated by irradiation with laser light, represented by m photons each of energy $h\nu$, according to equation (8.12). The gain in this process over that of equation (8.11) was considered by Neumann *et al.* (1983), who found that the capture rate could be enhanced by a factor of 100. As reviewed by Wolf (1993), large laser-induced enhancements of the radiative recombination rates to various levels of hydrogen, using merged electron and proton beams, have been demonstrated in two separate experiments (Schramm *et al.*, 1991 and Yousif *et al.*, 1991).

Antihydrogen production by this method was initially conceived as taking place in one of the straight sections of LEAR itself, by the addition of a small storage ring to recirculate the positrons. Poth (1987) gave a comprehensive account of the relevant antiproton intensities and the positron production, accumulation and recirculation scenarios which would maximize $R_{\bar{H}}$. Here the antihydrogen retains the velocity of the antiproton and is readily separated from the stored beam at one of the corners of the storage ring. Detection of this high energy collimated antihydrogen beam can thus be accomplished. Further discussion of a possible realization of this reaction was given by Meshkov and Skrinsky (1997). Detailed accounts of potential physics experiments with an antihydrogen beam have been given by Neumann (1987) and Meshkov (1997).

As outlined by Wolf (1993) and Holzscheiter *et al.* (1996), similar considerations to those described above also apply to recombination in traps, and in particular the nested Penning-trap scheme (see below), again with appropriate assumptions regarding the speed distributions of the trapped positrons and antiprotons and their degree of spatial overlap. As an example, Holzscheiter *et al.* (1996) argued that the recombination rates are of the order of one per second (though dependent upon E_e) for 10^6 positrons and 10^5 antiprotons trapped in a volume of 1 cm^3.

Use of the charge-exchange mechanism, reaction (8.13), to produce antihydrogen was first proposed by Deutch *et al.* (1986), and subsequently it was shown that the cross section for this process could be obtained by applying the charge conjugation and time reversal operators to the process of positronium formation in positron–hydrogen collisions (Humberston *et al.*, 1987, and see section 4.2). Under time reversal, the positronium formation process equation (4.5) becomes

$$\mathrm{Ps} + \mathrm{p} \rightarrow \mathrm{H} + \mathrm{e}^+, \tag{8.22}$$

with a cross section σ_{H}. The further application of charge conjugation then yields reaction (8.13), for which the cross section is also σ_{H}, assuming charge conjugation invariance. Time reversal invariance implies the symmetry of the S-matrix, from which the relationship between σ_{H} and σ_{Ps} can be derived. Consider a positron colliding with a hydrogen atom, in a state with principal and orbital angular momentum quantum numbers n_{H} and l_{H} respectively, and producing positronium with corresponding quantum numbers n_{Ps} and l_{Ps}. If the cross section for this process is $\sigma_{\mathrm{Ps}}(n_{\mathrm{Ps}}, l_{\mathrm{Ps}}; n_{\mathrm{H}}, l_{\mathrm{H}})$, then the cross section for the time-reversed process is $\sigma_{\mathrm{H}}(n_{\mathrm{H}}, l_{\mathrm{H}}; n_{\mathrm{Ps}}, l_{\mathrm{Ps}})$, and

$$\begin{aligned} \sigma_{\bar{\mathrm{H}}}(n_{\bar{\mathrm{H}}}, l_{\bar{\mathrm{H}}}; n_{\mathrm{Ps}}, l_{\mathrm{Ps}}) &= \sigma_{\mathrm{H}}(n_{\mathrm{H}}, l_{\mathrm{H}}; n_{\mathrm{Ps}}, l_{\mathrm{Ps}}) \\ &= \frac{k^2(2l_{\mathrm{H}}+1)}{\kappa^2(2l_{\mathrm{Ps}}+1)}\sigma_{\mathrm{Ps}}(n_{\mathrm{Ps}}, l_{\mathrm{Ps}}; n_{\mathrm{H}}, l_{\mathrm{H}}), \end{aligned} \tag{8.23}$$

where k and κ are the wave numbers of the positron and positronium respectively. Energy conservation gives the relationship between them as

$$E = \frac{k^2}{2} - \frac{1}{2n_{\bar{\mathrm{H}}}^2} = \frac{\kappa^2}{4} - \frac{1}{4n_{\mathrm{Ps}}^2}. \tag{8.24}$$

If the initial positronium and the residual antihydrogen are both in the ground state ($n_{\bar{\mathrm{H}}} = 1$ and $n_{\mathrm{Ps}} = 1$) the relationship between the cross sections reduces to

$$\sigma_{\bar{\mathrm{H}}} = \sigma_{\mathrm{H}} = \frac{k^2}{\kappa^2}\sigma_{\mathrm{Ps}}. \tag{8.25}$$

The simple conversion factor k^2/κ^2 was applied by Humberston *et al.* (1987) to the positronium formation cross sections of Brown and Humberston (1984) within the Ore gap, the distorted-wave Born approximation results of Shakeshaft and Wadehra (1980) from the top of the Ore gap to 200 eV, and the Born results of Omidvar (unpublished) beyond 200 eV, to yield the first estimates of the antihydrogen formation cross section. These results relate only to the formation of ground state antihydrogen in collisions of antiprotons with ground state positronium, but Darewych (1987) and Nahar and Wadehra (1988), using the first Born approximation, calculated the cross sections for the formation of antihydrogen in various excited states with $n_{\bar{\mathrm{H}}} \leq 3$. The latter authors also attempted to include contributions to the total formation cross section from antihydrogen states with $n_{\bar{\mathrm{H}}} > 3$: they exploited the fact that, at sufficiently high energies, the Born approximation predicts the cross section for the formation of antihydrogen in a state with principal quantum number $n_{\bar{\mathrm{H}}}$ to be $\sigma_{\bar{\mathrm{H}}}(n_{\bar{\mathrm{H}}}) \propto 1/n_{\bar{\mathrm{H}}}^3$. However, this scaling law is almost certainly not valid at the low energies of greatest experimental interest here. Nahar and Wadehra (1988) also used the Born approximation to calculate the cross sections for ground state antihydrogen formation in collisions of antiprotons with positronium in the $n_{\mathrm{Ps}} = 2$ state. They then used the above form of scaling law to estimate the contributions to the total ground state antihydrogen formation cross section from antiproton collisions with positronium in higher excited states.

All these contributions add up to a total antihydrogen formation cross section of approximately 2×10^{-15} cm^2 in the antiproton energy range 2–10 keV where the charge-exchange production mechanism is likely to be most effective. This value is consistent with results obtained by Ermolaev, Bransden and Mandal (1987), who used the classical trajectory Monte Carlo method, and also with the results of a recent experiment (Merrison *et al.*, 1997) which measured the hydrogen atom formation cross section via reaction (8.22).

Further investigations have been made of antihydrogen formation in collisions of protons with excited state positronium. Igarashi, Toshima and Shirai (1994), using a hyperspherical coupled-channel method, found that at an energy of approximately 0.04 ryd the formation cross sections for antiprotons in collision with positronium in the 2S and 2P states were each approximately $300\pi a_0^2$, more than ten times the magnitude of the formation cross section from ground state positronium. They surmised that the formation cross sections would be even larger for positronium in higher excited states. Mitroy and Stelbovics (1994), using the unitarized Born approximation, obtained results that were qualitatively similar but rather smaller in magnitude.

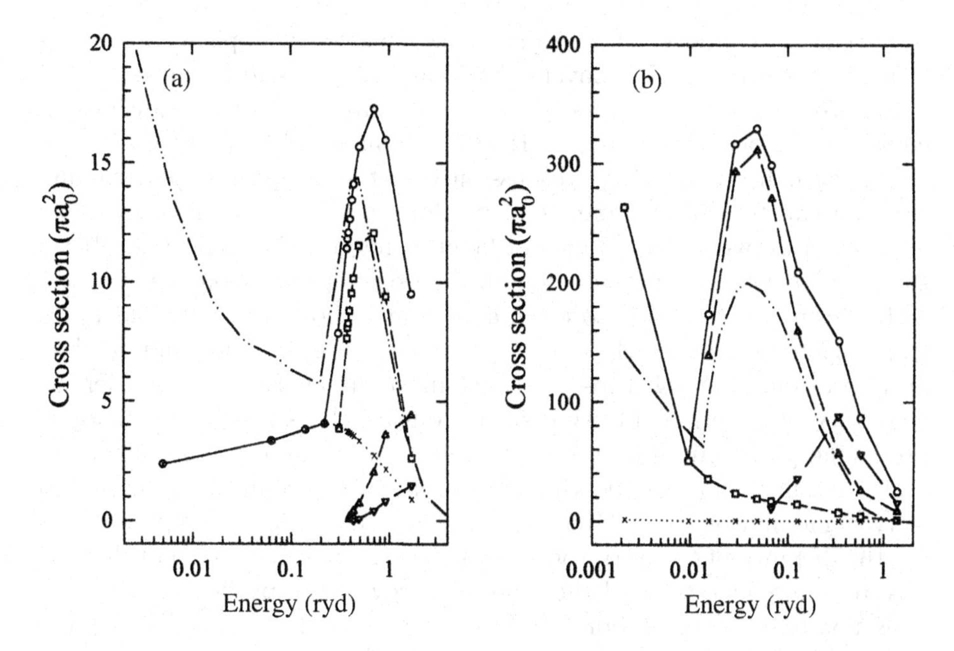

Fig. 8.10. Cross sections for antihydrogen formation in collisions of stationary antiprotons with positronium atoms (from Igarashi, Toshima and Shirai, 1994). (a) is for 1S positronium and (b) is for the 2P state (note the changes in scale). Key (same for both figures): dotted curve with crosses, formation into the $n_{\bar{H}} = 1$ state; short-broken line plus squares, formation into the $n_{\bar{H}} = 2$ state; long-broken line plus triangles, formation into the $n_{\bar{H}} = 3$ state; very-long-broken line plus inverted triangles, formation into the $n_{\bar{H}} = 4$ states. The solid curve with circles is the total cross section summed over all $n_{\bar{H}}$ states and the double chain curve is this quantity as calculated by Mitroy and Stelbovics (1994).

The possibility that total antihydrogen production (i.e. integrated over all states) might be markedly enhanced by antiproton collisions with excited state positronium was first suggested by Charlton (1990), who argued from a classical standpoint that $\sigma_{\bar{H}}$ should increase as the area of the positronium atom, giving an n_{Ps}^4 scaling. He also argued that as n_{Ps} was increased, the relative speed at which the cross section would be a maximum would fall as n_{Ps}^{-1}. The justification for this assumption was based on the Massey criterion, which states that the capture probability is highest when the speed of the projectile is matched to the quasi-classical orbital speed of the positron in the positronium. Both these features are demonstrated qualitatively by the theoretical results shown in Figure 8.10. Mitroy (1995) investigated antihydrogen formation from positronium states up to $n_{Ps} = 4$, using the coupled-state method, and found that the scaling for $\sigma_{\bar{H}}$ is actually closer to n_{Ps}^3 than to the fourth power suggested above. Other work of relevance to this topic has been performed by Mitroy and Ratnavelu (1995) and Mitroy and Ryzhikh (1997). An interesting method of realizing excited state positronium–antiproton collisions has been suggested by Hessels, Homan and Cavagnero (1998).

It is clear from the above discussion that the antiproton–positronium reaction can be a useful means of producing antihydrogen in low-lying atomic states over a wide range of kinetic energies. Deutch *et al.* (1988) suggested a possible method for antihydrogen production via reaction (8.13), using antiprotons stored in a circular radio-frequency quadrupole trap. However, given the advances in trapping, cooling and holding antiprotons in Penning traps described above, it is natural to consider the feasibility of experiments using such a device. It is possible to imagine producing positronium in a Penning trap using a target material such as SiO_2 (see subsection 1.5.3), which has been shown to emit positronium in a cryogenic environment (Mills *et al.*, 1989a). Here, the crucial parameters are the distance of the antiprotons from the shaped electrode housing the positronium converter and the kinetic energy of the positronium (which does not have to correspond to the same temperature as that of the target). At present only conceptual designs of such an arrangement exist (Deutch *et al.*, 1993; Surko, Greaves and Charlton 1997).

Deutch *et al.* (1993) also noted that the effective recoil temperature, $T_{\bar{H}}$, of the antihydrogen is given from conservation of energy and momentum, ignoring the initial energies of the two reactants, as

$$T_{\bar{H}} = \frac{2\Delta E\, m}{3kM} = 28.6 \left(\frac{2}{n_{\bar{\mathrm{H}}}^2} - \frac{1}{n_{\mathrm{Ps}}^2} \right) \quad \text{(K)}, \tag{8.26}$$

where m and M are the positron and antiproton masses respectively and ΔE is the difference between the binding energies of the antihydrogen and positronium states, governed predominantly by $n_{\bar{\mathrm{H}}}$ and n_{Ps}. The most favourable condition for capture, so-called resonant charge transfer, occurs (formally) when $\Delta E = 0$ ($n_{\bar{\mathrm{H}}} = \sqrt{2} n_{\mathrm{Ps}}$) and thus has $T_{\bar{\mathrm{H}}} = 0$. As an extension to the analysis of Deutch *et al.* (1993) and in an effort to simulate a real experimental situation, Cassidy *et al.* (1999) performed a Monte Carlo simulation of the antiproton–positronium reaction. This was done by making assumptions concerning the temperature distribution of the trapped antiprotons and the energy distribution of the positronium atoms, taken to be that typically resulting from positron impact on a surface; they used as input the cross sections, differential and total, of Mitroy and Ryzhikh (1997) and Mitroy (1997, private communication) for the antiproton–positronium reaction. This analysis found the antihydrogen formation rate in the ground state to be rather low, of the order of 100 per hour for 10^7 trapped antiprotons and with a positron flux of 3×10^7 s^{-1}. (These figures are close to the maximum which can presently be expected from a laboratory-based positron beam and for antiprotons held in a Penning trap.) Of these anti-atoms, only approximately 1% had a temperature below 1 K, which is around the maximum well depth that can be achieved using magnetic-gradient traps (see e.g. Hess *et al.*, 1987, and Gomer *et al.*, 1997, for discussions of this type of device) with currently available technology. Further discussion of the details of antihydrogen formation by the antiproton–positronium reaction have been given by Charlton (1996, 1997).

We now turn to reaction (8.14), the trapped plasmas combination reaction, in which the excess energy is removed by an extra positron as shown in Figure 8.9(c). The use of this ternary reaction for antihydrogen formation was first suggested by Gabrielse *et al.* (1988), who noted that its matter equivalent had been studied, mainly in relation to high temperature plasmas, for some time. These authors were therefore able to write the antihydrogen production rate, which corresponds to the recombination rate in conventional plasma physics, as

$$R_{\bar{\mathrm{H}}} = 6 \times 10^{-12} \left(\frac{4.2}{T}\right)^{9/2} n_{\mathrm{e}+}^2. \tag{8.27}$$

This formula for the rate per trapped antiproton is notable for its very strong positron-temperature dependence, $T^{-9/2}$ (in degrees K), and the presence of the positron density, $n_{\mathrm{e}+}$ (in cm^{-3}), to the second power. The latter dependence arises because two positrons are needed in the reaction. In their analysis, Gabrielse *et al.* (1988) assumed that a plasma of 10^7 positrons per cm^3 would be produced at 4.2 K, which yielded $R_{\bar{\mathrm{H}}} \sim 600$ s^{-1} per trapped antiproton. Glinsky and O'Neil (1991) re-examined this problem from a plasma physics viewpoint, and found that the combination rate given by equation (8.27), which is actually that pertaining to zero field, $B = 0$, is reduced by an order of magnitude in a highly magnetized plasma ($B \approx \infty$). In essence, this is caused by the constraint imposed on the positron orbits, since they cannot cross the magnetic field lines.

In order to promote reaction (8.14) it is necessary to trap both antiprotons and positrons and merge the two swarms. A schematic illustration of a sequence of traps to achieve this, the so-called nested Penning trap arrangement (Gabrielse *et al.*, 1988), is shown in Figure 8.11. Again the radial confinement of the two clouds is provided by large axial magnetic fields with appropriate voltage elements that produce potential wells to trap the opposite charges. The positron well is nested within empty wells into which the antiprotons can be moved, and the basic idea is to adjust the depth of the wells so that the antiprotons just pass through the positron cloud. The nested-trap scheme was developed further by Quint *et al.* (1993), who reported the results of preliminary measurements on

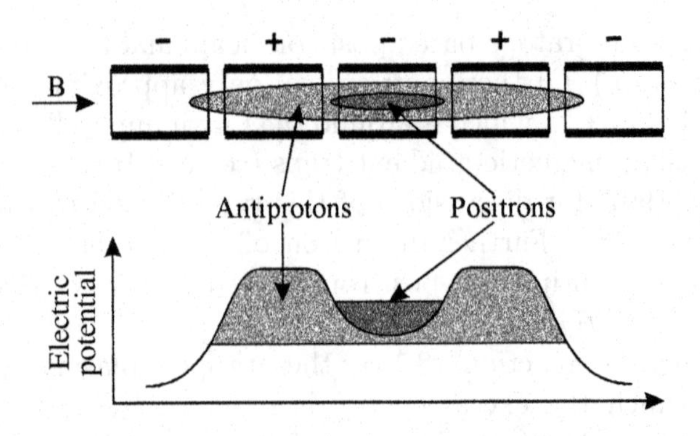

Fig. 8.11. Illustration of the principle of nested Penning traps for holding antiprotons and positrons in close proximity, in order to promote antihydrogen formation.

merged, trapped swarms of protons and electrons. Hall and Gabrielse (1996) demonstrated the electron cooling of trapped protons using a nested-well apparatus. Haarsma, Abdullah and Gabrielse (1995) reported progress in accumulating positrons under ultra-high vacuum conditions typical of those in antiproton traps in quantities sufficient to promote reaction (8.14). Experiments carried out at CERN by Gabrielse and coworkers (reported by Gabrielse *et al.*, 1999) during the last days of the operation of LEAR managed to confine antiprotons and positrons together in a multi-electrode apparatus for the first time, though the effects observed when the clouds were merged could not be attributed to antihydrogen formation.

One of the potential drawbacks of reaction (8.14) is that the antihydrogen will initially be produced in a Rydberg state with a high value of $n_{\bar{\rm H}}$, since the energy removed by the extra positron is of the order of $k_{\rm B}T$. This has been confirmed by the analysis of Pajek and Schuch (1997), who studied the time-reversed process (impact ionization) to derive a recombination coefficient for reaction (8.14). Their numerical values are in accord with equation (8.27), but they found that the coefficient was proportional to $n_{\bar{\rm H}}^6$, thus dramatically favouring initial capture into high-lying Ryberg states. Such atoms have long radiative lifetimes, and Glinsky and O'Niel (1991) and Fedichev (1997) have pointed out mechanisms which can lead to collisional stabilization. These involve so-called replacement collisions at higher values of $n_{\bar{\rm H}}$, whereby once the atom has been formed, another positron travelling on a path closer to the antiproton takes the place of the first positron, which leaves, carrying the excess energy. At lower values of $n_{\bar{\rm H}}$ this process becomes inefficient at the plasma densities envisaged, but another process, transverse collisional drift, occurs. Here the bound positron drifts across the field lines towards the antiproton under the influence of distant positron–atom collisions.

Both of these analyses rely on the fact that, to produce stable antihydrogen, the relaxation processes must occur before the atom leaves the plasma, otherwise it will be field ionized. Alternatively, as discussed by Wolf (1993), it may be possible to laser-stimulate transitions to more tightly bound levels to avoid the field-ionization problem.

We now consider the final method listed above for antihydrogen production. Reactions (8.15) were first suggested by Yamazaki (1992), with further details provided by Ito, Widmann and Yamazaki (1993). The method is based upon the discovery of Iwasaki *et al.* (1991) that up to 3% of antiprotons stopping in helium gas can form a metastable antiprotonic helium atom ($\alpha e\bar{p}$) with a lifetime of the order of 3 μs. The basic idea that such an effect could occur was proposed by Condo (1964). The antiproton, on stopping in the medium before annihilation, forms this exotic atom when its kinetic energy falls below the binding energy of the electrons in the helium (around 25 eV). The captured antiproton then falls into a state with principal quantum number, n_o, close to $(M^*/m)^{1/2}$, where M^* is the reduced mass of the antiproton–nucleus system. In this case $n_o \approx 38$, and the energy level spacing is ≈ 2 eV. If the antiproton is captured into a state with large orbital angular momentum, the transition to the lower levels will most likely occur through a series of radiative transitions, with a concomitantly long period (hence the appellation 'metastable') before annihilation. The atomic nature of these states was confirmed by Morita *et al.* (1994) in a study involving a laser-induced resonant transition between two of them. Further discussion of the $\alpha e\bar{p}$ 'atomcule' has been given by Kartavstev (1996) and Eades and Hartmann (1999).

The proposed scenario for antihydrogen production involves stopping a bunch of approximately 10^9 antiprotons in dense helium; this is followed shortly afterwards by the injection of a bunch of approximately 10^8 positrons into the same medium. A detailed discussion was given by Ito *et al.* (1993); this included a treatment of the differences between the positrons and the positronium atoms which result from the positron injection, in terms of the known behaviour of these species in helium gas; see Chapters 6 and 7. Under optimum conditions, the number of antihydrogen atoms formed could be as great as 10^3–10^4 per antiproton bunch.

Ito, Widmann and Yamazaki (1993) also addressed the problem of the detection of antihydrogen under these conditions. They estimated that most of the antihydrogen will be formed tens of nanoseconds after the positron injection into the 37 atmospheres of helium gas, envisaged as the stopping medium, and will probably be destroyed within 1 ns after formation. Thus the antihydrogen signature will be a small spike on the

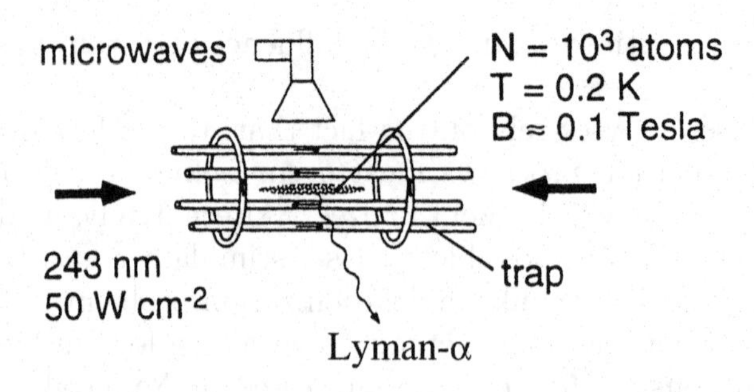

Fig. 8.12. A possible scenario for trapped antihydrogen spectroscopy. The microwaves quench the $2S_{1/2}$ antihydrogen via the $2P_{3/2}$ state, which spontaneously decays by emission of a Lyman-α photon.

$\alpha e\bar{p}$ atomcule annihilation spectrum, at a time governed mainly by the time difference between the injection of the positrons and the antiprotons and the diffusion coefficients of the lighter species. Ito, Widmann and Yamazaki (1993) noted that such an observation would, strictly speaking, indicate only that some of the antiprotons had annihilated more rapidly with the positrons present and not necessarily that antihydrogen had been formed. To prove the latter they envisaged using lasers to ionize the antihydrogen and to observe the associated antiproton annihilation. For this purpose they asserted that the antihydrogen has to survive for around 10 ns in the gas, which suggests that the pressure will have to be lower than the value assumed above.

It should be noted that even if antihydrogen were to be formed by this technique, the prospects for precision studies of the kind outlined in subsection 8.3.1, which might offer a challenge to some of the basic postulates of physics, are not likely to be easy.

4 Prospects for antihydrogen physics

Advances in trapping and cooling antiprotons make it likely that antihydrogen will soon be produced with kinetic energies suitable for trapping and also, possibly, in beams. Almost all the production scenarios outlined above rely on using a source of low energy antiprotons, such as LEAR and associated machinery and, although the former unique facility closed at the end of 1996, continuation of a low energy antiproton programme at CERN in 1999 and beyond has been assured by the approval of the Antiproton Decelerator project (Maury, 1997).

A possible glimpse of the future is provided by Figure 8.12, which shows a schematic illustration of a magnetic-gradient trap for anti-atom spectroscopy. The scenario described below is due to Hänsch and Zimmermann (1993), who assumed that a sample of 10^3 antihydrogen atoms could be held in such a trap at a temperature of 0.2 K. A uniform magnetic field of around 0.1 T (for antiproton and positron confinement) is superimposed upon the field produced by the quadrupole coils, which ensures that a local field minimum exists in the trap centre. Such a field configuration is well known to produce a shallow trap for neutral species with an appropriate magnetic moment. The schematic shows the anti-atoms, initially in the 1S state, raised into the 2S state by a Doppler-free two photon (243 nm) transition and further pumped by microwaves into the 2P state, which then decays back to the ground state with the emission of a Lyman-α photon. This resonance fluorescence is used as the signature of the transition. Hänsch and Zimmermann (1993) also described how this cycle can be operated on trapped anti-atoms to prevent spin flips and keep them stored for longer periods. Holzscheiter and Charlton (1999) have described an alternative scheme of observing the 1S–2S transition which does not rely on the detection of the Lyman-α photon.

As described briefly in subsection 8.3.1, the ultimate goal for a hydrogen–antihydrogen comparison for this transition is a fractional frequency difference in the region of 10^{-18} and as such would provide a stringent test of CPT invariance and of the WEP. The achievement of such precision is, however, a distant goal, most probably to be realized through a long series of steps, with much development work still to be done with ordinary matter reactions.

Appendix
Positron conference proceedings

This is a list of the proceedings of the most important conference series devoted to topics in positron physics discussed in this book.

A International conferences on positron annihilation

ICPA-1 *Proceedings of the First International Conference on Positron Annihilation* (held at Detroit, USA, 1965) edited by A.T. Stewart and L.O. Roellig, Academic Press, New York (1967)

ICPA-2 *Proceedings of the Second International Conference on Positron Annihilation* (held at Kingston, Canada, 1971) edited by A.T. Stewart, B.T.A. McKee and C.H. Markham

ICPA-3 *Proceedings of the Third International Conference on Positron Annihilation* (held at Helsinki, Finland, 1973) edited by P. Hautojarvi and A. Seeger, Springer-Verlag, Berlin (1975)

ICPA-4 *Proceedings of the Fourth International Conference on Positron Annihilation* (held at Lyngby, Denmark, 1976) edited by G. Trumpy

ICPA-5 *Proceedings of the Fifth International Conference on Positron Annihilation* (held at Lake Yamanaka, Japan, 1979) edited by R.R. Hasiguti and K. Fujiwara, The Japan Institute of Metals, Sendai, Japan (1979)

ICPA-6 *Proceedings of the Sixth International Conference on Positron Annihilation* (held at Arlington, Texas, USA, 1982) edited by P.G. Coleman, S.C. Sharma and L.M. Diana, North-Holland, Amsterdam (1982)

ICPA-7 *Proceedings of the Seventh International Conference on Positron Annihilation* (held at New Delhi, India, 1985) edited by P.C. Jain, R.M. Singru and K.P. Gopinathan, World Scientific, Singapore (1985)

ICPA-8 *Proceedings of the Eighth International Conference on Positron Annihilation* (held at Gent, Belgium, 1988) edited by L. Dorikens-Vanpraet, M. Dorikens and D. Segers, World Scientific, Singapore (1989)

ICPA-9 *Proceedings of the Ninth International Conference on Positron Annihilation* (held at Szombathely, Hungary, 1991) edited by Z. Kajcsos and S. Szeles, Trans Tech Publications, Switzerland (1992) published in *Materials Science Forum* **105–110** (1992)

ICPA-10 *Proceedings of the Tenth International Conference on Positron Annihilation* (held at Beijing, China, 1994) edited by Y-J He, B-S Cao and Y.C. Jean, Trans Tech Publications, Switzerland (1995) published in *Materials Science Forum* **175–178** (1995)

ICPA-11 *Proceedings of the Eleventh International Conference on Positron Annihilation* (held at Kansas City, USA, 1997) edited by Y.C. Jean, M. Eldrup, D.M. Schrader and R.N. West, Trans Tech Publications, Switzerland (1997) published in *Materials Science Forum* **255–257** (1997)

B International workshops on positron collisions in gases

1. *Proceedings of the International Conference on Positron Scattering and Annihilation in Gases* (held at Toronto, Canada, 1981) edited by J.W. Darewych, J.W. Humberston, R.P. McEachran, A.D. Stauffer and D.A.L. Paul, *Canadian Journal of Physics* **60** (1982)
2. *Proceedings of the International Workshop on Positron Scattering in Gases* (held at Egham, UK, 1983) edited by J.W. Humberston and M.R.C. McDowell, Plenum Press, New York (1984)
3. *Proceedings of the Third International Workshop on Positron (Electron)–Gas Scattering* (held at Detroit, Michigan, USA, 1985) edited by W.E. Kauppila, T.S. Stein and J.M. Wadehra, World Scientific, Singapore (1986)
4. *Proceedings of the International Workshop on Atomic Physics with Positrons* (held at London, UK, 1987) edited by J.W. Humberston and E.A.G. Armour, Plenum Press, New York (1987)
5. *Proceedings of the International Workshop on Annihilation in Gases and Galaxies* (held at Greenbelt, Maryland, USA, 1989) edited by R.J. Drachman, NASA Conference Publication 3058 (1990)
6. *Proceedings of the International Workshop on Positron Interactions with Gases* (held at Sydney, Australia, 1991) edited by L.A. Parcell, published in *Hyperfine Interactions* **73** (1992)
7. *Proceedings of the International Workshop on Positron Interactions with Atoms, Molecules and Clusters* (held at Bielefeld, Germany, 1993) edited by W. Raith and R.P. McEachran, published in *Hyperfine Interactions* **89** (1994)
8. *Proceedings of the 1995 Positron Workshop* (held at Vancouver, Canada, 1995) edited by R.P. McEachran and A.D. Stauffer, published in *Canadian Journal of Physics* **74** (1996)

9. *Proceedings of the International Workshop on Low Energy Positron and Positronium Physics* (held at Nottingham, UK, 1997) edited by H.H. Andersen, E.A.G. Armour, J.W. Humberston and G. Laricchia, published in *Nuclear Instruments and Methods in Physics Research, Section B* **143** (1998)

C International conferences on slow positron-beam techniques

SLOPOS-1 *Proceedings of the International Workshop on Slow Positrons in Surface Science* (held at Helsinki, Finland, 1984) edited by A. Vehanen, Helsinki University of Technology Laboratory of Physics Report 135 (1984)

SLOPOS-2 *Proceedings of the MURR Slow Positron Beam Workshop* (held at Columbia, Missouri, USA, 1985) edited by D.C. Reichel and W.B. Yelon

SLOPOS-3 *Proceedings of the International Workshop on Slow Positron Beams for Solids and Surfaces* (held at Norwich, UK, 1986) edited by P.G. Coleman and A.B. Walker

SLOPOS-4 *Proceedings of the International Workshop on Positron Beams for Solids and Surfaces* (held at London, Ontario, Canada, 1990) edited by P.J. Schultz, G.R. Massoumi and P.J. Simpson, American Institute of Physics Conference Proceedings 218, New York (1990)

SLOPOS-5 *Proceedings of the Fifth International Workshop on Slow Positron Beam Techniques for Solids and Surfaces* (held at Jackson Hole, Wyoming, USA, 1992) edited by E. Ottewitte and A.H. Weiss, American Institute of Physics Conference Proceedings 303, New York (1993)

SLOPOS-6 *Proceedings of the Sixth International Workshop on Slow Positron Beam Techniques for Solids and Surfaces* (held at Makuhari, Japan, 1994) edited by M. Doyama, T. Akahane and M. Fujinami, published in *Applied Surface Science* **85** (1995)

SLOPOS-7 *Proceedings of the Seventh International Workshop on Slow Positron Beam Techniques for Solids and Surfaces* (held at Unterægeri, Switzerland, 1996) edited by W.B. Waeber, M. Shi and A.A. Manuel, published in *Applied Surface Science* **116** (1997)

SLOPOS-8 *Proceedings of the Eighth International Workshop on Slow Positron Beam Techniques for Solids and Surfaces* (held at Cape Town, South Africa, 1998) edited by D.T. Britton and M. Härtig, published in *Applied Surface Science* **149** (1999)

D Positron and positronium chemistry

Proceedings of the International Workshop on Positron and Positronium Chemistry edited by D.M. Schrader and Y.C. Jean, Elsevier Science, Amsterdam (1988)

Proceedings of the International Workshop on Positron Annihilation Studies of Fluids edited by S.C. Sharma, World Scientiffc, Singapore (1988)

Proceedings of the International Workshop on Positron and Positronium Chemistry edited by Y.C. Jean, World Scientific, Singapore (1990)

Proceedings of the Fourth International Workshop on Positron and Positronium Chemistry (held at Le Mont Saint-Odile, France, 1993) edited by I. Billard, G. Duplatre and J.Ch. Abbe, published in *Journal de Physique* **IV**, Coll. C4, supp. *Journal de Physique* **II** 3 (1993)

Proceedings of the Fifth Intgernational Workshop on Positron and Positronium Chemistry edited by Zs. Kajcsos, B. Levay and K. Suvegh, published in *Journal of Radioanalytical and Nuclear Chemistry* **210**, **211** (1996)

E Others

Proceedings of the Workshop on Industrial Applications of Positron Annihilation, Europhysics Industrial Workshop EIW-12 (held at Oisterwijk, The Netherlands, 1994) edited by A. van Veen, C. Corbel and P.E. Mijnarends, published in *Journal de Physique* **IV**, Coll. C1, supp. *Journal de Physique* **III** 1 (1995)

References

Abdel-Raouf, M.A. (1988). Inelastic collisions of positrons with lithium and sodium atoms. *J. Phys. B: At. Mol. Opt. Phys.* **21** 2331–2352.

Adachi, S., Chiba, M., Hirose, T., Nagayama, S., Nakamitsu, Y., Sato, T. and Yamada, T. (1994). Precise measurements of e^+e^- annihilation at rest into four photons and the search for exotic particles. *Phys. Rev. A* **49** 3201–3208.

Adkins, G.S. (1983). Radiative corrections to positronium decay. *Ann. Phys.* **146** 78–128.

Adkins, G.S., Salahuddin, A.A. and Schalm, K.E. (1992). Order-α corrections to the decay rate of orthopositronium in the Fried–Yennie gauge. *Phys. Rev. A* **45** 7774–7781.

Aharonov, Y., Avignone III, F.T., Brodzinski, R.L., Collar, J.I., García, E., Miley, H.S., Morales, A., Nussinov, S., Ortiz de Solórzano, A., Puimedón, J., Reeves, J.H., Sáenz, C., Salinas, A., Sarsa, M.L. and Villar, J.A. (1995). New laboratory bounds on the stability of the electron. *Phys. Rev. D* **52** 3785–3792.

Alekseev, A.I. (1958). Two-photon annihilation of positronium in the P-state. *Sov. Phys. JETP* **34** 826–830.

Alekseev, A.I. (1959). Three-photon annihilation of positronium in the P-state. *Sov. Phys. JETP* **36** 1312–1315.

Allison, S.K. (1958). Experimental results on charge-changing collisions of hydrogen and helium ions of kinetic energies above 0.2 keV. *Rev. Mod. Phys.* **30** 1137–1168.

Al-Ramadhan, A.H. and Gidley, D.W. (1994). New precision measurement of the decay rate of singlet positronium. *Phys. Rev. Lett.* **72** 1632–1635.

Amusia, M.Ya., Cherepkov, N.A., Chernysheva, L.V. and Shapiro, S.G. (1976). Elastic scattering of slow positrons by helium. *J. Phys. B: At. Mol. Phys.* **9** L531–L534.

Andersen, L.H., Hvelplund, P., Knudsen, H., Møller, S.P., Sørensen, A., Elsener., K., Rensfelt and K.-G., Uggerhøj, E. (1987). Multiple ionization of He, Ne and Ar by fast protons and antiprotons. *Phys. Rev. A* **36** 3612–3629.

Andersen, L.H., Hvelplund, P., Knudsen, H., Møller, S.P., Pedersen, J.O.P., Tang-Petersen, S., Uggerhøj, E., Elsener, K. and Morenzoni, E. (1990a). Single ionization of helium by 40–3000 keV antiprotons. *Phys. Rev. A* **41** 6536–6539.

Andersen, L.H., Hvelplund, P., Knudsen, H., Møller, S.P., Pedersen, J.O.P., Tang-Petersen, S., Uggerhøj, E., Elsener, K. and Morenzoni, E. (1990b). Non-dissociative and dissociative ionisation of H_2 by 50–2000 keV antiprotons. *J. Phys. B: At. Mol. Opt. Phys.* **23** L395–L400.

Anderson, C. D. (1933). The positive electron. *Phys. Rev.* **43** 491–494.

Arbic, B.K., Hatamian, S., Skalsey, M., Van House, J. and Zheng, W. (1988). Angular-correlation test of CPT in polarised positronium. *Phys. Rev. A* **37** 3189–3194.

Archer, B.J., Parker, G.A. and Pack, R.T. (1990). Positron–hydrogen–atom S-wave coupled-channel scattering at low energies. *Phys. Rev. A* **41** 1303–1310.

Armour, E.A.G. (1978). Justification of the absence of a positron–hydrogen bound state. *J. Phys. B: At. Mol. Phys.* **11** 2803–2811.

Armour, E.A.G. (1982). New method for taking into account finite nuclear mass in the determination of absence of bound states: application to e^+H. *Phys. Rev. Lett.* **48** 1578–1581.

Armour, E.A.G. (1983). Application to μ^+e–e^+ and $p\mu$–e^+ of a new method for taking into account finite nuclear mass in the determination of the absence of bound states. *J. Phys. B: At. Mol. Phys.* **16** 1295–1302.

Armour, E.A.G. (1984). Application of a generalisation of the Kohn variational method to the calculation of cross sections for low-energy positron–hydrogen molecule scattering. *J. Phys. B: At. Mol. Opt. Phys.* **17** L375–L382.

Armour, E.A.G. (1988). The theory of low-energy positron collisions with molecules. *Phys. Rep.* **169** 1–98.

Armour, E.A.G. and Humberston, J.W. (1991). Methods and programs in collisions of positrons with atoms and molecules. *Phys. Rep.* **204** 165–251.

Armour, E.A.G., Baker, D.J. and Plummer, M. (1990). The theoretical treatment of low-energy e^+–H_2 scattering using the Kohn variational method. *J. Phys. B: At. Mol. Opt. Phys.* **23** 3057–3074.

Armstead, R.L. (1968). Electron–hydrogen scattering calculation. *Phys. Rev.* **171** 91–93.

Asai, S., Orito, O. and Shinohara, N. (1995). New measurement of the orthopositronium decay rate. *Phys. Lett. B* **357** 475–480.

Asai, S., Orito, S., Sanuki, T., Yasuda, M. and Yokoi, T. (1991). Direct search for orthopositronium decay into two photons. *Phys. Rev. Lett.* **66** 1298–1301.

Asai, S., Shigekuni, K., Sanuki, T. and Orito, S. (1994). Search for short-lived neutral bosons in orthopositronium decay. *Phys. Lett. B* **323** 90–94.

Ashley, P.N., Moxom, J. and Laricchia, G. (1996). Near-threshold ionization of He and H_2 by positron impact. *Phys. Rev. Lett.* **77** 1250–1253.

Au, C.K. and Drachman, R.J. (1986). van der Waals force between positronium and hydrogenic atoms: finite-mass corrections. *Phys. Rev. Lett.* **56** 324–327.

Aulenkamp, H., Heiss, P. and Wichmann, E. (1974). A calculation for the scattering of low-energy positrons by helium atoms. *Z. Phys.* **268** 213–215.

Azbel, M.Ya. and Platzman, P.M. (1981). Evidence for a positron mobility edge in gaseous helium. *Solid State Commun.* **39** 679–681.

Baille, P., Darewych, J.W. and Lodge, J.G. (1974). Elastic scattering and rotational excitation of molecular hydrogen by low energy positrons. *Can. J. Phys.* **52** 667–677.

Bandyopadhyay, A., Roy, K., Mandal, P. and Sil, N.C. (1994). Ionization of atomic hydrogen by positron impact. *J. Phys. B: At. Mol. Opt. Phys.* **27** 4337–4347.

Barker, M.I. and Bransden, B.H. (1968). The quenching of orthopositronium by helium. *J. Phys. B: At. Mol. Phys.* **1** 1109–1114 and corrigendum (1969) *J. Phys B: At. Mol. Phys.* **2** 730.

Bartschat, K., McEachran, R.P and Stauffer, A.D. (1988). Optical potential approach to positron and electron scattering from noble gases: I. Argon. *J. Phys. B: At. Mol. Opt. Phys.* **21** 2789–2800.

Basu, M. and Ghosh, A.S. (1988). Positronium formation in the $n = 1$ and 2 states in e^+–H scattering. *J. Phys. B: At. Mol. Opt. Phys.* **21** 3439–3447.

Basu, M. and Ghosh, A.S. (1991). Positron–lithium scattering at low and intermediate energies. *Phys. Rev. A* **43** 4746–4750.

Basu, M., Mazumdar, P.S. and Ghosh, A.S. (1985). Ionisation cross sections in positron–helium scattering. *J. Phys. B: At. Mol. Phys.* **18** 369–377.

Basu, M., Mukherjee, M. and Ghosh, A.S. (1990). Positron–hydrogen scattering below the first excitation threshold. *J. Phys. B: At. Mol. Opt. Phys.* **23** 2641–2648.

Baur, G., Boero, G., Brauksiepe, S., Buzzo, A., Eyrich, W., Geyer, R., Grzonka, D., Hauffe, J., Kilian, K., LoVetere, M., Macri, M., Moosburger, M., Nellen, R., Oelert, W., Passagio, S., Pozzo, A., Röhrich, K., Sachs, K., Schepers, G., Sefzick, T., Simon, R.S., Stratmann, R., Stinzing, F. and Wolke, M. (1996). Production of antihydrogen. *Phys. Lett. B* **368** 251–258.

Baz, A.I. (1958). The energy dependence of a scattering cross section near the threshold of a reaction. *Sov. Phys. JETP* **6** 709–713.

Beling, C.D. and Charlton, M. (1987). Low-energy positron beams – origins, developments and applications. *Contemp. Phys.* **28** 241–266.

Beling, C.D., Simpson, R.I., Charlton, M. Jacobsen, F.M., Griffith, T.C., Moriarty, P. and Fung, S. (1987). A field-assisted moderator for low-energy positron beams. *Appl. Phys. A* **42** 111–116.

Berakder, J. and Klar, H. (1993). Structures in triply and doubly differential ionization cross sections of atomic hydrogen. *J. Phys. B: At. Mol. Opt. Phys.* **26** 3891–3913.

Bergstrom Jr., P.M., Kissel, L. and Pratt, R.H. (1996). Production or annihilation of positrons with bound electrons. *Phys. Rev. A* **53** 2865–2868.

Berko, S. and Pendleton, H.N. (1980). Positronium. *Ann. Rev. Nucl. Part. Sci.* **30** 543–581.

Bethe, H.A. and Salpeter, E.E. (1977). *Quantum Mechanics of One- and Two-Electron Atoms* (Springer-Verlag).

Beverini, N., Lagomarsino, V., Manuzio, G., Scuri, F. and Torelli, G. (1988). Possible measurements of the gravitational acceleration with neutral antimatter. *Hyperfine Interactions* **44** 357–364.

Bhabha, H.J. (1936). The scattering of positrons by electrons with exchange on Dirac's theory of the positron. *Proc. Roy. Soc. Lond. A* **154** 195–206.

Bhabha, H.J. and Hulme, H.R. (1934). The annihilation of positrons by electrons in the K-shell. *Proc. Roy. Soc. Lond. A* **146** 723–736.

Bhatia, A.K. and Drachman, R.J. (1983). New calculation of the properties of the positronium ion. *Phys. Rev. A* **28** 2523–2525.

Bhatia, A.K. and Drachman, R.J. (1985). Photodetachment of the positronium negative ion. *Phys. Rev. A* **32** 3745–3747.

Bhatia, A.K., Temkin, A. and Eiserike, H. (1974). Rigorous precision p-wave positron–hydrogen scattering calculation. *Phys. Rev. A* **9** 219–222.

Bhatia, A.K., Temkin, A., Drachman, R.J. and Eiserike, H. (1971). Generalized Hylleraas calculation of positron–hydrogen scattering. *Phys. Rev. A* **3** 1328–1335.

Bhattacharyya, P.K. and Ghosh, A.S. (1975). Elastic positron–hydrogen-molecule scattering using the eikonal approximation. *Phys. Rev. A* **12** 1881–1884.

Biswas, P.K., Mukherjee, T. and Ghosh, A.S. (1991). Ground-state Ps formation in e^+–H_2 scattering using the FBA. *J. Phys. B: At. Mol. Opt. Phys.* **24** 2601–2607.

Biswas, P.K., Basu, M., Ghosh, A.S. and Darewych, J.W. (1991). Positronium formation in the $n = 2$ states in e^+–H_2 scattering. *J. Phys. B: At. Mol. Opt. Phys.* **24** 3507–3515.

Blaauw, H.J., de Heer, F.J., Wagenaar, R.W. and Barends, D.H. (1977). Total cross sections for electron scattering from N_2 and He. *J. Phys. B: At. Mol. Phys.* **10** L299–L303.

Blackett, P.M.S. and Occhialini, G.P.S. (1933). Some photographs of the tracks of penetrating radiation. *Proc. Roy. Soc. Lond. A* **139** 699–718.

Blanford, G., Christian, D.C., Gollwitzer, K., Mandelkern, M., Munger, C.T., Schultz, J. and Zioulas, G. (1998). Observation of atomic antihydrogen. *Phys. Rev. Lett.* **80** 3037–3040.

Bluhme, H., Knudsen, H., Merrison, J.P. and Poulsen, M.R. (1998). Strong suppression of the positronium channel in double ionization of noble gases by positron impact. *Phys. Rev. Lett.* **81** 73–76.

Bodwin, G.T. and Yennie, D.R. (1978). Hyperfine splitting in positronium and muonium. *Phys. Rep.* **43** 267–303.

Böse, N., Paul, D.A.L. and Tsai, J.-S. (1981). Positron drift in molecular hydrogen. *J. Phys. B: At. Mol. Phys.* **14** L227–L232.

Brandes, G.R., Mills Jr., A.P. and Zuckerman, D.M. (1992). Positron workfunction of diamond C(100) surfaces. *Materials Science Forum* **105–110** 1363–1366.

Brandes, G.R., Canter, K.F., Horsky, T.N., Lippel, P.H. and Mills Jr., A.P. (1988). Scanning positron microbeam. *Rev. Sci. Inst.* **59** 228–232.

Brandt, W. and Paulin, R. (1968). Positronium diffusion in solids. *Phys. Rev. Lett.* **21** 193–195.

Brandt, W. and Paulin, R. (1977). Positron implantation profile effects in solids. *Phys. Rev. B* **15** 2511–2518.

Bransden, B.H. and Hutt, P.K. (1975). Electron and positron scattering by helium and neon. *J. Phys. B: At. Mol. Phys.* **8** 603–611.

Bransden, B.H. and Jundi, Z. (1967). Positronium formation by positron impact on hydrogen. *Proc. Phys. Soc.* **92** 880–888.

Bransden, B.H., Hutt, P.K. and Winters, K.H. (1974). Total cross sections for the scattering of positrons by helium. *J. Phys. B: At. Mol. Phys.* **7** L129–L131.

Brauner, M. and Briggs, J.S. (1986). Ionisation to the projectile continuum by positron and electron collisions with neutral atoms. *J. Phys. B: At. Mol. Phys.* **19** L325–L330.

Brauner, M. and Briggs, J.S. (1991). Structures in differential cross sections for positron impact ionization of hydrogen. *J. Phys. B: At. Mol. Opt. Phys.* **24** 2227–2236.

Brauner, M. and Briggs, J.S. (1993). Structure in differential cross sections for positron and electron impact ionization of hydrogen. *J. Phys. B: At. Mol. Opt. Phys.* **26** 2451–2461.

Brauner, M., Briggs, J.S. and Klar, H. (1989). Triply-differential cross sections for ionisation of hydrogen atoms by electrons and positrons. *J. Phys. B: At. Mol. Opt. Phys.* **22** 2265–2287.

Bray, I. and Stelbovics, A.T. (1993). Convergent close-coupling calculation of low-energy positron–atomic hydrogen scattering. *Phys. Rev. A* **48** 4787–4789.

Bray, I. and Stelbovics, A.T. (1994). Calculation of the total and total ionization cross sections for positron scattering on atomic hydrogen. *Phys. Rev. A* **49** R2224–R2226.

Breit, G. (1957). Energy dependence of reactions at thresholds. *Phys. Rev.* **107** 1612–1615.

Brenton, A.G., Dutton, J. and Harris, F.M. (1978). Total cross sections for the scattering of positrons by neon and argon. *J. Phys. B: At. Mol. Phys.* **11** L15–L19.

Brenton, A.G., Dutton, J., Harris, F.M., Jones, R.A. and Lewis, D.M. (1977). Experimental determination of total scattering cross sections for positron–helium collisions. *J. Phys. B: At. Mol. Phys.* **10** 2699–2710.

Brewer, D.F.C., Newell, W.R., Harper, S.F.W. and Smith, A.C.H. (1981). Elastic scattering of low-energy electrons by neon atoms. *J. Phys. B: At. Mol. Phys.* **14** L749–L754.

Briggs, J.S. (1989). Cusps, dips and peaks in differential cross-sections for fast three-body Coulomb collisions. *Comm. At. Mol. Phys.* **13** 155–174.

Britton, D.T., Huttunen, P.A., Mäkinen, J., Soininen, E. and Vehanen, A. (1989). Positron reflection from the surface potential. *Phys. Rev. Lett.* **62** 2413–2416.

Brown, B.L. (1985). Creation of monoenergetic positronium (Ps) in a gas. *Bull. Am. Phys. Soc.* **30** 614.

Brown, B.L. (1986). Creation of a monoenergetic positronium beam in a gas and measurement of positron survival fractions in a gas. In *Positron(Electron)–Gas Scattering*, eds. W.E. Kauppila, T.S. Stein and J.M. Wadehra (World Scientific) pp. 212–221.

Brown, C.J. and Humberston, J.W. (1984). Positronium formation in p-wave positron–hydrogen scattering. *J. Phys. B: At. Mol. Phys.* **17** L423–L426.

Brown, C.J. and Humberston, J.W. (1985). Positronium formation in positron–hydrogen scattering. *J. Phys. B: At. Mol. Phys.* **18** L401–L406.

Buckley, B.D. and Walters, H.R.J. (1975). Second Born approximation to electron and positron impact excitations of the 1^1S $\to$ 2^1S transition in helium. *J. Phys. B: At. Mol. Phys.* **8** 1693–1715.

Budker, G.I. and Skrinsky, A.N. (1978). Electron cooling and new possibilities in elementary particle physics. *Sov. Phys. Usp.* **21** 277–296.

Burke, P.G. and Robb, W.D. (1975). The R-matrix theory of atomic processes. *Adv. At. Mol. Phys.* **11** 143–214.

Burke, P.G., Berrington, K.A. and Sukumar, C.V. (1981). Electron–atom scattering at intermediate energies. *J. Phys. B: At. Mol. Phys.* **14** 289–305.

Bussard, R.W., Ramaty, R. and Drachman, R.J. (1979). The annihilation of galactic positrons. *Astroph. J.* **228** 928–934.

Byron Jr., F.W. (1982). Some comments on positron–atom scattering at intermediate energy. *Can. J. Phys.* **60** 558–564.

Byron Jr., F.W. and Joachain, C.J. (1973). Elastic electron–atom scattering at intermediate energies. *Phys. Rev. A* **8** 1267–1282.

Byron Jr., F.W. and Joachain, C.J. (1977a). Elastic scattering of electrons and positrons by atomic hydrogen and helium at intermediate and high energies. *J. Phys. B: At. Mol. Phys.* **10** 207–226.

Byron Jr., F.W. and Joachain, C.J. (1977b). Eikonal theory of electron– and positron–atom collisions. *Phys. Rep.* **34** 233–324.

Byron Jr., F.W. and Joachain, C.J. (1981). A third order optical potential theory for elastic scattering of electrons and positrons by atomic hydrogen. *J. Phys. B: At. Mol. Phys.* **14** 2429–2448.

Byron Jr., F.W., Joachain, C.J. and Potvliege, R.M. (1981). Unitarisation of the eikonal-Born series method for electron– and positron–atom collisions. *J. Phys. B: At. Mol. Phys.* **14** L609–L615.

Byron Jr., F.W., Joachain, C.J. and Potvliege, R.M. (1982). Unitarisation of the eikonal-Born series method for electron– and positron–atom collisions. *J. Phys. B: At. Mol. Phys.* **15** 3915–3943.

Byron Jr., F.W., Joachain, C.J. and Potvliege, R.M. (1985). Elastic and inelastic scattering of electrons and positrons by atomic hydrogen at intermediate and high energies in the unitarised eikonal-Born method. *J. Phys. B: At. Mol. Phys.* **18** 1637–1660.

Campbell, C.P., McAlinden, M.T., Kernoghan, A.A. and Walters, H.R.J. (1998a). Positron collisions with one- and two-electron atoms. *Nuc. Inst. Meth. B* **143** 41–56.

Campbell, C.P., McAlinden, M.T., MacDonald, F.G.R.S. and Walters, H.R.J. (1998b). Scattering of positronium by atomic hydrogen. *Phys. Rev. Lett.* **80** 5097–5100.

Campeanu, R.I. (1977). The scattering of low energy positrons by helium. Ph.D. Thesis, University of London.

Campeanu, R.I. (1981). On the theoretical and experimental cross sections for low-energy positron–rare-gas scattering. *J. Phys. B: At. Mol. Phys.* **14** L157–L160.

Campeanu, R.I. (1982). Positron diffusion in krypton and xenon. *Can. J. Phys.* **60** 615–617.

Campeanu, R.I. and Beu, T. (1983). Hydrogen–antihydrogen interaction potential. *Phys. Lett. A* **93** 223–226.

Campeanu, R.I. and Dubau, J. (1978). Positron–neon elastic scattering. *J. Phys. B: At. Mol. Phys.* **11** L567–L570.

Campeanu, R.I. and Humberston, J.W. (1975). The scattering of p-wave positrons by helium. *J. Phys. B: At. Mol. Phys.* **8** L244–L248.

Campeanu, R.I., and Humberston, J.W. (1977a). The scattering of s-wave positrons by helium. *J. Phys. B: At. Mol. Phys.* **10** L153–L158.

Campeanu, R.I. and Humberston, J.W. (1977b). Diffusion of positrons in helium gas. *J. Phys. B: At. Mol. Phys.* **10** 239–250.

Campeanu, R.I., McEachran, R.P. and Stauffer, A.D. (1996). Positron impact ionization of helium, neon and argon. *Can. J. Phys.* **74** 544–547.

Campeanu, R.I., Fromme, D., Kruse, G., McEachran, R.P., Parcell, L.A., Raith, W., Sinapius, G. and Stauffer, A.D. (1987). Partitioning of the positron–helium total scattering cross section. *J. Phys. B: At. Mol. Phys.* **20** 3557–3570.

Canter, K.F. (1984). Low energy positron and positronium diffraction. In *Positron Scattering in Gases*, eds. J.W. Humberston and M.R.C. McDowell (Plenum Press) pp. 219–225.

Canter, K.F. (1986). Slow positron optics. In *Positron Studies of Solids, Surfaces and Atoms*, eds. A.P. Mills Jr., W.S. Crane and K.F. Canter (World Scientific) pp. 102–120.

Canter, K.F. and Roellig, L.O. (1970). Critical behaviour of positrons in low temperature gaseous helium. *Phys. Rev. Lett.* **25** 328–330.

Canter, K.F. and Roellig, L.O. (1975). Positron annihilation in low-temperature rare gases. II. Argon and neon. *Phys. Rev. A* **12** 386–395.

Canter, K.F., Mills Jr., A.P. and Berko, S. (1975). Observation of positronium Lyman-α radiation. *Phys. Rev. Lett.* **34** 177–180 and erratum p. 848.

Canter, K.F., Coleman, P.G., Griffith, T.C. and Heyland, G.R. (1972). Measurement of total cross-sections for low energy positron–helium collisions. *J. Phys. B: At. Mol. Phys.* **5** L167–L169.

Canter, K.F., Coleman, P.G., Griffith, T.C. and Heyland, G.R. (1973). The measurement of total cross sections for positrons of energies 2–400 eV in He, Ne, Ar, and Kr. *J. Phys. B: At. Mol. Phys.* **6** L201–L203.

Canter, K.F., Fishbein, M., Fox, R.A., Gyasi, K. and Steinman, J.F. (1980). Is there a positron mobility edge in gaseous helium? *Solid State Commun.* **34** 773–776.

Canter, K.F., Lippel, P.H., Crane, W.S. and Mills Jr., A.P. (1986). Modified Soa immersion lens positron gun. In *Positron Studies of Solids, Surfaces and Atoms*, ed. A.P. Mills Jr., W.S. Crane and K.F. Canter (World Scientific) pp. 102–120.

Canter, K.F., Brandes, G.R., Horsky, T.N., Lippel P.H. and Mills Jr., A.P. (1987). The high brightness beam at Brandeis. In *Atomic Physics with Positrons*, eds. J.W. Humberston and E.A.G Armour (Plenum Press) pp. 153–160.

Cassidy, D.B., Merrison, J.P., Charlton, M., Mitroy, J. and Ryzhikh, G. (1999). Antihydrogen from positronium impact with cold antiprotons: a Monte Carlo simulation. *J. Phys. B: At. Mol. Opt. Phys.* **32** 1923–1932.

Caswell, W.E. and Lepage, G.P. (1979). $O(\alpha^2 \ln \alpha^{-1})$ corrections in positronium: hyperfine splitting and decay rate. *Phys. Rev. A* **20** 36–43.

Caswell, W.E., Lepage, G.P. and Sapirstein, J. (1977). $O(\alpha)$ corrections to the decay rate of orthopositronium. *Phys. Rev. Lett.* **38** 488–491.

Cesar, C.L., Fried, D.G., Killian, T.C., Polcyn, A.D., Sandberg, J.C., Yu, I.A., Greytak, T.J. and Kleppner, D. (1996). Two-photon spectroscopy of trapped atomic hydrogen. *Phys. Rev. Lett.* **77** 255–258.

Chamberlain, O., Segrè, E., Wiegand, C. and Ypsilantis, T. (1956). Observation of antiprotons. *Phys. Rev.* **100** 947–949.

Chan, Y.F. and Fraser, P.A. (1973). S-wave positron scattering by hydrogen atoms. *J. Phys. B: At. Mol. Phys.* **6** 2504–2515.

Chan, Y.F. and McEachran, R.P. (1976). Inelastic p-wave positron–hydrogen scattering. *J. Phys. B: At. Mol. Phys.* **9** 2869–2875.

Chang, T., Tang, H and Yaoqing, L. (1985). The gamma ray energy spectrum in orthopositronium 3 gamma decay. In *Positron Annihilation*, eds. P.C. Jain, R.M. Singru and K.P. Gopinathan (World Scientific) pp. 212–214.

Charlton, M. (1985a). Experimental studies of positron scattering in gases. *Rep. Prog. Phys.* **48** 737–793.

Charlton, M. (1985b). A determination of positron mobilities in low-density gases. *J. Phys. B: At. Mol. Phys.* **18** L667–L671.

Charlton, M. (1990). Antihydrogen production in collisions of antiprotons with excited states of positronium. *Phys. Lett. A* **143** 143–146.

Charlton, M. (1996). Possibilities for antihydrogen formation by antiproton–positronium collisions. *Can. J. Phys.* **73** 483–489.

Charlton, M. (1997). Possible antihydrogen formation by antiproton–positronium reactions; where we've been and where we're going. *Hyperfine Interactions* **109** 269–278.

Charlton, M. and Curry, P.J. (1985). Effect of an electric field on Ps formation in gaseous CO_2. *Il Nuovo Cimento* **6D** 17–24.

Charlton, M. and Jacobsen, F.M. (1987). Applications of accelerator-based low-energy positron beams in atomic physics. *Appl. Phys. A* **43** 235–245.

Charlton, M. and Laricchia, G. (1986). Positronium formation and positron drift experiments. In *Positron(Electron)–Gas Scattering*, eds. W.E. Kauppila, T.S. Stein and J.M. Wadhera (World Scientific) pp. 73–84.

Charlton, M. and Laricchia, G. (1991). Collision phenomena involving positronium. *Comm. At. Mol. Phys.* **26** 253–267.

Charlton, M., Griffith, T.C., Heyland, G.R. and Twomey, T.R. (1980a). Total scattering cross sections for low-energy electrons in helium and argon. *J. Phys. B: At. Mol. Phys.* **13** L239–L244.

Charlton, M., Griffith, T.C., Heyland, G.R. and Wright, G.L. (1980b). Total scattering cross sections for intermediate-energy positrons in the molecular gases H_2, O_2, N_2, CO_2 and CH_4. *J. Phys. B: At. Mol. Phys.* **13** L353–L356.

Charlton, M., Griffith, T.C., Heyland, G.R., Lines, K.S. and Wright, G.L. (1980c). The energy dependence of positronium formation in gases. *J. Phys. B: At. Mol. Phys.* **13** L757–L760.

Charlton, M., Griffith, T.C., Heyland, G.R. and Wright, G.L. (1983a). Total cross sections for low-energy positrons in the molecular gases H_2, N_2, CO_2, O_2 and CH_4. *J. Phys. B: At. Mol. Phys.* **16** 323–341.

Charlton, M., Clark, G., Griffith, T.C. and Heyland, G.R. (1983b). Positronium formation cross sections in the inert gases. *J. Phys. B: At. Mol. Phys.* **16** L465–L470.

Charlton, M., Laricchia, G., Griffith, T.C., Wright, G.L. and Heyland, G.R. (1984). Measurements of low-energy $e^{\pm}$–Ne and e^{+}–Ar total scattering cross sections. *J. Phys. B: At. Mol. Phys.* **17** 4945–4951.

Charlton, M., Andersen, L.H., Brun-Nielsen, L., Deutch, B.I., Hvelplund, P., Jacobsen, F.M., Knudsen, H., Laricchia, G., Poulsen, M.R. and Pedersen, J.O.P. (1988). Positron and electron impact double ionisation of helium. *J. Phys. B: At. Mol. Opt. Phys.* **21** L545–L549.

Charlton, M., Brun-Nielsen, L., Deutch, B.I., Hvelplund, P., Jacobsen, F.M., Knudsen, H., Laricchia, G. and Poulsen, M.R. (1989). Double ionisation of noble gases by positron and electron impact. *J. Phys. B: At. Mol. Opt. Phys.* **22** 2779–2788.

Charlton, M., Eades, J., Horváth, D., Hughes, R.J. and Zimmermann, C. (1994). Antihydrogen physics. *Phys. Rep.* **241** 65–117.

Chaudhury, J., Ghosh, A.S. and Sil, N.C. (1974). e^{-}–H scattering for a wider range of energy. *Phys. Rev. A* **10** 2257–2263.

Chen, X.X., Chen, J. and Kuang, J. (1992). Ionization cross sections in e^{+}–H_2 scattering. *J. Phys. B: At. Mol. Opt. Phys.* **25** 5489–5494.

Chen, Z. and Msezane, A.Z. (1994). Calculation of the cross section for positron- and proton-impact ionization of helium. *Phys. Rev. A* **49** 1752–1756.

Cherry, W.H. (1958). Secondary electron emission produced from surfaces by positron bombardment. Ph.D. Dissertation, Princeton University.

Christophorou, L.G. (1971). *Atomic and Molecular Radiation Physics* (Wiley-Interscience).

Chu, S. and Mills Jr., A.P. (1982). Excitation of the positronium 1^3S_1–2^3S_1 two-photon transition. *Phys. Rev. Lett.* **48** 1333–1337.

Chu, S., Mills Jr., A.P. and Hall, J.S. (1984). Measurement of the positronium 1^3S_1–2^3S_1 interval by Doppler-free two-photon spectroscopy. *Phys. Rev. Lett.* **52** 1689–1692.

Chu, S., Mills Jr., A.P. and Murray, C.A. (1981). Thermodynamics of positronium thermal desorption from surfaces. *Phys. Rev. B* **23** 2060–2064.

Clark, G. (1984). An experimental study of positronium formation in gases. Ph.D. thesis, University of London.

Cody, W.J., Lawson, J., Massey, H.S.W. and Smith, K. (1964). The elastic scattering of slow positrons by hydrogen atoms. *Proc. Roy. Soc. Lond. A* **278** 479–489.

Coleman, P.G. (1979). The distortion of TAC–MCA spectra by the measuring process. *J. Phys. E.* **12** 590–592.

Coleman, P.G. and Hutton, J.T. (1980). Excitation of helium atoms by positron impact. *Phys. Rev. Lett.* **45** 2017–2020.

Coleman, P.G. and McNutt, J.D. (1979). Measurement of differential cross sections for the elastic scattering of positrons by argon atoms. *Phys. Rev. Lett.* **42** 1130–1133.

Coleman, P.G., Griffith, T.C. and Heyland, G.R. (1973). A time of flight method of investigating the emission of low energy positrons from metal surfaces. *Proc. Roy. Soc. Lond. A* **331** 561–569.

Coleman, P.G., Griffith, T.C. and Heyland, G.R. (1974). The analysis of data obtained with time to amplitude converter and multichannel analyser systems. *Appl. Phys.* **5** 223–230.

Coleman, P.G., Griffith, T.C. and Heyland, G.R. (1981). Rotational excitation and momentum transfer in slow positron–molecule collisions. *J. Phys. B: At. Mol. Phys.* **14** 2509–2517.

Coleman, P.G., Griffith, T.C., Heyland, G.R. and Killeen, T.L. (1974). A simple positron lifetime system suitable for low density gases. *Appl. Phys.* **3** 271–273.

Coleman, P.G., Griffith, T.C., Heyland, G.R. and Killeen, T.L. (1975a). Positronium formation in the noble gases. *J. Phys. B: At. Mol. Phys.* **8** L185–L189.

Coleman, P.G., Griffith, T.C., Heyland, G.R. and Killeen, T.L. (1975b). Positron lifetime spectra for the noble gases. *J. Phys. B: At. Mol. Phys.* **8** 1734–1743.

Coleman, P.G., Griffith, T.C., Heyland, G.R. and Killeen, T.L. (1976a). Slow positrons in nitrogen gas. In *Proc. Fourth Int. Conf. on Positron Annihilation, Helsingor* pp. 62–67.

Coleman, P.G., Griffith, T.C., Heyland, G.R. and Twomey, T.R. (1976b). Measurement of positron–rare gas total cross-sections at intermediate energies. *Appl. Phys.* **11** 321–325.

Coleman, P.G., McNutt, J.D., Diana, L.M. and Burciaga, J.R. (1979). Scattering of low-energy positrons by helium and neon atoms. *Phys. Rev. A* **20** 145–153.

Coleman, P.G., McNutt, J.D., Diana, L.M. and Hutton, J.T. (1980a). Measurement of total cross sections for the scattering of positrons by argon and xenon atoms. *Phys. Rev. A* **22** 2290–2292.

Coleman, P.G., McNutt, J.D., Hutton, J.T., Diana, L.M. and Fry, J.L. (1980b). A time-of-flight spectrometer for the measurement of angular distributions of scattered positrons and electrons. *Rev. Sci. Inst.* **51** 935–943.

Coleman, P.G., Hutton, J.T., Cook, D.R. and Chandler, C.A. (1982). Inelastic scattering of slow positrons by helium, neon, and argon atoms. *Can. J. Phys.* **60** 584–590.

Coleman, P.G., Johnston, K.A., Cox, A.M.G., Goodyear, A. and Charlton, M. (1992). Elastic positron–helium scattering near the positronium formation threshold. *J. Phys. B: At. Mol. Opt. Phys.* **25** L585–L588.

Coleman, P.G., Rayner, S., Jacobsen, F.M., Charlton, M. and West, R.N. (1994). Angular correlation studies of positron annihilation in the noble gases. *J. Phys. B: At. Mol. Opt. Phys.* **27** 981–991.

Condo, G.T. (1964). On the absorption of negative pions by liquid helium. *Phys. Lett.* **9** 65–66.

Conti, R.S., Hatamian, S., Lapidus, L., Rich, A. and Skalsey, M. (1993). Search for C-violating and P-conserving interactions and observation of 2^3S_1 to 2^1P_1 transitions in positronium. *Phys. Lett. A* **177** 43–48.

Costello, D.G., Groce, D.E., Herring, D.F. and McGowan, J.W. (1972). (Positron, He) total scattering. *Can. J. Phys.* **50** 23–33.

Crompton, R.W. and Elford, M.T. (1973). The drift velocity of electrons in oxygen at 293 K. *Aust. J. Phys.* **26** 771–782.

Curry, P.J. and Charlton, M. (1985). Evidence for Ps formation in spurs in high-density gaseous CO_2. *Chem. Phys.* **95** 313–320.

Czarnecki, A., Melnikov, K. and Yelkhovsky, A. (1999). Positronium hyperfine splitting: analytical value at $O(m\alpha^6)$. *Phys. Rev. Lett.* **82** 311–314.

Dababneh, M.S., Kauppila, W.E., Downing, J.P., Laperriere, F., Pol, V., Smart, J.H. and Stein, T.S. (1980). Measurements of total scattering cross sections for low-energy positrons and electrons colliding with krypton and xenon. *Phys. Rev. A* **22** 1872–1877.

Dahm, J., Ley, R., Niebling, K.D., Schwarz, R. and Werth, G. (1988). Electro-produced slow positrons. *Hyperfine Interactions* **44** 151–166.

Dalba, G., Fornasini, P., Lazzizzera, I., Ranieri, G. and Zecca, A. (1980). Measurements of total absolute cross sections for 0.2–100 eV electrons on H_2. *J. Phys. B: At. Mol. Phys.* **13** 2839–2848.

Dalgarno, A. and Kingston, A.E. (1960). The refractive indices and Verdet constants of the inert gases. *Proc. Roy. Soc. Lond. A* **259** 424–429.

Dalgarno, A. and Lynn, N. (1957). An exact calculation of second order long range forces. *Proc. Phys. Soc. A* **70** 223–225.

Danby, G. and Tennyson, J. (1988). Positron–HF collisions: prediction of a weakly bound state. *Phys. Rev. Lett.* **61** 2737–2739.

Danby, G. and Tennyson, J. (1990). Differential cross sections for elastic positron–H_2 collisions using the R-matrix method. *J. Phys. B: At. Mol. Opt. Phys.* **23** 1005–1016 and corrigendum, **23** 2471.

Daniel, T.B. and Stump, R. (1959). Positron annihilation in helium. *Phys. Rev.* **115** 1599–1600.

Danzmann, K., Fee, M.S. and Chu, S. (1989). Doppler-free laser spectroscopy of positronium and muonium: reanalysis of the 1S–2S measurements. *Phys. Rev. A* **39** 6072–6073.

Darewych, J.W. (1982). Elastic scattering and annihilation of low-energy positrons by molecular nitrogen. *J. Phys. B: At. Mol. Phys.* **15** L415–L419.

Darewych, J.W. (1987). Formation of antihydrogen in excited states in antiproton–positronium collisions. *J. Phys. B: At. Mol. Phys.* **20** 5917–5924.

Darewych, J.W. and Baille, P. (1974). Interaction of low-energy positrons with molecular nitrogen. *J. Phys. B: At. Mol. Phys.* **7** L1–L4.

Davies, S.A., Charlton, M. and Griffith, T.C. (1989). Free positron annihilation in gases under the influence of a static electric field. *J. Phys. B: At. Mol. Opt. Phys.* **22** 327–340.

Day, D.J. (1993). An electrostatic positron beam and its use in an experimental investigation of the first excited state of positronium. Ph.D. thesis, University of London.

Day, D.J., Charlton, M. and Laricchia, G. (2000). On the formation of excited state positronium in vacuum by positron impact on untreated surfaces. In preparation.

Deb, N.C., Crothers, D.S.F and Fromme, D. (1990). Positronium formation in positron–helium collision. *J. Phys. B: At. Mol. Opt. Phys.* **23** L483–L489.

Deb, N.C., McGuire, J.H. and Sil, N.C. (1987). Evaluation of cross section for electron capture by positrons. *Phys. Rev. A* **36** 3707–3714.

DeBenedetti, S., Cowan, C.E. and Konneker, W.E. (1949). Angular distribution of annihilation radiation. *Phys. Rev.* **76** 440.

DeBenedetti, S., Cowan, C.E., Konneker, W.R. and Primakoff, H. (1950). On the angular distribution of two-photon annihilation radiation. *Phys. Rev.* **77** 205–212.

de Heer, F.J. and Jansen, R.H.J. (1977). Total cross sections for electron scattering by He. *J. Phys. B: At. Mol. Phys.* **10** 3741–3758.

de Heer, F.J., Jansen, R.H.J. and van der Kaay, W. (1979). Total cross sections for electron scattering by Ne, Ar, Kr and Xe. *J. Phys. B: At. Mol. Phys.* **12** 979–1002.

de Heer, F.J., McDowell, M.R.C. and Wagenaar, R.W. (1977). Numerical study of the dispersion relation for e$^-$–H scattering. *J. Phys. B: At. Mol. Phys.* **10** 1945–1953.

Deuring, A., Floeder, K., Fromme, D., Raith, W., Schwab, A., Sinapius, G., Zitzewitz, P.W. and Krug, J. (1983). Total cross section measurements for positron and electron scattering on molecular hydrogen between 8 and 400 eV. *J. Phys. B: At. Mol. Phys.* **16** 1633–1656.

Deutch, B.I., Jensen, A.S., Miranda, A. and Oades, G.C. (1986). Antiproton capture in flight. In *Proc. Fermilab Workshop in Antimatter Physics at Low Energies.* p. 371.

Deutch, B.I., Andersen, L.H., Hvelplund, P., Jacobsen, F.M., Knudsen, H., Holzscheiter, M.H., Charlton, M. and Laricchia, G. (1988). Antihydrogen by positronium–antiproton collisions. *Hyperfine Interactions* **44** 271–286.

Deutch, B.I., Charlton, M., Holzscheiter, M.H., Hvelplund, P., Jørgensen, L.V., Knudsen, H., Laricchia, G., Merrison, J.P. and Poulsen, M.R. (1993). Antihydrogen synthesis by the reaction of antiprotons with excited state positronium atoms. *Hyperfine Interactions* **76** 153–161.

Deutsch, M. (1951). Evidence for the formation of positronium in gases. *Phys. Rev.* **82** 455–456.

Deutsch, M. and Brown, S.C. (1952). Zeeman effect and hyperfine splitting of positronium. *Phys. Rev.* **85** 1047–1048.

Deutsch. M. and Dulit, E. (1951). Short range interaction of electrons and fine structure of positronium. *Phys. Rev.* **84** 601–602.

Dewangan, D.P. (1980). On higher order Born approximations in atomic scattering calculations. *J. Phys. B: At. Mol. Phys.* **13** L595–L598.

Dewangan, D.P. and Walters, H.R.J. (1977). The elastic scattering of electrons and positrons by helium and neon: the distorted-wave second Born approximation. *J. Phys. B: At. Mol. Phys.* **10** 637–661.

Diana, L.M., Fornari, L.S., Sharma, S.C., Pendleton, P.K. and Coleman, P.G. (1985). Measurements of total ionization cross sections for positrons. In *Positron Annihilation*, eds. P.C. Jain, R.M. Singru and K.P. Gopinathan (World Scientific) pp. 342–343.

Diana, L.M., Brooks, D.L., Coleman, P.G., Pendleton, P.K., Norman, D.M., Seay, B.E. and Sharma, S.C. (1986a). Total cross sections for positronium formation in molecular hydrogen, krypton and xenon. In *Positron (Electron)–Gas Scattering*, eds. W.E. Kauppila, T.S. Stein and J.M. Wadehra (World Scientific) pp. 293–295.

Diana, L.M., Coleman, P.G., Brooks, D.L., Pendleton, P.K. and Norman, D.M. (1986b). Positronium formation cross sections in He and H_2 at intermediate energies. *Phys. Rev. A* **34** 2731–2737.

Diana, L.M., Coleman, P.G., Brooks, D.L., Pendleton, P.K., Norman, D.M., Seay, B.E. and Sharma, S.C. (1986c). Measurement of total positronium formation cross section in argon to 441.3 eV. In *Positron(Electron)–Gas Scattering*, eds. W.E. Kauppila, T.S. Stein and J.M. Wadehra (World Scientific) pp. 296–298.

Dirac, P.A.M. (1930). A theory of electrons and protons. *Proc. Roy. Soc. Lond. A* **126** 360–365.

Dirks, J.F. and Hahn, Y. (1971). Generalized variational bounds on the positron–hydrogen reaction matrix. II Effective distortions. *Phys. Rev. A* **3** 310–319.

Doolen, G., McCartor, G., McDonald, F.A. and Nuttall, J. (1971). s-wave elastic positron–hydrogen scattering in the ionization region. *Phys. Rev. A* **4** 108–111.

Dou, L., Kauppila, W.E., Kwan, C.K., Pryzybyla, D., Smith, S.J. and Stein, T.S. (1992a). Evidence for resonances and absorption effects in positron–krypton differential-elastic-scattering measurements. *Phys. Rev. A* **46** R5327–R5330.

Dou, L., Kauppila, W.E., Kwan, C.K. and Stein, T.S. (1992b). Observation of structure in intermediate-energy positron–argon differential elastic scattering. *Phys. Rev. Lett.* **68** 2913–2916.

Drachman, R.J. (1965). Positron–hydrogen scattering at low energies. *Phys. Rev.* **138A** 1582–1585.

Drachman, R.J. (1966a). Theory of low-energy positron–helium scattering. *Phys. Rev.* **144** 25–28.

Drachman, R.J. (1966b). Positron annihilation in helium. *Phys. Rev.* **150** 10–14.

Drachman, R.J. (1968). Variational bounds in positron–atom scattering. *Phys. Rev.* **173** 190–201.

Drachman, R.J. (1972). The method of models in variational scattering calculations. *J. Phys. B: At. Mol. Phys.* **5** L30–L32.

Drachman, R.J. (1987). Theoretical aspects of positronium collisions. In *Atomic Physics with Positrons*, eds. J.W. Humberston and E.A.G. Armour (Plenum Press) pp. 203–214.

Drachman, R.J. and Houston, S.K. (1975). Positronium–hydrogen elastic scattering. *Phys. Rev. A* **12** 885–890.

Drachman, R.J. and Houston, S.K. (1976). Positronium–hydrogen elastic scattering: the electronic $S = 1$ state. *Phys. Rev. A* **14** 894–896.

Drachman, R.J. and Sucher, J. (1979). Annihilation in positron–atom collisions: a new approach. *Phys. Rev. A* **20** 442–444.

DuBois, R.D. and Rudd, M.E. (1975). Absolute differential cross sections for 20–800 eV electrons elastically scattered from argon. *J. Phys. B: At. Mol. Phys.* **8** 1474–1483.

DuBois, R.D. and Rudd, M.E. (1978). Absolute doubly differential cross sections for ejection of secondary electrons from gases by electron impact. II 100–500 eV electrons on neon, argon, molecular hydrogen and molecular nitrogen. *Phys. Rev. A* **17** 843–848.

DuMond, J.M.W., Lind, D.A. and Watson, B.B. (1949). Precision measurement of the wavelength and spectral profile of the annihilation radiation from ^{64}Cu with the two-meter focusing curved crystal spectrometer. *Phys. Rev.* **75** 1226–1239.

Dutton, J., Evans, C.J. and Mansour, H.L. (1982). Total cross-sections for positron scattering in nitrogen at energies from 20 to 3000 eV. In *Positron Annihilation*, eds. P.G. Coleman, S.C. Sharma and L.M. Diana (North-Holland) pp. 82–84.

Dzuba, V.A., Flambaum, V.V., Gribakin, G.F. and King, W.A. (1995). Bound states of positrons and neutral atoms. *Phys. Rev. A* **52** 4541–4546.

Eades, J. and Hartmann, F.J. (1999). Forty years of antiprotons. *Rev. Mod. Phys.* **71** 373–419.

Ebel, F., Faust, W., Hahn, C., Rückert, M., Schneider, H., Singe, A. and Tobehn, I. (1989). Ratio of inner-shell ionization by low energy electron and positron impact. *Phys. Lett. A* **140** 114–116.

Ebel, F., Faust, W., Hahn, C., Rückert, M., Schneider, H., Singe, A. and Tobehn, I. (1990). Research at the positron source TEPOS. *Nuc. Inst. Meth. B* **50** 328–330.

Egan, P.O., Hughes, V.W. and Yam, M.H. (1977). Precision determination of the fine-structure interval in the ground state of positronium. IV. Measurement of positronium fine-structure density shifts in noble gases. *Phys. Rev. A* **15** 251–260.

Ehrhardt, H., Knoth, G., Schlemmer, P. and Jung, K. (1985). Absolute H(e, 2e)p cross section measurements: comparison with first and second order theory. *Phys. Lett. A* **110** 92–94.

Eldrup, M., Vehanen, A., Schultz, P.J. and Lynn, K.G. (1985). Positronium formation and diffusion in crystalline and amorphous ice using a variable-energy positron beam. *Phys. Rev. B* **32** 7048–7064.

El-Gogary, M.H.H., Abdel-Raouf, M.A., Hassan, M.Y.M. and Saleh, S.A. (1995). Variational treatment of positronium molecules. *J. Phys. B: At. Mol. Opt. Phys.* **28** 4927–4945.

Ermolaev, A.M., Bransden, B.H. and Mandal, C.R. (1987). Theoretical cross sections for formation of antihydrogen in $\bar{\mathrm{p}}$–Ps collisions in the antiproton energy range 2–100 keV lab. *Phys. Lett. A* **125** 44–46.

Falke, T., Brandt, T., Kühl, O., Raith, W. and Weber, M. (1997). Differential Ps-formation and impact-ionization cross sections for positron scattering on Ar and Kr atoms. *J. Phys. B: At. Mol. Opt. Phys.* **30** 3247–3256.

Falke, T., Raith, W., Weber, M. and Wesskamp, U. (1995). Differential positronium formation in positron–argon collisions. *J. Phys. B: At. Mol. Opt. Phys.* **28** L505–L509.

Farazdel, A. (1986). Confirmation of the positron mobility edge in gaseous helium by Monte Carlo simulation. *Phys. Rev. Lett.* **57** 2664–2666.

Farazdel, A. and Cade, P.E. (1977). The electronic structure and positron annihilation characteristics of positronium halides $[X^-e^+]$. II. Two-photon annihilation. *J. Chem. Phys.* **66** 2612–2620.

Farazdel, A. and Epstein, I.R. (1977). Monte Carlo studies of positrons in matter. Method and application to annihilation spectra in helium gas. *Phys. Rev. A* **16** 518–524.

Farazdel, A. and Epstein, I.R. (1978). Monte Carlo studies of positrons in matter. Temperature and electric field effects on lifetime spectra in low-temperature, high-density helium gas. *Phys. Rev. A* **17** 577–586.

Faust, W., Hahn, C., Rückert, M., Schneider, H., Singe, A. and Tobehn, I. (1991). New results at the Giessen positron sources TEPOS. *Nuc. Inst. Meth. B* **56/57** 575–577.

Fedichev, P.O. (1997). Formation of antihydrogen atoms in an ultra-cold positron–antiproton plasma. *Phys. Lett. A* **226** 289–292.

Fee, M.S., Chu, S., Mills Jr., A.P., Chichester, R.J., Zuckerman, D.M., Shaw, E.D. and Danzmann, K. (1993a). Measurement of the positronium 1^3S_1–2^3S_1 interval by continuous-wave two-photon excitation. *Phys. Rev. A* **48** 192–219.

Fee, M.S., Mills Jr., A.P., Chu, S., Shaw, E.D., Danzmann, K., Chichester, R.J. and Zuckerman, D.M. (1993b). Measurement of the positronium 1^3S_1–2^3S_1 interval by continuous-wave two-photon excitation. *Phys. Rev. Lett.* **70** 1397–1400.

Fell, R.N. (1992). Order $\alpha^4 \ln \alpha^{-1} f_{\mathrm{RYD}}$ corrections to the $n = 1$ and $n = 2$ energy levels of positronium. *Phys. Rev. Lett.* **68** 25–28.

Feng, X., Holzscheiter, M.H., Charlton, M., Hangst, J., King, N.S.P., Lewis, R.A., Rochet, J. and Yamazaki, Y. (1997). Capture and cooling of antiprotons. *Hyperfine Interactions* **109** 145–152.

Ferch, J., Masche, C. and Raith, W. (1981). Total cross section measurement for e–CO_2 scattering down to 0.07 eV. *J. Phys. B: At. Mol. Phys.* **14** L97–L100.

Ferch, J., Raith, W. and Schröder, K. (1980). Total cross section measurements for electron scattering from molecular hydrogen at very low energies. *J. Phys. B: At. Mol. Phys.* **13** 1481–1490.

Feshbach, H. (1962). A unified theory of nuclear reactions. II. *Ann. Phys.* **19** 287–313.

Ficocelli Varracchio, E. (1990). Positron excitation of He: a random phase approximation analysis of experimental results. *J. Phys. B: At. Mol. Opt. Phys.* **23** L779–L785.

Ficocelli Varracchio, E. and Parcell, L.A. (1992). Positron impact excitation of the $n = 2$ and $n = 3$ manifolds of He in the RPA formulation. *J. Phys. B: At. Mol. Opt. Phys.* **25** 3037–3048.

Finch, R.M., Kövér, Á., Charlton, M. and Laricchia, G. (1996a). Positron–argon elastic, positronium formation and ionizing collisions at 60°. *J. Phys. B: At. Mol. Opt. Phys.* **29** L667–L672.

Finch, R.M., Kövér, Á., Charlton, M. and Laricchia, G. (1996b). Differential cross-section measurements in positron–argon collisions. *Can. J. Phys.* **74** 505–508.

Floeder, K., Höner, P., Raith, W., Schwab, A., Sinapius, G. and Spicher, G. (1988). Differential elastic scattering of positrons from argon atoms at low energies. *Phys. Rev. Lett.* **60** 2363–2366.

Fornari, L.S., Diana, L.M. and Coleman, P.G. (1983). Positronium formation in collisions of positrons with He, Ar and H_2. *Phys. Rev. Lett.* **51** 2276–2279.

Fox, R.A. and Canter, K.F. (1978). Positronium lifetimes in helium–oxide-powder mixtures. *J. Phys. B: At. Mol. Phys.* **11** L255–L258.

Fox, R.A., Canter, K.F. and Fishbein, M. (1977). Positron and orthopositronium decay rates in helium at high densities. *Phys. Rev. A* **15** 1340–1343.

Fraser, P.A. (1961a). The scattering of low-energy ortho-positronium by hydrogen atoms. *Proc. Phys. Soc.* **78** 329–347.

Fraser, P.A. (1961b). The scattering of low-energy ortho-positronium by helium atoms. *Proc. Phys. Soc.* **79** 721–731.

Fraser, P.A. (1968). Positrons and positronium in gases. *Adv. At. Mol. Phys.* **4** 63–107.

Fraser, P.A. and Kraidy, M. (1966). The pick-off quenching of ortho-positronium in helium. *Proc. Phys. Soc.* **89** 533–539.

Frolov, A.M and Smith Jr., V.H. (1997). Ground state of positronium hydride. *Phys. Rev. A* **56** 2417–2420.

Frolov, A.M. and Yeremin, A.Yu. (1989). Ground states in two-electron systems with $Z = 1$. *J. Phys. B: At. Mol. Opt. Phys.* **22** 1263–1268.

Fromme, D., Kruse, G., Raith, W. and Sinapius, G. (1986). Partial cross-section measurements for ionization of helium by positron impact. *Phys. Rev. Lett.* **57** 3031–3034.

Fromme, D., Kruse, G., Raith, W. and Sinapius, G. (1988). Ionisation of molecular hydrogen by positrons. *J. Phys. B: At. Mol. Opt. Phys.* **21** L262–L265.

Gabrielse, G. (1988). Trapped antihydrogen for spectroscopy and gravitation studies: is it possible? *Hyperfine Interactions* **44** 349–356.

Gabrielse, G., Haarsma, L. and Rolston, S.L. (1989). Open-endcaps Penning traps for precision experiments. *Int. J. Mass. Spec. Ion Processes* **88** 319–332.

Gabrielse, G., Fei, X., Helmerson, K., Rolston, S.L., Tjoekler, R., Trainor, T.A., Kalinowsky, H., Haas, J. and Kells, W. (1986). First capture of antiprotons in a Penning trap: a kiloelectronvolt source. *Phys. Rev. Lett.* **57** 2504–2507.

Gabrielse, G., Rolston, S.L., Haarsma, L. and Kells, W. (1988). Possible antihydrogen production using trapped plasmas. *Hyperfine Interactions* **44** 287–294.

Gabrielse, G., Fei, X., Orozco, L.A., Kalinowsky, H., Trainor, T., Haas, J. and Kells, W. (1990). Thousand-fold improvement in the measured antiproton mass. *Phys. Rev. Lett.* **65** 1317–1320.

Gabrielse, G., Hall, D.S, Roach, T., Yelsey P., Khabbaz, A., Estrada, J., Heimann, C. and Kalinowsky, H. (1999). The ingredients of cold antihydrogen: simultaneous confinement of antiprotons and positrons at 4 K. *Phys. Lett. B* **455** 311–315.

Gailitis, M. (1965). Extremal properties of approximate methods of collision theory in the presence of inelastic processes. *Sov. Phys. JETP* **20** 107–111.

Garner, A.J., Laricchia, G. and Özen, A. (1996). Ps beam production and scattering from gaseous targets. *J. Phys. B: At. Mol. Opt. Phys.* **29** 5961–5968.

Garner, A.J., Özen, A. and Laricchia, G. (1998). Positronium beam scattering from atoms and molecules. *Nuc. Inst. Meth. B* **143** 155–161.

Ghosh, A.S., Mazumder, P.S. and Basu, M. (1985). The sensitivity of the final state wavefunction in positron impact ionization. *J. Phys. B: At. Mol. Phys.* **18** 1881–1886.

Ghosh, A.S., Sil, N.C. and Mandal, P. (1982). Positron–atom and positron–molecule collisions. *Phys. Rep.* **87** 313–406.

Gianturco, F.A. and Mukherjee, T. (1997). Dynamical coupling effects in the vibrational excitation of H_2 and N_2 colliding with positrons. *Phys. Rev. A* **55** 1044–1055.

Gianturco, F.A. and Paioletti, P. (1997). Elastic collisions and rotational excitation in positron scattering from CO_2 molecules. *Phys. Rev. A* **55** 3491–3503.

Gidley, D.W. and Zitzewitz, P.W. (1978). Measurement of the vacuum decay rate of orthopositronium formed in an MgO-lined cavity. *Phys. Lett. A* **69** 97–99.

Gidley, D.W., Marko, K. and Rich, A. (1976). Precision measurement of the decay rate of ortho-positronium in SiO_2 powders. *Phys. Rev. Lett.* **36** 395–398.

Gidley, D.W., Nico, J.S. and Skalsey, M. (1991). Direct search for two-photon decay modes of orthopositronium. *Phys. Rev. Lett.* **66** 1302–1305.

Gidley, D.W., Rich, A., Sweetman, E. and West, D. (1982). New precision measurement of the decay rates of singlet and triplet positronium. *Phys. Rev. Lett.* **49** 525–528.

Gidley, D.W., Mayer, R., Frieze, W.E. and Lynn, K.G. (1987). Glancing-angle scattering and neutralization of a positron beam at metal surfaces. *Phys. Rev. Lett.* **58** 595–598.

Gien, T.T. (1994). Structures at intermediate energies in positron–hydrogen scattering. *J. Phys. B: At. Mol. Opt. Phys.* **27** L25–L31.

Gien, T.T. (1997). Coupled-state calculations of positron–hydrogen scattering. *Phys. Rev. A* **56** 1332–1337.

Gillespie, E.S. and Thompson, D.G. (1975). Positron scattering by molecular nitrogen. *J. Phys. B: At. Mol. Phys.* **8** 2858–2868.

Gillespie, E.S. and Thompson, D.G. (1977). Positronium formation in neon and argon. *J. Phys. B: At. Mol. Phys.* **10** 3543–3549.

Glinsky, M.E. and O'Niel, T.M. (1991). Guiding center atoms: three-body recombination in a strongly magnetized plasma. *Phys. Fluids B* **3** 1279–1293.

Gninenko, S.N. (1994). Limit on 'disappearance' of orthopositronium in vacuum. *Phys. Lett. B* **326** 317–319.

Gninenko, S.N., Klubakov, Yu.M., Poblaguev, A.A. and Postoev, V.E. (1990). A search for a keV pseudoscalar in the two-body decay of orthopositronium. *Phys. Lett. B* **237** 287–290.

Goldanskii, V.I. (1968). Physical chemistry of the positron and positronium. *Atomic Energy Review* **6** 3–148.

Goldanskii, V.I. and Mokrushin, A.D. (1968). Chemical interaction of positronium atoms with chlorine. *High Energy Chem.* **2** 77–81.

Gomer, V., Harms, O., Haubrich, D., Schadwinkel, H., Strauch, F., Ueberholz, B., aus der Wiesche, S. and Meschede, D. (1997). Magnetostatic traps for charged and neutral particles. *Hyperfine Interactions* **109** 281–292.

Greaves, R.G. and Surko, C.M. (1996). Solid neon moderator for positron trapping experiments. *Can. J. Phys.* **74** 445–448.

Greaves, R.G., Tinkle, M.D. and Surko, C.M. (1994). Creation and uses of positron plasmas. *Phys. Plasmas* **1** 1439–1446.

Gribakin, G.F. and King, W.A. (1996). Positron scattering from Mg atoms. *Can. J. Phys.* **74** 449–459.

Griffith, T.C. and Heyland, G.R. (1978). Experimental aspects of the study of the interaction of low-energy positrons with gases. *Phys. Rep.* **39** 169–277.

Griffith, T.C., Heyland, G.R., Lines, K.S. and Twomey, T.R. (1978a). A reappraisal of the experimental cross sections for low-energy positron–helium scattering. *J. Phys. B: At. Mol. Phys.* **11** L635–L637.

Griffith, T.C., Heyland, G.R., Lines, K.S. and Twomey, T.R. (1978b). The decay rate of ortho-positronium in vacuum. *J. Phys. B: At. Mol. Phys.* **11** L743–L748.

Griffith, T.C., Heyland, G.R., Lines, K.S. and Twomey, T.R. (1979a). Total cross-sections for the scattering of positrons by helium, neon, and argon at intermediate energies. *Appl. Phys.* **19** 431–437.

Griffith, T.C., Heyland, G.R., Lines, K.S. and Twomey, T.R. (1979b). Inelastic scattering of positrons by helium atoms at intermediate energies. *J. Phys. B: At. Mol. Phys.* **12** L747–L753.

Griffith, T.C., Charlton, M., Clark, G., Heyland, G.R. and Wright, G.L. (1982). Positrons in gases – a progress report. In *Positron Annihilation*, eds. P.G. Coleman, S.C. Sharma and L.M. Diana (North-Holland) pp. 61–70.

Groce, D.E., Costello, D.G., McGowan, J.W. and Herring, D.F. (1968). Time-of-flight observation of low-energy positrons. *Bull. Am. Phys. Soc.* **13** 1397.

Gryziński, M. and Kowalski, M. (1993). Theory of inner shell ionization by positrons. *Phys. Lett. A* **183** 196–200.

Guha, S. and Ghosh, A.S. (1981). Positron–lithium–atom collisions using the two-state approximation. *Phys. Rev. A* **23** 743–750.

Guha, S. and Mandal, P. (1980). Model potential approach for positron–atom collisions I. Positronium formation in ground state in alkali atoms Na, K, Rb and Cs using the distorted-wave approximation. *J. Phys. B: At. Mol. Phys.* **13** 1919–1935.

Gullikson, E.M. and Mills Jr., A.P. (1986). Positron dynamics in rare-gas solids. *Phys. Rev. Lett.* **57** 376–379.

Gullikson, E.M., Mills Jr., A.P., Crane, W.S. and Brown, B.L. (1985). Absence of energy loss in positron emission from metal surfaces. *Phys. Rev. B* **32** 5484–5486.

Haarsma, L.H., Abdullah, K. and Gabrielse, G. (1995). Extremely cold positrons accumulated electronically in ultrahigh vacuum. *Phys. Rev. Lett.* **75** 806–809.

Hagena, D., Ley, R., Weil, D., Werth, G., Arnold, W. and Schneider, H. (1993). Precise measurement of $n = 2$ positronium fine-structure intervals. *Phys. Rev. Lett.* **71** 2887–2890.

Hall, D.S. and Gabrielse, G. (1996). Electron cooling of protons in a nested Penning trap. *Phys. Rev. Lett.* **77** 1962–1965.

Hänsch, T.W. and Zimmermann, C. (1993). Laser spectroscopy of hydrogen and antihydrogen. *Hyperfine Interactions* **76** 47–57.

Hansen, H. and Flammersfeld, A. (1966). Messung des wirkungsquerschnitts fur K-ionisierung durch stoss niederenergetischer negatonen und positonen. *Nucl. Phys.* **79** 135–144.

Hansen, H., Weigmann, H. and Flammersfeld, A. (1964). Messung des wirkungsquerschnitts fur K-ionisierung durch negatonen- und positonenstoß. *Nucl. Phys.* **58** 241–253.

Hara, S. (1974). Scattering of slow positrons by hydrogen molecules. *J. Phys. B: At. Mol. Phys.* **7** 1748–1755.

Hara, S. and Fraser, P.A. (1975). Low-energy ortho-positronium scattering by hydrogen atoms. *J. Phys. B: At. Mol. Phys.* **8** L472–L476.

Harris, F.E. (1967). Expansion approach to scattering. *Phys. Rev. Lett.* **19** 173–175.

Harting, E. and Read, F.H. (1976). *Electrostatic Lenses* (Elsevier).

Hasbach, P., Hilkert, G., Klempt, E. and Werth, G. (1987). Experimental determination of the ortho-positronium lifetime in vacuum. *Il Nuovo Cimento A* **97** 419–425.

Hatamian, S., Conti, R.S. and Rich, A. (1987). Measurements of the 2^3S_1–2^3P_J $(J = 0, 1, 2)$ fine-structure splittings in positronium. *Phys. Rev. Lett.* **58** 1833–1836.

Haugen, H.K., Andersen, L.H., Hvelplund, P. and Knudsen, H. (1982). Multiple ionization of noble gases by fully stripped ions. *Phys. Rev. A* **26** 1962–1974.

Hautojärvi, P. and Rytsölä, K. (1979). Positron-induced cluster and positronium bubble in low temperature helium. In *Positron Annihilation*, eds. R.R. Hasiguti and K. Fujiwara (Tokyo Inst. Metals) pp. 807–813.

Hautojärvi, P. and Vehanen, A. (1979). Introduction to positron annihilation. In *Positrons in Solids (Topics in Current Physics, Vol. 12)*, ed. P. Hautojärvi (Springer-Verlag) pp. 1–23.

Hautojärvi, P., Rytsölä, K., Tuovinen, P., Vehanen, A. and Jauho, P. (1977). Microscopic gas–liquid-like phase transition around the positron in helium gases. *Phys. Rev. Lett.* **38** 842–844.

Heddle, D.W.O. (1979). Excitation of atoms by electron impact. *Adv. At. Mol. Phys.* **15** 381–421.

Heinberg, M. and Page, L.A. (1957). Annihilation of positrons in gases. *Phys. Rev.* **107** 1589–1600.

Heitler, W. (1954). *The Quantum Theory of Radiation* (Oxford University Press).

Helms, S., Brinkmann, U., Deiwiks, J., Schneider, H. and Hippler, R. (1994a). Double ionisation of rare gas atoms by positron impact. *Hyperfine Interactions* **89** 395–400.

Helms, S., Brinkmann, U., Deiwiks, J., Hippler, R., Schneider, H. Segers, D. and Paridaens, J. (1994b). Inner shell contributions to multiple ionization of argon by positron impact. *J. Phys. B: At. Mol. Opt. Phys.* **27** L557–L562.

Helms, S., Brinkmann, U., Deiwiks, J., Hippler, R., Schneider, H. Segers, D. and Paridaens, J. (1995). Multiple ionization of argon, krypton and xenon atoms by positron impact. *J. Phys. B: At. Mol. Opt. Phys.* **28** 1095–1103.

Hess, H.F., Kochanski, G.P., Doyle, J.M., Masuhara, N., Kleppner, D. and Greytak, T.J. (1987). Magnetic trapping of spin-polarized atomic hydrogen. *Phys. Rev. Lett.* **59** 672–675.

Hessels, E.A., Homan, D.M. and Cavagnero, M.J. (1998). Two-state Rydberg charge exchange: an efficient method for production of antihydrogen. *Phys. Rev. A* **57** 1668–1671.

Hewitt, R.N., Noble, C.J. and Bransden, B.H. (1990). Positronium formation in positron–hydrogen scattering. *J. Phys. B: At. Mol. Opt. Phys.* **23** 4185–4195.

Hewitt, R.N., Noble, C.J. and Bransden, B.H. (1991). Intermediate energy resonances in positron–hydrogen scattering. *J. Phys. B: At. Mol. Opt. Phys.* **24** L635–L639.

Hewitt, R.N., Noble, C.J. and Bransden, B.H. (1992a). He(2^1S,2^1P) excitation and positronium formation in positron–helium collisions at intermediate energies. *J. Phys. B: At. Mol. Opt. Phys.* **25** 557–570.

Hewitt, R.N., Noble, C.J. and Bransden, B.H. (1992b). The effect of positronium formation on close-coupling calculations of positron–lithium collisions. *J. Phys. B: At. Mol. Opt. Phys.* **25** 2683–2695.

Hewitt, R.N., Noble, C.J. and Bransden, B.H. (1993). Positron collisions with alkali atoms at low and intermediate energies. *J. Phys. B: At. Mol. Opt. Phys.* **26** 3661–3677.

Hewitt, R.N., Noble, C.J. and Bransden, B.H. (1994). Coupled-channel calculations of positron–atom scattering at intermediate energies. *Hyperfine Interactions* **89** 195–207.

Heyland, G.R., Charlton, M., Davies, S.A. and Griffith, T.C. (1986). On the annihilation rate of thermalised free positrons in gases. *Phys. Lett. A* **119** 289–292.

Heyland, G.R., Charlton, M., Griffith, T.C. and Clark, G. (1985). The temperature dependence of free positron lifetimes and positronium fractions in gaseous CO_2 and SF_6. *Chem. Phys.* **95** 157–163.

Heyland, G.R., Charlton, M., Griffith, T.C. and Wright, G.L. (1982). Positron lifetime spectra for gases. *Can. J. Phys.* **60** 503–516.

Higgins, K. and Burke, P.G. (1991). Positron scattering by atomic hydrogen including positronium formation. *J. Phys. B: At. Mol. Opt. Phys.* **24** L343–L349.

Higgins, K. and Burke, P.G. (1993). Positron scattering by atomic hydrogen including positronium formation. *J. Phys. B: At. Mol. Opt. Phys.* **26** 4269–4288.

Higgins, K., Burke, P.G. and Walters, H.R.J. (1990). Positron scattering by atomic hydrogen at intermediate energies. *J. Phys. B: At. Mol. Opt. Phys.* **23** 1345–1357.

Ho, Y.K. (1983). Positron annihilation in the positronium negative ion. *J. Phys. B: At. Mol. Phys.* **16** 1503–1509.

Ho, Y.K. (1984). Doubly excited states of positronium negative ions. *Phys. Lett. A* **102** 348–350.

Ho, Y.K. (1986a). Binding energy of positronium molecules. *Phys. Rev. A* **33** 3584–3587.

Ho, Y.K. (1986b). Positron annihilation in positronium hydrides. *Phys. Rev. A* **34** 609–611.

Ho, Y.K. (1989). Resonant states of positronium molecules. *Phys. Rev. A* **39** 2709–2711.

Ho, Y.K. (1993). Variational calculation of the ground-state energy of the positronium negative ion. *Phys. Rev. A* **48** 4780–4783.

Ho, Y.K. and Fraser, P.A. (1976). S-wave collisions of positrons with helium. *J. Phys. B: At. Mol. Phys.* **9** 3213–3224.

Hodges, C.H. and Stott, M.J. (1973). Work functions for positrons in metals. *Phys. Rev. B* **7** 73–79.

Hoffman, K.R., Dababneh, M.S., Hsieh, Y.-F., Kauppila, W.E., Pol, V., Smart, J.H. and Stein, T.S. (1982). Total-cross-section measurements for positrons and electrons colliding with H_2, N_2, and CO_2. *Phys. Rev. A* **25** 1393–1403.

Hofmann, A., Falke, T., Raith, W., Weber, M., Becker, D.P. and Lynn, K.G. (1997). Ionization of atomic hydrogen by positrons. *J. Phys. B: At. Mol. Opt. Phys.* **30** 3297–3303.

Holzscheiter, M.H. and Charlton, M. (1999). Ultra-low energy antihydrogen. *Rep. Prog. Phys.* **62** 1–60.

Holzscheiter, M.H., Feng, X., Goldman, T., King, N.S.P., Lewis, R.A., Nieto, M.M. and Smith, G.A. (1996). Are antiprotons forever? *Phys. Lett. A* **214** 279–284.

Horbatsch, M. and Darewych, J.W. (1983). Model potential description of low-energy e^+CO_2 scattering. *J. Phys. B: At. Mol. Phys.* **16** 4059–4064.

Houston, S.K. (1973). Variational calculations of the positron–helium scattering length using approximate target wavefunctions. *J. Phys. B: At. Mol. Phys.* **6** 136–145.

Houston, S.K. and Drachman, R.J. (1971). Positron–atom scattering by Kohn and Harris methods. *Phys. Rev. A* **3** 1335–1342.

Howell, R.H., Alvarez, R.A. and Stanek, M. (1982). Production of slow positrons with a 100-MeV electron linac. *Appl. Phys. Lett.* **40** 751–752.

Howell, R.H., Rosenberg, I.J. and Fluss, M. (1986). Production of energetic positronium at metal surfaces. *Phys. Rev. B* **34** 3069–3075.

Huang, W.-T. (1973). Binding energy of exitonic molecules in isotropic semiconductors. *Phys. Stat. Sol. B* **60** 309–317.

Hughes, R.J. (1993a). Antihydrogen and fundamental symmetries. *Hyperfine Interactions* **76** 3–16.

Hughes, R.J. (1993b). The equivalence principle. *Contemporary Physics* **34** 177–191.

Hughes, R.J. and Deutch, B.I. (1992). Electric charges of positrons and antiprotons. *Phys. Rev. Lett.* **69** 578–581.

Hughes, R.J. and Holzscheiter, M.H. (1992). Tests of the weak equivalence principle with trapped antimatter. *J. Mod. Opt.* **39** 263–278.

Hughes, V.W. (1998). High precision spectroscopy of positronium and muonium. *Adv. Quant. Chem.* **30** 99–123.

Humberston, J.W. (1973). The scattering of low-energy positrons by helium. *J. Phys. B: At. Mol. Phys.* **6** L305–L308.

Humberston, J.W. (1974). Annihilation in positron–helium scattering. *J. Phys. B: At. Mol. Phys.* **7** L286–L289.

Humberston, J.W. (1978). A comparison of experimental and theoretical total cross-sections for low-energy positron–helium scattering. *J. Phys. B: At. Mol. Phys.* **11** L343–L346.

Humberston, J.W. (1979). Theoretical aspects of positron collisions in gases. *Adv. At. Mol. Phys.* **15** 101–133.

Humberston, J.W. (1982). Positronium formation in s-wave positron–hydrogen scattering. *Can. J. Phys.* **60** 591–596.

Humberston, J.W. (1984). Positronium formation in s-wave positron–hydrogen scattering. *J. Phys. B: At. Mol. Phys.* **17** 2353–2361.

Humberston, J.W. (1986). Positronium – its formation and interaction with simple systems. *Adv. At. Mol. Phys.* **22** 1–36.

Humberston, J.W. and Campeanu, R.I. (1980). The calculation of p-wave phaseshifts for positron–helium scattering. *J. Phys. B: At. Mol. Phys.* **13** 4907–4917.

Humberston, J.W. and Van Reeth, P. (1996). Positronium formation in low-energy positron–helium scattering. *Can. J. Phys.* **74** 335–342.

Humberston, J.W and Wallace, J.B.G. (1972). The elastic scattering of positrons by atomic hydrogen. *J. Phys. B: At. Mol. Phys.* **5** 1138–1148.

Humberston, J.W. and Watts, M.S.T. (1994). Positron–lithium scattering with the inclusion of positronium formation. *Hyperfine Interactions* **89** 47–55.

Humberston, J.W., Charlton, M., Jacobsen, F.M. and Deutch, B.I. (1987). Antihydrogen formation in collisions of antiprotons with positronium. *J. Phys. B: At. Mol. Phys.* **20** L25–L29.

Humberston, J.W., Van Reeth, P., Watts, M.S.T. and Meyerhof, W.E. (1997). Positron–hydrogen scattering in the vicinity of the positronium formation threshold. *J. Phys. B: At. Mol. Opt. Phys.*, **30** 2477–2493.

Hutchins, S.M., Coleman, P.G., Stone, R.J. and West, R.N. (1986). A low-distortion slow positron filter. *J. Phys. E* **19** 282–283.

Huxley, L.G.H. and Crompton, R.W. (1974). *The Diffusion and Drift of Electrons in Gases* (Wiley-Interscience).

Hvelplund, P., Knudsen, H., Mikkelsen, U., Morenzoni, E., Møller, S.P., Uggerhøj, E. and Worm, T. (1994). Ionization of helium and molecular hydrogen by slow antiprotons. *J. Phys. B: At. Mol. Opt Phys.* **27** 925–934.

Hyder, G.M.A., Dababneh, M.S., Hseih, Y.-F., Kauppila, W.E., Kwan, C.K., Mahdavi-Hezaveh, M. and Stein, T.S. (1986). Positron differential elastic-scattering cross-section measurements for argon. *Phys. Rev. Lett.* **57** 2252–2255.

Hylleraas, E.A. and Ore, A. (1947). Binding energy of the positronium molecule. *Phys. Rev.* **71** 493–496.

Hyodo, T. (1992). ACAR study of positronium–gas molecule interactions. *Material Science Forum* **105–110** pp. 281–288.

Hyodo, T., Kakimoto, M., Chang, T.B., Deng, J., Akahane, T., Chiba, T., McKee, B.T.A. and Stewart, A.T. (1989). Relaxation of the momentum distribution of free positronium atoms interacting with silica fine particles. In *Positron Annihilation*, eds. L. Dorikens-Vanpraet, M. Dorikens and D. Segers (World Scientific) pp. 878–880.

Iakubov, I.T. and Khrapak, A.G. (1982). Self-trapped states of positrons and positronium in dense gases and liquids. *Rep. Prog. Phys.* **45** 697–751.

Igarashi, A. and Toshima, N. (1992). Positronium formation in positron–helium collisions at intermediate energies. *Phys. Lett. A* **164** 70–72.

Igarashi, A. and Toshima, N. (1994). Hyperspherical coupled-channel study of positronium formation. *Phys. Rev. A* **50** 232–239.

Igarashi, A., Toshima, N. and Shirai, T. (1994). Hyperspherical coupled-channel calculation for antihydrogen formation in antiproton–positronium collisions. *J. Phys. B: At. Mol. Opt. Phys.* **27** L497–L501.

Ihra, W., Macek, J.H., Mota-Furtado, F. and O'Mahoney, P.F. (1997). Threshold law for positron impact ionization of atoms. *Phys. Rev. Lett.* **78** 4027–4030.

Inokuti, M. and McDowell, M.R.C. (1974). Elastic scattering of fast electrons by atoms I. Helium to neon. *J. Phys. B: At. Mol. Phys.* **7** 2382–2395.

Ito, S., Shimizu, S., Kawaratani, T. and Kubuta, K. (1980). Inner-shell ionization of silver by 100–400 keV electrons and positrons. *Phys. Rev. A* **22** 407–412.

Ito, Y., Widmann, E. and Yamazaki, T. (1993). Possible formation of antihydrogen atoms from metastable antiprotonic helium atoms and positrons/positroniums. *Hyperfine Interactions* **76** 163–173.

Itoh, Y., Lee, K.H., Nakajyo, T., Goto, A., Nakanishi, N., Kase, M., Kanazawa, I., Yamamoto, Y., Oshima, N. and Ito, Y. (1995). Slow positron production using the RIKEN AVF cyclotron. *Applied Surface Science* **85** 165–171.

Iwasaki, M., Nakamura, S.N., Shigaki, K., Shimizu, Y., Tamura, H., Ishikawa, T., Hayano, R.S., Takada, E., Widmann, E., Outa H., Aoki, M., Kitching, P. and Yamazaki, T. (1991). Discovery of antiproton trapping by long-lived metastable states in liquid helium. *Phys. Rev. Lett.* **67** 1246–1249.

Iwata, K., Greaves, R.G. and Surko, C.M. (1997). γ-ray spectra from positron annihilation on atoms and molecules. *Phys. Rev. A* **55** 3586–3604.

Iwata, K., Greaves, R.G., Murphy, T.J., Tinkle, M.D. and Surko, C.M. (1995). Measurements of positron annihilation rates on molecules. *Phys. Rev. A* **51** 473–487.

Iwata, K., Gribakin, G., Greaves, R.G., Kurz, C. and Surko, C.M. (2000). Positron annihilation on large molecules. *Phys. Rev. A* **61** 22 719–22 735.

Jackson, J.D. (1975). *Classical Electrodynamics* (Wiley).

Jacobsen, F.M. (1984). Positronium formation in gases and liquids. In *Positron Scattering in Gases*, eds. J.W. Humberston and M.R.C. McDowell (Plenum Press) pp. 85–97.

Jacobsen, F.M. (1986). On Ps formation in moderately dense molecular gases. *Chem. Phys.* **109** 455–464.

Jacobsen, F.M. and Lynn, K.G. (1996). Positron quantum reflection in thin metal films and efficient generation of high brightness low energy positron beams at 4.2 K. *Phys. Rev. Lett.* **76** 4262–4264.

Jacobsen, F.M., Charlton, M. and Laricchia, G. (1986). Comment on the density and electric-field dependence of the positronium fraction in gaseous methane. *J. Phys. B: At. Mol. Phys.* **19** L111–L114.

Jacobsen, F.M., Frandsen, N.P., Knudsen, H. and Mikkelsen, U. (1995a). Non-dissociative single ionization of molecular hydrogen by electron and positron impact. *J. Phys. B: At. Mol. Opt. Phys.* **28** 4675–4689.

Jacobsen, F.M., Frandsen, N.P., Knudsen, H., Mikkelsen, U. and Schrader, D.M. (1995b). Single ionization of He, Ne and Ar by positron impact. *J. Phys. B: At. Mol. Opt. Phys.* **28** 4691–4695.

Jaduszliwer, B. and Paul, D.A.L. (1973). Positron–helium scattering cross sections and phase shifts below 19 eV. *Can. J. Phys.* **51** 1565–1572.

Jaduszliwer, B. and Paul, D.A.L. (1974). Elastic scattering of positrons in neon and argon and phase shift analysis from 4 eV to inelastic thresholds. *Can. J. Phys.* **52** 272–277.

Jaduszliwer, B., Nakashima, A. and Paul, D.A.L. (1975). The scattering of positrons by helium: total cross sections up to 270 eV. *Can. J. Phys.* **53** 962–967.

Jain, A. and Thompson, D.G. (1983). The scattering of slow positrons by CH_4 and NH_3. *J. Phys. B: At. Mol. Phys.* **16** 1113–1123.

Janev, R.K. and Solov'ev, E.A. (1998). Application of hidden crossing theory to positron–hydrogen collisions. In *Photonic, Electronic and Atomic Collisions*, eds. F. Aumayr and H. Winter (World Scientific) pp. 393–398.

Joachain, C.J. (1987). Recent developments in the theory of fast positron–atom collisions. In *Atomic Physics with Positrons*, eds. J.W. Humberston and E.A.G. Armour (Plenum Press) pp. 71–82.

Joachain, C.J. and Potvliege, R.M. (1987). Importance of absorption effects on fast positron–argon differential cross sections. *Phys. Rev. A* **35** 4873–4875.

Joachain, C.J., Vanderpoorten, R., Winters, K.H. and Byron Jr., F.W. (1977). Optical model theory of elastic electron– and positron–argon scattering at intermediate energies. *J. Phys. B: At. Mol. Phys.* **10** 227–238.

Jones, G.O., Charlton, M., Slevin, J. Laricchia, G., Kövér, Á., Poulsen, M.R. and Nic Chormaic, S. (1993). Positron impact ionization of atomic hydrogen. *J. Phys. B: At. Mol. Opt. Phys.* **26** L483–L488.

Kakimoto, M. and Hyodo, T. (1988). Evidence for normal formation of positronium in gaseous Xe. *J. Phys. B: At. Mol. Opt. Phys.* **21** 2977–2987.

Kakimoto, M., Hyodo, T. and Chang, T.B. (1990). Conversion of ortho-positronium in low density oxygen gas. *J. Phys. B: At. Mol. Opt Phys.* **23** 589–597.

Kakimoto, M., Hyodo, T., Chiba, T., Akahane, T. and Chang, T.B. (1987). Observation of triplet–singlet conversion of positronium via inelastic scattering by oxygen. *J. Phys. B: At. Mol. Opt Phys.* **20** L107–L113.

Kakimoto, M., Nagashima, Y., Hyodo, T., Fujiwara, K. and Chang, T.B. (1989). Slowing-down of positronium in gases. In *Positron Annihilation,* eds. L. Dorikens-Vanpraet, M. Dorikens and D. Segers (World Scientific) pp. 737–739.

Kara, V. (1999) Positron impact ionization studies. Ph.D. thesis, University of London.

Kara, V., Paludan, K., Moxom, J., Ashley, P. and Laricchia, G. (1997a). Single and double ionization of neon, krypton and xenon by positron impact. *J. Phys. B: At. Mol. Opt. Phys.* **30** 3933–3949.

Kara, V., Paludan, K., Moxom, J., Ashley, P. and Laricchia, G. (1997b). Positron impact ionisation of atoms. *Nuc. Inst. Meth. B* **143** 94–99.

Karl, M.W., Nakanishi, H. and Schrader, D.M. (1984). Chemical stability of positronic complexes with atoms and atomic ions. *Phys. Rev. A* **30** 1624–1628.

Karplus, R. and Klein, A. (1952). Electrodynamic displacement of atomic energy levels. III. The hyperfine structure of positronium. *Phys. Rev.* **87** 848–858.

Kartavstev, O.I. (1996). Variational calculations of antiprotonic helium atoms. *Russian J. Nucl. Phys.* **59** 1541–1550.

Katayama, Y., Sueoka, O. and Mori, S. (1987). Inelastic cross section measurement for slow positron–O_2 collisions. *J. Phys. B: At. Mol. Opt. Phys.* **20** 1645–1657.

Kauppila, W.E. and Stein, T.S. (1982). Positron–gas cross section measurements. *Can. J. Phys.* **60** 471–493.

Kauppila, W.E., Stein, T.S. and Jesion, G. (1976). Direct observation of a Ramsauer–Townsend effect in positron–argon collisions. *Phys. Rev. Lett.* **36** 580–584.

Kauppila, W.E., Stein, T.S., Jesion, G., Dababneh, M.S. and Pol, V. (1977). Transmission experiment for measuring total positron–atom collision cross sections in a curved, axial magnetic field. *Rev. Sci. Inst.* **48** 822–828.

Kauppila, W.E., Stein, T.S., Smart, J.H., Dababneh, M.S., Ho, Y.K., Downing, J.P. and Pol, V. (1981). Measurements of total scattering cross sections for intermediate-energy positrons and electrons colliding with helium, neon, and argon. *Phys. Rev. A* **24** 725–742.

Kauppila, W.E., Kwan, C.K., Stein, T.S. and Zhou, S. (1994). Evidence for channel-coupling effects in positron scattering by sodium and potassium atoms. *J. Phys. B: At. Mol. Opt. Phys.* **27** L551–L555.

Kauppila, W.E., Kwan, C.K., Przybyla, D., Smith, S.J. and Stein, T.S. (1996). Positron–inert gas atom elastic DCS measurements. *Can. J. Phys.* **74** 474–482.

Kawaratani, T., Nakayama, Y. and Mizogawa, T. (1985). Density and temperature dependences of ortho-positronium annihilation rates in low temperature gaseous N_2. *Phys. Lett. A* **108** 75–79.

Kennerly, R.E. (1980). Absolute total electron scattering cross sections for N_2 between 0.5 and 50 eV. *Phys. Rev. A* **21** 1876–1883.

Kernoghan, A.A. (1996). Positron scattering by atomic hydrogen and the alkali metals. Ph.D. thesis, Queens's University of Belfast.

Kernoghan, A.A., McAlinden, M.T. and Walters, H.R.J. (1994a). Scattering of low energy positrons by lithium. *J. Phys. B: At. Mol. Opt. Phys.* **27** L625–L631.

Kernoghan, A.A., McAlinden, M.T. and Walters, H.R.J. (1994b). Resonances above the ionization threshold in positron–hydrogen scattering. *J. Phys. B: At. Mol. Opt. Phys.* **27** L543–L549.

Kernoghan, A.A., McAlinden, M.T. and Walters, H.R.J. (1995). An 18-state calculation of positron–hydrogen scattering. *J. Phys. B: At. Mol. Opt. Phys.* **28** 1079–1094.

Kernoghan, A.A., McAlinden, M.T. and Walters, H.R.J. (1996). Positron scattering by rubidium and caesium. *J. Phys. B: At. Mol. Opt. Phys.* **29** 3971–3987.

Kernoghan, A.A., Robinson, D.J.R., McAlinden, M.T. and Walters, H.R.J. (1996). Positron scattering by atomic hydrogen. *J. Phys. B: At. Mol. Opt. Phys.* **29** 2089–2102.

Khan, P. and Ghosh, A.S. (1983). Positronium formation in positron–helium scattering. *Phys. Rev. A* **28** 2181–2189.

Khan, P., Dutta, S. and Ghosh, A.S. (1987). Positron–lithium scattering using the eigenstate expansion method. *J. Phys. B: At. Mol. Phys.* **20** 2927–2935.

Khan, P., Mazumdar, P.S. and Ghosh, A.S. (1985). Positronium formation in the $n = 2$ level in positron scattering from hydrogen and helium atoms. *Phys. Rev. A* **31** 1405–1414.

Khatri, R. Charlton, M., Sferlazzo, P., Lynn, K.G., Mills Jr., A.P. and Roellig, L.O. (1990). Improvement of rare-gas solid moderators by using conical geometry. *Appl. Phys. Lett.* **57** 2374–2376.

Khriplovich, I.B. and Yelkhovsky, A.S. (1990). On the radiative correction $\alpha^2 \ln \alpha$ to the positronium decay rate. *Phys. Lett. B* **246** 520–522.

Khriplovich, I.B., Milstein, A.I. and Yelkhovsky, A.S. (1992). Corrections of $O(\alpha^6 \log \alpha)$ in the two-body QED problem. *Phys. Lett. B* **282** 237–242.

Kiefl, R.F. (1982). Temperature dependence of spin conversion of o-Ps by O_2 in SiO_2 powder. In *Positron Annihilation*, eds. P.G. Coleman, S.C. Sharma and L.M. Diana (North-Holland) pp. 690–692.

Kimura, M., Sueoka, O., Hamada, A. and Itikawa, Y. (2000). A comparative study of electron– and positron–polyatomic molecule scattering. *Adv. Chem. Phys.* **111** 537–622.

Kinghorn, D.B. and Poshusta, R.D. (1993). Nonadiabatic variational calculation on dipositronium using explicitly correlated Gaussian basis functions. *Phys. Rev. A* **47** 3671–3681.

Kirkegaard, P., Pederson, N.J. and Eldrup, M. (1989). *PATFIT-88: A Data-processing System for Positron Annihilation Spectra on Mainframe and Personal Computers* (Risø National Laboratory, Denmark, Risø-M-2740).

Klar, H. (1981). Threshold ionisation of atoms by positrons. *J. Phys. B: At. Mol. Phys.* **14** 4165–4170.

Klobuchar, R.L. and Karol, P.J. (1980). Positronium formation and quenching in argon–oxygen mixtures. *J. Chem. Phys.* **84** 483–488.

Knoll, G.F. (1989). *Radiation Detection and Measurement*, 2nd edition (Wiley Interscience).

Knudsen, H. and Reading, J.F. (1992). Ionization of atoms by particle and antiparticle impact. *Phys. Rep.* **212** 107–222.

Knudsen, H., Brun-Nielsen, L., Charlton, M. and Poulsen, M.R. (1990). Single ionization of H_2, He, Ne and Ar by positron impact. *J. Phys. B: At. Mol. Opt. Phys.* **23** 3955–3976.

Knudsen, H., Mikkelsen, U., Paludan, K., Kirsebom, K., Møller, S.P., Uggerhøj, E., Slevin, J., Charlton, M. and Morenzoni, E. (1995a). Non-dissociative and dissociative ionization of N_2, CO, CO_2 and CH_4 by impact of 50–6000 keV protons and antiprotons. *J. Phys. B: At. Mol. Opt. Phys.* **28** 3569–3592.

Knudsen, H., Mikkelsen, U., Paludan, K., Kirsebom, K., Møller, S.P., Uggerhøj, E., Slevin, J., Charlton, M. and Morenzoni, E. (1995b). Ionization of atomic hydrogen by 30–1000 keV antiprotons. *Phys. Rev. Lett.* **74** 4627–4630.

Kövér, Á. and Laricchia, G. (1998). Triply differential study of positron impact ionization of H_2. *Phys. Rev. Lett.* **80** 5309–5312.

Kövér, Á., Laricchia, G. and Charlton, M. (1993). Ionization by positrons and electrons at zero degrees. *J. Phys. B: At. Mol. Opt. Phys.* **26** L575–L580.

Kövér, Á., Laricchia, G. and Charlton, M. (1994). Doubly differential cross sections for collisions of 100 eV positrons and electrons with argon atoms. *J. Phys. B: At. Mol. Opt. Phys.* **27** 2409–2416.

Kövér, Á., Finch, R.M., Charlton, M. and Laricchia, G. (1997). Double-differential cross sections for collisions of positrons with argon atoms. *J. Phys. B: At. Mol. Opt. Phys.* **30** L507–L512.

Kozlowski, P.M. and Adamowicz, L. (1993). Nonadiabatic calculations for the ground state of the positronium molecule. *Phys. Rev. A* **48** 1903–1908.

Kragh, H. (1990). From 'Electrum' to positronium. *J. Chem. Ed.* **67** 196–197.

Kraidy, M. and Fraser, P.A. (1967). Scattering of positrons by helium and the rate of positron annihilation in helium. In *Fifth Int. Conf. on Physics of Electronic and Atomic Collisions, Leningrad*, Abstracts pp. 110–116.

Krishnakumar, E. and Srivastava, S.K. (1988). Ionisation cross sections of rare-gas atoms by electron impact. *J. Phys. B: At. Mol. Opt. Phys.* **21** 1055–1082.

Kruit, P. and Read, F.H. (1983). Magnetic field paralleliser for 2π electron-spectrometer and electron-image magnifier. *J. Phys. E* **16** 313–324.

Kruse, G., Quermann, W., Raith, W., Sinapius, G. and Weber, M. (1991). Multiple ionization of xenon by positron impact. *J. Phys. B: At. Mol. Opt. Phys.* **24** L33–L37.

Kuang, Y.R. and Gien, T.T. (1997). Positron–hydrogen collisions at low energies. *Phys. Rev. A* **55** 256–264.

Kumar, M., Srivastava, R. and Tripathi, A.N. (1985). Systematic approach for discrete excitation of helium in the Coulomb–Born model. *Phys. Rev. A* **31** 652–658.

Kurz, C., Greaves, R.G. and Surko, C.M. (1996). Temperature dependence of positron annihilation rates in noble gases. *Phys. Rev. Lett.* **77** 2929–2932.

Kvitsinsky, A.A., Carbonell, J. and Gignoux, C. (1992). Faddeev calculation of e–Ps scattering lengths. *Phys. Rev. A* **46** 1310–1315.

Kvitsinsky, A.A., Carbonell, J. and Gignoux, C. (1995). S-wave positron–hydrogen scattering via Faddeev equations: elastic scattering and positronium formation. *Phys. Rev. A* **51** 2997–3004.

Kvitsinsky, A.A., Wu, A. and Hu, C.-H. (1995). Scattering of electrons and positrons on hydrogen using the Faddeev equations. *J. Phys. B: At. Mol. Opt. Phys.* **28** 275–285.

Kwan, C.K., Hsieh, Y.-F., Kauppila, W.E., Smith, S.J., Stein, T.S. and Uddin, M.N. (1984). Total-scattering measurements and comparisons for collisions of electrons and positrons with N_2O. *Phys. Rev. Lett.* **16** 1417–1420.

Kwan, C.K., Kauppila, W.E., Lukaszew, R.A., Parikh, S.P., Stein, T.S., Wan, Y.J. and Dababneh, M.S. (1991). Total cross-section measurements for positrons and electrons scattered by sodium and potassium atoms. *Phys. Rev. A* **44** 1620–1635.

Kwan, C.K., Kauppila, W.E., Parikh, S.P., Stein, T.S. and Zhou, S. (1994). Measurements of total and positronium formation cross sections for the scattering of positrons by alkali atoms. *Hyperfine Interactions* **89** 33–45.

Lahmam-Bennani, A. (1991). Recent developments and new trends in (e, 2e) and (e, 3e) studies. *J. Phys. B: At. Mol. Opt. Phys.* **24** 2401–2442.

Lahtinen, J., Vehanen, A., Huomo, H., Mäkinen, J., Huttunen, P., Rytsölä, K., Bentzon, M. and Hautojärvi, P. (1986). High-intensity variable-energy positron beam for surface and near-surface studies. *Nucl. Inst. Meth. B* **17** 73–80.

Laricchia, G. (1995a). Positron and positronium collisions. *Nuc. Inst. Meth. B* **99** 363–367.

Laricchia, G. (1995b). Positronium beams and surfaces. In *Positron Spectroscopy of Solids, Proc. Int. School of Physics 'Enrico Fermi', Vol. 125*, eds. A. Dupasquier and A.P. Mills Jr. (IOS) pp. 401–418.

Laricchia, G. and Moxom, J. (1993). Ionization of CO_2 by positron impact. *Phys. Lett. A* **174** 255–257.

Laricchia, G. and Wilkin, C. (1997). Semiempirical approach to positron annihilation in molecules. *Phys. Rev. Lett.* **79** 2241–2244.

Laricchia, G., Charlton, M. and Griffith, T.C. (1988). Observation of simultaneous positronium formation and excitation in positron–CO_2 scattering. *J. Phys. B: At. Mol. Opt. Phys.* **21** L227–L232.

Laricchia, G., Moxom, J. and Charlton, M. (1993). Near threshold effects in positron–O_2 scattering. *Phys. Rev. Lett.* **70** 3229–3230.

Laricchia, G., Charlton, M., Clark, G. and Griffith, T.C. (1985). Excited state positronium formation in low density gases. *Phys. Lett. A* **109** 97–100.

Laricchia, G., Charlton, M. Griffith, T.C. and Jacobsen, F.M. (1986). Preliminary results on the angular dependence of Ps emission in positron–gas collisions. In *Positron(Electron)–Gas Scattering*, eds. W.E. Kauppila, T.S. Stein and J.M. Wadehra (World Scientific) pp. 303–306.

Laricchia, G., Charlton, M., Beling, C.D. and Griffith, T.C. (1987a). Density dependence of positronium formation in H_2 gas at temperatures between 77 and 297 K. *J. Phys. B: At. Mol. Phys.* **20** 1865–1874.

Laricchia, G., Charlton, M., Davies, S.A., Beling, C.D. and Griffith, T.C. (1987b). The production of collimated beams of o-Ps atoms using charge exchange in positron–gas collisions. *J. Phys. B: At. Mol. Phys.* **20** L99–L105.

Laricchia, G., Davies, S.A., Charlton, M. and Griffith, T.C. (1988). The production of a timed tunable beam of positronium atoms. *J. Phys. E* **21** 886–888.

Lee, G.F. and Jones, G. (1974). Electric field and temperature dependence of positron annihilation in argon gas. *Can. J. Phys.* **52** 17–28.

Lee, G.F., Orth, P.H.R. and Jones, G. (1969). Electric field dependence of the annihilation rate of positrons in helium. *Phys. Lett. A* **28** 674–675.

Lee, M.A., Vashista, P. and Kalia, R.K. (1983). Ground state excitonic molecules by the Green's function Monte Carlo method. *Phys. Rev. Lett.* **51** 2422–2425.

Lennard, W.N., Schultz, P.J., Massoumi, G.R. and Logan, L.R. (1988). Observation of the difference between e^-–e^- and e^-–e^+ interactions. *Phys. Rev. Lett.* **61** 2428–2430.

Leung, C.Y. and Paul, D.A.L. (1969). Positron diffusion and annihilation in helium. *J. Phys. B: At. Mol. Phys.* **2** 1278–1292.

Ley, R., Niebling, K.D., Werth, G., Hahn, C., Schneider, H. and Tobehn, I. (1990). Energy dependence of excited positronium formation at a molybdenum surface. *J. Phys. B: At. Mol. Opt. Phys.* **23** 3437–3442.

Liang, E.P. and Dermer, C.D. (1988). Laser cooling of positronium. *Optics Communications* **65** 419–424.

Liang, S.C. (1955). On the calculation of thermal transpiration. *Can. J. Chem.* **33** 279–285.

Lodge, J.G., Darewych, J.W. and McEachran, R.P. (1971). Elastic scattering of positrons by hydrogen molecules. *Can. J. Phys.* **49** 13–19, and erratum (1973) *Can. J. Phys.* **51** 779.

Lynn, K.G. (1979). Observation of surface traps and vacancy trapping with slow positrons. *Phys. Rev. Lett.* **43** 391–394.

Lynn, K.G. (1980). Observation of the amorphous-to-crystalline surface transition in Al-Al_xO_y using slow positrons. *Phys. Rev. Lett.* **44** 1330–1333.

Lynn, K.G., Weber, M., Roellig, L.O., Mills Jr., A.P. and Moodenbaugh, A.R. (1987). A high intensity positron beam at the Brookhaven reactor. In *Atomic Physics with Positrons*, eds. J.W. Humberston and E.A.G. Armour (Plenum Press) pp. 161–174.

MacKenzie, I.K. (1983). Experimental methods of annihilation time and energy spectrometry. In *Positron Solid-State Physics, Proc. Int. School of Physics 'Enrico Fermi', Course 83*, eds. W. Brandt and A. Dupasquier (North-Holland) pp. 196–264.

MacKenzie, I.K., Shulte, C.W., Jackman, T. and Campbell, J.L. (1973). Positron transmission and scattering measurements using superposition of annihilation line shapes: backscatter coefficients. *Phys. Rev. A* **7** 135–145.

Madansky, L. and Rasetti, F. (1950). An attempt to detect thermal energy positrons. *Phys. Rev.* **79** 397.

Madison, D.H. and Winters, K.H. (1983). A second-order distorted-wave model for the excitation of the 2^1P state of helium by electron and positron impact. *J. Phys. B: At. Mol. Phys.* **16** 4437–4450.

Mandal, P., Ghosh, A.S. and Sil, N.C. (1975). e^+–He scattering using integral form of the close-coupling approximation. *J. Phys. B: At. Mol. Phys.* **8** 2377–2389.

Mandal, P., Guha, S. and Sil, N.C. (1979). Positronium formation in positron scattering from hydrogen and helium atoms: the distorted-wave approximation. *J. Phys. B: At. Mol. Phys.* **12** 2913–2924.

Mandal, P., Guha, S. and Sil, N.C. (1980). Positron–helium collisions: positronium formation into an arbitrary excited S state. *Phys. Rev. A* **22** 2623–2629.

Manninen, M. (1987). Private communication to Tuomisaari *et al.* (1988).

Manson, J.R. and Ritchie, R.H. (1985). Completely quantal treatment of the van der Waals forces between atoms: application to positronium. *Phys. Rev. Lett.* **54** 785–788.

Marder, S., Hughes, V.W., Wu, C.S. and Bennett, W. (1956). Effect of an electric field on positronium formation in gases: experimental. *Phys. Rev.* **103** 1258–1265.

Martin, D.W. and Fraser, P.A. (1980). The van der Waals force between positronium and light atoms. *J. Phys. B: At. Mol. Phys.* **13** 3383–3387.

Massey, H.S.W. (1969). *Electronic and Ionic Impact Phenomena. Vol. II* (Oxford University Press).

Massey, H.S.W. (1975). *Electronic and Ionic Impact Phenomena. Vol. V* (Oxford University Press).

Massey, H.S.W. and Burhop, E.H.S. (1938). The probability of annihilation of positrons without emission of radiation. *Proc. Roy. Soc. Lond. A* **167** 53–61.

Massey, H.S.W. and Mohr, C.B.O. (1954). Gaseous reactions involving positronium. *Proc. Phys. Soc. A* **67** 695–704.

Massey, H.S.W. and Moussa, A.H. (1958). The elastic scattering of positrons by atoms and molecules. *Proc. Phys. Soc. A* **71** 38–44.

Massey, H.S.W. and Moussa, A.H. (1961). Positronium formation in helium. *Proc. Phys. Soc. A* **77** 811–816.

Massey, H.S.W., Lawson, J. and Thompson, D.G. (1966). Collision of slow positrons with atoms. In *Quantum Theory of Atoms, Molecules and the Solid State*, ed. P.-O. Löwdin (Academic Press) pp. 203–215.

Matsumoto, T., Chiba, M., Hamatsu, R., Hirose, T., Yang, J. and Yu, J. (1996). Measurement of five-photon decay in orthopositronium. *Phys. Rev. A* **54** 1947–1951.

Maury, S. (1997). The antiproton decelerator: AD. *Hyperfine Interactions* **109** 43–52.

McAlinden, M.T. and Walters, H.R.J. (1992). Positron scattering by the noble gases. *Hyperfine Interactions* **73** 65–83.

McAlinden, M.T. and Walters, H.R.J. (1994). Differential cross sections for elastic scattering and positronium formation for positron collisions with Ne, Ar, Kr and Xe. *Hyperfine Interactions* **89** 407–418.

McAlinden, M.T., Kernoghan, A.A. and Walters, H.R.J. (1994). Cross-channel coupling in positron–atom scattering. *Hyperfine Interactions* **89** 161–194.

McAlinden, M.T., Kernoghan, A.A. and Walters, H.R.J. (1996). Positron scattering by potassium. *J. Phys. B: At. Mol. Opt. Phys.* **29** 555–569.

McAlinden, M.T., MacDonald, F.G.R.S. and Walters, H.R.J. (1996). Positronium–atom scattering. *Can. J. Phys.* **74** 434–444.

McAlinden, M.T., Kernoghan, A.A. and Walters, H.R.J. (1997). Positron scattering by lithium. *J. Phys. B: At. Mol. Opt. Phys.* **30** 1543–1561.

McCallion, P., Shah, M.B. and Gilbody, H.B. (1992). A crossed beam study of the multiple ionization of argon by electron impact. *J. Phys. B: At. Mol. Opt. Phys.* **25** 1061–1071.

McDaniel, E.W. (1989). *Atomic Collisions – Electron & Photon Projectiles* (Wiley Interscience).

McEachran, R.P. and Fraser, P.A. (1965). The elastic scattering of low energy positrons by atomic hydrogen. *Proc. Phys. Soc.* **86** 369–373.

McEachran, R.P. and Stauffer, A.D. (1986). Differential cross-sections for positron noble gas scattering. In *Positron(Electron)-Gas Scattering*, eds. W.E. Kauppila, T.S. Stein and J.M. Wadehra (World Scientific) pp. 122–130.

McEachran, R.P., Horbatsch, M. and Stauffer, A.D. (1991). Positron scattering from rubidium. *J. Phys. B: At. Mol. Opt. Phys.* **24** 1107–1113.

McEachran, R.P., Ryman, A.G. and Stauffer, A.D. (1978). Positron scattering from neon. *J. Phys. B: At. Mol. Phys.* **11** 551–561.

McEachran, R.P., Ryman, A.G. and Stauffer, A.D. (1979). Positron scattering from argon. *J. Phys. B: At. Mol. Phys.* **12** 1031–1041.

McEachran, R.P., Stauffer, A.D. and Campbell, L.E.M. (1980). Positron scattering from krypton and xenon. *J. Phys. B: At. Mol. Phys.* **12** 1281–1292.

McEachran, R.P., Morgan, D.L., Ryman, A.G. and Stauffer, A.D. (1977). Positron scattering from noble gases. *J. Phys. B: At. Mol. Phys.* **10** 663–677 and corrigendum and addendum, (1978), *J. Phys. B: At. Mol. Phys.* **11** 951–953.

McGuire, J.H. (1982). Double ionization of helium by protons and electrons at high velocities. *Phys. Rev. Lett.* **49** 1153–1157.

McGuire, J.H. (1986). Differences in atomic collisions by high velocity protons and positrons. In *Positron (Electron)–Gas Scattering.*, eds. W.E. Kauppila, T.S. Stein, and J.M. Wadehra (World Scientific) pp. 222–231.

McGuire, J.H. (1992). Multiple-electron excitation, ionization, and transfer in high-velocity atomic and molecular collisions. *Adv. At. Mol. Opt. Phys.* **29** 217–323.

McGuire, J.H., Berrah, N., Bartlett, R.J., Samson, J.A.R., Tanis, J.A., Cocke, C.L. and Schlachter, A.S. (1995). The ratio of cross sections for double to single ionization of helium by high energy photons and charged particles. *J. Phys. B: At. Mol. Opt. Phys.* **28** 913–940.

McIntyre, D.H. and Hänsch, T.W. (1986). Absolute calibration of the $^{130}Te_2$ reference line for positronium 1^3S_1–2^3S_1 spectroscopy. *Phys. Rev. A* **34** 4504–4507.

McNutt, J.D. and Sharma, S.C. (1978). Dependence of orthopositronium annihilation rates on density fluctuations in methane gas. *J. Chem. Phys.* **68** 130–133.

McNutt, J.D., Sharma, S.C. and Brisbon, R.D. (1979). Positron annihilation in gaseous hydrogen and hydrogen–neon mixtures. I. Low-energy positrons. *Phys. Rev. A* **20** 347–356.

McNutt, J.D., Summerour, V.B., Ray, A.D. and Huang, P.H. (1975). Complex dependence of the positron annihilation rate on methane gas density and temperature. *J. Chem. Phys.* **62** 1777–1789.

McNutt, J.D., Sharma, S.C., Franklin, M.H. and Woodall II, M.A. (1979). Positron annihilation in gaseous hydrogen and hydrogen–neon mixtures. II. Positronium. *Phys. Rev. A* **20** 357–363.

Merrison, J.P., Bluhme, H., Chevallier, J., Deutch, B.I., Hvelplund, P., Jørgensen, L.V., Knudsen, H., Poulsen M. and Charlton, M. (1997). Hydrogen formation by proton impact on positronium. *Phys. Rev. Lett.* **78** 2728–2731.

Merrison, J.P., Charlton, M., Deutch, B.I. and Jørgensen, L.V. (1992). Field assisted moderation by surface charging of rare gas solids. *J. Phys. Condens. Matter* **4** L207–L212.

Meshkov, I.N. (1997). Experimental studies of antihydrogen and positronium physics: problems and possibilities. *Phys. Part. Nucl.* **28** 198–215.

Meshkov, I.N. and Skrinsky, A.N. (1997). The antihydrogen and positronium generation and studies using storage rings. *Nuc. Inst. Meth. A* **391** 205–209.

Meyerhof, W.E. (1963). Threshold effects in average cross sections according to R-matrix theory. *Phys. Rev.* **129** 692–702.

Meyerhof, W.E., Laricchia, G., Moxom, J., Watts, M.S.T. and Humberston, J.W. (1996). Positron scattering on atoms and molecules near the positronium threshold. *Can. J. Phys.* **74** 427–433.

Mikhailov, A.I. and Porsev, S.G. (1992). Double ionization of the atomic K shell in the annihilation of a positron with a bound K electron. *J. Phys. B: At. Mol. Opt. Phys.* **25** 1097–1101.

Miller, B.N., Reese, T.L. and Worrell, G.A. (1996). Virial expansion of the positron and positronium decay rate in a dilute gas: the linear contribution. *Can. J. Phys.* **74** 548–553.

Mills Jr., A.P. (1979). Thermal activation measurement of positron binding energies at surfaces. *Solid State Commun.* **31** 623–626.

Mills Jr., A.P. (1980). Brightness enhancement of slow positron beams. *Appl. Phys.* **23** 189–191.

Mills Jr., A.P. (1981). Observation of the positronium negative ion. *Phys. Rev. Lett.* **46** 717–720.

Mills Jr., A.P. (1983a). Experimentation with low-energy positron beams. In *Positron Solid-State Physics, Proc. Int. School of Physics 'Enrico Fermi', Course 83*, eds. W. Brandt and A. Dupasquier (North-Holland) pp. 432–509.

Mills Jr., A.P. (1983b). Measurement of the decay rate of the positronium negative ion. *Phys. Rev. Lett.* **50** 671–674.

Mills Jr., A.P. (1983c). Line-shape effects in the measurement of the positronium hyperfine interval. *Phys. Rev. A* **27** 262–267.

Mills Jr., A.P. (1984). Techniques for studying systems containing many positrons. In *Positron Scattering in Gases*, eds. J.W. Humberston and M.R.C. McDowell (Plenum Press) pp. 121–138.

Mills Jr., A.P. (1993). Spectroscopy with few-atom samples. *Hyperfine Interactions* **76** 233–248.

Mills Jr., A.P. and Bearman, G.H. (1975). New measurement of the positronium hyperfine interval. *Phys. Rev. Lett.* **34** 246–250.

Mills Jr., A.P. and Crane, W.S. (1985). Beam–foil production of fast positronium. *Phys. Rev. A* **31** 593–597.

Mills Jr., A.P. and Gullikson, E.M. (1986). Solid neon moderator for producing slow positrons. *Appl. Phys. Lett.* **49** 1121–1123 .

Mills Jr., A.P. and Pfeiffer, L. (1979). Desorption of surface positrons: a source of free positronium at thermal velocities. *Phys. Rev. Lett.* **43** 1961–1964.

Mills Jr., A.P. and Pfeiffer, L. (1985). Velocity spectrum of positronium thermally desorbed from an Al(111) surface. *Phys. Rev. B* **32** 53–57.

Mills Jr., A.P. and Zuckerman, D.M. (1990). Search for spatial anisotropy in orthopositronium annihilation. *Phys. Rev. Lett.* **64** 2637–2639.

Mills Jr., A.P., Shaw, E.D., Chichester, R.J. and Zuckerman, D.M. (1989a). Positronium thermalisation in SiO_2 powder. *Phys. Rev. B* **40** 2045–2052.

Mills Jr., A.P., Shaw, E.D., Chichester, R.J. and Zuckerman, D.M. (1989b). Production of slow positron bunches using a microtron accelerator. *Rev. Sci. Inst.* **60** 825–830.

Mills Jr., A.P., Friedman, P.G. and Zuckerman, D.M. (1990). Decay rate and other properties of the positronium negative ion. In *Annihilation in Gases and Galaxies, NASA Conference Publication 3058*, ed. R.J. Drachman pp. 213–221.

Mil'stein, A.I. and Khriplovich, I.B. (1994). Large relativistic corrections to the positronium decay probability. *JETP* **79** 379–383.

Mitroy, J. (1993). Another s-wave resonance in positron–hydrogen scattering. *J. Phys. B: At. Mol. Opt. Phys.* **26** L625–L631.

Mitroy, J. (1995). Formation of antihydrogen by the charge-transfer reaction. *Phys. Rev. A* **52** 2859–2864.

Mitroy, J. (1996). An L^2 calculation of positron–hydrogen scattering at intermediate energies. *J. Phys. B: At. Mol. Opt. Phys.* **29** L263–L269.

Mitroy, J. and Stelbovics, A.T. (1994). Formation of antihydrogen from antiproton collisions with positronium. *J. Phys. B: At. Mol. Opt. Phys.* **27** L79–L82.

Mitroy, J. and Ratnavelu, K. (1995). The positron–hydrogen system at low energies. *J. Phys. B: At. Mol. Opt. Phys.* **28** 287–306.

Mitroy, J. and Ryzhikh, G. (1997). The formation of antihydrogen by the charge transfer reaction. *J. Phys. B: At. Mol. Opt. Phys.* **30** L371–L375.

Mitroy, J., Berge, L. and Stelbovics, A. (1994). Positron–hydrogen scattering at low energies. *Phys. Rev. Lett.* **73** 2966–2969.

Mitsui, T., Fujimoto, R., Ishisaki, Y., Ueda, Y., Yamazaki, Y., Asai, S. and Orito, S. (1993). Search for invisible decay of orthopositronium. *Phys. Rev. Lett.* **70** 2265–2268.

Mittleman, M.H. (1966). Resonances in proton–hydrogen and positron–hydrogen scattering. *Phys. Rev.* **152** 76–78.

Mizogawa, T., Nakayama, Y., Kawaratani, T. and Tosaki, M. (1985). Precise measurements of positron–helium total cross sections from 0.6 to 22 eV. *Phys. Rev. A* **31** 2171–2179.

Mogensen, O.E. (1974). Spur reaction model of positronium formation. *J. Chem. Phys.* **60** 998–1004.

Mogensen, O.E. (1979). Solvated positron chemistry. II. The reaction of hydrated positrons with Cl^-, Br^- and I^- ions. *Chem. Phys.* **37** 139–158.

Mogensen, O.E. (1982). Positronium formation in condensed matter and high-density gases. In *Positron Annihilation*, eds. P.G. Coleman, S.C. Sharma and L.M. Diana (North-Holland) pp. 763–772.

Mogensen, O.E. (1995). *Positron Annihilation in Chemistry* (Springer-Verlag).

Mogensen, O.E. and Shantarovitch, V.P. (1974). Solvated positron chemistry. The reaction of hydrated positrons with chloride ions. *Chem. Phys.* **6** 100–108.

Mohorovičić, S. (1934). Möglichkeit neuer elemente und ihre bedeutung für die astrophysik. *Astron. Nachr.* **253** 93–108.

Møller, C. (1932). Zur theorie des durchgangs schneller elecktronen durch materie. *Ann. Phys. (Leipzig)* **14** 531–585.

Montague, R.G., Harrison, M.F.A. and Smith, A.C.H. (1984). A measurement of the the cross section for ionisation of helium by electron impact using a fast crossed beam technique. *J. Phys. B: At. Mol. Phys.* **17** 3295–3310.

Montgomery, R.E. and LaBahn, R.W. (1970). Annihilation of positrons in helium, neon, and argon. *Can. J. Phys.* **48** 1288–1303.

Moores, D.L. (1998). Positron impact ionisation of rare gas atoms by a distorted wave method with close coupled target states. *Nuc. Inst. Meth. B* **143** 105–111.

Mori, S. and Sueoka, O. (1984). Excitation of He, Ne and Ar atoms by positron impact. *At. Coll. Res. Japan* **10** 8–11.

Mori, S. and Sueoka, O. (1994). Excitation and ionization cross sections of He, Ne and Ar by positron impact. *J. Phys. B: At. Mol. Opt. Phys.* **27** 4349–4364.

Morita, N., Kumakura, M., Yamazaki, T., Widmann, E., Masuda, H., Sugai, I., Hayano, R.S., Maas, F.E., Torii, H.A., Hartmann, F.J., Daniel, H., von Egidy, T., Ketzer, B., Müller, W., Schmidt, W., Horváth, D. and Eades, J. (1994). First observation of laser-induced resonant annihilation in metastable antiprotonic helium atoms. *Phys. Rev. Lett.* **72** 1180–1183.

Morrison, M.A. (1986). Perspectives on polarization in positron-molecule collisions. In *Positron(Electron)-Gas Scattering*, eds. W.E. Kauppila, T.S. Stein and J.M. Wadehra (World Scientific) pp. 100–109.

Morrison, M.A., Gibson, T.L. and Austin, D. (1984). Polarisation potentials for positron–molecule collisions: positron–H_2 scattering. *J. Phys. B: At. Mol. Phys.* **17** 2725–2745.

Mott, N.F. and Davis, E.A. (1979). *Electronic Processes in Non-crystalline Materials* (Oxford University Press).

Mott, N.F. and Massey, H.S.W. (1965). *The Theory of Atomic Collisions* (Oxford University Press).

Moxom, J., Ashley, P. and Laricchia, G. (1996). Single ionization by positron impact. *Can. J. Phys.* **74** 367–372.

Moxom, J., Laricchia, G. and Charlton, M. (1993). Total ionisation cross sections of He, H_2 and Ar by positron impact. *J. Phys. B: At. Mol. Opt. Phys.* **26** L367–L372.

Moxom, J., Laricchia, G. and Charlton, M. (1995a). A gated positron beam incorporating a scattering cell and novel ion extractor. *Appl. Surf. Sci.* **85** 118–123.

Moxom, J., Laricchia, G. and Charlton, M. (1995b). Ionization of He, Ar and H_2 by positron impact at intermediate energies. *J. Phys. B: At. Mol. Opt. Phys.* **28** 1331–1347.

Moxom, J., Laricchia, G., Charlton, M., Jones, G.O. and Kövér, Á. (1992). Ejected-electron energy spectra in low energy positron–argon collisions. *J. Phys. B: At. Mol. Opt. Phys.* **25** L613–L619.

Moxom, J., Laricchia, G., Charlton, M., Kövér, Á. and Meyerhof, W.E. (1994). Threshold effects in positron scattering on noble gases. *Phys. Rev. A* **50** 3129–3133.

Moxom, J., Xu, J., Laricchia, G., Hulett, L.D., Schrader, D.M., Kobayashi, Y., Somieski, B. and Lewis, T. (1998). Fragmentation and ionization of CH_3F by positron and electron impact. *Nuc. Inst. Meth. B* **143** 112–120.

Mukherjee, A. and Sural, D.P. (1982). Elastic scattering of positrons by hydrogen and helium atoms in a second-order potential model. *J. Phys. B: At. Mol. Phys.* **15** 1121–1130.

Mukherjee, M., Basu, M. and Ghosh, A.S. (1990). Positron–hydrogen scattering at intermediate energies using the close-coupling approximation. *J. Phys. B: At. Mol. Opt. Phys.* **23** 757–766.

Mukherjee, K.K., Singh, N.R. and Mazumdar, P.N. (1989). Ionisation from the ground state of atomic hydrogen by positron impact. *J. Phys. B: At. Mol. Opt. Phys.* **22** 99–103.

Müller, A. and Wolf, A. (1997). Production of antihydrogen by recombination of p with e^+: what can we learn from electron–ion studies? *Hyperfine Interactions* **109** 233–267.

Müller, B. and Thoma, M.H. (1992). Vacuum polarization and the electric charge of the positron. *Phys. Rev. Lett.* **69** 3432–3434.

Munger, C.T., Brodsky, S.J. and Schmidt, I. (1994). Production of relativistic antihydrogen atoms by pair production with positron capture. *Phys. Rev. D* **49** 3228–3235.

Murphy, T.J. and Surko, C.M. (1990). Annihilation of positrons in xenon gas. *J. Phys. B: At. Mol. Opt. Phys.* **23** L727–L732.

Murphy, T.J. and Surko, C.M. (1991). Annihilation of positrons on organic molecules. *Phys. Rev. Lett.* **67** 2954–2957.

Murphy, T.J. and Surko, C.M. (1992). Positron trapping in an electrostatic well by inelastic collisions with nitrogen molecules. *Phys. Rev. A* **46** 5696–5705.

Nagashima, Y., Kakimoto, M., Hyodo, T., Fujiwara, K., Ichimura, A., Chang, T., Deng, J., Akahane, T., Chiba, T., Suzuki, K., McKee, B.T.A. and Stewart, A.T. (1995). Thermalization of free positronium atoms by collisions with silica-powder grains, aerogel grains and gas molecules. *Phys. Rev. A* **52** 258–265.

Nagashima, Y., Hyodo, T., Fujiwara, K. and Ichimura, A. (1998). Momentum-transfer cross section for slow positronium–He scattering. *J. Phys. B: At. Mol. Opt. Phys.* **31** 329–339.

Nahar, S.N. and Wadehra, J.M. (1987). Elastic scattering of positrons and electrons by argon. *Phys. Rev. A* **35** 2051–2064.

Nahar, S.N. and Wadehra, J.M. (1988). Formation of ground and excited states of antihydrogen. *Phys. Rev. A* **37** 4118–4124.

Nakanishi, H. and Schrader, D.M. (1986a). Polarization potentials for positron– and electron–atom systems. *Phys. Rev. A* **34** 1810–1822.

Nakanishi, H. and Schrader, D.M. (1986b). Simple but accurate calculations on the elastic scattering of electrons and positrons from neon and argon. *Phys. Rev. A* **34** 1823–1840.

Neumann, R. (1987). Possible experiments with antihydrogen. In *Fundamental Symmetries, Proc. Int. School of Physics with Low Energy Antiprotons, Erice, Sicily, 1986*, eds. P. Bloch, P. Pavlopoulos and R. Klapisch (Plenum Press) pp. 95–114.

Neumann, R., Poth, H., Winnacker, A. and Wolf, A. (1983). Laser-induced electron–ion capture and antihydrogen formation. *Z. Phys. A* **313** 253–262.

Newell, W.R., Brewer, D.F.C. and Smith, A.C.H. (1981). Elastic scattering of electrons in helium. *J. Phys. B: At. Mol. Phys.* **14** 3209–3226.

Newton, R.G. (1959). Threshold properties of scattering and reaction cross sections. *Phys. Rev.* **114** 1611–1618.

Newton, R.G. (1982). *Scattering Theory of Waves and Particles* (Springer-Verlag).

Nico, J.S., Gidley, D.W., Rich, A. and Zitzewitz, P.W. (1990). Precision measurement of the ortho-positronium decay rate using the vacuum technique. *Phys. Rev. Lett.* **65** 1344–1347.

Nico, J.S., Gidley, D.W., Skalsey, M. and Zitzewitz, P.W. (1992). Measurement of the decay rate and 2-gamma branching ratio of orthopositronium. *Material Science Forum* **105–110** 401–410.

Nieminen, R M. (1980). Nonlinear density dependence of the positron decay rate in helium. *Phys. Rev. A* **21** 1347–1349.

Obenshain, F.E. and Page, L.A. (1962). Strong electric field experiments on positronium in gases. *Phys. Rev.* **125** 573–581.

Ohsaki, A., Watanabe, T., Nakanishi, K. and Iguchi, K. (1985). Classical-trajectory Monte Carlo calculation for collision processes of $e^+ + (e^- p)$ and $\mu^+ + (\mu^+ p)$. *Phys. Rev. A* **32** 2640–2644.

O'Malley, T.F., Spruch, L. and Rosenberg, L. (1961). Modification of effective-range theory in the presence of a long range (r^{-4}) potential. *J. Math. Phys.* **2** 491–498.

O'Malley, T.F., Spruch, L. and Rosenberg, L. (1962). Low-energy scattering of a charged particle by a neutral polarizable system. *Phys. Rev.* **125** 1300–1310.

Omidvar, K. (1975). Asymptotic form of the charge-exchange cross section in three-body rearrangement collisions. *Phys. Rev. A* **12** 911–926.

Onsager, L. (1938). Initial recombination of ions. *Phys. Rev.* **54** 554–557.

Ore, A. (1949). Annihilation of positrons in gases. *Naturvitenskapelig Rekke* **9** 1–15.

Ore, A. and Powell, J.L. (1949). Three-photon annihilation of an electron–positron pair. *Phys. Rev.* **75** 1696–1699.

Orito, S., Yoshimura, K., Haga, T., Minowa, M. and Tsuchiaki, M. (1989). New limits on exotic two-body decay of orthopositronium. *Phys. Rev. Lett.* **63** 597–600.

Orth, P.H.R., Falk, W.R., and Jones, G. (1968). Use of the maximum likelihood technique for fitting counting distributions. *Nuc. Inst. Meth.* **65** 301–306.

Orth, P.H.R. and Jones, G. (1969). Annihilation of positrons in argon II. Theoretical. *Phys. Rev.* **183** 16–22.

Osmon, P.E. (1965). Positron lifetime spectra in molecular gases. *Phys. Rev.* **140** A8–A11.

Overton, N., Mills, R.J. and Coleman, P.G. (1993). The energy dependence of the positronium formation cross section in helium. *J. Phys. B: At. Mol. Opt. Phys.* **26** 3951–3957.

Pachucki, K. and Karshenboim, G. (1998). Complete results for positronium energy levels at order $m\alpha^6$. *Phys. Rev. Lett.* **80** 2101–2104.

Pack, R.T. and Parker, G.A. (1987). Quantum reactive scattering in three dimensions using hyperspherical (APH) coordinates. Theory. *J. Chem. Phys.* **87** 3888–3921.

Page, B.A.P. (1975). Calculation of electron– or positron–atom scattering lengths using approximate target wavefunctions. *J. Phys. B: At. Mol. Phys.* **8** 2486–2499.

Page, B.A.P. (1976). Positronium–hydrogen scattering lengths. *J. Phys. B: At. Mol. Phys.* **9** 1111–1114.

Page, B.A.P. and Fraser, P.A. (1974). The ground state of positronium hydride. *J. Phys. B: At. Mol. Phys.* **7** L389–L392.

Pajek, M. and Schuch, R. (1997). Three-body recombination of ions with electrons in cooler-storage rings. *Hyperfine Interactions* **108** 185–194.

Palathingal, J.C., Asoka-Kumar, P., Lynn, K.G. and Wu, X.Y. (1995). Nuclear-charge and positron-energy dependence of the single quantum annihilation of positrons. *Phys. Rev. A* **51** 2122–2130.

Paludan, K., Laricchia, G., Ashley, P., Kara, V., Moxom, J., Bluhme, H., Knudsen, H., Mikkelsen, U., Møller, S.P., Uggerhøj, E. and Morenzoni, E. (1997). Ionization of rare gases by particle–antiparticle impact. *J. Phys. B: At. Mol. Opt. Phys.* **30** L581–L587.

Parcell, L.A., McEachran, R.P. and Stauffer, A.D. (1983). Positron excitation of the 2^1S state of helium. *J. Phys. B: At. Mol. Phys.* **16** 4249–4257.

Parcell, L.A., McEachran, R.P. and Stauffer, A.D. (1987). Positron excitation of the 2^1P state of helium. *J. Phys. B: At. Mol. Phys.* **20** 2307–2315.

Parcell, L.A., McEachran, R.P. and Stauffer, A.D. (1990). Positron excitation of neon. In *Annihilation in Gases and Galaxies, NASA Conf. Pub. 3058*, ed. R.J. Drachman, pp. 109–111.

Parikh, S.P., Kauppila, W.E., Kwan, C.K., Lukaszew, R.A., Przybyla, D., Stein, T.S. and Zhou, S. (1993). Toward measurements of total cross sections for positrons and electrons scattered by potassium and rubidium atoms. *Phys. Rev. A* **47** 1535–1538.

Paul, D.A.L. and Böse, N. (1982). Positron drift and diffusion in gases. In *Electron and Ion Swarms*, ed. L.G. Christophorou (Pergamon Press) pp. 65–82.

Paul, D.A.L. and Leung, C.Y. (1968). On the thermalization times of positrons in polyatomic gases. *Can. J. Phys.* **46** 2779–2788.

Paul, D.A.L. and Tsai, J.-S. (1979). On the theory of positron drift in a uniform electric field. *Can. J. Phys.* **57** 1667–1671.

Paulin, R. and Ambrosino, G. (1968). Annihilation libre de l'ortho-positronium formé dans certaines poudres de grande surface spécifique. *Le Journal de Physique* **29** 263–270.

Peach, G., Saraph, H.E. and Seaton, M.J. (1988). Atomic data for opacity calculations: IX. The lithium isoelectronic sequence. *J. Phys. B: At. Mol. Opt. Phys.* **21** 3669–3683.

Peisert, A. and Sauli, F. (1984). Drift and diffusion of electrons in gases: a compilation. CERN Report 84-08.

Peterkop, R. and Rabik, L. (1971). Effect of inaccuracy of atomic wavefunction in variational calculations of scattering length. *J. Phys. B: At. Mol. Phys.* **4** 1440–1449.

Petelenz, P. and Smith Jr., V.H. (1987). Binding energies of the muonium and positronium negative ions. *Phys. Rev. A* **36** 5125–5126.

Pirenne, J. (1946). Le champ propre et l'interaction des particules de Dirac suivant l'électrodynamique quantique. *Arch. Sci. Phys. Nat.* **28** 233–272.

Phillips, T.J. (1997). Antimatter gravity studies with interferometry. *Hyperfine Interactions* **109** 357–365.

Platzman, P.M. and Mills Jr., A.P. (1994). Possibilities for Bose condensation of positronium. *Phys. Rev. B* **49** 454–458.

Poth, H. (1987). Antihydrogen formation through positron–antiproton radiative combination. *Appl. Phys. A* **43** 287–293.

Poulsen, M.R., Frandsen, N.P. and Knudsen, H. (1994). Non-dissociative and dissociative ionization of molecules by positron impact. *Hyperfine Interactions* **89** 73–81.

Poulsen, M.R., Charlton, M., Chevallier, J., Deutch, B.I., Jørgensen, L.V. and Laricchia, G. (1991). Thermal activation of positronium from thin Ag(100) films in backscattering and transmission geometries. *J. Phys. Condens. Matter* **3** 2849–2858.

Puckett, L.J. and Martin, D.W. (1970). Analysis of recoil He^+ and He^{++} ions produced by fast protons in helium gas. *Phys. Rev. A* **1** 1432–1439.

Quint, W., Kaiser, R., Hall, D.S. and Gabrielse, G. (1993). (Anti)hydrogen recombination studies in a nested Penning trap. *Hyperfine Interactions* **76** 181–188.

Raith, W. (1987). Positron-impact ionization and positronium formation. In *Atomic Physics with Positrons*, eds. J.W. Humberston and E.A.G. Armour (Plenum Press) pp. 1–14.

Raith, W., Hofmann, A., Lynn, K.G. and Weber, M. (1996). The e^+–H experiment of the Bielefeld–Brookhaven collaboration. *Can. J. Phys.* **74** 361–366.

Rapp, D. and Englander-Golden, P. (1965). Total cross sections for ionization and attachment in gases by electron impact I. Positive ionization. *J. Chem. Phys.* **43** 1464–1479.

Rapp, D., Englander-Golden, P. and Briglia, D.D. (1965). Cross section for dissociative ionization of molecules by electron impact. *J. Chem. Phys.* **42** 4081–4085.

Ratnavelu K. (1991). Positron impact ionisation of H and He atoms: the continuum model. *Aust. J. Phys.* **44** 265–270.

Ray, H. and Ghosh, A.S. (1996). Positronium–hydrogen atom scattering using the static exchange model. *J. Phys. B: At. Mol. Opt. Phys.* **29** 5505–5511 and corrigendum (1997) **30** 3745–3746.

Ray, A., Ray, P.P. and Saha, B.C. (1980). Positronium formation in positron–hydrogen-molecule collisions. *J. Phys. B: At. Mol. Phys.* **13** 4509–4519.

Read, F.H. (1985). Threshold behaviour of ionization cross sections. In *Electron Impact Ionization*, eds. T.D. Mark and G.H. Dunn (Springer-Verlag) pp. 42–88.

Register, D. and Poe, R.T. (1975). Algebraic variational method – a quantitative assessment in $e^{\pm}$–H scattering. *Phys. Lett. A* **51** 431–433.

Rich, A. (1981). Recent experimental advances in positronium research. *Rev. Mod. Phys.* **53** 127–165.

Risley, J.S. (1972). Design parameters for the cylindrical mirror analyser. *Rev. Sci. Inst.* **43** 95–103.

Rødbro, M. and Andersen, F.D. (1979). Charge transfer to the continuum for 15 to 1500 keV H^+ in He, Ne, Ar and H_2 gases under single-collision conditions. *J. Phys. B: At. Mol. Phys.* **12** 2883–2903.

Rodionov, S.N., Sannikov, B.P. and Solodov, E.P. (1969). Electric drift of positrons in helium. *JETP Lett.* **10** 325–326.

Rodionov, S.N., Sannikov, B.P. and Solodov, E.P. (1971). The energy-dependence of the slow positron elastic scattering cross-section in helium. *Phys. Lett. A* **35** 297–298.

Roellig, L.O. and Kelly, T.M. (1965). Positron lifetimes in low-temperature helium gas. *Phys. Rev. Lett.* **15** 746–748.

Rost, J.M. and Heller, E.J. (1994). Ionization of hydrogen by positron impact near the fragmentation threshold. *Phys. Rev. A* **49** R4289–R4292.

Rost, J.M. and Pattard, T. (1997). Analytical parameterization for the shape of atomic ionization cross sections. *Phys. Rev. A* **55** R5–R7.

Ruark, A.E. (1945). Positronium. *Phys. Rev.* **68** 278.

Ruttenberg, A.H., Tawel, R. and Canter, K.F. (1985). Experimental investigation of a DC electric field on positron self-trapping in helium gas. *Solid State Commun.* **53** 63–66.

Rytsölä, K., Rantapuska, K. and Hautojärvi, P. (1984). Positron annihilation and cluster formation in nitrogen fluid. *J. Phys. B: At. Mol. Phys.* **17** 299–317.

Ryzhikh, G. and Mitroy, J. (1997). Positronic lithium, an electronically stable Li–e^+ ground state. *Phys. Rev. Lett.* **79** 4124–4126.

Ryzhikh, G., Mitroy, J. and Varga, K. (1998a). The stability of the ground state for positronic sodium. *J. Phys. B: At. Mol. Opt. Phys.* **31** L265–L271.

Ryzhikh, G., Mitroy, J. and Varga, K. (1998b). The structure of exotic atoms containing positrons and positronium. *J. Phys. B: At. Mol. Opt. Phys.* **31** 3965–3996.

Saigusa, T. and Shimizu, S. (1994). A note on nuclear excitation by positron annihilation with K-shell electrons. *Hyperfine Interactions* **89** 445–451.

Sarkar, N.K. and Ghosh, A.S. (1997). Ps–He scattering using a static exchange model. *J. Phys. B: At. Mol. Opt. Phys.* **30** 4591–4597.

Sarkar, N.K., Basu, M. and Ghosh, A.S. (1992). Positronium formation in the $n = 1$ and $n = 2$ states in e^+–He scattering. *Phys. Rev. A* **45** 6887–6890.

Sarkar, N.K., Basu, M. and Ghosh, A.S. (1993). Resonances just above the Ps-formation threshold in positron–hydrogen scattering. *J. Phys. B: At. Mol. Opt. Phys.* **26** L799–L802.

Sauder, W.C. (1968). Thermalisation by elastic collisions: positronium in a rare gas moderator. *J. Res. Natl. Bur. Stand. A* **72** 91–93.

Saxena, S., Gupta, G.P. and Mathur, K.C. (1984). Excitation of the hydrogen atom from its ground and metastable states by positron and proton impact at intermediate energies. *J. Phys. B: At. Mol. Phys.* **17** 3743–3762.

Schmitt, A., Cerny, U., Möller, H., Raith, W. and Weber, M. (1994). Positron–atom doubly differential ionization cross sections. *Phys. Rev. A* **49** R5–R7.

Schneibel, U., Bentz, E., Müller, A., Salzborn, E. and Tawara, H. (1976). Comparison of inner shell ionization by relativistic electron and positron impact. *Phys. Lett. A* **59** 274–276.

Schneider, H., Tobehn, I. and Hippler, R. (1991). L-shell ionization by positron and electron impact. *Phys. Lett. A* **156** 303–306.

Schneider, H., Tobehn, I. and Hippler, R. (1992). Inner-shell ionization by positron and electron impact. *Hyperfine Interactions* **73** 17–26.

Schneider, H., Tobehn, I., Ebel, F. and Hippler, R. (1993). Absolute cross sections for inner shell ionization by lepton impact. *Phys. Rev. Lett.* **71** 2707–2709.

Schoepf, D.C., Berko, S., Canter, K.F. and Sferlazzo, P. (1992). Observation of Ps(n =2) from well-characterized metal surfaces in ultrahigh vacuum. *Phys. Rev. A* **45** 1407–1411.

Schrader, D.M. (1979). Semiempirical polarization potential for low-energy positron–atom and positron–atomic-ion interactions. I. Theory: hydrogen and the noble gases. *Phys. Rev. A* **20** 918–939.

Schrader, D.M. (1998). Bound states of positrons with atoms and molecules: theory. *Nuc. Inst. Meth. B* **143** 209–217.

Schrader, D.M. and Svetic, R.E. (1982). The interaction of positrons with large atoms and molecules: theory. *Can. J. Phys.* **60** 517–542.

Schrader, D.M., Laricchia, G. and Horsky, T.N. (1993). Crossed beam measurement of binding energies of positronium compounds: a preliminary study. *J. de Phys. IV, C4 supplement to J. de Phys. II* 3 61–67.

Schrader, D.M., Jacobsen, F.M., Frandsen, N.-P. and Mikkelsen, U. (1992). Formation of positronium hydride. *Phys. Rev. Lett.* **69** 57–60.

Schramm, U., Berger, J., Grieser, M., Habs, D., Jaeschke, E., Kilgus, G., Schwalm, D., Wolf, A. and Schuch, R. (1991). Observation of laser-induced recombination in merged electron and proton beams. *Phys. Rev. Lett.* **67** 22–25.

Schultz, D.R. and Olson, R.E. (1988). Single-electron-removal processes in collisions of positrons and protons with helium at intermediate velocities. *Phys. Rev. A* **38** 1866–1876.

Schultz, D.R. and Reinhold, C.O. (1990). Electron capture to the continuum and binary ridge structures in positron–hydrogen collisions. *J. Phys. B: At. Mol. Opt. Phys.* **23** L9–L14.

Schultz, D.R., Olson, R.E. and Reinhold, C.O. (1991). Recent advances in the comparison of matter– and antimatter–atom collisions. *J. Phys. B: At. Mol. Opt. Phys.* **24** 521–558.

Schultz, D.R., Reinhold, C.O. and Olson, R.E. (1989). Large-angle scattering in positron–helium and positron–krypton collisions. *Phys. Rev. A* **40** 4947–4958.

Schulz, G.J. (1973). Resonances in electron impact on atoms. *Rev. Mod. Phys.* **45** 378–422.

Schultz, P.J. (1988). A variable-energy positron beam for low to medium energy research. *Nuc. Inst. Meth. B* **30** 94–104.

Schultz, P.J. and Campbell, J.L. (1985) Cross-section ratio for K-shell ionization of copper by low energy electrons and positrons. *Phys. Lett. A* **112** 316–218.

Schultz, P.J. and Lynn, K.G. (1988). Interaction of positron beams with surfaces, thin films and interfaces. *Rev. Mod. Phys.* **60** 701–779.

Schwartz, C. (1961a). Lamb shift in the helium atom. *Phys. Rev.* **123** 1700–1705.

Schwartz, C. (1961b). Electron scattering from hydrogen. *Phys. Rev.* **124** 1468–1471.

Seif el Nasr, S.A.H., Berényi, D. and Bibok, Gy. (1974). Positron impact inner shell ionization. *Z. Phys.* **271** 207–210.

Shah, M.B., Elliot, D.S. and Gilbody, H.B. (1987). Pulsed crossed-beam study of the ionisation of atomic hydrogen by electron impact. *J. Phys. B: At. Mol. Phys.* **20** 3501–3514.

Shakeshaft, R. and Wadehra, J.M. (1980). Distorted-wave Born approximation for inelastic collisions: application to electron capture by positrons from hydrogen atoms. *Phys. Rev. A* **22** 968–978.

Sharma, R.R. (1968). Binding energy of the positronium molecule. *Phys. Rev.* **171** 36–42.

Sharma, S.C. (1988). Positron and positronium annihilation in gases. In *Positron and Positronium Chemistry, Studies in Physical and Theoretical Chemistry, Vol. 57*, eds. D.M. Schrader and Y.C. Jean, pp. 193–239.

Sharma, S.C. (1992). Positronium localization in spontaneous density fluctuations in molecular gases. *Material Science Forum*, **105–110** pp. 451–458.

Sharma, S.C. and McNutt, J.D. (1978). Positron annihilation in gaseous nitrogen and nitrogen–neon mixtures at 77 K. *Phys. Rev. A* **18** 1426–1434.

Sharma, S.C., Eftekhari, A. and McNutt, J.D. (1982). Sensitivity of orthopositronium annihilation rates to density fluctuations in ethane gas. *Phys. Rev. Lett.* **48** 953–956.

Sharma, S.C., Kafle, S.R. and Hart, J.S. (1984). New features in the behaviour of orthopositronium annihilation rates near the vapour-liquid critical point of ethane. *Phys. Rev. Lett.* **52** 2233–2236.

Sharma, S.C., Hyatt, S.D., Ward, M.H. and Dark, C.A. (1985). Effects of density and electric field on positron annihilation in methane gas. *J. Phys. B: At. Mol. Phys.* **18** 3245–3253.

Shimizu, S., Mukoyama, T. and Nakayama, Y. (1965). Search for radiationless annihilation of positrons. *Phys. Lett.* **17** 295–296.

Shimizu, S., Mukoyama, T. and Nakayama, Y. (1968). Radiationless annihilation of positrons in lead. *Phys. Rev.* **173** 405–416.

Shizgal, B. and Ness, K. (1987). Thermalisation and annihilation of positrons in helium and neon. *J. Phys. B: At. Mol. Phys.* **20** 847–865.

Simon, B. (1974). Absence of positive eigenvalues in a class of multiparticle quantum systems. *Math. Ann.* **207** 133–138.

Simon, B. (1978). Resonances and complex scaling: a rigorous overview. *Int. J Quantum Chem.* **XIV** 529–542.

Sinapius, G., Raith, W. and Wilson, W.G. (1980). Scattering of low-energy positrons from noble-gas atoms. *J. Phys. B: At. Mol. Phys.* **13** 4079–4090.

Singh, V. (1989). Computer simulation of positron annihilation and diffusion characteristics in Kr and Xe. *Aust. J. Phys.* **42** 187–196.

Singh, V. and Grover, P.S. (1987). Computer-aided analysis of electric and magnetic field effects on positron annihilation in argon. *J. Phys. B: At. Mol. Phys.* **20** 403–412.

Sinha, K.V. and Grover, P.S. (1987). Simulation of time-dependent positron behaviour in neon gas. *Phys. Rev. A* **35** 3309–3312.

Sinha, P.K., Chaudhury, P. and Ghosh, A.S. (1997). Ps–H scattering using three-state positronium close-coupling approximation. *J. Phys. B: At. Mol. Opt. Phys.* **30** 4643–4652.

Skalsey, M., Engebrecht, J.J., Bithell, R.K., Vallery, R.S. and Gidley, D.W. (1998). Thermalization of positronium in gases. *Phys. Rev. Lett.* **80** 3727–3730.

Slevin, J. and Stirling, W. (1981). Radio frequency atomic hydrogen beam source. *Rev. Sci. Inst.* **52** 1780–1782.

Smith, A.J., Read, F.H. and Imhof, R.E. (1975). Measurement of the lifetimes of ionic excited states using the inelastic electron–photon delayed coincidence technique. *J. Phys. B: At. Mol. Phys.* **8** 2869–2879.

Smith, S.J., Hyder, G.M.A., Kauppila, W.E., Kwan, C.K. and Stein, T.S. (1990). Evidence for absorption effects in positron elastic scattering by argon. *Phys. Rev. Lett.* **64** 1227–1230.

Sparrow, R.A. and Olson, R.E. (1994). Projectile and electron spectra resulting from positron-argon collisions. *J. Phys. B: At. Mol. Opt. Phys.* **27** 2647–2655.

Spektor, D.M. and Paul, D.A.L. (1975). Determination of positronium-atom collision cross sections. *Can. J. Phys.* **53** 13–22.

Sperber, W., Becker, D., Lynn, K.G., Raith, W., Schwab, A., Sinapius, G., Spicher, G. and Weber, M. (1992). Measurement of positronium formation in positron collisions with hydrogen atoms. *Phys. Rev. Lett.* **68** 3690–3693.

Spicher, G., Olsson, B., Raith, W., Sinapius, G. and Sperber, W. (1990). Ionization of atomic hydrogen by positron impact. *Phys. Rev. Lett.* **64** 1019–1022.

Spruch, L. and Rosenberg, L. (1960). Low-energy scattering by a compound system: positrons on atomic hydrogen. *Phys. Rev.* **117** 143–151.

Srivastava, R., Kumar, M and Tripathi, A.N. (1986). Excitation of helium by positron: a distorted wave polarized orbital approach. *J. Chem. Phys.* **84** 4715–4717.

Srivastava, S.K., Tanaka, H., Chutjian, A. and Trajmar, S. (1981). Elastic scattering of intermediate-energy electrons by Ar and Kr. *Phys. Rev. A* **23** 2156–2166.

Srivastava, S.K. and Vušković, L. (1980). Elastic and inelastic scattering of electrons by Na *J. Phys. B: At. Mol. Phys.* **13** 2633–2643.

Steiger, T.D. and Conti, R.S. (1992). Formation of $n = 2$ positronium from untreated metal surfaces. *Phys. Rev. A* **45** 2744–2752.

Stein, J. and Sternlicht, R. (1972). Inelastic s-wave positron–hydrogen scattering. *Phys. Rev. A* **6** 2165–2169.

Stein, T.S. and Kauppila, W.E. (1982). Positron–gas scattering experiments. *Adv. At. Mol. Phys.* **18** 53–96.

Stein, T.S., Kauppila, W.E. and Roellig, L.O. (1974). Production of a monochromatic, low-energy positron beam using the ^{11}B(p, n) ^{11}C reaction. *Rev. Sci. Inst.* **45** 951–953.

Stein, T.S., Kauppila, W.E., Pol, V., Smart, J.H. and Jesion, G. (1978). Measurements of total scattering cross sections for low-energy positrons and electrons colliding with helium and neon atoms. *Phys. Rev. A* **17** 1600–1608.

Stein, T.S., Gomez, R.D., Hsieh, Y.-F., Kauppila, W.E., Kwan, C.K. and Wan, Y.J. (1985). Total-cross-section measurements for positrons and electrons colliding with potassium. *Phys. Rev. Lett.* **55** 488–491.

Stein, T.S., Kauppila, W.E., Kwan, C.K., Lukaszew, R.A., Parikh, S.P., Wan, Y.J., Zhou, S. and Dababneh, M.S. (1990). Scattering of positrons and electrons by alkali atoms. In *Annihilation in Gases and Galaxies, NASA Conference Publication 3058*, ed. R.J. Drachman, pp. 13–27.

Stein, T.S., Jiang, J., Kauppila, W.E., Kwan, C.K., Li, H., Surdutovich, A. and Zhou, S. (1996). Measurements of total and (or) positronium-formation cross sections for positrons scattered by alkali, magnesium and hydrogen atoms. *Can. J. Phys.* **74** 313–333.

Stewart, A.T., Briscoe, C.V. and Steinbacher, J.J. (1990). Positron annihilation in simple condensed gases. *Can. J. Phys.* **68** 1362–1376.

Straton, J.C. (1987). Reduced-mass Fock–Tani representations for $a^+ + (b^+c^-) \to (a^+c^-) + b^+$ and first-order results for $\{abc\} = \{ppe, epe, \mu p\mu, \mu d\mu$, and $\mu t\mu\}$. *Phys. Rev. A* **35** 3725–3740.

Sueoka, O. (1982). Excitation and ionisation of He atom by positron impact. *J. Phys. Soc. Japan* **51** 3757–3758.

Sueoka, O. and Mori, S. (1994). Partitioning of the total cross sections of positrons scattered on He, Ne and Ar atoms. *J. Phys. B: At. Mol. Opt. Phys.* **20** 5083–5088.

Sueoka, O., Mori, S. and Katayama, Y. (1987). Total cross sections for positron and electron collisions with NH_3 and H_2O molecules. *J. Phys. B: At. Mol. Phys.* **20** 3237–3246.

Sural, D.P. and Mukherjee, S.C. (1970). Electron capture by positrons from molecular hydrogen. *Physica* **49** 249–260.

Surdutovich, A., Jiang, J., Kauppila, W.E., Kwan, C.K., Stein, T.S. and Zhou, S. (1996). Measurement of positronium-formation cross sections for positrons scattering by Rb atoms. *Phys. Rev. A* **53** 2861–2864.

Surko, C.M., Greaves, R.G. and Charlton, M. (1997). Stored positrons for antihydrogen production. *Hyperfine Interactions* **109** 181–188.

Surko, C.M., Leventhal, M., Crane, W.S., Passner, A., Wysocki, F., Murphy, T.J., Strachan, J. and Rowan, W.L. (1986). Use of positrons to study transport in tokamak plasmas. *Rev. Sci. Inst.* **57** 1862–1867.

Szmytkowski, C. and Zubek, M. (1978). Absolute total electron scattering cross section of CO, CO_2 and OCS in the low energy region. *Chem Phys. Lett.* **57** 105–108.

Tang, S. and Surko, C.M. (1993). Angular dependence of positronium formation in molecular hydrogen. *Phys. Rev. A* **47** R743–R746.

Tang, S., Tinkle, M.D., Greaves, R.G. and Surko, C.M. (1992). Annihilation gamma-ray spectra from positron–molecule interactions. *Phys. Rev. Lett.* **68** 3793–3796.

Tao, S.J. (1965). Resonance annihilation of positrons in chlorine and argon. *Phys. Rev. Lett.* **14** 935–936.

Tao, S.J. (1970). Annihilation of positrons in nitrogen. *Phys. Rev. A* **2** 1669–1675.

Tawel, R. and Canter, K.F. (1986). Observation of a positron mobility threshold in gaseous helium. *Phys. Rev. Lett.* **56** 2322–2325.

Temkin, A. (1957). Polarization and exchange effects in the scattering of electrons from atoms with applications to oxygen. *Phys. Rev.* **107** 1004–1012.

Temkin, A. (1959). A note on the scattering of electrons from atomic hydrogen. *Phys. Rev.* **116** 358–363.

Temkin, A. (1982). Threshold law for electron–atom impact ionization. *Phys. Rev. Lett.* **49** 365–368.

Temkin, A. and Lamkin, J.C. (1961). Application of the method of polarized orbitals to the scattering of electrons from hydrogen. *Phys. Rev.* **121** 788–794.

Tennyson, J. (1986). Low-energy, elastic positron–molecule collisions using the R-matrix method: e^+–H_2 and e^+–N_2. *J. Phys. B: At. Mol. Phys.* **19** 4255–4263.

Tennyson, J. and Danby, G. (1987). The *ab initio* inclusion of polarisation effects in low-energy positron–molecule collisions using the R-matrix method. In *Atomic Physics with Positrons*, eds. J.W. Humberston and E.A.G. Armour (Plenum Press) pp. 111–121.

Tennyson, J. and Morgan, L. (1987). Rotational and polarisation effects in low-energy positron–CO collisions using the R-matrix method. *J. Phys. B: At. Mol. Phys.* **20** L641–L646.

Theriot Jr., E.D., Beers, R.H., Hughes, V.W. and Ziock, K.O.H. (1970). Precision redetermination of the fine-structure interval of the ground state of positronium and a direct measurement of the decay rate of parapositronium. *Phys. Rev. A* **2** 707–721.

Thomas, M.A. and Humberston, J.W. (1972). The polarizability of helium. *J. Phys. B: At. Mol. Phys.* **5** L229–L232.

Tisenko, Yu. A. (1981). Decay of the positronium molecule into a positronium atom and photons. *Russian Physics J.* **24** 99–102.

Tong, B.Y. (1972). Negative work function of thermal positrons in metals. *Phys. Rev. B* **5** 1436–1439.

Toumisaari, M., Rytsölä, K. and Hautojärvi, P. (1985). Localized state of positron in argon. *Phys. Lett. A* **122** 279–282.

Toumisaari, M., Rytsölä, K. and Hautojärvi, P. (1988). Positron annihilation in xenon. *J. Phys. B: At. Mol. Opt. Phys.* **21** 3917–3928.

Tsai, J.-S., Lebow, L. and Paul, D.A.L. (1976). Measurement of total cross sections (e^+, Ne) and (e^+, Ar). *Can. J. Phys.* **54** 1741–1748.

Turner, D.W. (1969). *Molecular Photoelectron Spectroscopy* (Wiley).

Ujc, H. and Stauffer, A.D. (1985). Calculation of positron annihilation in helium using a global operator. *J. Phys. B: At. Mol. Phys.* **18** 2087–2092.

Vallery, R.S., Leanhardt, A.E, Skalsey, M. and Gidley, D.W. (2000). Temperature dependence of positronium decay rates in gases. *J. Phys. B: At. Mol. Opt. Phys.* **33** 1047–1055.

Van Dyck Jr., R.S., Schwinberg, P.B. and Dehmelt, H.G. (1987). New high-precision comparison of electron and positron g factors. *Phys. Rev. Lett.* **59** 26–29.

Van Reeth, P. and Humberston, J.W. (1995a). The use of inexact wavefunctions in positron–helium scattering. *J. Phys. B: At. Mol. Opt. Phys.* **28** L23–L28.

Van Reeth, P. and Humberston, J.W. (1995b). Positronium formation in low energy s-wave positron–helium scattering. *J. Phys. B: At. Mol. Opt. Phys.* **28** L511–L517.

Van Reeth, P. and Humberston, J.W. (1997). A partial-wave analysis of positronium formation in positron–helium scattering. *J. Phys. B: At. Mol. Opt. Phys.* **30** L95–L100.

Van Reeth, P. and Humberston, J.W. (1998). The energy dependence of the annihilation rate in positron–atom scattering. *J. Phys. B: At. Mol. Opt. Phys.* **31** L231–L238.

Van Reeth, P. and Humberston, J.W. (1999a). A significant feature in the total cross section for positron–helium scattering at the positronium formation threshold. *J. Phys. B: At. Mol. Opt. Phys.* **32** L103–L106.

Van Reeth, P. and Humberston, J.W. (1999b). Elastic scattering and positronium formation in low energy positron–helium collisions. *J. Phys. B: At. Mol. Opt. Phys.*, in press.

Van Reeth, P., Humberston J.W., Iwata, K., Greaves, R.G. and Surko, C.M. (1996). Annihilation in low-energy positron–helium scattering. *J. Phys. B: At. Mol. Opt. Phys.* **29** L465–L471.

Van Wingerden, B., Wagenaar, R.W. and de Heer, F.J. (1980). Total cross sections for electron scattering by molecular hydrogen. *J. Phys. B: At. Mol. Phys.* **13** 3481–3491.

Vlasov, R.A., Gadomskii, O.N. and Shageev, M.G. (1990). Annihilation superradiance in system of positronium atoms. *Sov. Phys. Dokl.* **35** 369–370.

Wadehra, J.M., Stein, T.S. and Kauppila, W.E. (1981). Analysis of low-energy positron–helium total cross sections. *J. Phys. B: At. Mol. Phys.* **14** L783–L787.

Wakid, S.E.A. (1973). Positron–hydrogen scattering for energies up to the $n = 2$ threshold of hydrogen. *Phys. Rev. A* **8** 2456–2462.

Wakid, S.E.A. and LaBahn, R.W. (1972). Positronium formation in positron–hydrogen collisions. *Phys. Rev. A* **6** 2039–2049.

Wakiya, K. (1978). Differential and integral cross sections for the electron impact excitation of O_2 I. Optically allowed transitions from the ground state. *J. Phys. B: At. Mol. Phys.* **11** 3913–3930.

Walraven, J.T.M. (1993). Trapping and cooling of (anti)hydrogen. *Hyperfine Interactions* **76** 205–220.

Walters, H.R.J. (1988). Positron scattering by atomic hydrogen at intermediate energies: 1s–1s, 1s–2s and 1s–2p transitions. *J. Phys. B: At. Mol. Opt. Phys.* **21** 1893–1906.

Wannier, G.H. (1953). The threshold law for single ionization of atoms or ions by electrons. *Phys. Rev.* **90** 817–825.

Ward, S.J., Humberston, J.W. and McDowell, M.R.C. (1985). The scattering of low-energy s-wave electrons by positronium. *J. Phys. B: At. Mol. Phys.* **18** L525–L530.

Ward, S.J., Humberston, J.W. and McDowell, M.R.C. (1987). Elastic scattering of electrons (or positrons) from positronium and the photodetachment of the positronium negative ion. *J. Phys. B: At. Mol. Phys.* **20** 124–149.

Ward, S.J., Macek, J.H. and Ovchinnikov, S.Yu. (1998). Hidden crossing theory applied to positronium formation. *Nuc. Inst. Meth. B* **143** 175–183.

Ward, S.J., Horbatsch, M., McEachran, R.P. and Stauffer, A.D. (1988). Close-coupling approach to positron scattering from potassium. *J. Phys. B: At. Mol. Opt. Phys.* **21** L611–L616.

Ward, S.J., Horbatsch, M., McEachran, R.P. and Stauffer, A.D. (1989). Close-coupling approach to positron scattering for lithium, sodium and potassium. *J. Phys. B: At. Mol. Opt. Phys.* **22** 1845–1861.

Weber, M., Tang, S., Berko, S., Brown, B.L., Canter, K.F., Lynn, K.G., Mills Jr., A.P., Roellig, L.O. and Viescas, A.J. (1988). Observation of positronium specular reflection from LiF. *Phys. Rev. Lett.* **61** 2542–2545.

Weber, M., Schwab, A., Becker, D. and Lynn, K.G. (1992). Solid neon moderated electrostatic or magnetic positron beam. *Hyperfine Interactions* **73** 147–157.

Weber, M., Hofmann, A., Raith, W., Sperber, W., Jacobsen, F.M. and Lynn, K.G. (1994). Results of the Bielefeld–Brookhaven e^+–H experiment. *Hyperfine Interactions* **89** 221–242.

West, R.N., Mayers, J. and Walters, P.A. (1981). A high-efficiency two-dimensional angular correlation spectrometer for positron studies. *J. Phys. E* **14** 478–488.

Westbrook, C.I., Gidley, D.W., Conti, R.S. and Rich, A. (1987). New precision measurement of the orthopositronium decay rate: a discrepancy with theory. *Phys. Rev. Lett.* **58** 1328–1331.

Westbrook, C.I., Gidley, D.W., Conti, R.S. and Rich, A. (1989). Precision measurement of the orthopositronium vacuum decay rate using the gas technique. *Phys. Rev. A* **40** 5489–5499.

Wetmore, A.E. and Olson, R.E. (1986). Ionization of H and He^+ by electrons and positrons colliding at near-threshold energies. *Phys. Rev. A* **34** 2822–2829.

Wheeler, J.A. (1946). Polyelectrons. *Ann. N.Y. Acad. Sci.* **48** 219–238.

Wigner, E.P. (1948). On the behaviour of cross sections near thresholds. *Phys. Rev.* **73** 1002–1009.

Wikander, G. and Mogensen, O.E. (1982). A study of picosecond dehalogenation of chlorobenzene anions in liquids by positronium inhibition measurements. *Chem. Phys.* **72** 407–423.

Williams, J.F. (1979). A phaseshift analysis of the experimental angular distributions of electrons elastically scattered from He, Ne and Ar over the range 0.5 to 20 eV. *J. Phys. B: At. Mol. Phys.* **12** 265–282.

Willis, S.L. and McDowell, M.R.C. (1982). Pseudostate effects on positron–helium excitation cross sections at intermediate energies. *J. Phys. B: At. Mol. Phys.* **15** L31–L35.

Wilson, W.G. (1978). Scattering of low-energy positrons from helium atoms. *J. Phys. B: At. Mol. Phys.* **11** L629–L633.

Winick, J.R. and Reinhardt, W.P. (1978a). Moment T-matrix approach to e^+–H scattering. I. Angular distribution and total cross section for energies below the pickup threshold. *Phys. Rev. A* **18** 910–924.

Winick, J.R. and Reinhardt, W.P. (1978b). Moment T-matrix approach to e^+–H scattering. II. Elastic scattering and total cross section at intermediate energies. *Phys. Rev. A* **18** 925–934.

Wolf, A. (1993). Laser-stimulated formation and stabilization of antihydrogen atoms. *Hyperfine Interactions* **76** 189–201.

Wolfenstein, L. and Ravenhall, D.G. (1952). Some consequences of invariance under charge conjugation. *Phys. Rev.* **88** 279–282.

Worrell, G.A., Miller, B.N. and Reese, T.L. (1996). Virial expansion of a quantum particle in a classical gas: application to the orthopositronium decay rate. *Phys. Rev. A* **53** 2101–2107.

Wright, G.L. (1982). Positron and electron interactions with gaseous media. Ph.D. thesis, University of London.

Wright, G.L., Charlton, M., Clark, G., Griffith, T.C. and Heyland, G.R. (1983). Positron lifetime parameters in H_2, CO_2 and CH_4. *J. Phys. B: At. Mol. Phys.* **16** 4065–4088.

Wright, G.L., Charlton, M., Griffith, T.C. and Heyland, G.R. (1985). The annihilation of positrons and positronium formation in gaseous Kr and Xe. *J. Phys. B: At. Mol. Phys.* **18** 4327–4347.

Yamazaki, T. (1992). A possible way to promote antihydrogen formation via metastable antiprotonic helium atoms. *Z. Phys. A* **341** 223–225.

Yang, C.N. (1950). Selection rules for the dematerialization of a particle into two photons. *Phys. Rev.* **77** 242–245.

Yang, J., Chiba, M., Hamatsu, R., Hirose, T., Matsumoto, T. and Yu, J. (1996). Four-photon decay of orthopositronium: a test of charge-conjugation invariance. *Phys. Rev. A* **54** 1952–1956.

Yousif, F.B., Van der Donk, P., Kucherovsky, Z., Reis, J., Brannen, E., Mitchell, J.B.A. and Morgan, T.J. (1991). Experimental observation of laser-stimulated radiative recombination. *Phys. Rev. Lett.* **67** 26–29.

Zafar, N., Laricchia, G., Charlton, M. and Garner, A. (1996). Positronium–argon scattering. *Phys. Rev. Lett.* **76** 1595–1598.

Zafar, N., Laricchia, G., Charlton, M. and Griffith, T.C. (1991). Diagnostics of a positronium beam. *J. Phys. B: At. Mol. Opt. Phys.* **24** 4661–4670.

Zafar, N., Laricchia, G., Charlton, M. and Griffith, T.C. (1992). The new magnetically confined positron beam at UCL. *Hyperfine Interactions* **73** 213–215.

Zhang, Z. and Ito, Y. (1990). A new model of positronium formation: resonant positronium formation. *J. Chem. Phys.* **93** 1021–1029.

Zhou, S., Kauppila, W.E., Kwan, C.K. and Stein, T.S. (1994a). Measurements of total cross sections for positrons and electrons colliding with atomic hydrogen. *Phys. Rev. Lett.* **72** 1443–1446.

Zhou, S., Parikh, S.P., Kauppila, W.E., Kwan, C.K., Lin, D., Surdutovich, A. and Stein, T.S. (1994b). Measurements of positronium formation cross sections for positron scattering by K, Na and Ar atoms. *Phys. Rev. Lett.* **73** 236–239.

Zhou, S., Li, H., Kauppila, W.E., Kwan, C.K. and Stein, T.S. (1997). Measurements of total and positronium formation cross sections for positrons and electrons scattered by hydrogen atoms and molecules. *Phys. Rev. A* **55** 361–368.

Zhou, Y. and Lin, C.D. (1994). Hyperspherical close-coupling calculation of positronium formation cross sections in positron–hydrogen scattering at low energies. *J. Phys. B: At. Mol. Opt. Phys.* **27** 5065–5081.

Zhou, Y. and Lin, C.D. (1995a). On the non-existence of resonances above the ionization threshold in positron–hydrogen scattering. *J. Phys. B: At. Mol. Opt. Phys.* **28** L519–L523.

Zhou, Y. and Lin, C.D. (1995b). Hyperspherical close-coupling calculation of positronium formation and excitation cross sections in positron–hydrogen scattering at energies below the $H(n = 4)$ threshold. *J. Phys. B: At. Mol. Opt. Phys.* **28** 4907–4925.

Ziock, K.P., Dermer, C.D., Howell, R.H., Magnotta, F. and Jones, K.M. (1990a). Optical saturation of the 1^3S–2^3P transition in positronium. *J. Phys. B: At. Mol. Opt. Phys.* **23** 329–336.

Ziock, K.P., Howell, R.H., Magnotta, F., Failor, R.A. and Jones, K.M. (1990b). First observation of resonant excitation of high-n states of positronium. *Phys. Rev. Lett.* **64** 2366–2369.

Zitzewitz, P.W., Van House, J.C., Rich, A. and Gidley, D.W. (1979). Spin polarization of low-energy positron beams. *Phys. Rev. Lett.* **43** 1281–1284.

Index